PROGRESS IN

Molecular Biology and Translational Science

Volume 93

PROGRESS IN

Molecular Biology and Translational Science

Glycosaminoglycans in Development, Health and Disease

edited by

Lijuan Zhang

Department of Pathology and Immunology
Washington University Medical School
St. Louis, Missouri, USA

Volume 93

ELSEVIER

AMSTERDAM • BOSTON • HEIDELBERG • LONDON
NEW YORK • OXFORD • PARIS • SAN DIEGO
SAN FRANCISCO • SINGAPORE • SYDNEY • TOKYO
Academic Press is an imprint of Elsevier

Academic Press is an imprint of Elsevier
32 Jamestown Road, London, NW1 7BY, UK
Radarweg 29, PO Box 211, 1000 AE Amsterdam, The Netherlands
30 Corporate Drive, Suite 400, Burlington, MA 01803, USA
525 B Street, Suite 1900, San Diego, CA 92101-4495, USA

This book is printed on acid-free paper. ∞

Library of Congress Cataloging-in-Publication Data
A catalog record for this book is available from the Library of Congress

British Library Cataloguing in Publication Data
A catalogue record for this book is available from the British Library

ISBN: 978-0-12-381282-7
ISSN: 1877-1173

For information on all Academic Press publications
visit our website at elsevierdirect.com

Printed and bound by CPI Group (UK) Ltd, Croydon, CR0 4YY

Transferred to digital print 2013

10 11 12 13 10 9 8 7 6 5 4 3 2 1

Contents

Congenital Disorders of Glycosylation with Emphasis on loss of Dermatan-4-Sulfotransferase 289

Lijuan Zhang, Thomas Müller, Jacques U. Baenziger, and Andreas R. Janecke

Heparan Sulfate Proteoglycans in Amyloidosis. 309

Xiao Zhang and Jin-Ping Li

Heparin as an Inhibitor of Cancer Progression 335

Lubor Borsig

Contributors

Numbers in parentheses indicate the pages on which the authors' contributions begin.

Rafael S. Aquino, Division of Respiratory Diseases, Children's Hospital, Harvard Medical School, Boston, Massachusetts, USA (373)

Jacques U. Baenziger, Department of Pathology, Washington University School of Medicine, St. Louis, Missouri, USA (289)

Lubor Borsig, Institute of Physiology, and Zürich Center for Integrative Human Physiology, University of Zürich, Switzerland (335)

Ariane de Agostini, Department of Gynaecology and Obstetrics, Laboratory of Reproductive Biology, and Department of Genetic and Laboratory Medicine, Service of Clinical Pathology, Geneva University Hospitals and University of Geneva, Geneva, Switzerland (153)

Jeffrey D. Esko, Department of Cellular and Molecular Medicine, Glycobiology Research and Training Center, University of California, San Diego, La Jolla, California, USA (213)

Erin M. Foley, Department of Cellular and Molecular Medicine, Glycobiology Research and Training Center, University of California, San Diego, La Jolla, California, USA (213)

Mark M. Fuster, Department of Medicine, Division of Pulmonary and Critical Care, University of California San Diego, and VA San Diego Healthcare System, La Jolla, California, USA (179)

Hiroko Habuchi, Research Complex for the Medicine Frontiers, Aichi Medical University, Nagakute, Aichi, Japan (79)

Masaru Inatani, Sanford Children's Health Research Center, Sanford-Burnham Medical Research Institute, La Jolla, California, USA, and Department of Ophthalmology and Visual Science, Kumamoto University Graduate School of Medical Sciences, Kumamoto, Japan (133)

Fumitoshi Irie, Sanford Children's Health Research Center, Sanford-Burnham Medical Research Institute, La Jolla, California, USA (133)

Tomomi Izumikawa, Department of Biochemistry, Kobe Pharmaceutical University, Kobe, Japan (19)

Andreas R. Janecke, Department of Pediatrics II, Innsbruck Medical University, Anichstrasse 35, Innsbruck, Austria (289)

Koji Kimata, Research Complex for the Medicine Frontiers, Aichi Medical University, Nagakute, Aichi, Japan (79)

Hiroshi Kitagawa, Department of Biochemistry, Kobe Pharmaceutical University, Kobe, Japan (19)

Lena Kjellén, Department of Medical Biochemistry and Microbiology, The Biomedical Center, Uppsala University, Uppsala, Sweden, and Department of Biomedicine, University of Bergen, Bergen, Norway (35)

Michael Klüppel, Human Molecular Genetics Program, Children's Memorial Research Center, and Department of Pediatrics; and Robert H. Lurie Comprehensive Cancer Center, Northwestern University, Feinberg School of Medicine, Chicago, Illinois, USA (113)

Takashi Kobayashi, Department of Medicine, Dartmouth Medical School, Hanover, New Hampshire, USA (153)

Eui Seung Lee, Division of Respiratory Diseases, Children's Hospital, Harvard Medical School, Boston, Massachusetts, USA (373)

Jin-Ping Li, Department of Medical Biochemistry and Microbiology, University of Uppsala, Uppsala, Sweden (59, 309)

Hong Lu, Department of Pathology and Immunology, Washington University Medical School, St. Louis, Missouri, USA (395, 423, 443, 473)

Thomas Müller, Department of Pediatrics II, Innsbruck Medical University, Anichstrasse 35, Innsbruck, Austria (289)

Yoshihiro Matsumoto, Sanford Children's Health Research Center, Sanford-Burnham Medical Research Institute, La Jolla, California, USA, and Department of Orthopaedic Surgery, Graduate School of Medical Sciences, Kyushu University, Fukuoka, Japan (133)

Junko Ogawa, Sanford Children's Health Research Center, Sanford-Burnham Medical Research Institute, La Jolla, California, USA (133)

Jing Pan, Department of Pathology and Immunology, Washington University Medical School, St. Louis, Missouri, USA (395, 423, 443, 473)

Pyong Woo Park, Division of Respiratory Diseases, Children's Hospital, Harvard Medical School, Boston, Massachusetts, USA (373)

Yi Qian, Department of Pathology and Immunology, Washington University Medical School, St. Louis, Missouri, USA (395, 423, 443,473)

Maria Ringvall, Department of Medical Biochemistry and Microbiology, The Biomedical Center, Uppsala University, Uppsala, Sweden (35)

Barbara P. Schick, Department of Medicine, Cardeza Foundation for Hematologic Research, Thomas Jefferson University, Philadelphia, Pennsylvania, USA (235)

Fei F. Shih, Department of Pediatrics, Washington University Medical School, St. Louis, Missouri, USA (443)

Nicholas W. Shworak, Department of Medicine, and Department of Pharmacology and Toxicology, Dartmouth Medical School, Hanover, New Hampshire, USA (153)

Nicole C. Smits, Department of Medicine, Dartmouth Medical School, Hanover, New Hampshire, USA (153)

Daniel R. Studelska, Department of Pathology and Immunology, Washington University Medical School, St. Louis, Missouri, USA (443, 473)

Douglas M. Tollefsen, Hematology Division, Washington University Medical School, St. Louis, Missouri, USA (351)

Lianchun Wang, Department of Biochemistry and Molecular Biology, Complex Carbohydrate Research Center, University of Georgia, Athens, Georgia, USA (179)

Peter Weiser, Department of Pathology and Immunology, Washington University Medical School, St. Louis, Missouri, USA (395, 423, 443, 473)

Yu Yamaguchi, Sanford Children's Health Research Center, Sanford-Burnham Medical Research Institute, La Jolla, California, USA (133)

Lijuan Zhang, Department of Pathology and Immunology, Washington University Medical School, St. Louis, Missouri, USA (1, 289, 395, 423, 443, 473)

Xiao Zhang, Department of Public Health and Caring Sciences, Molecular Geriatrics, University of Uppsala, Uppsala, Sweden (309)

Xiaodong Zhou, Department of Pathology and Immunology, Washington University Medical School, St. Louis, Missouri, USA (395, 423, 443, 473)

Preface

All animal cells produce two major types of sulfated linear polysaccharide glycosaminoglycans (GAGs): heparan sulfate and chondroitin sulfate. Due to their structural diversity, GAGs have been claimed to be the most information-dense biopolymers found in nature. GAGs bind to numerous protein ligands and receptors involved in diverse biological processes, including cell division, growth control, signal transduction, cell adhesion, hemostasis, and lipid metabolism (Chapters 1 and 2). Emerging from its roots in classical chemistry and biochemistry, the progress in the field of GAGs has greatly accelerated during the past 20 years through genetic analysis. Transgenic and knockout animal data (Chapters 2–13) provide compelling evidence that this structural diversity is a component of a sugar/sulfation-GAG code. The sugar/sulfation-GAG code imparts unique and specific biological functions during development, health, and disease (Chapters 2–16), which makes GAG an essential component of modern molecular biology and human physiology.

Heparin, the mostly sulfated heparan sulfate made by mast cells, has the highest anticoagulation activity by inhibiting thrombin generation and thrombin activities among different GAGs and has been used as an anticoagulant drug for over 70 years. Crude heparin isolated from animal tissues consists of ~50% heparin and ~50% less sulfated GAGs including heparan sulfate, chondroitin sulfate, and dermatan sulfate. Heparin manufacturers purchase crude heparin from small vendors and produce heparin from crude heparin by removing the low-sulfated GAGs using sophisticated techniques. In 2007 and 2008, contaminated heparin was associated with hundreds of anaphylactic reactions and at least 149 anaphylactic reaction-associated deaths in the United States. It turns out that the heparin contaminants are chemically sulfated/modified low-sulfated GAGs.

The contaminated heparin associated anaphylactic reactions are a result of up-regulated immune system by oversulfated GAGs through contact system activation. The contact system was first discovered as an *in vitro* thrombin generation system where artificial surfaces induce thrombin generation and clotting. Injury-associated thrombin generation in animals instantly leads to immune system up-regulation. However, the *in vivo* role of contact system activation-generated thrombin and immune system up-regulation has been overlooked during the past 50 years. Chapters 17–20 demonstrate that contact system activation-generated thrombin and immune system up-regulation induced by abnormal GAG/protein aggregates are the outcomes of different autoimmune diseases, including Lupus, rheumatoid arthritis, psoriasis, heparin-induced thrombocytopenia/thrombosis (HIT), and different kinds of human cancers.

The Editor is indebted to many others who made this book possible. Special thanks are due to Dr. Michael Conn, the editor of *Progress in Molecular Biology and Translational Science* book series, who projected this book volume; and to Lisa Tickner, Delsy Retchagar, Malathi Samayan, and Sunita Sundararajan at Elsevier for keeping us on track and converting our efforts into a product. Last but not least, we acknowledge the support and hard work of our families and lab members in producing each chapter in the book. It now remains for the reader to decide whether we have achieved our goals in compiling the GAG book.

LIJUAN ZHANG
Missouri, USA

Glycosaminoglycan (GAG) Biosynthesis and GAG-Binding Proteins

LIJUAN ZHANG

Department of Pathology and Immunology, Washington University Medical School, St. Louis, Missouri, USA

Two major types of glycosaminoglycan (GAG) polysaccharides, heparan sulfate and chondroitin sulfate, are polymerized and modified by enzymes that are encoded by more than 40 genes in animal cells. Because of the expression repertoire of the GAG assembly and modification enzymes, each heparan sulfate and chondroitin sulfate chain has a sulfation pattern, chain length, and fine structure that is potentially unique to each animal cell. GAGs interact with hundreds of proteins. Such interactions protect growth factors, chemokines, and cytokines against proteolysis. GAGs catalyze protease (such as thrombin) inhibition by serpins. GAGs regulate multiple signaling pathways including, but not limited to, fibroblast growth factor (FGF)/FGFR, hepatocyte growth factor (HGF)/c-Met, glial cell line-derived neurotrophic factor (GDNF)/c-Ret/GFRα1, vascular endothelial growth factor (VEGF)/VEGFR, platelet derived growth factor (PDGF)/PDGFR, BAFF/TACI, Indian hedgehog, Wnt, and BMP signaling pathways,where genetic studies have revealed an absolute requirement for GAGs in these pathways. Most importantly, protein/GAG aggregates induce thrombin generation and immune system upregulation by activating the contact system. Abnormal protein/GAG aggregates are associated with a variety of devastating human diseases including, but not limited to, Alzheimer's, diabetes, prion or transmissible spongiform encephalopathies, Lupus, heparin-induced thrombocytopenia/thrombosis, and different kinds of cancers. Therefore, GAGs are essential components of modern molecular biology and human physiology. Understanding GAG structure and function at

Progress in Molecular Biology
and Translational Science, Vol. 93
DOI: 10.1016/S1877-1173(10)93001-9

1877-1173/10 $35.00

molecular level with regard to development and health represents a unique opportunity in combating different kinds of human diseases.

Abbreviations: CS, chondroitin sulfate; GAG, glycosaminoglycan; HS, heparan sulfate

I. Glycosaminoglycans (GAGs)

GAGs are linear polysaccharides that are made by all animal cells. Two major types of GAGs are heparan sulfate (HS) and chondroitin sulfate (CS). HS and CS comprise repeating hexosamine-uronic acid disaccharides that are sulfated to varying degrees[1,2] (Fig. 1). Heparin is the most highly sulfated HS made by mast cells. DS is one type of CS containing IdoA residues and is made by many types of animal cells.

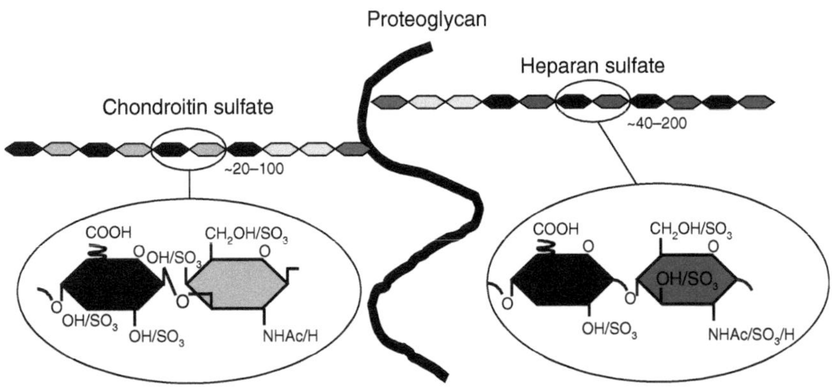

Fig. 1. Diagrammatic representation of GAG assembly on a proteoglycan core protein. Both HS and CS are attached to specific serine residues of proteoglycan core protein by the linkage tetrasaccharide GlcA (black)-Gal (yellow)-Gal (yellow)-Xyl (pink). Biosynthesis starts with the transfer of xylose from UDP-xylose to a serine residue of a core protein catalyzed by two xylosyltransferases. The linkage region is then synthesized by the sequential addition of two galactose residues (by galactosyltransferase I and II), and glucuronic acid (by glucuronosyltransferase I) from the corresponding UDP-sugars. After completion of the linkage tetrasaccharides, the addition of first HexNac residue occurs. Addition of GalNAc from UDP-GalNAc by *N*-acetylgalactosaminyl transferase I to the nonreducing terminal GlcA commits the intermediate to CS sysnthesis, which occurs subsequently through alternating addition of GlcA and GalNAc (green) by chondroitin synthase. If GlcNAc is added to the linkage tetrasaccharide instead by *N*-acetylglucosaminyltransferase I, HS synthesis occurs. Alternating GlcA and GlcNAc (red) residues are then added by HS copolymerases (EXT-1 and EXT-2) from their corresponding UDP-sugars. (For interpretation of the references to color in this figure legend, the reader is referred to the Web version of this chapter.) (See Color Insert.)

HS and CS are abundantly produced (~ 10^5–10^6 copies on the cell surface, ~ mg/ml concentrations in the extracellular matrix) by animal cells.[3] This synthesis occurs in the *Golgi* in the form of HS, CS, or HS/CS hybrid proteoglycans. In proteoglycans, anywhere between one (e.g., decorin) to over one hundred GAG chains (e.g., aggrecan) can be assembled on a proteoglycan core protein (Fig. 1). The number of repeating disaccharides of HS and CS varies depending on the source of cells or tissues used for GAG isolation. GAGs adopt an extended helical coil structure with a length ranging from 40 to 160 nm. Such abundance and size implies that GAGs are a dominant feature of the cell surface and are an important feature of the extracellular matrix.

II. Proteoglycan

All mammalian cells produce proteoglycans and can secrete them into the extracellular matrix, insert them into the plasma membrane, or store them in secretory granules. Over 50 proteoglycan cDNAs have been cloned (Table I). Some of the proteoglycan gene products were known as important functional proteins long before they were known to be proteoglycans. It is typical for a cell to express multiple types of HS and CS proteoglycans. For example, at least 11 different proteoglycans are expressed by human lung fibroblasts[4–10] and at least 23 types are expressed in the nervous system.[11] Most core proteins not only serve as GAG carriers but also have their own functional protein domains. In some instances, they may also contain GAG-binding domains. The core proteins usually determine the number of GAG chains, the type (HS or CS), and the ultimate destination (apical, luminal, intra- or extracellular) of the finished proteoglycan. GAG chains carried by proteoglycans are sometimes cleaved before they become biologically active. Extracellular heparanases (heparan sulfate endoglucuronidases),[12] sulfatases,[13,14] and free GAG chains have been extensively reported.[15–18]

The structural variation of proteoglycans in different cells or tissues is due to a number of factors. First, over 50 core proteins have been identified, and these can be substituted with CS or both HS and CS. Another source of variability lies in the stoichiometry of GAG chain substitution. For example, syndecan-1 has five GAG attachment sites, but not all of the sites are used equally. Thus, a preparation of syndecan-1 represents a diverse population of syndecan-1 molecules. Other proteoglycans, such as thrombomodulin, can be a "part time proteoglycan" that is, they may exist with or without a GAG chain or with only a truncated GAG chain. These characteristics, typical of all proteoglycans, create diversity that may facilitate the formation of binding sites of variable density and affinity for different ligands in a cell- and tissue-specific manner, beyond its GAG-binding specificity.

TABLE I
CLONED PROTEOGLYCANS

HS proteoglycans	CS proteoglycans		
Syndecan	Syndecan-1 (TM)	Lectican	Aggrecan
	Syndecan-2 (TM)		Versican/PG-M
	Syndecan-3 (TM)		Neurocan
	Syndecan-4 (TM)		Brevican (soluble or GPI-linked)
Glypican	Glypican-1 (GPI-Linked)	IPM	IPM150 (interphotoreceptor matrix PG, TM)
	Glypican-2 (GPI-Linked)		IPM 200 (TM)[92]
	Glypican-3 (GPI-Linked)		SPACRCAN (TM)[93]
	Glypican-4 (GPI-Linked)	SLRP	Decorin (small leucine-rich proteoglycans)
	Glypican-5 (GPI-Linked)		Biglycan
	Glypican-6 (GPI-Linked)		PG-lb
Testican	Testican-1	Others	CD44 (CS splicing form, TM)
	Testican-2		Thrombomodulin (TM)
	Testican-3		Invariant chain
Others	Betaglycan (TM)		APP (amyloid precursor protein)
	Tyrosine kinase receptor (TM)		APLP2
	PRPg (proline rich PG)		Lepecan[94]
	Perlecan		Bicunin
	Agrin		Chromogranin A
	Collagen XVIII		α2(IX) collagen
	CD44 (HS splicing form, TM)		Tanasin-C
	Serglycin		Laminin α4[95]
			CSF (colony stimulating factor)
			Phosphocan (3 splicing forms, soluble or TM)
			Claustrin
			NG2 (TM)
			Neuroglycan-C (TM)
			MCSP (melanoma-associated CSPG)[96]
			PRG4 (megakaryocyte stimulating factor)[97]
			Endoglycan[98]
			Endocan[99]
			Bamacan[100]

Accession numbers of nonreferenced genes can be found in Refs. 11,22.
TM, transmembrane protein.

III. GAG Biosynthesis

HS and CS are elaborated after the synthesis of the GAG-protein linkage region, GlcA-Gal-Gal-Xyl, which is attached to specific Ser residues on the proteoglycan core protein (Fig. 1). The synthesis of this region is initiated by the addition of a Xyl to Ser followed by the addition of two Gal residues and is completed by the addition of GlcA. The pathways of HS and CS synthesis diverge after formation of the linkage region. The addition of GlcNAc to the linkage region commits the intermediate proteoglycan molecule to the assembly of HS. Similarly, the addition of a GalNAc commits it to CS synthesis.

CS assembly on the linkage region represents a default pathway.[19–21] HS assembly requires amino acid determinants proximal to the linkage tetrasaccharide. The enzyme necessary for initiating HS synthesis prefers sites containing multiple acidic residues, one or more hydrophobic residues, and repetitive Ser-Gly units.[22] Later on, it was discovered that certain structural determinants exist in glypican-1 that prevent CS formation. These may act by blocking access of the initiating β-GalNAc transferase. The proportion of HS to CS carried on a HS proteoglycan is cell type- (or tissue-) dependent. For example, the only "absolute" HS proteoglycan, glypican-1, carries 90% HS and 10% CS when expressed in COS cells and 80% HS and 20% CS when expressed in CHO cells.[23] Very low percentages of HS have been detected on certain CS proteoglycans, for example biglycan, and aggrecan.[24,25] The enzymes responsible for making essential GAG building blocks and linkage region are summarized in Table II.

The fine structure of HS is determined by the step-wise action of multiple enzymes and enzyme isoforms (Table III). The two subunits of the HS copolymerase are each encoded by genes that are known to be tumor suppressor genes (EXT-1 and EXT-2).[26,27] EXT genes in humans are associated with multiple hereditary exostoses, a syndrome characterized by neonatal growth plate tumors.

As shown in Table III, several HS modification enzymes are present in multiple isoforms. The expression of each isoform, including those of sulfotransferases NST, 6-OST and 3-OST, is tissue specific.[1] This is thought to be important for generating HS with unique saccharide sequences. In this regard, it is instructive to point out that 3-OST-1 modified HS binds antithrombin and has anticoagulant activity,[1,28] whereas 3-OST-3 modified HS binds to herpes envelope protein gD and serves as an entry receptor for the herpes simplex virus-1.[29,30] In contrast, 3-OST-5 modified HS binds to both antithrombin and herpes envelope protein gD and has both anticoagulant and entry receptor properties.[31]

In summary, at least 20 genes are involved in HS polymerization and modification. Combinatorial expression of these genes means that the structure of HS can be altered to modulate a wide variety of tissue-specific functions.

TABLE II

Cloned Enzymes for Making GAG Building Blocks and for GAG Initiation

DTDST sulfate transporter 1	UDP-GlcA transporter	UDP-glucose dehydrogenase	Galactosyltransferase I
PAPS synthetase-1	UDP-GlcNAc transporter	UDP-GlcA decarboxylase	Galactosyltransferase II
PAPS synthetase-2	UDP-GalNAc transporter	Xylosyltransferase I	Glucuronosyltransferase I (knockout mice; chapter "Mice Deficient in Glucuronyltransferase-I")
PAPS transporter	UDP-Gal transporter	Xylosyltransferase II	Glucuronosyltransferase II

Accession numbers of corresponding human genes can be found in Ref. 1. Three glucuronosyltransferases for linkage tetrasaccharide biosynthesis in *Drosophila* have been identified.[101]

TABLE III
CLONED ENZYMES FOR HS FORMATION

αGlcNAcTIA (EXTL1)	NDST-1	Epimerase	6-OST-3	3-OST-3B
αGlcNAcTIB (EXTL2)	NDST-2	2-OST	3-OST-1	3-OST-4
Copolymerase EXT1	NDST-3	6-OST-1	3-OST-2	3-OST-5
Copolymerase EXT2	NDST-4	6-OST-2	3-OST-3A	3-OST-6

Accession numbers of corresponding human genes can be found in Ref. 1.

EXT1 conditional knock out: chapter "Roles of Heparan Sulfate in Mammalian Brain Development: Current Views Based on the Findings from Ext1 Conditional Knockout Studies."

NDST-1 knockout and conditional knockout mice: chapters "Mice Deficient in Heparan Sulfate *N*-Deacetylase/*N*-Sulfotransferase 1, Endothelial Heparan Sulfate in Angiogenesis, and Hepatic Heparan Sulfate Proteoglycans and Endocytic Clearance of Triglyceride-Rich Lipoproteins."

Epimerase knockout mice: chapter "Mice Deficient in Glucuronyl C5-Epimerase—An Enzyme Converting Glucuronic Acid to Iduronic Acid in Heparan Sulfate/Heparin Biosynthesis."

6-OST-1 knockout mice: chapter "Mice Deficient in Heparan Sulfate 6-O-Sulfotransferase-1."

3-OST-1 knockout mice: chapter "Anticoagulant Heparan Sulfate: To Clot or Not."

TABLE IV
TYPES OF CS DISACCHARIDES

Major disaccharides	Other disaccharides found	
CS-A: GlcA-GalNAc4S	GlcA-GalNAc	GlcA2S-GalNAc
CS-B: IdoA-GalNAc4S	IdoA2S-GalNAc4S	GlcA3S-GalNAc
CS-C: GlcA-GalNAc6S	IdoAGalNAc4S6S	GlcA3S-GalNAc4S
CS-D: GlcA2S-GalNAc6S	IdoA2SGalNAc4S6S	GlcA3S-GalNAc4S6S
CS-E: GlcA-GalNAc4S6S	IdoA2S-GalNAc	GlcA3S-GalNAc6S

CS has been classified as CS-A, CS-B (dermatan sulfate), CS-C, CS-D, and CS-E (Table IV) according to the major constituent of the repeating disaccharides. However, all CSs are hybrid structures that contain more than two types of disaccharides. CS-B, or dermatan sulfate, is distinguished because it contains IdoA residues.

The fine structure of CS also depends on the temporal and tissue-specific expression of a variety of modifying enzymes and enzyme isoforms (Table V).

Some sulfation appears to be specific to CS or HS. GlcA 3-O-sulfation has been detected only in CS.[32,33] Conversely, N-sulfation occurs only in HS but not in CS. Because the HS epimerase requires N-sulfated residues for its activity and CS is devoid of N-sulfation, epimerization of GlcA to IdoA residues of CS are catalyzed by two distinct CS epimerases.[34]

TABLE V
Enzymes for CS Formation

GalNAc I	GlcA 3-OST	4OST3[102]	6OST3
GalNAc II	Epimerase I and II	IdoA-GalNAc 4-OST	6OST4
Chondroitin synthase	4-OST1	6OST1	GalNAc4S 6-OST
GlcA/IdoA 2-OST	4-OST2	6OST2	IdoA-GalNAc-IdoA 6OST

Accession numbers of non-referenced genes can be found in Ref. 2.

4-OST1 knockout mice: chapter "Roles of Heparan Sulfate in Mammalian Brain Development: Current Views Based on the Findings from Ext1 Conditional Knockout Studies."

IdoA-GalNAc 4-OST deficiency in human: chapters "Congenital Disorders of Glycosylation with Emphasis on Loss of Dermatan-4-Sulfotransferase."

In contrast to CS, which tend to have long tracts of fully modified disaccharides, the modification reactions in HS biosynthesis occur in clusters along the chain, with regions devoid of sulfation separating the modified tracts. This arrangement gives rise to segments referred to as *N*-acetylated (NA), *N*-sulfated (NS), and mixed domains (NA/NS). In general, the sulfation reactions fail to go to completion, resulting in tremendous chemical heterogeneity within the modified regions in CS and HS.

In summary, HS and CS are characterized by a linear chain of 20–400 disaccharide units. The disaccharide repeat unit in HS can be modified by N- and O-sulfation (6-O- and 3-O-sulfation of the glucosamine and 2-O-sulfation of the uronic acid) and by epimerization of the glucuronic acid to iduronic acid. The disaccharide repeat unit in CS can be modified by 4-O- and 6-O-sulfation of the galactosamine and 2-O- and 3-O-sulfation of the uronic acid and by epimerization of glucuronic acid to iduronic acid. Together, the five different modifications for disaccharides in HS and CS give rise to $2^5 = 32$ combinations. With 23 disaccharides found in CS and 24 found in HS,[1] a HS or CS hexasaccharide could have several thousand possible sequences, thereby making HS and CS not only the most acidic but also the most information-dense biopolymers found in nature. Understanding how and in what order the cells assemble specific HS and CS sequences is one of the important areas in research on GAG.

IV. GAG-Binding Proteins

GAGs participate in a variety of physiological processes such as binding, activation, or immobilization of various protein ligands, such as growth factors, cytokines, chemokines, extracellular matrix components, proteases, protease inhibitors, and lipoprotein lipase.[1,3,35–42] These interactions depend, to a large extent, on the composition and fine structure of the GAG chains, which in turn depend on the substrate specificity of the various biosynthetic enzymes and regulatory factors.

Heparin interacts with 23% of the plasma proteins.[43] More than 200 GAG-binding proteins have been described in literature (Fig. 2). To a large extent, these studies have focused on protein interactions with heparin. This bias may reflect the commercial availability of heparin, which is frequently used for fractionation studies and heparin-Sepharose affinity chromatography. The binding of protein ligands to heparin is thought to mimic the physiological interaction of proteins with the HS that is present on cell surfaces and in the extracellular matrix.

Most animal tissues and organs contain more CS than HS.[44] CS acts as a biological activator in a number of instances, in apparent independence of its co-existence with HS. Examples include modulation of axon growth,[45,46] wound healing,[47] NFκB transcriptional activation of endothelial cells,[48] circumscribed release of short-lived kinin hormones from precursor depots to regulate local blood pressure and inflammatory responses,[49] heparin cofactor II inhibition of localized coagulation,[50] and low density lipoprotein (LDL) binding that affects intramural retention of atherogenic lipoproteins.[51] A variety of bacterial-, viral-, and parasite-proteins also bind to GAGs (chapter "Diverse Functions of Glycosaminoglycans in Infectious Diseases").

GAGs regulate many growth factor signaling pathways[52] including, but not limited to, fibroblast growth factor (FGF)/FGFR,[53] hepatocyte growth factor (HGF)/c-Met,[54–57] glial cell line-derived neurotrophic factor (GDNF)/c-Ret/GFRα1,[58,59] vascular endothelial growth factor (VEGF)/VEGFR,[60,61] platelet derived growth factor (PDGF)/PDGFR,[62] BAFF/TACI,[63] Indian hedgehog (Ihh), Wnt, and BMP signaling pathways, where genetic studies revealed an absolute requirement for GAGs.[64,65]

Heparin has been used for preventing and treating thromboembolic disorders for over 70 years as it inhibits thrombin generation and thrombin activities. Thromboembolic disorders are the leading cause of disabilities and deaths in a variety of unrelated human diseases, such as coronary heart disease,[66] cancer,[67] diabetes,[68] kidney failure,[69] autoimmune diseases,[70] and heparin-induced thrombocytopenia and thrombosis (HITT).[71] Thrombin is the only known enzyme that causes thrombus formation. Thrombin also plays multiple roles in development, tissue repair, inflammation, and hemostasis.[72] Conditional loss of prothrombin leads to the rapid death of adult mice,[73] suggesting that thrombin is a key survival factor, which is continuously generated in blood circulation.

In 2007 and 2008, contaminated heparin was associated with hundreds of anaphylactic reactions and at least 149 anaphylactic reaction-associated deaths in the United States. Heparin is contaminated with chemically sulfated or

Morphogens
- Activin
- BMP-2, -4
- Chordin
- Sonic hedgehog
- Frizzled-related peptides
- Sprouty peptides
- Wnt (1–13)

Growth factor binding proteins (BP)
- Follistatin
- IGF BP-3, -5
- TGF-β BP
- Noggin

Anti-angiogenic factors
- Angiostatin
- Endostatin
- PF4

Collectins
- SPA, D
- MBP

Antimicrobial peptides
- PR-39
- Bac 5, 7
- β defensin

Growth factors and receptors
- EGF family
  - Amphiregulin
  - Betacellulin
  - Heparin-binding-EGF
  - Neuregulin
- FGFs (1–23)/FGFRs
- PDGF/PDGFR
- GDNF/cRet
- VEGFs/VEGFRs
- HGF/cMet
- TGFβ1, -2

ECM components
- Fibrin
- Fibronectin
- Interstitial collagens
- Laminins
- Pleiotropin (HB-GAM)
- Tenascin
- Thrombospondin
- Vitronectin
- Fibrillin
- Tropoelastin

Energy balance
- Agouti signaling peptide
- Agouti-related protein
- ApoB, E
- Lipoprotein lipase
- Triglyceride lipases

Tissue remodeling
- Tissue plasminogen activator
- Plasminogen activator inhibitor-1
- Protease nexin I
- TIMP-3

Complement proteins (25)
Contact system protein (4)
Coagulation proteins
- Antithrombin III
- Heparin cofactor II
- Leuserpin
- Tissue factor pathway inhibitor
- Thrombin
- Factors IX, X, XI, and XII

Proteinases
- Neutrophil elastase
- CathepsinG
- MCP-4, -5
- Carboxypeptidase A

Unclassified
- Acetylcholinesterase
- HIP
- Thyroglobulin
- Cyclophilin A
- Superoxidase dismutase

Cell adhesion molecules
- E-, L-, P-selectins
- MAC-1
- N-CAM
- PECAM

Chemokines
- C–C, for example MIP-1a, RANTES
- CXC, for example IP-10, IL-8

Cytokines
- GM-CSF
- IL-2, -3, -4, -5, -7, -12
- Interferon γ
- Kininogen
- TNF-α

Viral and parasite
Coat proteins
- HIV-1 tat
- HIV-1 gp41, 120
- HSV gB, gC, gD
- HHV-6 gp65
- HHV-8 gK8.1A
- Circumsprorozoite

Fig. 2. GAG-binding proteins. The GAG-binding proteins reported in literature were originally cataloged and compiled by Dr. Merton Bernfield and were updated by the author.

modified low sulfated GAGs containing HS, CS, and DS.[74–76] Contaminated heparin-associated anaphylactic reactions are a result of an upregulated immune system by oversulfated GAGs through contact system activation.[74,77] The contact system was first discovered as an *in vitro* thrombin generation system, wherein artificial surfaces induce thrombin generation and clotting.[78] Injury associated-thrombin generation in animals instantly leads to immune system upregulation. However, the *in vivo* role of contact system activation-generated thrombin and immune system upregulation has been overlooked during the past 50 years.

Heparin-induced thrombocytopenia with thrombosis (HITT) is the most severe side effect of heparin administration. HITT patients may die or have permanent sequelae such as stroke or limb amputation. Contaminated heparin is associated with high incidence of HITT.[79,80] It turns out that whereas oversulfated GAGs induce thrombin activities in both normal and HIT patient plasmas through contact system activation, authentic heparin induces thrombin activities only in HIT patient plasmas containing autoantibodies against the protein/heparin complex.[81,82] These data suggest that the negatively charged protein/contaminated heparin or IgG/PF4/heparin complex activate the contact system and produce thrombin in human plasma. Thrombin partially activates the platelets, allowing subsequent platelet activation through IgG/Fc receptor II signaling.[82] In addition, kallikrein and thrombin generate potent immune system activators such as C3a, C5a, and bradykinin.[83–85]

The newly discovered molecular mechanism of contaminated heparin-induced anaphylactic reactions suggests that GAG/protein complexes initiate immune reactions through contact system activation. The contact system can recognize different protein complexes and foreign surfaces.[74,86] Indeed, over 20 different types of GAG-containing protein aggregates (amyloids) (chapter "Heparan Sulfate Proteoglycans in Amyloidosis") are associated with different human diseases, such as Alzheimer's, diabetes, and prion/transmissible spongiform encephalopathies.[87,88]

In summary, GAG/protein interactions play multiple roles in development, health, and disease through different molecular mechanisms. GAGs remain the most complex and the least understood components of modern molecular biology and human physiology.

V. Concluding Remarks

HS or CS GAG chains are assembled in the *Golgi* on a tetrasaccharide linkage group that is present in different sites of proteoglycan core proteins. Both HS and CS are polymerized and modified by enzymes that are encoded by more than 40 genes. Because of the expression repertoire of the GAG assembly enzymes, each HS and CS chain has a sulfation pattern, chain length, and fine structure that is potentially unique to each cell. Cell surface GAGs turn over within 1/8 to 1/3 of a cell cycle.[89] This means their structures are able to rapidly change in response to a variety of environmental factors. Therefore, different from RNA and protein, GAGs are not coded on a template but coded on a dynamic assembly line contributed by environment factor-regulated biosynthesis and modification enzymes produced by specific cells.

Indeed, GAG-protein interactions seem to be cell- and tissue-specific and independent of the specific proteoglycan core proteins that the GAG chains are covalently attached to.[90,91] This idea is further supported by the fact that many proteins require GAGs to facilitate their biological functions and the involvement of most of these proteins in processes that are fundamental to survival during development at the level of cell signaling and at the organismal level through immune system regulation and hemostasis control (dealt in remaining chapters). Therefore, understanding GAG-protein interactions at the molecular level will be key in using such knowledge to prevent and cure certain devastating human diseases, such as heparin-induced thrombocytopenia/thrombosis (chapter "Molecular Mechanism Underlines Heparin Induced Thrombocytopenia and Thrombosis"), rheumatoid arthritis (chapter "Chondroitin Sulfate and Abnormal Contact System in Rheumatoid Arthritis"), Lupus (chapter "Activated Contact System and Abnormal Glycosaminoglycans in Lupus and Other Auto- and Non-Autoimmune Diseases"), different kinds of cancers (chapter "Glycosaminoglycans and Activated Contact System in Cancer Patient Plasmas"), Alzheimer's, diabetes, and prion/transmissible spongiform encephalopathies.

References

1. Esko JD, Selleck SB. Order out of chaos: assembly of ligand binding sites in heparan sulfate. *Annu Rev Biochem* 2002;**71**:435–71.
2. Silbert JE, Sugumaran G. Biosynthesis of chondroitin/dermatan sulfate. *IUBMB Life* 2002;**54**:177–86.
3. Lander AD, Selleck SB. The elusive functions of proteoglycans: in vivo veritas. *J Cell Biol* 2000;**148**:227–32.
4. Marynen P, Zhang J, Cassiman JJ, Van den Berghe H, David G. Partial primary structure of the 48- and 90-kilodalton core proteins of cell surface-associated heparan sulfate proteoglycans of lung fibroblasts. Prediction of an integral membrane domain and evidence for multiple distinct core proteins at the cell surface of human lung fibroblasts. *J Biol Chem* 1989;**264**:7017–24.
5. David G, Van den Berghe H. Cell-surface heparan sulfate and heparan-sulfate/chondroitin-sulfate hybrid proteoglycans of mouse mammary epithelial cells. *Eur J Biochem* 1989;**178**:609–17.

6. David G, Lories V, Decock B, Marynen P, Cassiman JJ, Van den Berghe H. Molecular cloning of a phosphatidylinositol-anchored membrane heparan sulfate proteoglycan from human lung fibroblasts. *J Cell Biol* 1990;**111**:3165–76.
7. David G, van der Schueren B, Marynen P, Cassiman JJ, van den Berghe H. Molecular cloning of amphiglycan, a novel integral membrane heparan sulfate proteoglycan expressed by epithelial and fibroblastic cells. *J Cell Biol* 1992;**118**:961–9.
8. Lories V, Cassiman JJ, Van den Berghe H, David G. Differential expression of cell surface heparan sulfate proteoglycans in human mammary epithelial cells and lung fibroblasts. *J Biol Chem* 1992;**267**:1116–22.
9. Heremans A, Cassiman JJ, Van den Berghe H, David G. Heparan sulfate proteoglycan from the extracellular matrix of human lung fibroblasts. Isolation, purification, and core protein characterization. *J Biol Chem* 1988;**263**:4731–9.
10. Vogel KG, Peterson DW. Extracellular, surface, and intracellular proteoglycans produced by human embryo lung fibroblasts in culture (IMR-90). *J Biol Chem* 1981;**256**:13235–42.
11. Hartmann U, Maurer P. Proteoglycans in the nervous system–the quest for functional roles in vivo. *Matrix Biol* 2001;**20**:23–35.
12. Vlodavsky I, Friedmann Y. Molecular properties and involvement of heparanase in cancer metastasis and angiogenesis. *J Clin Invest* 2001;**108**:341–7.
13. Morimoto-Tomita M, Uchimura K, Werb Z, Hemmerich S, Rosen SD. Cloning and Characterization of Two Extracellular Heparin-degrading Endosulfatases in Mice and Humans. *J Biol Chem* 2002;**277**:49175–85.
14. Dhoot GK, Gustafsson MK, Ai X, Sun W, Standiford DM, Emerson Jr CP. Regulation of Wnt signaling and embryo patterning by an extracellular sulfatase. *Science* 2001;**293**:1663–6.
15. Kosir MA, Quinn CC, Wang W, Tromp G. Matrix glycosaminoglycans in the growth phase of fibroblasts: more of the story in wound healing. *J Surg Res* 2000;**92**:45–52.
16. Mack M, Pfirstinger J, Weber C, Weber KS, Nelson PJ, Rupp T, et al. Chondroitin sulfate A released from platelets blocks RANTES presentation on cell surfaces and RANTES-dependent firm adhesion of leukocytes. *Eur J Immunol* 2002;**32**:1012–20.
17. Piepkorn M, Hovingh P, Linker A. Evidence for independent metabolism and cell surface localization of cell surface localization of cellular proteoglycans and glycosaminoglycan free chains. *J Cell Physiol* 1988;**135**:189–99.
18. Piepkorn M, Hovingh P, Linker A. Glycosaminoglycan free chains. External plasma membrane components distinct from the membrane proteoglycans. *J Biol Chem* 1989;**264**:8662–9.
19. Zhang L, Esko JD. Amino acid determinants that drive heparan sulfate assembly in a proteoglycan. *J Biol Chem* 1994;**269**:19295–9.
20. Zhang L, David G, Esko JD. Repetitive Ser-Gly sequences enhance heparan sulfate assembly in proteoglycans. *J Biol Chem* 1995;**270**:27127–35.
21. Zhang L, Esko JD. Accumulation of a pentasaccharide terminating in alpha-N- acetylglucosamine in an animal cell mutant defective in heparan sulfate biosynthesis. *J Biol Chem* 1995;**270**:12557–62.
22. Esko JD, Zhang L. Influence of core protein sequence on glycosaminoglycan assembly. *Curr Opin Struct Biol* 1996;**6**:663–70.
23. Chen RL, Lander AD. Mechanisms underlying preferential assembly of heparan sulfate on glypican-1. *J Biol Chem* 2001;**276**:7507–17.
24. Govindraj P, West L, Koob TJ, Neame P, Doege K, Hassell JR. Isolation and identification of the major heparan sulfate proteoglycans in the developing bovine rib growth plate. *J Biol Chem* 2002;**277**:19461–9.
25. Kresse H, Seidler DG, Muller M, Breuer E, Hausser H, Roughley PJ, et al. Different usage of the glycosaminoglycan attachment sites of biglycan. *J Biol Chem* 2001;**276**:13411–6.

26. McCormick C, Leduc Y, Martindale D, Mattison K, Esford LE, Dyer AP, et al. The putative tumour suppressor EXT1 alters the expression of cell-surface heparan sulfate. *Nat Genet* 1998;**19**:158–61.
27. Lind T, Tufaro F, McCormick C, Lindahl U, Lidholt K. The putative tumor suppressors EXT1 and EXT2 are glycosyltransferases required for the biosynthesis of heparan sulfate. *J Biol Chem* 1998;**273**:26265–8.
28. Zhang L, Lawrence R, Schwartz JJ, Bai X, Wei G, Esko JD, et al. The effect of precursor structures on the action of 3-OST-1 and the biosynthesis of anticoagulant heparan sulfate. *J Biol Chem* 2001;**24**:24.
29. Shukla D, Liu J, Blaiklock P, Shworak NW, Bai X, Esko JD, et al. A novel role for 3-O-sulfated heparan sulfate in herpes simplex virus 1 entry. *Cell* 1999;**99**:13–22.
30. Liu J, Shriver Z, Pope RM, Thorp SC, Duncan MB, Copeland RJ, et al. Characterization of a heparan sulfate octasaccharide that binds to herpes simplex virus type 1 glycoprotein D. *J Biol Chem* 2002;**277**:33456–67.
31. Xia G, Chen J, Tiwari V, Ju W, Li JP, Malmstrom A, et al. Heparan sulfate 3-O-sulfotransferase isoform 5 generates both an antithrombin-binding site and an entry receptor for herpes simplex virus, type 1. *J Biol Chem* 2002;**277**:37912–9.
32. Sugahara K, Tanaka Y, Yamada S, Seno N, Kitagawa H, Haslam SM, et al. Novel sulfated oligosaccharides containing 3-O-sulfated glucuronic acid from king crab cartilage chondroitin sulfate K. Unexpected degradation by chondroitinase ABC. *J Biol Chem* 1996;**271**:26745–54.
33. Kinoshita A, Yamada S, Haslam SM, Morris HR, Dell A, Sugahara K. Novel tetrasaccharides isolated from squid cartilage chondroitin sulfate E contain unusual sulfated disaccharide units GlcA(3-O-sulfate)beta1-3GalNAc(6-O-sulfate) or GlcA(3-O-sulfate)beta1-3GalNAc. *J Biol Chem* 1997;**272**:19656–65.
34. Maccarana M, Kalamajski S, Kongsgaard M, Magnusson SP, Oldberg A, Malmstrom A. Dermatan sulfate epimerase 1-deficient mice have reduced content and changed distribution of iduronic acids in dermatan sulfate and an altered collagen structure in skin. *Mol Cell Biol* 2009;**29**:5517–28.
35. Rosenberg RD, Shworak NW, Liu J, Schwartz JJ, Zhang L. Heparan sulfate proteoglycans of the cardiovascular system. Specific structures emerge but how is synthesis regulated? *J Clin Invest* 1997;**100**:S67–75.
36. Esko JD, Lindahl U. Molecular diversity of heparan sulfate. *J Clin Invest* 2001;**108**:169–73.
37. Perrimon N, Bernfield M. Specificities of heparan sulphate proteoglycans in developmental processes. *Nature* 2000;**404**:725–8.
38. Selleck SB. Proteoglycans and pattern formation: sugar biochemistry meets developmental genetics. *Trends Genet* 2000;**16**:206–12.
39. Lindahl U. What else can 'Heparin' do? *Haemostasis* 1999;**29**(Suppl. S1):38–47.
40. Reizes O, Lincecum J, Wang Z, Goldberger O, Huang L, Kaksonen M, et al. Transgenic expression of syndecan-1 uncovers a physiological control of feeding behavior by syndecan-3. *Cell* 2001;**106**:105–16.
41. Iozzo RV. Heparan sulfate proteoglycans: intricate molecules with intriguing functions. *J Clin Invest* 2001;**108**:165–7.
42. Iozzo RV, San Antonio JD. Heparan sulfate proteoglycans: heavy hitters in the angiogenesis arena. *J Clin Invest* 2001;**108**:349–55.
43. Saito A, Munakata H. Analysis of plasma proteins that bind to glycosaminoglycans. *Biochim Biophys Acta* 2007;**1770**:241–6.
44. Zhang L. Amino acid sequences that regulate heparan sulfate biosynthesis. In: Zhang L, editor. *Biochemistry and molecular genetics*. Birmingham: University of Alabama at Birmingham; 1995. p. 201.

45. Margolis RU, Margolis RK. Chondroitin sulfate proteoglycans as mediators of axon growth and pathfinding. *Cell Tissue Res* 1997;**290**:343–8.
46. Inatani M, Honjo M, Otori Y, Oohira A, Kido N, Tano Y, et al. Inhibitory effects of neurocan and phosphacan on neurite outgrowth from retinal ganglion cells in culture. *Invest Ophthalmol Vis Sci* 2001;**42**:1930–8.
47. Penc SF, Pomahac B, Winkler T, Dorschner RA, Eriksson E, Herndon M, et al. Dermatan sulfate released after injury is a potent promoter of fibroblast growth factor-2 function. *J Biol Chem* 1998;**273**:28116–21.
48. Penc SF, Pomahac B, Eriksson E, Detmar M, Gallo RL. Dermatan sulfate activates nuclear factor-kappab and induces endothelial and circulating intercellular adhesion molecule-1. *J Clin Invest* 1999;**103**:1329–35.
49. Renne T, Muller-Esterl W. Cell surface-associated chondroitin sulfate proteoglycans bind contact phase factor H-kininogen. *FEBS Lett* 2001;**500**:36–40.
50. Maimone MM, Tollefsen DM. Structure of a dermatan sulfate hexasaccharide that binds to heparin cofactor II with high affinity. *J Biol Chem* 1990;**265**:18263–71.
51. Kaplan M, Aviram M. Macrophage plasma membrane chondroitin sulfate proteoglycan binds oxidized low-density lipoprotein. *Atherosclerosis* 2000;**149**:5–17.
52. Conrad HE. *Heparin-binding proteins.* San Diego: Academic Press; 1998.
53. McDowell LM, Frazier BA, Studelska DR, Giljum K, Chen J, Liu J, et al. Inhibition or activation of Apert syndrome FGFR2 (S252W) signaling by specific glycosaminoglycans. *J Biol Chem* 2006;**281**:6924–30.
54. Schwall RH, Chang LY, Godowski PJ, Kahn DW, Hillan KJ, Bauer KD, et al. Heparin induces dimerization and confers proliferative activity onto the hepatocyte growth factor antagonists NK1 and NK2. *J Cell Biol* 1996;**133**:709–18.
55. Zioncheck TF, Richardson L, Liu J, Chang L, King KL, Bennett GL, et al. Sulfated oligosaccharides promote hepatocyte growth factor association and govern its mitogenic activity. *J Biol Chem* 1995;**270**:16871–8.
56. Lyon M, Deakin JA, Rahmoune H, Fernig DG, Nakamura T, Gallagher JT. Hepatocyte growth factor/scatter factor binds with high affinity to dermatan sulfate. *J Biol Chem* 1998;**273**:271–8.
57. Lyon M, Deakin JA, Lietha D, Gherardi E, Gallagher JT. The interactions of hepatocyte growth factor/scatter factor and its NK1 and NK2 variants with glycosaminoglycans using a modified gel mobility shift assay: elucidation of the minimal size of binding and activatory oligosaccharides. *J Biol Chem* 2004;**279**:43560–7.
58. Barnett MW, Fisher CE, Perona-Wright G, Davies JA. Signalling by glial cell line-derived neurotrophic factor (GDNF) requires heparan sulphate glycosaminoglycan. *J Cell Sci* 2002;**115**:4495–503.
59. Davies JA, Yates EA, Turnbull JE. Structural determinants of heparan sulphate modulation of GDNF signalling. *Growth Factors* 2003;**21**:109–19.
60. Ashikari-Hada S, Habuchi H, Kariya Y, Kimata K. Heparin regulates vascular endothelial growth factor165-dependent mitogenic activity, tube formation, and its receptor phosphorylation of human endothelial cells: comparison of the effects of heparin and modified heparins. *J Biol Chem* 2005;**280**:31508–15.
61. Dougher AM, Wasserstrom H, Torley L, Shridaran L, Westdock P, Hileman RE, et al. Identification of a heparin binding peptide on the extracellular domain of the KDR VEGF receptor. *Growth Factors* 1997;**14**:257–68.
62. Rolny C, Spillmann D, Lindahl U, Claesson-Welsh L. Heparin amplifies platelet-derived growth factor (PDGF)-BB-induced PDGF alpha -receptor but not PDGF beta -receptor tyrosine phosphorylation in heparan sulfate-deficient cells. Effects on signal transduction and biological responses. *J Biol Chem* 2002;**277**:19315–21.

63. Bischof D, Elsawa SF, Mantchev G, Yoon J, Michels GE, Nilson A. Selective activation of TACI by syndecan-2. *Blood* 2005; 2005-2001-0256.
64. Ruppert R, Hoffmann E, Sebald W. Human bone morphogenetic protein 2 contains a heparin-binding site which modifies its biological activity. *Eur J Biochem* 1996;**237**:295–302.
65. Rubin JB, Choi Y, Segal RA. Cerebellar proteoglycans regulate sonic hedgehog responses during development. *Development* 2002;**129**:2223–32.
66. Duvall WL. Antithrombotic therapy. *Curr Mol Med* 2006;**6**:603–19.
67. Mousa SA. Role of current and emerging antithrombotics in thrombosis and cancer. *Timely Top Med Cardiovasc Dis* 2006;**10**:E19.
68. Bartnik M, Norhammar A, Ryden L. Hyperglycaemia and cardiovascular disease. *J Intern Med* 2007;**262**:145–56.
69. Grand'Maison A, Charest AF, Geerts WH. Anticoagulant use in patients with chronic renal impairment. *Am J Cardiovasc Drugs* 2005;**5**:291–305.
70. Swigris JJ, Fischer A, Gillis J, Meehan RT, Brown KK. Pulmonary and thrombotic manifestations of systemic lupus erythematosus. *Chest* 2008;**133**:271–80.
71. Linkins L, Warkentin TE. The approach to heparin-induced thrombocytopenia. *Semin Respir Crit Care Med* 2008;**1**:66–74.
72. Jenny NS, Lundblad RL, Mann KG. Thrombin. In: Colman RW, editor. *Hemostasis and thrombosis*. Philadelphia: Lippincott Williams & Wilkins; 2006. p. 193–213.
73. Mullins ES, Kombrinck KW, Talmage KE, Shaw MA, Witte DP, Ullman JM, et al. Genetic elimination of prothrombin in adult mice is not compatible with survival and results in spontaneous hemorrhagic events in both heart and brain. *Blood* 2009;**113**:696–704.
74. Pan J, Qi Y, Zhou X, Lu H, Ramacciotti E, Zhang L. Chemically oversulfated glycosaminoglycans are potent modulators of contact system activation and different cell signaling pathways. *J Biol Chem* 2010;doi:10.1074/jbc.M109.063735.
75. Pan J, Qian Y, Zhou X, Pazandak A, Frazier SB, Weiser P, et al. Oversulfated chondroitin sulfate is not the sole contaminant in heparin. *Nat Biotech* 2010;**28**:203–7.
76. Pan J, Qian Y, Zhou X, Pazandak A, Frazier SB, Weiser P, et al. Identification of chemically sulfated/desulfated glycosaminoglycans in contaminated heparins and development of a simple assay for the detection of most contaminants in heparin. *Glycobiol. Insight* 2010;**2**:1–12.
77. Kishimoto TK, Viswanathan K, Ganguly T, Elankumaran S, Smith S, Pelzer K, et al. Contaminated heparin associated with adverse clinical events and activation of the contact system. *N Engl J Med* 2008;**358**:2457–67.
78. Waaler BA. Contact activation in the intrinsic blood clotting system; studies on a plasma product formed on contact with glass and similar surfaces. *Scand J Clin Lab Invest* 1959;**11**:1–133.
79. Adiguzel C, Bansal V, Litinas E, Cunanan J, Iqbal O, Nelson K, et al. Increased prevalence of antiheparin platelet factor 4 antibodies in patients may be due to contaminated heparin. *Clin Appl Thromb Hemost* 2009;**15**:145–51.
80. Greinacher A, Warkentin TE. Contaminated heparin. *N Engl J Med* 2008;**359**:1291–2 author reply 1293.
81. Qian Y, Pan J, Zhou X, Weiser P, Hourcade D, Lizsewski KM, Atkinson JP, Lu H, Zhang L. Oversulfated heparin byproducts induce thrombin generation through contact system activation. *Clin Appl Thromb Hemost* 2010;**16**:244–50.
82. Qian Y, Pan J, Zhou X, Weiser P, Lu H, Shih FF, Porche-Sorbet R, Eby C, Zhang L. Heparin and oversulfated heparin byproduct induce thrombin generation through contact system activation in plasma of patients with HIT. *Clin Appl Thromb Hemost* 2010;**16**:251–60.
83. Clark A, Weymann A, Hartman E, Turmelle Y, Carroll M, Thurman JM, et al. Evidence for non-traditional activation of complement factor C3 during murine liver regeneration. *Mol Immunol* 2008;**45**:3125–32.

84. Colman RW. Contact activation (kallikrein-kinin) pathway: multiple physiologic and pathophysiologic activities. In: Colman RW, editor. *Hemostasis and thrombosis*Philadelphia: Lippincott Williams & Wilkins; 2006. p. 107–30.
85. Huber-Lang M, Sarma JV, Zetoune FS, Rittirsch D, Neff TA, McGuire SR, et al. Generation of C5a in the absence of C3: a new complement activation pathway. *Nat Med* 2006;**12**:682–7.
86. Maas C, Govers-Riemslag JW, Bouma B, Schiks B, Hazenberg BP, Lokhorst HM, et al. Misfolded proteins activate Factor XII in humans, leading to kallikrein formation without initiating coagulation. *J Clin Invest* 2008;**118**:3208–18.
87. Warner RG, Hundt C, Weiss S, Turnbull JE. Identification of the heparan sulfate binding sites in the cellular prion protein. *J Biol Chem* 2002;**277**:18421–30.
88. Supattapone S. Prion protein conversion in vitro. *J Mol Med* 2004;**82**:348–56.
89. Yanagishita M, Hascall VC. Cell surface heparan sulfate proteoglycans. *J Biol Chem* 1992;**267**:9451–4.
90. Sanderson RD, Turnbull JE, Gallagher JT, Lander AD. Fine structure of heparan sulfate regulates syndecan-1 function and cell behavior. *J Biol Chem* 1994;**269**:13100–6.
91. Shworak NW, Kojima T, Rosenberg RD. Isolation and characterization of ryudocan and syndecan heparan sulfate proteoglycans, core proteins, and cDNAs from a rat endothelial cell line. *Haemostasis* 1993;**23**(Suppl 1):161–76.
92. Kuehn MH, Hageman GS. Molecular characterization and genomic mapping of human IPM 200, a second member of a novel family of proteoglycans. *Mol Cell Biol Res Commun* 1999;**2**:103–10.
93. Acharya S, Foletta VC, Lee JW, Rayborn ME, Rodriguez IR, Young 3rd WS, et al. SPACRCAN, a novel human interphotoreceptor matrix hyaluronan-binding proteoglycan synthesized by photoreceptors and pinealocytes. *J Biol Chem* 2000;**275**:6945–55.
94. Wassenhove-McCarthy DJ, McCarthy KJ. Molecular characterization of a novel basement membrane-associated proteoglycan, leprecan. *J Biol Chem* 1999;**274**:25004–17.
95. Sasaki T, Mann K, Timpl R. Modification of the laminin alpha 4 chain by chondroitin sulfate attachment to its N-terminal domain. *FEBS Lett* 2001;**505**:173–8.
96. Pluschke G, Vanek M, Evans A, Dittmar T, Schmid P, Itin P, et al. Molecular cloning of a human melanoma-associated chondroitin sulfate proteoglycan. *Proc Natl Acad Sci USA* 1996;**93**:9710–5.
97. Ikegawa S, Sano M, Koshizuka Y, Nakamura Y. Isolation, characterization and mapping of the mouse and human PRG4 (proteoglycan 4) genes. *Cytogenet Cell Genet* 2000;**90**:291–7.
98. Sassetti C, Van Zante A, Rosen SD. Identification of endoglycan, a member of the CD34/podocalyxin family of sialomucins. *J Biol Chem* 2000;**275**:9001–10.
99. Bechard D, Gentina T, Delehedde M, Scherpereel A, Lyon M, Aumercier M, et al. Endocan is a novel chondroitin sulfate/dermatan sulfate proteoglycan that promotes hepatocyte growth factor/scatter factor mitogenic activity. *J Biol Chem* 2001;**276**:48341–9.
100. Wu RR, Couchman JR. cDNA cloning of the basement membrane chondroitin sulfate proteoglycan core protein, bamacan: a five domain structure including coiled-coil motifs. *J Cell Biol* 1997;**136**:433–44.
101. Kim BT, Tsuchida K, Lincecum J, Kitagawa H, Bernfield M, Sugahara K. Identification and characterization of three *Drosophila melanogaster* glucuronyltransferases responsible for the synthesis of the conserved glycosaminoglycan-protein linkage region of proteoglycans. Two novel homologs exhibit broad specificity towards oligosaccharides from proteoglycans, glycoproteins and glycosphingolipids. *J Biol Chem* 2003;**278**:9116–24.
102. Kang HG, Evers MR, Xia G, Baenziger JU, Schachner M. Molecular cloning and expression of an N-acetylgalactosamine-4-O- sulfotransferase that transfers sulfate to terminal and non-terminal beta1, 4-linked N-acetylgalactosamine. *J Biol Chem* 2001;**3**:3.

Mice Deficient in Glucuronyltransferase-I

Tomomi Izumikawa and Hiroshi Kitagawa

Department of Biochemistry, Kobe Pharmaceutical University, Kobe, Japan

Chondroitin sulfate (CS), dermatan sulfate (DS), heparan sulfate (HS), and heparin (Hep) are a class of glycosaminoglycans (GAGs) that are distributed on the surface of virtually all cells and in the extracellular matrices. CS/DS and HS/Hep chains share a common carbohydrate–protein linkage region structure, GlcAβ1-3Galβ1-3Galβ1-4Xylβ1-*O*-Ser. Glucuronyl transfer to the Gal residue, the final biosynthetic step in the common linkage region, is catalyzed by a key enzyme, β1,3-glucuronyltransferase, which is termed glucuronyltransferase I (GlcAT-I). As it has been reported that the expression level of GlcAT-I correlates well with the amount of GAGs, GlcAT-I is thought to regulate the expression of GAGs. In fact, a defect in the *squashed vulva 8* (*sqv-8*) gene which encodes GlcAT-I in *Caenorhabditis elegans* eliminates the expression of GAGs and the mutant worms show not only a perturbation in vulval invagination but also a defect in the cytokinesis in fertilized eggs, resulting in alternating cell division and cell fusion. Here, we summarize the recent knowledge on the roles of GlcAT-I in mammalian GAG biosynthesis and embryonic cell division using GlcAT-I knock-out mice.

Progress in Molecular Biology and Translational Science, Vol. 93
DOI: 10.1016/S1877-1173(10)93002-0

I. Introduction

Chondroitin sulfate (CS), dermatan sulfate (DS), heparan sulfate (HS), and heparin (Hep) are a class of glycosaminoglycans (GAGs) that are distributed on the surface of virtually all cells and in the extracellular matrices. CS/DS and HS/Hep are covalently linked to a specific serine residue in the core protein and occur as CS/DS proteoglycans (PGs) and HS-PGs. Many of the physiological roles of CS/DS-PGs and HS-PGs are thought to be due to the CS/DS and HS side chains with the core proteins largely playing the role of a scaffold in order to make CS/DS and HS functionally available for binding to a variety of ligands. In fact, gene-targeting technology in vertebrates and invertebrates has led to elucidation of the physiological functions of HS during development and morphogenesis in addition to their regulation of signaling molecules. In contrast to the series of model organisms that are deficient in HS, we have generated a model that is lacking the CS backbone biosynthesis in *Caenorhabditis elegans* only.[1] A study of these worms revealed that nonsulfated chondroitin is required for normal cell division and cytokinesis at an early developmental stage[2,3]; whereas HS is essential for embryonic morphogenesis in the later stages of development.[4] These observations suggested that, in *C. elegans*, even though the structure of chondroitin is similar to that of HS, the function of chondroitin is different from that of HS.[4] In mice, although the deficiency of an enzyme that synthesizes the HS backbones leads to neonatal lethality with not only abnormal organogenesis but also with aberrations in the signaling pathways of morphogens and growth factors,[5,6] little is known about the roles of CS, mainly due to the unexpected redundancy of CS-synthesizing enzymes, thereby making the functional analysis of CS all the more difficult.[1]

CS/DS and HS/Hep chains are synthesized onto a common carbohydrate–protein linkage region structure, GlcAβ1-3Galβ1-3Galβ1-4Xylβ1-*O*-Ser.[7] The linkage region tetrasaccharide is formed by the sequential stepwise addition of monosaccharide residues by their respective specific glycosyltransferases, that is, xylosyltransferase, galactosyltransferase-I, galactosyltransferase-II, and glucuronyltransferase-I (GlcAT-I).[8] The repeating disaccharide region, [(-4GlcAβ1-4GlcNAcα1-)$_n$], of HS/Hep is synthesized on the linkage region by the HS copolymerase complex of EXT1 and EXT2.[9,10] In contrast, the repeating disaccharide region [(-4GlcAβ1-3GalNAcβ1-)$_n$] of CS/DS is formed on the linkage region by any two combinations of chondroitin synthases-1,[11] -2,[12] -3,[13] and the chondroitin polymerizing factor.[14] Also, the functionally redundant multiple glycosyltransferases involved in CS/DS synthesis have been cloned.[15,16] Thus, as previously mentioned, this redundancy makes it difficult to investigate the specific role of CS/DS during early embryogenesis in mammals.

To clarify the functions of CS/DS during mammalian early embryogenesis, we focused on GlcAT-I. As GlcAT-I transfers GlcA from UDP-GlcA to the trisaccharide-serine (Galβ1-3Galβ1-4Xylβ1-*O*-Ser) and thus finalizes the formation of the common linkage region,[17,18] *GlcAT-I* knockout mice would show complete elimination of CS/DS as well as HS/Hep. In addition, transfection of CHO cell mutants (defective in GlcAT-I) with GlcAT-I cDNA augments GAG synthesis to levels that are approximately twofold higher than those observed in wild-type cells,[19] suggesting that GlcAT-I activity is rate-limiting in GAG biosynthesis. Thus, we generated *GlcAT-I* knockout mice and attempted to analyze *in vivo* functions of GlcAT-I and CS/DS at an early developmental stage.[20] This chapter will discuss the significant recent advances in our understanding of the roles of GlcAT-I in mouse GAG biosynthesis and embryonic cell division. In particular, the importance of CS in embryonic cell division and cytokinesis will be presented.

II. *GlcAT-I* Gene Structure

The human *GlcAT-I* gene is located on chromosome 11q12.3, and spans approximately 7 kb of human genomic DNA.[21] The transcript is encoded by five exons, and all exons contain the coding sequences indicated in Fig. 1A. In addition, exon V contains the 3′-untranslated region, which includes the polyadenylation signal AATAAA. Notably, the human genome also contains a related, processed pseudogene that covers approximately 1.4 kb of the genomic sequence and shares 95.3% nucleotide identity with exons 1–5 of *GlcAT-I*.[21] The pseudogene is localized to chromosome 3. The mouse *GlcAT-I* locus is located on chromosome 19-A, is organized in a very similar fashion (Fig. 1B), and spans approximately 8.5 kb of the mouse genomic sequence. In contrast to the human genome, there is no *GlcAT-I* pseudogene in the mouse genome. So far, orthologues have been found at least in the zebrafish, *Drosophila*, and *C. elegans* genomes,[22] suggesting the existence of conserved functions in different organisms.

III. cDNA and Protein Structure of GlcAT-I

GlcAT-I activity was first detected in a chick embryo cartilage extract[23] and was subsequently partially purified from embryonic chick brain[24] and mouse mastocytoma cells[25]; however, attempts to purify GlcAT-I to homogeneity have not been successful due to its low concentration and difficulty in solubilizing the enzyme. Previously, the cDNA encoding a glucuronyltransferase involved in the biosynthesis of the HNK-1 carbohydrate epitope on glycoproteins (GlcAT-P) was cloned.[26] GlcAT-P is a β1,3-glucuronyltransferase that utilizes glycoprotein acceptor substrates having the terminal Galβ1-4GlcNAc

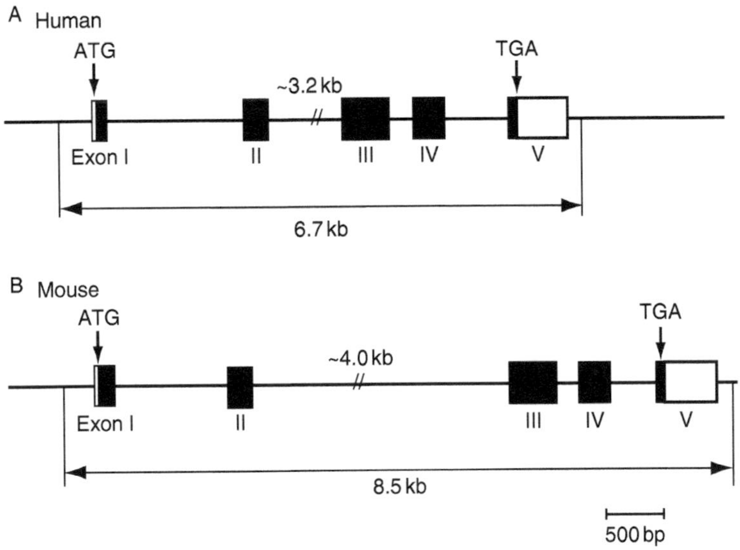

FIG. 1. Comparison of the genomic organization of human and mouse GlcAT-I. (A) Schematic illustration of the human *GlcAT-I* locus on chromosome 11q12.3. *GlcAT-I* gene consists of five exons and spans approximately 7 kb of human genomic sequence. (B) Schematic illustration of the mouse *GlcAT-I* locus on chromosome 19-A. *GlcAT-I* gene consists of five exons and spans approximately 8.5 kb of mouse genomic sequence. Its genomic organization is very similar to that of human GlcAT-I. Exon regions are denoted by boxes. Closed boxes represent the coding sequence, and open boxes denote 5′- and 3′-untranslated sequences. Translation initiation (ATG) and termination (TGA) codons are also shown. Black horizontal bars denote introns.

sequence. Mixed-substrate experiments have indicated that GlcAT-I and GlcAT-P are distinct enzymes, although the two enzymatic reactions are similar.[27] In view of the resemblance in the reactions catalyzed by GlcAT-I and GlcAT-P, we cloned the cDNA encoding GlcAT-I from human placenta based upon information on the amino acid sequence alignment of rat GlcAT-P with the putative proteins in *C. elegans* and *Schistosoma mansoni.*[17] Subsequently, based on the sequence homology to human GlcAT-I (GeneBank TM accession number AB009598), the mouse cDNA and genes were identified (GeneBank TM accession number AB019523). The human GlcAT-I cDNA had a single open reading frame of 1005-bp coding for a 335 amino acid protein showing the characteristics of a type II transmembrane protein, with one prominent hydrophobic segment of 18 amino acid residues in the N-terminal region, and a Golgi transmembrane domain.[17,28] Mouse GlcAT-I cDNA also encodes a 355 amino acid, type II transmembrane protein that was 95% identical to human GlcAT-I. The overall fold of the catalytic domain (Thr-76-Val-335) of human GlcAT-I can be divided into two subdomains, the N-terminal cofactor binding domain and the C-terminal substrate-binding domain.[29] Both human and mouse proteins contain a DDD sequence in their C-terminal, which appear to correspond to the conserved DXD motif found in most glycosyltransferases.[30] One or two N-linked glycosylation sites are present in the C-terminal part of human and mouse proteins, respectively.[17] The predicted size of the human or the mouse protein is approximately 37 kDa; however, both, human and mouse, recombinant protein A-tagged GlcAT-I showed two bands. Removal of the *N*-glycan from both the recombinant GlcAT-I proteins by enzymatic digestion resulted in the expected single band, suggesting that one potential N-linked glycosylation site of GlcAT-I is partly utilized.

IV. Enzymatic Activities

GlcAT-I transfers GlcA from UDP-GlcA to the trisaccharide-serine, Galβ1-3Galβ1-4Xylβ1-*O*-Ser, thus finalizing the formation of the common linkage region, GlcAβ1-3Galβ1-3Galβ1-4Xylβ1-*O*-Ser. Although GlcAT-I was first identified in 1969, the properties of GlcAT-I were not described in detail until the cDNA encoding GlcAT-I was cloned.[17,28] The properties and substrate specificities of recombinant GlcAT-I, which was expressed in COS-1 cells as a soluble, protein A chimera and purified using immunoglobulin G (IgG)-Sepharose, have now been determined.[18,28]

Recombinant human GlcAT-I utilizes only a linkage region trisaccharide derivative, such as Galβ1-3Galβ1-4Xyl and Galβ1-3Galβ1-4Xylβ1-*O*-Ser, and little GlcA incorporation is observed with other substrates containing a terminal GalNAc or Gal residue (i.e., polymer chondroitin), longer oligosaccharide–serines derived from the linkage region, *N*-acetyllactosamine, lactose, asialoorosomucoid, Galβ1-3GlcNAc, Galβ1-3GalNAc, and, notably, Galβ1-3Galβ1-*O*-benzyl.[17,18] In addition, recombinant mouse GlcAT-I utilizes a linkage region trisaccharide derivative, such as Galβ1-3Galβ1-4Xyl. Thus, GlcAT-I is distinct from the enzyme termed glucuronyltransferase II, which is involved in the formation of the repeating disaccharide units of CS.[11–13] The above mentioned substrate specificity indicates that the minimum structural requirement for the acceptor substrate of GlcAT-I is the trisaccharide sequence, Galβ1-3Galβ1-4Xyl (K_m 80.4 μM), and that the enzyme recognizes up to the third saccharide residue (Xyl) from the nonreducing end.[18] This deduced trisaccharide recognition disagrees with the finding of Wei et al.,[28] who claimed that Galβ1-3Galβ1-*O*-napthalenemethanol and Galβ1-3Galβ1-*O*-benzyl served as acceptors of recombinant hamster GlcAT-I obtained by expression cloning. X-ray crystallographic analysis revealed the crystal structure of recombinant human GlcAT-I in the presence of both the donor-substrate product UDP and the acceptor substrate analog Galβ1-3Galβ1-4Xyl, and helped the identification of the key residues involved in catalysis and the binding of the two Gal residues.[29]

The Xyl and Gal residues in the linkage region can be modified by phosphorylation and sulfation, respectively.[31–36] Enzyme assays showed that the synthetic molecules, Galβ1-3Galβ1-4Xyl(2-*O*-phosphate)-*O*-Ser and Galβ1-3Gal(6-*O*-sulfate)β1-4Xyl(2-*O*-phosphate)-*O*-Ser, were better substrates for the truncated form of the recombinant human GlcAT-I than the unmodified trisaccharide serine, whereas Gal(6-*O*-sulfate)β1-3Galβ1-4Xyl(2-*O*-phosphate)-*O*-Ser exhibited no acceptor activity.[37] The crystal structure of the catalytic domain of GlcAT-I with UDP and bound Galβ1-3Gal(6-*O*-sulfate)β 1-4Xyl(2-*O*-phosphate)-*O*-Ser revealed that the Xyl(2-*O*-phosphate)-*O*-Ser is disordered and that the 6-*O*-sulfate interacts with Gln-318 from the second GlcAT-I monomer in the dimeric enzyme.[29,37] The above results indicate the possible involvement of such modifications in the processing and maturation of the growing linkage-region-oligosaccharide required for the assembly of GAG chains. In addition, the sulfotransferase responsible for the 6-O-sulfation of Gal residues and the kinase responsible for the 2-O-phosphorylation of the Xyl residue in the linkage region have been identified.[38,39]

V. Homologous Proteins

GlcAT-I belongs to a group of Golgi-associated β1,3-glucuronyltransferases whose founding member is the human HNK-1 glucuronyltransferase.[26,40,41] Phylogenetic analysis showed that GlcAT-I is closely related to the homologous proteins GlcAT-P and GlcAT-S. The amino acid sequence of GlcAT-I has a proline-rich domain (from Pro-30 to Pro-75) next to the transmembrane region, as is seen in several other glycosyltransferases, including GlcAT-P and GlcAT-S, which synthesizes the precursor structure GlcAβ1-3Galβ1-4GlcNAc-R for the HNK-1 carbohydrate epitope GlcA(3-*O*-sulfate)β1-3Galβ1-4GlcNAc-R.[26,40,41] Database searches indicate that the amino acid sequence of GlcAT-I displays 43% and 46% sequence identity to GlcAT-P and GlcAT-S, respectively. The highest sequence identity is found in the C-terminal catalytic domain, which follows the proline-rich region (252 amino acids between Pro-68 and Glu-319 overlap with about 60% identity) [17,28] and contains the four previously identified highly conserved motifs (I–IV) of putative glucuronyltransferases among animal species.[26,39]

VI. Expression Pattern of *GlcAT-I*

Northern blot analysis for GlcAT-1 mRNA demonstrated a single band of ~1.5 kb in all human tissues examined.[21] The *GlcAT-I* gene exhibited ubiquitous but differential expression in the human tissues examined. Notably, expression was particularly abundant in the placenta, followed by the pancreas, brain, and heart. In addition, Northern blot analysis of mouse mRNA demonstrated a single band of ~1.8 kb (Fig. 2). The mouse *GlcAT-I* gene was expressed from embryonic day 7 (Fig. 2B). In adult tissues, its expression was detected in virtually all the tissues examined, with the highest levels found in the liver, brain, and kidney (Fig. 2A). These expression patterns of mouse GlcAT-I have also been previously described.[28]

Additionally, to establish the expression pattern of *GlcAT-I* during mouse embryogenesis, *in situ* hybridization was carried out. At embryonic day 8, *GlcAT-I* was expressed in the telencephalon, hindbrain, and in the frontonasal prominence. At embryonic day 16, *GlcAT-I* was prominently expressed in the teeth, thymus, kidney, and cartilage primordia (unpublished results). These results indicate a highly specific temporal and spatial expression pattern of *GlcAT-I* during embryogenesis.

VII. GlcAT-I in Mouse Early Embryogenesis

A. Targeted Disruption of GlcAT-I Results in Early Embryonic Lethality

To examine the functions of *GlcAT-I* in mammalian early embryogenesis, the *GlcAT-I* gene was inactivated via homologous recombination in mouse ES cells.[20] The targeting vector was constructed by inserting a neomycin resistance cassette into exon II (Fig. 3).

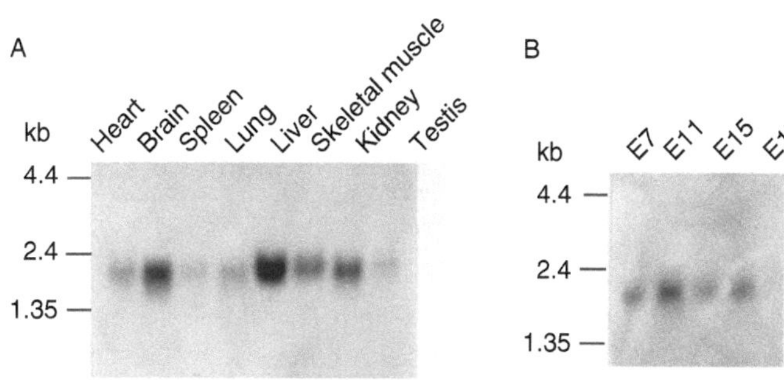

FIG. 2. Northern blot analysis of mouse *GlcAT-I*. Northern blots with mRNAs from various mouse tissues (A) or embryonic stages (B) were hybridized with a probe for mouse GlcAT-I cDNA. (A) *Lane 1*, heart; *lane 2*, brain; *lane 3*, spleen; *lane 4*, lung; *lane 5*, liver; *lane 6*, skeletal muscle; *lane 7*, kidney; *lane 8*, testis. (B) *Lane 1*, embryonic day 7; *lane 2*, embryonic day 11; *lane 3*, embryonic day 15; *lane 4*, embryonic day 17.

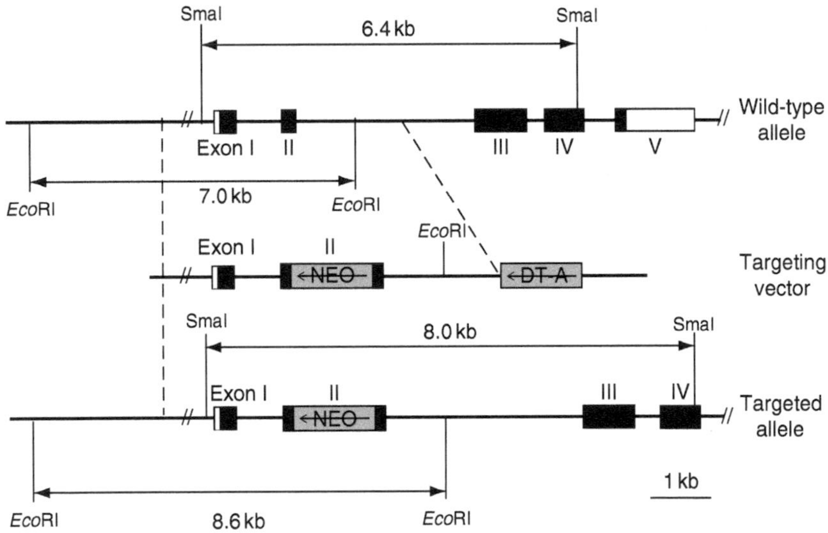

FIG. 3. Targeted disruption of mouse *GlcAT-I* gene.[20] The neomycin resistance cassette was inserted into exon II of *GlcAT-I* gene. Coding and noncoding exons of *GlcAT-I* gene are shown by closed and open boxes, respectively, and the PGKneobpA cassette (NEO) and diphtheria toxin A fragment gene cassette (DT-A) are represented by gray boxes.

GlcAT-I$^{+/-}$ mice had an apparently normal phenotype and were born at a largely Mendelian frequency. They were intercrossed and more than 300 offspring were genotyped by PCR. No *GlcAT-I*$^{-/-}$ neonates and embryos after E6.5 could be detected, indicating that the mutant embryos died during early embryogenesis (Table I). In fact, even at E2.5 (8-cell stage), only 2% of the *GlcAT-I*$^{-/-}$ embryos were detected, suggesting that most *GlcAT-I*$^{-/-}$ mutants died before E2.5 (8-cell stage).[20]

B. GlcAT-I is Essential for Embryonic Cytokinesis and Cell Division

To further analyze the lethalilty and phenotypes of the *GlcAT-I*$^{-/-}$ embryos during early development, 2-cell stage embryos, derived from heterozygous intercrosses (*GlcAT-I*$^{+/-}$ × *GlcAT-I*$^{+/-}$) or other matings (*GlcAT-I*$^{+/-}$ × *GlcAT-I*$^{+/+}$ or *GlcAT-I*$^{+/+}$ × *GlcAT-I*$^{+/+}$), were cultured until the blastocyst stage (Table II). The results of the *in vitro* culture experiments showed that 13% of the embryos from the *GlcAT-I*$^{+/-}$ heterozygous intercrosses died between the 2-cell and 8-cell stage, although all embryos from *GlcAT-I*$^{+/-}$ × *GlcAT-I*$^{+/+}$ or *GlcAT-I*$^{+/+}$ × *GlcAT-I*$^{+/+}$ were viable (Table II). Notably, of the embryos from

TABLE I
GENOTYPE ANALYSIS OF PROGENY FROM *GLCAT-I* HETEROZYGOUS INTERCROSSES[20]

	No. of progeny with genotype[a]				
Day	+/+	+/–	–/–	No. resorbed[b]	No. total
Neonate	46 (38%)	74 (62%)	0 (0%)	–	120
E8.5	22 (39%)	31 (56%)	0 (0%)	3 (5%)	56
E7.5	12 (19%)	43 (67%)	0 (0%)	9 (14%)	64
E6.5	17 (39%)	18 (41%)	0 (0%)	9 (20%)	44
E2.5	19 (45%)	22 (53%)	1 (2%)	–	42

Percentages of different genotypes appear in parentheses.
[a]Genotyping of each developmental stage was performed by PCR.
[b]Resorbed embryos were not genotyped.

the heterozygous intercrosses, only 7% *GlcAT-I*$^{-/-}$ embryos could be identified at the implantation stage, while the fraction of the *GlcAT-I*$^{+/+}$ and the *GlcAT-I*$^{+/-}$ embryos that were viable was within Mendelian expectations (1:2), confirming that *GlcAT-I* inactivation is lethal before the 8-cell stage. Moreover, reversion of cell division was observed in embryos only from the *GlcAT-I*$^{+/-}$ heterozygous intercrosses. Figure 4 shows a representative example of the reversion of cell division in one embryo from the *GlcAT-I*$^{+/-}$ heterozygous intercrosses. The 2-cell (E1.5) embryo divided into a 4-cell embryo, and then insufficient cytoplasmic division seemed to force the embryonic cell compartments to revert to an unusual 3-cell embryo with four nuclei. This unusual embryo eventually died, most likely due to incomplete cytokinesis (Fig. 4A and B). These results indicate that *GlcAT-I* function is essential for embryonic cytokinesis and cell division.

C. Expression of CS and HS in Mouse Embryos

As mentioned earlier, because GlcAT-I transfers GlcA from UDP-GlcA to the trisaccharide-serine, Galβ1-3Galβ1-4Xylβ1-*O*-Ser, thus finalizing the formation of the common GAG–protein linkage region GlcAβ1-3Galβ1-3Galβ1-4Xylβ1-*O*-Ser,[17] it was expected that inactivation of *GlcAT-I* would abolish both CS and HS chains in mouse embryos. Immunocytochemistry with wild-type mouse 2-cell embryos and blastocysts was first performed using an anti-CS (LY111) or anti-HS (Hepss-1) monoclonal antibodies. As expected, fluorescent signals were detected in all 2-cell embryos (Fig. 5C and D) and blastocysts with

TABLE II

Genotype Analysis of Embryos Cultured *in vitro* from 2-Cell-Stage Embryos to Blastocyst Implantation Stages[20]

Parental genotype	No. of progeny with genotype[a]			No. of dead embryos[b]					
	+/+	+/–	–/–	2-cell to 8-cell	8-cell to morula	Morula to blastocyst	Blastocyst to hatched blastocyst	Total dead embryos	No. total
GlcAT-I$^{+/-}$ × *GlcAT-I*$^{+/-}$	32 (29%)	55 (49%)	8 (7%)	15	0	0	2	17 (15%)	112
GlcAT-I$^{+/-}$ × *GlcAT-I*$^{+/+}$	23 (59%)	14 (36%)	–	0	0	0	2	2 (5%)	39
GlcAT-I$^{+/+}$ × *GlcAT-I*$^{+/+}$	42 (98%)	–	–	0	1	0	0	1 (2 %)	43

Percentages of different genotypes appear in parentheses.

[a]Genotyping was performed by PCR.

[b]Dead embryos were not genotyped, but their lethal stages were determined.

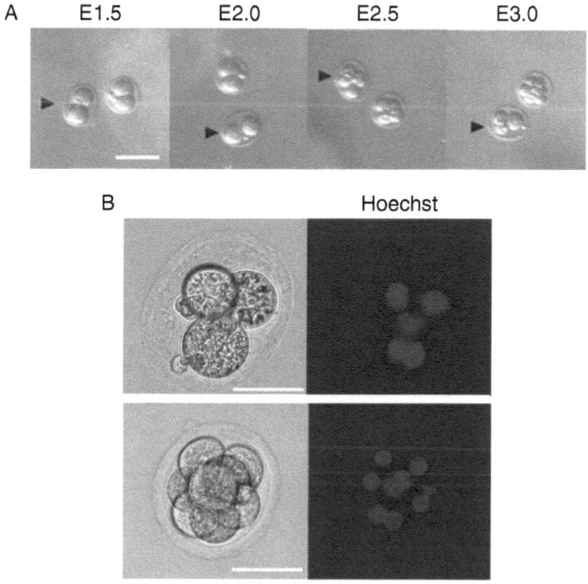

FIG. 4. Reversion of cytokinesis in embryos from *GlcAT-I* heterozygous intercrosses.[20] (A) E1.5 embryos were isolated from heterozygous crosses and cultured. Representative features are depicted. Reversal of cytokinesis was observed in one embryo from *GlcAT-I* heterozygous intercrosses (arrowhead). No reversal of cytokinesis was detected in embryos from crosses of wild-type and heterozygous mice. (B) The two embryos shown in panel (A) were stained with Hoechst 33342. The abnormal embryo (arrowheads in panel (A)) failed to complete cytokinesis and a double nucleated cell appeared (upper), whereas cell division of the other embryo continued normally (lower).

either of these antibodies; and the corresponding signals were eliminated by chondroitinase ABC (CSase) or heparitinases (HSase), respectively, indicating that both CS and HS were produced in mouse 2-cell embryos and blastocysts.[20]

D. Inactivation of GlcAT-I Results in Defective CS and HS

Next, double immunostaining of *GlcAT-I*$^{-/-}$ and *GlcAT-I*$^{+/-}$ embryos as well as *GlcAT-I*$^{+/+}$ embryos using anti-CS (LY111) and anti-ΔHS (3G10) monoclonal antibodies was carried out. *GlcAT-I*$^{+/+}$ and *GlcAT-I*$^{+/-}$ embryos were stained by both anti-CS and anti-ΔHS monoclonal antibodies, whereas the *GlcAT-I*$^{-/-}$ embryos were not stained.[20] These findings suggest that *GlcAT-I*$^{-/-}$ embryos seem to lack both the CS and HS chains.

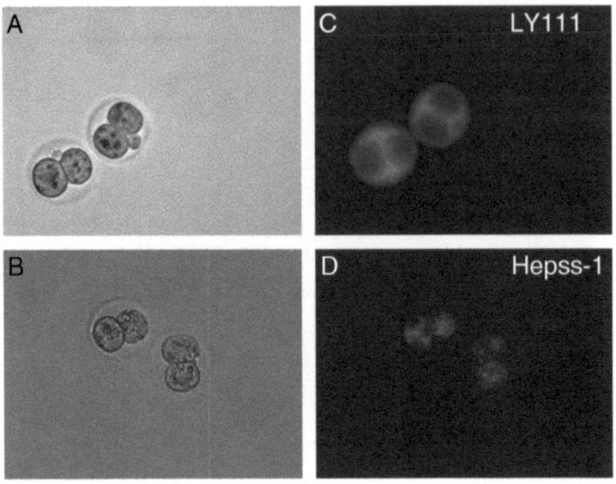

FIG. 5. Immunocytohistochemistry of mouse 2-cell embryos using an anti-CS or anti-HS monoclonal antibody.[20] Wild-type embryos at E1.5 were collected and stained. Left panels (A, B) show embryos examined by phase-contrast microscopy. Immunofluorescent staining by anti-CS (LY111) or anti-HS (Hepss-1) antibody is shown in right panels (C, D). (See Color Insert.)

E. CS Chains are Involved in Controlling Embryonic Cell Division and Cytokinesis

As described above, *GlcAT-I*$^{-/-}$ embryos showed loss of synthesis of both CS and HS and died before the 8-cell stage due to failure of cytokinesis. However, it is not clear whether the embryonic cell death observed in the *GlcAT-I*$^{-/-}$ embryos is due to deficiency of CS, HS, or both. If CS or HS is indispensable for proper embryonic cytokinesis and cell division, digestion of CS or HS at the embryonic cell surface might also induce abnormal cell division. The treatment of 2-cell embryos with CSase showed that 67% of treated embryos died between the 2-cell and 8-cell stages. In contrast, most embryos treated with heat-inactivated CSase/HSase and 65% of embryos treated with HSase developed normally in to blastocysts (Table III, Fig. 6). These results indicate that CS chains, but not HS chains, are involved in controlling embryonic cell division and cytokinesis.

VIII. Concluding Remarks

It has been revealed that in *C. elegans*, nonsulfated chondroitin is required for normal cell division and cytokinesis at an early developmental stage, while HS is essential for embryonic morphogenesis in the later stages of development.[2–4] In addition, prior studies on *EXT1* or *EXT2* in mice demonstrated that

TABLE III
ANALYSIS OF LETHAL STAGES OF *IN VITRO* CULTURED EMBRYOS AFTER TREATMENT WITH CSASE OR HSASE[20]

	No. of dead embryos				
	2-cell to 8-cell	8-cell to morula	Morula to blastocyst	No. of viable embryos to blastocyst stage	No. total
Control[a]	0	0	2	29	31
CSase treatment	22	5	0	6	33
HSase treatment	2	5	1	15	23

[a]Embryos were treated with heat-inactivated CSase/HSase.

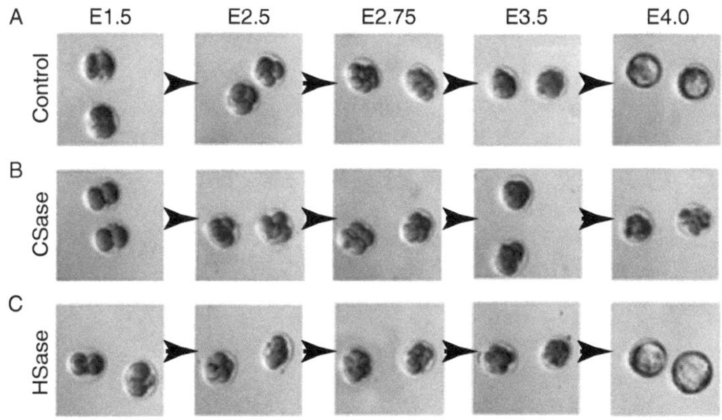

FIG. 6. Depletion of CS results in cytokinesis defects.[20] Wild-type embryos at E1.5 were collected and incubated with heat-inactivated CSase/HSase (A, control), CSase (B), or HSase (C), respectively. Treatment of embryos with CSase showed that these two embryos died from 2-cell to 8-cell stages (B), although control (A) and HSase-treated (C) embryos developed normally to blastocysts. Representative features are depicted.

these genes are essential for HS synthesis and early development.[5,6] Notably, *EXT1*- or *EXT2*-null embryos developed normally until around E6.5, when they became growth-arrested and failed to gastrulate. In addition, the marked reduction of HS in ES cells from *EXT1*- or *EXT2*-deficient mice has been reported.[5,6] To clarify the roles of CS in early embryogenesis in mammals, *GlcAT-I* knockout mice were generated by gene targeting.[20] Mice with a deletion of *GlcAT-I* showed marked reduction of the synthesis of CS and HS and embryonic lethality before the 8-cell stage due to failed cytokinesis.

In addition, treatment of wild-type 2-cell embryos with CSase had marked effects on cell division, although many HSase-treated embryos normally developed to blastocysts. These results suggest that CS in mammals, as with nonsulfated chondroitin in *C. elegans*, is indispensable for embryonic cell division. So far, there are no reports on the role of CS or CSPG in cytokinesis and cell division in mice. To gain more insight into the role of CSPG in mammalian cell division, future studies on the identification of core proteins modified with CS, that are involved in mouse embryonic cell division, are needed.

Moreover, although the *GlcAT-I* gene is ubiquitously expressed in virtually every human and mouse tissue examined, it exhibits marked differential expression among tissues.[17,28] Thus, conditional knockout of the gene in mice would provide essential information regarding the biological function of sulfated GAGs in each tissue.

References

1. Uyama T, Kitagawa H, Sugahara K. Biosynthesis of glycosaminoglycans and proteoglycans. In: Kamerling J, editor. *Comprehensive glycoscience, vol. 3*. Amsterdam: Elsevier; 2007. p. 79–104.
2. Mizuguchi S, Uyama T, Kitagawa H, Nomura K, Dejima K, Gengyo-Ando K, Mitani S, Sugahara K, Nomura K. Chondroitin proteoglycans are involved in cell division of *Caenorhabditis elegans*. *Nature* 2003;**423**:443–8.
3. Izumikawa T, Kitagawa H, Mizuguchi S, Nomura KH, Nomura K, Tamura J, Gengyo-Angyo K, Mitani K, Sugahara K. Nematode chondroitin polymerizing factor showing cell- /organ-specific expression is indispensable for chondroitin synthesis and embryonic cell division. *J Biol Chem* 2004;**279**:53755–61.
4. Kitagawa H, Izumikawa T, Mizuguchi S, Dejima K, Nomura KH, Egusa N, Taniguchi F, Tamura K, Gengyo-Ando K, Mitani S, Nomura K, Sugahara K. Expression of rib-1, a *Caenorhabditis elegans* homolog of the human tumor suppressor EXT genes, is indispensable for heparan sulfate synthesis and embryonic morphogenesis. *J Biol Chem* 2007;**282**:8533–44.
5. Lin X, Wei G, Shi Z, Dryer L, Esko JD, Wells DE, Matzuk MM. Disruption of gastrulation and heparan sulfate biosynthesis in EXT1-deficient mice. *Dev Biol* 2000;**224**:299–311.
6. Stickens D, Zak BM, Rougier N, Esko JD, Werb Z. Mice deficient in Ext2 lack heparan sulfate and develop exostoses. *Development* 2005;**132**:5055–68.
7. Lindahl U, Rodén L. In: Gottschalk A, editor. *Glycoprotein*. Amsterdam: Elsevier; 1972. p. 491–517.
8. Sugahara K, Kitagawa H. Recent advances in the study of the biosynthesis and functions of sulfated glycosaminoglycans. *Curr Opin Struct Biol* 2000;**10**:518–27.
9. McCormick C, Leduc Y, Martindale D, Mattison K, Esford LE, Dyer AP, Tufaro F. The putative tumor suppressor EXT1 alters the expression of cell-surface heparan sulfate. *Nat Genetics* 1998;**19**:158–61.
10. Lind T, Tufaro F, McCormick C, Lindahl U, Lidholt K. The putative tumor suppressor EXT1 alters the expression of cell-surface heparan sulfate. *J Biol Chem* 1998;**273**:26265–8.
11. Kitagawa H, Uyama T, Sugahara K. Molecular cloning and expression of a human chondroitin synthase. *J Biol Chem* 2001;**276**:38721–6.
12. Izumikawa T, Uyama T, Okuura Y, Sugahara K, Kitagawa H. Involvement of chondroitin sulfate synthase-3 (chondroitin synthase-2) in chondroitin polymerization through its interaction with chondroitin synthase-1 or chondroitin-polymerizing factor. *Biochem J* 2007;**403**:545–52.

13. Izumikawa T, Koike T, Shiozawa S, Sugahara K, Tamura J, Kitagawa H. Identification of chondroitin sulfate glucuronyltransferase as chondroitin synthase-3 involved in chondroitin polymerization: chondroitin polymerization is achieved by multiple enzyme complexes consisting of chondroitin synthase family members. *J Biol Chem* 2008;**283**:11396–406.
14. Kitagawa H, Izumikawa T, Uyama T, Sugahara K. Molecular cloning of a chondroitin polymerizing factor that cooperates with chondroitin synthase for chondroitin polymerization. *J Biol Chem* 2003;**278**:23666–71.
15. Uyama T, Kitagawa H, Tamura J, Sugahara K. Molecular cloning and expression of human chondroitin *N*-acetylgalactosaminyltransferase: the key enzyme for chain initiation and elongation of chondroitin/dermatan sulfate on the protein linkage region tetrasaccharide shared by heparin/heparan sulfate. *J Biol Chem* 2002;**277**:8841–6.
16. Uyama T, Kitagawa H, Tanaka J, Tamura J, Ogawa T, Sugahara K. Molecular cloning and expression of a second chondroitin *N*-acetylgalactosaminyltransferase involved in the initiation and elongation of chondroitin/dermatan sulfate. *J Biol Chem* 2003;**278**:3072–8.
17. Kitagawa H, Tone Y, Tamura J, Neumann KW, Ogawa T, Oka S, Kawasaki T, Sugahara K. Molecular cloning and expression of glucuronyltransferase I involved in the biosynthesis of the glycosaminoglycan-protein linkage region of proteoglycans. *J Biol Chem* 1998;**273**:6615–8.
18. Tone Y, Kitagawa H, Oka S, Kawasaki T, Sugahara K. Characterization of recombinant human glucuronyltransferase I involved in the biosynthesis of the glycosaminoglycan-protein linkage region of proteoglycans. *FEBS Lett* 1999;**459**:415–20.
19. Bei X, Wei G, Sinha A, Esko JD. Chinese hamster ovary cell mutants defective in glycosaminoglycan assembly and glucuronosyltransferase I. *J Biol Chem* 1999;**274**:13017–24.
20. Izumikawa T, Kanagawa N, Watamoto Y, Okada M, Saeki M, Sakano M, Sugahara K, Sugihara M, Asano M, Kitagawa H. Impairment of embryonic cell division and glycosaminoglycan biosynthesis in glucuronyltransferase-I-deficient mice. *J Biol Chem* 2010;**285**:12190–6.
21. Kitagawa H, Taoka M, Tone Y, Sugahara K. Human glycosaminoglycan glucuronyltransferase I gene and a related processed pseudogene: genomic structure, chromosomal mapping and characterization. *Biochem J* 2001;**358**:539–46.
22. Bulik DA, Wei G, Toyoda H, Kinoshita-Toyoda A, Waldrip WR, Esko JD, Robbins PW, Selleck SB. *sqv-3*, *-7*, and *-8*, a set of genes affecting morphogenesis in *Caenorhabditis elegans*, encode enzymes required for glycosaminoglycan biosynthesis. *Proc Natl Acad Sci USA* 2000;**97**:10838–43.
23. Helting T, Rodén L. Biosynthesis of chondroitin sulfate II. Glucuronosyl transfer in the formation of the carbohydrate–protein linkage region. *J Biol Chem* 1969;**244**:2799–805.
24. Brandt AE, Distler J, Jourdian GW. Biosynthesis of the chondroitin sulfate-protein linkage region: purification and properties of a glucuronosyltransferase from embryonic chick brain. *Proc Natl Acad Sci USA* 1969;**64**:374–80.
25. Helting T. Biosynthesis of heparin: solubilization and partial purification of uridine diphosphate glucuronic acid: acceptor glucuronosyltransferase from mouse mastocytoma. *J Biol Chem* 1972;**247**:4327–32.
26. Terayama K, Oka S, Seiki T, Miki Y, Nakamura A, Kozutsumi Y, Takio K, Kawasaki T. Cloning and functional expression of a novel glucuronyltransferase involved in the biosynthesis of the carbohydrate epitope HNK-1. *Proc Natl Acad Sci USA* 1997;**94**:6093–8.
27. Curenton T, Ekbrog G, Rodén L. Glucuronosyl transfer to galactose residues in the biosynthesis of HNK 1 antigens and xylose-containing glycosaminoglycans: one or two transferases? *Biochem Biophys Res Commun* 1991;**179**:416–22.
28. Wei G, Bai X, Sarkat AK, Esko JD. Formation of HNK-1 determinants and the glycosaminoglycan tetrasaccharide linkage region by UDP-GlcUA:Galactose β1,3-glucuronosyltransferases. *J Biol Chem* 1999;**274**:7857–64.

29. Pedersen LC, Tsuchida K, Kitagawa H, Sugahara K, Darden TA, Negishi M. Heparan/chondroitin sulfate biosynthesis: structure and mechanism of human glucuronyltransferase I. *J Biol Chem* 2000;**275**:34580–5.
30. Wiggins S, Munro S. Activity of the yeast MNN1 alpha-1, 3-mannosyltransferase requires a motif conserved in many other families of glycosyltransferases. *Proc Natl Acad Sci USA* 1998;**95**:7945–50.
31. Sugahara K, Yamashina I, De Waard P, Van Halbeek H, Vliegenthart JFG. Structural studies on sulfated glycopeptides from the carbohydrate–protein linkage region of chondroitin 4-sulfate proteoglycans of swarm rat chondrosarcoma. Demonstration of the structure Gal(4-*O*-sulfate) beta 1-3Gal beta 1-4XYL beta 1-*O*-Ser. *J Biol Chem* 1988;**263**:10168–74.
32. Sugahara K, Ohi Y, Harada T, de Waard P, Vliegenthart JF. Structural studies on sulfated oligosaccharides derived from the carbohydrate–protein linkage region of chondroitin 6-sulfate proteoglycans of shark cartilage. I. Six compounds containing 0 or 1 sulfate and/or phosphate residues. *J Biol Chem* 1992;**267**:6027–35.
33. de Waard P, Vliegenthart JF, Harada T, Sugahara K. Structural studies on sulfated oligosaccharides derived from the carbohydrate–protein linkage region of chondroitin 6-sulfate proteoglycans of shark cartilage. II. Seven compounds containing 2 or 3 sulfate residues. *J Biol Chem* 1992;**267**:6036–44.
34. Sugahara K, Ohkita Y, Shibata Y, Yoshida K, Ikegami A. Structural studies on the hexasaccharide alditols isolated from the carbohydrate–protein linkage region of dermatan sulfate proteoglycans of bovine aorta: demonstration of iduronic acid-containing components. *J Biol Chem* 1995;**270**:7204–12.
35. Lauder RM, Huckerby TN, Nieduszynski IA. Increased incidence of unsulphated and 4-sulphated residues in the chondroitin sulphate linkage region observed by high-pH anion-exchange chromatography. *Biochem J* 2000;**347**:339–48.
36. De Beer T, Inui A, Tsuda H, Sugahara K, Vliegenthart JF. Polydispersity in sulfation profile of oligosaccharide alditols isolated from the protein-linkage region and the repeating disaccharide region of chondroitin 4-sulfate of bovine nasal septal cartilage. *Eur J Biochem* 1996;**240**:789–97.
37. Tone Y, Pedersen LC, Yamamoto T, Izumikawa T, Kitagawa H, Nishihara J, Tamura J, Negishi K, Sugahara K. 2-*O*-phosphorylation of xylose and 6-*O*-sulfation of galactose in the protein linkage region of glycosaminoglycans influence the glucuronyltransferase-I activity involved in the linkage region synthesis. *J Biol Chem* 2008;**283**:16801–7.
38. Kitagawa H, Tsutsumi K, Ikegami-Kuzuhara A, Nadanaka S, Goto F, Ogawa T, Sugahara K. Sulfation of the galactose residues in the glycosaminoglycan-protein linkage region by recombinant human chondroitin 6-*O*-sulfotransferase-1. *J Biol Chem* 2008;**283**:27438–43.
39. Koike T, Izumikawa T, Tamura J, Kitagawa H. FAM20B is a kinase that phosphorylates xylose in the glycosaminoglycan-protein linkage region. *Biochem J* 2009;**421**:157–62.
40. Seiki T, Oka S, Terayama K, Imiya K, Kawasaki T. Molecular cloning and expression of a second glucuronyltransferase involved in the biosynthesis of the HNK-1 carbohydrate epitope. *Biochem Biophys Res Commun* 1999;**255**:182–7.
41. Shimoda Y, Tajima Y, Nagase T, Harii K, Osumi N, Sanai Y. Cloning and expression of a novel galactoside β1,3-glucuronyltransferase involved in the biosynthesis of HNK-1 epitope. *J Biol Chem* 1999;**274**:17115–22.

Mice Deficient in Heparan Sulfate *N*-Deacetylase/*N*-Sulfotransferase 1

Maria Ringvall* and
Lena Kjellén*,†

*Department of Medical Biochemistry and Microbiology, The Biomedical Center, Uppsala University, Uppsala, Sweden
†Department of Biomedicine, University of Bergen, Bergen, Norway

Ndsts (*N*-deacetylase/*N*-sulfotransferases) are enzymes responsible for *N*-sulfation during heparan sulfate (HS) and heparin biosynthesis. In this review, basic features of the Ndst1 enzyme are covered and a brief description of HS biosynthesis and its regulation is presented. Effects of Ndst1 deficiency on embryonic development are described. These include immature lungs, craniofacial dysplasia and eye developmental defects, branching defect during lacrimal gland development, delayed mineralization of the skeleton, and reduced pericyte recruitment during vascular development. A brief account of the effects of Ndst1 deficiency in selective cell types in adult mice is also given.

Progress in Molecular Biology
and Translational Science, Vol. 93
DOI: 10.1016/S1877-1173(10)93003-2

I. Introduction

The first *in vivo* evidence for the requirement of heparan sulfate (HS) during developmental processes came from studies of the *Drosophila melanogaster* mutants *sugarless* (also called *suppenkasper* or *kiwi*)[1–3] and *sulfateless*.[4,5] These studies also gave the first *in vivo* evidence for direct involvement of HS in specific signaling events. The *sugarless* gene is a UDP-glucose dehydrogenase necessary for production of UDP-glucuronic acid while *sulfateless* encodes the only known HS glucosaminyl Ndst in *Drosophila*. The *sulfateless* mutant suffers from disturbed wingless (*Wg*), hedgehog (*Hh*) and fibroblast growth factor (*fgf*) signaling.[4–6] Since then, the importance of regulated HS biosynthesis in physiological and pathological events has been demonstrated both *in vitro* and *in vivo* in different settings.

Ndst1 is one out of four mammalian Ndsts.[7–9] It is a bifunctional enzyme acting on *N*-acetylglucosamine residues in the growing HS chain during biosynthesis, removing *N*-acetyl groups which are replaced by *N*-sulfate groups. This is the first modification step and a key event during biosynthesis, determining the overall sulfation pattern of the polysaccharide. Subsequent modification enzymes recognize the *N*-sulfated glucosamine residues and act largely in their vicinity (see below; Fig. 1).

The first mammalian model organism with proven HS structural defects was the $Ndst1^{-/-}$ mouse developed in our lab in 2000.[10] We had previously produced an *Ndst2* knockout mouse strain with no detectable alterations in HS biosynthesis but with a complete lack of sulfated heparin[11] and a severe mast cell defect.[11,12] Complete lack of Ndst1 results in a decrease of HS *N*-sulfation levels to approximately one-third of that in wild-type HS while the total sulfation is reduced to half.[10,13] These HS structural alterations affect several processes in the developing embryo and a complete systemic inactivation of *Ndst1* turned out to be incompatible with life *ex utero*.

II. Gene and Protein Structure

The murine *Ndst1* gene is found on chromosome 18, location D2. It contains 14 exons and spans a region of 27.4 kb. The corresponding human gene is present on chromosome 5. The protein contains 882 amino acids. It was first purified from rat liver as an *N*-sulfotransferase.[14] Rat *Ndst1* was cloned in 1992,[15] and the human and mouse counterparts some years later.[16,17] During the same period, the homologue *Ndst2* was purified, cloned and characerized.[18–20] It was shown that this protein contained both *N*-deacetylase and *N*-sulfotransferase activities[18,21], which was also demonstrated for Ndst1.[22]

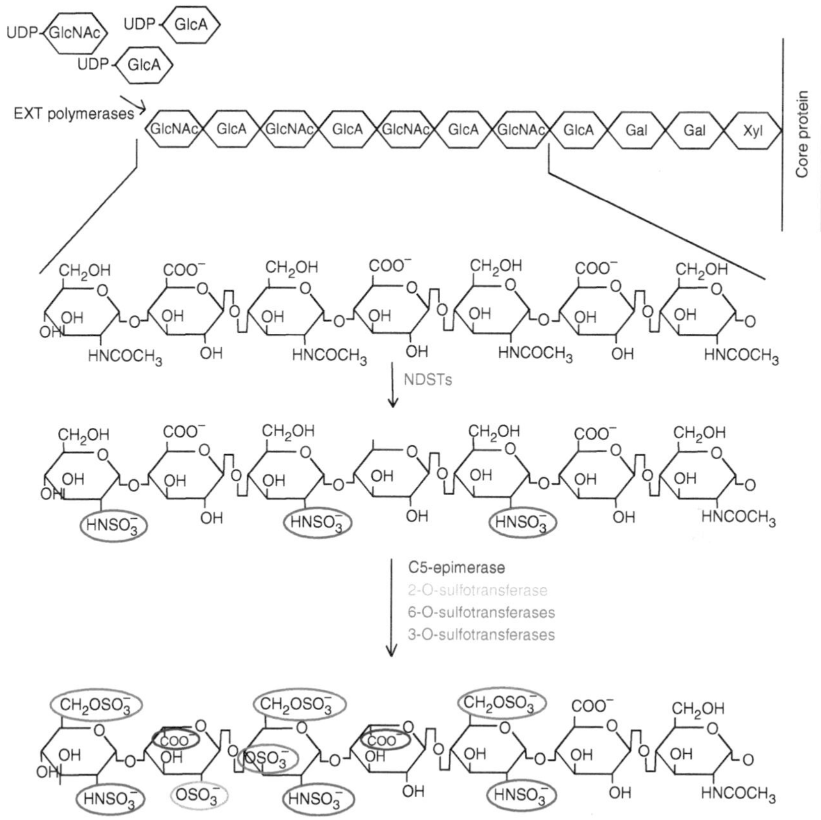

FIG. 1. Heparan sulfate biosynthesis. The growing polysaccharide is attached to a serine residue in a core protein. Different UDP-sugars are used as substrates and each individual reaction step is catalyzed by a specific enzyme. The HS-polymerase complex composed of EXT1 and EXT2 adds alternating units of glucuronic acid and *N*-acetylglucosamine to the nonreducing end of the chain. At the same time, as the chain is growing a series of modifications takes place, beginning with *N*-deacetylation and *N*-sulfation of the original *N*-acetylglucosamine units. *N*-Deacetylation/*N*-sulfation is followed by epimerization of glucuronic acid to iduronic acid and finally ending with stepwise *O*-sulfation of the sugars, including 2-*O*-sulfation of the uronic acid and 6-*O*- and 3-*O*-sulfation of the glucosamine residue.

The *N*-sulfotransferase domain of human Ndst1 (res 580–880) has been crystallized,[23] but the structure of the *N*-deacetylase domain is so far unknown. However, the segment comprising residues A66-P604 of Ndst2 has been expressed in *Escherichia coli* with retained *N*-deacetylase activity,[24] demonstrating that the *N*-decatylase active site is located between the Golgi membrane and the *N*-sulfotransferase domain. Cysteine 486 has been demonstrated to be located close to the *N*-deacetylase active site but does not take part in the catalytic action.[25] Mutation of this residue to an arginine or a tryptophan abolishes enzyme activity while mutation to an alanine residue instead results in enhanced *N*-deacetylase activity. Studies on cells expressing Ndst1 mutants lacking either *N*-deacetylase or *N*-sulfotransferase activity have shown that the two reaction steps do not have to be performed by the same Ndst molecule. Instead, one Ndst enzyme can perform the *N*-deacetylation reaction and another can add the sulfate group.[26]

III. Homologues

The four mammalian Ndst isoforms have been identified and cloned, Ndst1–Ndst4.[15,17,19,20,27,28] Mouse *Ndst1* and *Ndst2* transcripts are found in most tissues both during the embryonic stage and in adult mice whereas *Ndst3* and *Ndst4* show more restricted mRNA expression.[17,28,29] However, transcription levels may not necessarily correlate with translation levels. In fact, expression of the Ndst isoforms may be translationally regulated.[30,31]

Mouse and human Ndsts are highly similar on the protein level, the same isoforms showing 93–98% similarity. Comparing Ndst1 and Ndst2, the two proteins are 70% identical,[17] while Ndst3 and Ndst4 are more similar to each other than to Ndst1 and Ndst2.[28] The sulfotransferase domain of the Ndsts also shows homology with the 3-*O*-sulfotransferases.[32] Similar to *Ndst2* knockout mice, mice lacking Ndst3 are healthy and fertile with only minor hematological and behavioral abnormalities.[29] *Ndst4* knockout mice have so far not been generated.

As mentioned above, a single *Ndst*, *sulfateless*, is expressed in *Drosophila*. The nematode *Caenorhabditis elegans* harbors single orthologs for each of the mammalian sulfotransferase genes including the single *Ndst* gene, *hst-1*.[33] No data has yet been presented on the effects of impaired *hst-1* HS modifications but knockdown of *hst-2*, the *2-Ost* ortholog, in *C. elegans* results in multiple phenotypes that can be correlated to compromised cell migration such as defective gonad development, egg laying, and axonal guidance.[33] In zebrafish five *Ndst* genes are present (Carlsson, Filipek-Gorniok, Kjellén, Ledin, unpublished data). Impaired vascular development was reported as an effect of lowering the concentration of one of them using a splice donor morpholino.[34]

IV. Enzymatic Activities

All Ndst isoforms have both *N*-deacetylase and *N*-sulfotransferase activities. Attempts to compare the two activities in recombinant preparations have demonstrated that Ndst3 is a much better *N*-deacetylase than *N*-sulfotransferase, while the opposite is true for Ndst4.[28] The difference between Ndst1 and Ndst2 in this aspect is smaller, but Ndst2 is stronger on *N*-deacetylation than Ndst1 while Ndst1 is a better *N*-sulfotransferase.[28,35]

In addition to *N*-sulfated and *N*-acetylated glucosamine residues, low levels of *N*-unsubstituted glucosamine can also be detected in HS/heparin preparations.[36] Such residues have been suggested to be generated by the *N*-deacetylase activity of Ndst enzymes without subsequent *N*-sulfation. *N*-Unsubstituted glucosamine residues have been identified in mouse embryonic stem cells lacking both *Ndst1* and *Ndst2*[37] as well as in tissues of *Ndst3* knockout mice.[29] As Ndst4 is a strong *N*-sulfotransferase, it seems less likely that this isoform would miss out on *N*-sulfation. However, if any of the Ndst enzymes would encounter its HS substrate in a Golgi compartment lacking the sulfate donor 3′-phosphoadenosine 5′-phosphosulfate (PAPS), enzyme catalysis would result in the generation of $GlcNH_2$ residues. Whether such a mechanism is operating in cells is not known.

Studies of recombinant Ndst1 and Ndst2 demonstrated that both enzymes work in a processive manner *in vitro* if the sulfate donor PAPS is present,[38] the concentration of PAPS influencing the length of the NS-domains. Without PAPS, the substrate is attacked in a more random manner resulting in deacetylated units scattered along the polysaccharide chain.

A. The GAGosome Hypothesis

Little is still known about regulation of HS biosynthesis and how the specific sulfation patterns, characteristic for each cell type and tissue are formed.[13] It has been suggested that the biosynthesis enzymes may assemble into large complexes, GAGosomes, which in addition to enzymes may contain other components that influence the biosynthesis machinery.[39] In support of the GAGosome hypothesis several enzymes have been shown to interact, at least in pairs; The exostosin (EXT)1 and EXT2 proteins together form the functional HS-polymerase,[40,41] the association between the C5 epimerase and the 2-*O*-sulfotransferase has been shown to be required for both epimerase stability and translocation to the Golgi,[42] and the interaction between Ndst1 and EXT2 greatly influences the structure of HS produced.[43] In addition, early work demonstrated an interaction between the xylosyl-transferase and galactosyl-transferase 1, enzymes taking part in biosynthesis of the HS linkage region.[44]

The GAGosome hypothesis has also been used to tentatively explain why HS biosynthesis in most cells is unaffected by knockout of the Ndst2 enzyme.[45] If Ndst2 has a lower affinity than Ndst1 for EXT2, it will not be incorporated into the GAGosome when Ndst1 is present in sufficient amounts. Therefore, only cells with high expression of Ndst2 compared to Ndst1, such as connective tissue-type mast cells,[17] will be sensitive to lack of Ndst2.

V. Heparan Sulfate Structure in Ndst1 Deficient Embryos

What is the consequence of Ndst1 deficiency for HS structure? Using compositional disaccharide analysis, HS purified from $Ndst1^{-/-}$ embryonic E18.5 liver, lung, brain, and skeletal muscle have been analyzed as well as HS in whole E15.5 and E16.6 embryos.[13,29,45–47] As each cell type synthesize HS with a structure characteristic for that cell, the disaccharide composition of the various HS wild-type preparations also differed from each other. However, in HS from Ndst1 deficient mice a general decrease in *N*-sulfation to 30–50% of wild-type levels was observed together with a decrease in 2-*O*-sulfation of similar magnitude. In contrast, the overall 6-*O*-sulfation was only marginally decreased, mainly through the reduction of the trisulfated disaccharide containing *N*-sulfate, 2-*O*-sulfate, and 6-*O*-sulfate groups. In liver and brain, 6-*O*-sulfation of *N*-acetylglucosamine residues was even slightly increased in the absence of Ndst1.[13] Using ion exchange chromatography in combination with descending paper chromatography of disaccharides from E18.5 liver $Ndst1^{-/-}$ HS, metabolically labeled with ^{3}H-glucosamine, a dramatic decrease in iduronic acid-containing disaccharides could also be demonstrated.[45] Interestingly, here it was demonstrated that the increased amount of 6-*O*-sulfated *N*-acetylglucosamine residues was present in glucuronic acid-containing disaccharides, whereas those containing iduronic acid were decreased.

6-*O*-sulfate groups were also identified in HS from embryonic stem cells completely lacking *N*-sulfation of HS through ablation of both the *Ndst1* and *Ndst2* genes,[37] demonstrating that *N*-sulfation is not an absolute prerequisite for this modification.

The domain structure of HS, where so called NS-domains with a high degree of modification (including *O*-sulfation and epimerization of glucuronic acid residues to iduronic acid) alternate with nonsulfated stretches, is a characteristic feature of HS chains. Spacing of the NS-domains has been suggested to be of major importance for the ability of the HS chain to interact with several growth factors and morphogens.[48] Little is known about the effect of Ndst1 deficiency on the length and distribution of the NS-domains. However, on the basis of gel chromatography after specific cleavage of $Ndst1^{-/-}$ E18.5 liver HS chains at *N*-sulfated residues (HNO_2 at pH 1.5), we could conclude that very long stretches of the chain seemed to be devoid of *N*-sulfate groups.[45]

VI. Effects of Ndst1 Deficiency During Development

A. The Different Ndst1 Deficient Mouse Strains

Three complete systemic Ndst1 deficient mouse strains have been described.[10,47,49] Both our $Ndst1^{-/-}$ strain[10] and the one generated by Fan *et al.*[49] are classical knockout strains while the strain presented by Grobe *et al.*[47]

is a conditional knockout. Fan *et al.* inactivated the *Ndst1* gene by insertion of a neomycine (neo) cassette into exon 2, the first coding exon of *Ndst1*, 19 bp downstream of the ATG start codon. Our targeting vector was constructed in a slightly different way with a 2 kb deletion of exon 2 that includes the ATG start codon. The deleted part was substituted for by the 2 kb neo cassette that was inserted in the opposite direction relative to the transcription of the *Ndst1* gene to prevent any read through activity. The conditional *Ndst1* knockout was created from ES cells with a conditionally deletable floxed *Ndst1* allele with loxP sites inserted on both sides of exon 2. The conditional approach makes it possible to control the inactivation of the floxed gene in a spatiotemporal manner by introduction of Cre-expression in specific cells and/or at a certain time point. The $Ndst1^{f/f}$ allele has also been used for cell specific deletion, which makes it possible to study effects of Ndst1 unmodified HS in adult animals as a systemic lack of Ndst1 causes perinatal death. All three strains were generated from 129/Sv ES cells introduced into C57BL/6 blastocysts.

Perinatal death has been reported as a consequence for all of the three systemic *Ndst1* knockout mouse stains.[10,47,49] We found that while about two-thirds of the Ndst1 deficient pups die perinatally, one-third die earlier, between E14.5 and 18.5. Also Grobe *et al.* noted arrested development before E18.5, while this was not studied by Fan *et al.* Live-born Ndst1 deficient pups turned cyanotic and died within hours due to respiratory failure resembling the respiratory distress syndrome seen in premature infants.[10,49] The analysis of the conditional systemic *Ndst1* knockout mouse confirmed the perinatal death and the HS structural alterations.

In addition to the fully penetrant lung phenotype, other features, such as skeletal abnormalities and brain defects including eye development, have been reported (see Table I). The severity of the different phenotypical traits of Ndst1 deficiency are highly variable between individuals and not all are fully penetrant. This is a phenomenon often seen in knockout mouse strains and may have several reasons. Phenotypic analysis on a mixed genetic background is one of the reasons, and we have noticed (unpublished data) that head and eye defects are more frequent on a C57Bl/6 background than with the highly mixed C57Bl/6/129SVJ background from the first generations of congenic mice. On the contrary, the frequency of embryonic lethality is drastically lowered on a C57Bl/6 background. When it comes to the eye defects, this phenomenon is not too surprising as C57Bl/6 mice are associated with a congenital predisposition for microphthalmia.

$Ndst1^{+/-}$ animals have none of the described $Ndst1^{-/-}$ phenotypes and are viable, and as healthy and fertile as their wild-type littermates. We have not detected any alterations in HS sulfation in our $Ndst1^{+/-}$ mice while Grobe *et al.* reported a slight reduction of *N*-sulfation in their heterozygous mice. In light of the many possible tasks for HS in a developing embryo, it is remarkable how

TABLE I
NDST1 Knockout Developmental Phenotypes

Affected region	Defect	Reference
Nervous system	Neural tube closure	50
	Cerebral patterning and hypoplasia	47
	Eye and lens	
	Coloboma	51
	Microphthalmia	
	Anophthalmia	
Respiratory system	Respiratory distress syndrome	10,49,52
	Surfactant insufficiency	
	Immature type II pneumocytes	
	Hyperplasia	
	Thick alveolar septa	
	Less dilated terminal sacks	
	Elevated number of type II pneumocytes	
Skeletal system	Delayed ossification	47,50,53
	Partially split sternum	
	Hypoplastic or missing neural crest derived elements	
Vascular system (Hind brain primary vascular plexus)	Delayed pericyte recruitment	54
	Pericyte detachment	
Midline	Median cleft face, lip and palate	47,50
	Partially split sternum	50
Lacrimal gland[a]	Impaired lacrimal gland induction and budding decreased cell proliferation	55
Skeletal muscle—myotubes (*in vitro*)	Ca^{2+} kinetics	46

[a]Studies performed on Ndst1$^{f/f}$ systemical, lacrimal epitelial and mesenchymal specfic knockout embryos.

well some Ndst1 deficient individuals develop. Without close examination or genotyping, occasional *Ndst1*$^{-/-}$ embryos, cannot easily be distinguished from their wild-type and *Ndst1*$^{+/-}$ littermates.

B. Immature Lungs

Two types of pneumocytes, type I and type II, are present in the alveoli of lungs. While type I pneumocytes function in gas exchange, the type II pneumocytes in the respiratory epithelium secrete pulmonary surfactant consisting of a mixture of lipids and proteins with the main function to reduce surface tension at the air/liquid interface in the postnatal lung.[56–58] The two types of pneumocytes are considered to arise from a common progenitor cell[59] and type II pneumocytes can differentiate into type I both *in vitro* and *in vivo*.[60] At the end of mammalian gestation, the normal fetal lung differentiates for adaptation to postnatal life and the type II pneumocytes in the respiratory epithelium start to produce and secrete surfactant.

1. Neonatal Respiratory Distress Syndrome

The two classical *Ndst1* knockout mouse strains were both reported to suffer from neonatal respiratory distress syndrome,[10,49] which is characterized by lack of surfactant in the airways and affects premature infants. The newborn $Ndst1^{-/-}$ pups died within hours after birth and their lungs appeared atelectatic and hyperplastic with thick alveolar septa and less dilated terminal sacks than in wild-type lungs.[10,49] Several HS interacting signaling molecules, such as FGF9, FGF10, Sonic Hedgehog (SHh), Wnt7b, and bone morphogenetic protein (BMP) 4, are known to be important in development of the lung.[61–63] In view of this large number of potentially affected signaling pathways, Ndst1 deficient mice exhibit surprisingly little disturbances in lung development; for example, no defects in branching morphogenesis or vascularization of the lungs have been reported for these mice.

2. HS Sulfation Important for Type II—Type I Pneumocyte Transdifferentiation

There are reasons to speculate that Ndst1 deficiency may confer difficulties for lung epithelial cells to develop into type I pneumocytes. Differences in sulfation status of the basement membrane underlying the lung epithelial cells might be decisive for the differentiation into type I and type II pneumocytes; the sulfation levels of the HS in the basement membrane underlying the two pneumocyte types have been shown to differ, with a more highly sulfated HS in the basement membrane underlying type I cells.[64,65] Li *et al.* have shown that highly sulfated Ndst1-modified HS is necessary for *in vitro* transdifferentiation of rat type II to type I pneumocytes.[66] As shown by Newman *et al.* who studied rat alveolar type II pneumocytes, undersulfated HS can also cause enhanced proliferation of type II pneumocytes.[67]

3. Defective Production and/or Secretion of Surfactant

While the newborn $Ndst1^{-/-}$ pups of our mouse strain appeared to have no problem with the production of surfactant,[10] Hu *et al.* could barely detect any surfactant protein (SP)-staining in mutant lung from the other classical knockout strain, whereas wild-type lungs showed SP-A staining throughout the epithelium and SP-C specifically stained the distal epithelium at both E16.5 and 18.5.[52] At E18.5 there was also a reduction in SP-C producing cells in

Ndst1$^{-/-}$ lungs of these pups. In our mouse strain, stainings of *Ndst1*$^{-/-}$ E18.5 embryos showed production of both SP-A and -B but no or very little secretion, while immunostainings of lung sections from E18.5 wild-type embryos showed that both SP-A and -B were produced and secreted into the airways in large amounts. In our mutants, also the number of surfactant producing type II pneumocytes seemed to be elevated.[10] Lack of secreted surfactant and the elevated number of surfactant-producing cells are both signs of immature lungs. Electron microscopic pictures presented by Fan *et al.* revealed several signs of immature type II pneumocytes in mutant lungs such as glycogen accumulation, reduced numbers of surfactant containing lamellar bodies, and attenuated microvilli.[49] While both strains thus showed reduced levels of secreted surfactant, resulting in difficulties to breathe, they differed in synthesis of SPs. The reason for this discrepancy is not known, but could be due to differences in the mixed C57BL/6/129/Sv genetic background.

4. Enhaced BMP Signaling in Ndst1$^{-/-}$ Lungs

Hu *et al.* could also show that at E17.5, all four *Ndst* isoforms were expressed in normal lung and that the expression of *Ndst2-4* transcripts was upregulated in Ndst1 deficient lungs.[52] The hyperplastic lungs of *Ndst1*$^{-/-}$ mice also had more proliferative cells at both E16.5 and E18.5 than their wild-type littermates. As the lung phenotype in *Ndst1*$^{-/-}$ mice resembles the one seen in *Grem1*$^{-/-}$ mice,[68] which is caused by BMP dysregulation, Hu *et al.* also examined BMP signaling in Ndst1 deficient lungs *in vivo* and *in vitro*. They found an upregulation of BMP signaling that suggested Ndst1 modified HS to be essential for proper regulation of BMP signaling in the developing lung. At E18.5, the binding of BMP2 and BMP4 was less in mutant lungs than in normal littermates. Heparitinase treatment of wild-type lung epithelial cells also reduced BMP binding to the cell surface, further indicating that BMP binding is dependent on HS. While BMP-binding to heteromeric preformed receptor complexes preferably results in phosphorylation of Smads,[69] binding of BMP to HS rather causes internalization of the BMP–HS complex. In the developing lung, the binding of BMP to HS therefore inhibits BMP signaling. The lowered binding capacity of BMP to the undersulfated HS produced in Ndst1 deficient lungs therefore enhances BMP signaling via BMP receptors as a result of less internalization, and thereby lowered clearance of BMP. The internalization of BMP was restored in cultured *Ndst1*$^{-/-}$ lung epithelial cells by addition of heparin. Exogenous heparin also restored phosphorylation of Smad1 and elevated the expression of the two lung maturation markers, SP-C and aquaporin-5, in lung explants from *Ndst1*$^{-/-}$ embryos.

It cannot be excluded that Ndst1 modified HS also has other roles to play in surfactant secretion. As an example, the exocytosis mechanism of lamellar bodies is dependent of cytoskeletal components[70] and rearrangements of the cytoskeleton can be influenced by signaling through the syndecan-HSproteoglycans.[71] Elevated cytoplasmic Ca^{2+} levels are known to trigger surfactant secretion at least *in vitro*.[72] It has been shown that cultured myotubes from Ndst1 deficient mice show disturbed Ca^{2+} kinetics.[46] Tentatively, disturbed Ca^{2+} signaling in type II pneumocytes could influence surfactant secretion. It is also interesting to note that it has been suggested that type I pneumocytes can regulate surfactant secretion. Type I pneumocytes supposedly receive external stimuli and thereafter transduce a Ca^{2+} signal to type II pneumocytes for surfactant release.[72] As Hu *et al.* found that both type II and type I pneumocytes were immature, lack of enough numbers of fully differentiated type I cells may be a reason for disturbed secretion of surfactant by type II pneumocytes.

C. Branching Defect During Lacrimal Gland Development

Even though the airway branching morphogenesis in the *Ndst1* systemic knockout seems to be unaffected, a branching defect has been revealed during lacrimal gland development in these mice. In a study by Pan *et al.*[55], lacrimal gland development was studied in the systemic *Ndst1* mutant and in two tissue specific *Ndst1* knockouts. While newborn wild-type mice had extensively branched lacrimal glands, the systemic knockout completely lacked these glands at birth. The lacrimal glands are built up of both endothelial cells and neural-crest derived mesenchyme. By inactivating the *Ndst1* gene in one of these cell types at a time, it could be demonstrated that correct HS production in endothelial cells was sufficient for lacrimal gland development while the mesenchymal HS production did not influence the process. In compound mutants where *Ndst1* was deleted in both epithelial and mesenchymal cells, the lacrimal glands were missing in about half of the newborn pups, while the other half showed much reduced glands. By crossing the *Ndst1*$^{-/-}$ compound mutants with the *Ndst2* knockout mouse strain, it was further established that the incomplete penetrance was caused by *Ndst2* redundancy. Development of lacrimal glands was completely absent in the *Ndst1/2* double deficient compound mice. It was also shown that FGF-receptor (FGFR) 2 in the lacrimal gland epithelium was needed for development, thus demonstrating that in this process HS appears to function in cis with FGFR2.

D. Craniofacial Dysplasia and Eye Developmental Defects

Lack of Ndst1 affects formation of the head region, and severe brain malformations have been suggested to be a reason for the prenatal death observed in Ndst1 deficient mice.[47] The severity of the developmental disturbances in the head region is highly variable between individuals; however, defects in the developing forebrain were present in all mutant embryos with disturbed patterning, hypoplastic or absent anterior and hippocampal commissures, hypoplastic pituitary, and absence of olfactory bulbs. The cerebral cortex was generally smaller than in wild type and there was also a high frequency of eye defects of variable severity; from coloboma, lack of lens and/or retina, uni- or bilateral microphthalmia to complete lack of one or both eyes. Interestingly, unilateral defects were restricted to the right side, suggesting that HS is of importance for the left-right symmetry process. All out of 18 examined *Ndst1*$^{-/-}$ embryos at E9.5–E17.5 had a more or less pronounced eye-developmental defect. The majority of the embryos retained the retina while the lens was absent.

1. FGF Signaling Disturbed in Ndst1$^{-/-}$ Lens Development

In a separate study of early lens development in *Ndst1*$^{-/-}$ embryos, Pan *et al.* examined the BMP, Wnt, and FGF pathways that are important for the early events in lens formation.[51] Out of these three signaling pathways, only FGF–FGFR interactions and signaling seemed to be abrogated. BMP signaling was not apparently affected by loss of *Ndst1* and absence of genetic interaction between *Ndst1* and *Bmp* was further corroborated by crossing *Ndst1* mice with a *Bmp4*-mutant strain. Neither was canonical Wnt signaling notably altered in *Ndst1* mutant lenses.

The first morphological alterations in lens development in Ndst1 deficient embryos appeared to take place around E10. The lens placode was morphologically indistinguishable between mutant and wild type at the 24 somite stage (21–29 somites at E9.5; 30–34 somites at E10; 35–39 somites at E10.5; see Ref. 73). At the 30-somite stage, the wild-type placode had already formed a lens pit while *Ndst1* mutants exhibited less advanced indentation. At the 35-somite stage, the lens vesicles were reduced in size or absent in mutants. Transcripts from all four *Ndst* isoforms were present in the retina at E12.5. On the contrary, in the developing lens, *Ndst1* and *Ndst2* appeared to be the only expressed isoforms, with *Ndst1* as the predominant transcript species.

2. Sonic Hedgehog and Head Development

A smaller fraction (around 14%) of the *Ndst1*$^{-/-}$ embryos examined by Grobe *et al.*[47] and by Pallerla *et al.*[50] showed a more severe head phenotype with an extremely reduced size of the forebrain and a partial lack of frontonasal and mandibular/maxillary processes, agnathia or lack of defined skull. Some *Ndst1* mutants also suffered from a midline closure defect where fusion of the primary and secondary palatal shelves and nasal primordium failed, sometimes visible as cleft lip. There was also a rare neural tube closure defect displayed as exencephaly and incompletely fused neural tube in the thoracic/lumbar region.

Assessment of apoptosis and proliferation was performed on wild-type and mutant brain and revealed a significant enhancement of apoptosis both in the neopallial cortex and in the subventricular zone in E15.5 *Ndst1*$^{-/-}$ embryos while reduced proliferation was only detected in the lateral areas of the developing E17.5 cortex. In addition, a reduced number of glial cells were present in the mutant telencephalon. The proliferative pattern found in wild-type brains much resembled the distribution pattern of HS.

The head developmental phenotype resembles those seen in mouse embryos with either neural crest cells unresponsive to hedgehog[74] or embryos where *Fgf8* has been specifically deleted from the first branchial arch ectoderm.[75] It has also been demonstrated that chick embryos where retinoic signaling has been locally disrupted in the rostral head fail to synchronize development of the forebrain and the frontonasal process due to terminated *Shh* and *Fgf8* expression.[76] To search for functional disturbances in the hedgehog and FGF signaling pathways in *Ndst1* mutants, Pallerla *et al.* used different strategies.[50] One strategy was to generate *Ndst1:Shh* compound mutants. Mice heterozygous for *Ndst1* and *Shh* showed the same, but enhanced, developmental defects as *Ndst1*$^{-/-}$ mutants and surviving compound heterozygous *Ndst1: Shh* mice generally had small eyes and snouts. Even more severe defects were seen in *Ndst1*$^{-/-}$*:Shh*$^{+/-}$ mutants where both mid- and forebrain were collapsed. Only one extremely small and malformed *Ndst1*$^{-/-}$*:Shh*$^{-/-}$ embryo was found (at E10.5) while mere *Shh*$^{+/-}$ mice seemed to be completely unaffected. Additionally, recombinant SHh turned out to have reduced binding capacity to mutant HS and the SHh receptor patched (Ptch) expression was strongly reduced in the developing face at E15.5 in *Ndst1*$^{-/-}$ embryos. These results indicated that HS influences SHh signaling by a direct interaction between SHh and HS and maybe also by a more indirect interaction between HS and some factor that influences SHh signaling. Therefore, many of the developmental defects observed in *Ndst1*$^{-/-}$ embryos may originate from impaired SHh signaling.

3. FGF Signaling Also Important

Ndst1 unmodified HS was also suggested to be less effective as FGF coreceptor as FGF2 signaling was affected in cultured facial mesenchymal *Ndst1*$^{-/-}$ fibroblasts. Stimulation of the fibroblast cultures with FGF2 resulted in strong enhancement of ERK1/2 phosphorylation in wild-type cells while fibroblasts isolated from *Ndst1*$^{-/-}$ embryos were largely unresponsive to the FGF2 stimulation. Additionally, it was shown that FGF2 had a lower capacity to bind to the basement membranes in brain tissue sections prepared from *Ndst1*$^{-/-}$ embryos than to those from wild-type embryos.[50] Less FGF2 also bound to mutant HS than to wild-type HS in an affinity chromatography assay. The low number of glia found in the mutant telencephalon might also be a result of impaired FGF and SHh signaling as both pathways have been suggested to be important for oligodendrocyte specification. It has previously been shown that cultured oligodendrocyte-type2-astrocyte precursor cells from the rat optic nerve produce a highly sulfated heparin-like HS.[77] Therefore, there might be a general need for highly sulfated HS for survival or differentiation of some glial precursor cell lineages also *in vivo*.

Possible disturbances in Wnt signaling were examined by use of a lacZ based Wnt signaling reporter system. Reduced canonical Wnt signaling was detected in the presumptive mid- and hindbrain and spinal cord in one-third of the examined *Ndst1*$^{-/-}$ embryos whereas Wnt signaling seemed to be unaffected in the facial region of mutant embryos.[50]

E. Delayed Mineralization of the Skeleton

Two independent publications on skeletal developmental defects in *Ndst1*$^{-/-}$ mice were presented in early 2007.[50,53] The phenotype involves the whole skeleton including a general delay in mineralization. While Pallerla *et al.* suggested lowered Indian hedgehog (Ihh) signaling as the main molecular mechanism for the delayed mineralization in endochondral bones,[50] Hu *et al.* instead favored enhanced BMP-signaling.[53] The two soluble HS-binding ligands are interconnected through the parathyroid hormone-related protein (PTHrP) in the same regulative loop, which controls cartilage proliferation and differentiation. With a high probability, both IHh and BMP signaling are affected by the undersulfated HS produced by *Ndst1*$^{-/-}$ cells.

1. Importance of Indian Hedgehog

The importance of HS in the ossification process in endochondral bones was first shown by Koziel *et al.* in a report from 2004[79] where it was reported that an *Ext1* hypomorphic mutation in mice results in delayed chondrocyte hypertrophic differentiation. Koziel *et al.* suggested that the establishment of an IHh concentration gradient is dependent on HS concentration and that the decreased production of HS in the *Ext1* hypomorphic mice resulted in an extended range of IHh signaling as well as an elevation of IHh signaling during chondrocyte differentiation.

Pallerla *et al.* described several skeletal anomalies, which included less mineralization of longbones and vertebrae, invertebral disc hypoplasia with hypoplastic or missing nucleus pulposus, partially split sternum, as well as hypoplastic or missing neural crest derived skeletal elements of the skull.[50] To further explore the potential role of HS in mineralization, pluripotent mouse mesenchymal precursor cells (C3H10T1/2) were used as a model system for early stages of endochondral bone formation. Upon stimulation with hedgehogs these precursor cells differentiate into alkaline phosphatase producing mineralizing osteoblasts.[78] Treatment with either sodium chlorate, which lowers HS sulfation, or addition of exogenous heparin to the cell cultures significantly impaired alkaline phosphatase production when stimulated with SHh, demonstrating that HS can influence the differentiation process. RT-PCR analysis on differentiated and undifferentiated cells further showed that *Ndst1* is the prevailing *Ndst* transcript at both stages, which points to an important role of Ndst1 in the process of differentiation into the osteogenic lineage. shRNAi knockdown of *Ndst1* in C3H10T1/2 cells consequently resulted in a complete inhibition of alkaline phosphatase expression. The outcome of these assays and, again, the fact that recombinant SHh had significantly reduced binding capacity to mutant HS,[47] suggested that differentiation from precursors into mineralizing osteoblasts is dependent on sufficiently sulfated HS, not produced by Ndst1 deficient cells. Immunostaining additionally revealed downregulation of the hedgehog receptor Ptch in the periosteal collar. As there are also phenotypic similarities between *Ndst1* mutants and mice where the hedgehog dependent transcription factor Gli2 has been deleted and as *Ndst2* or *Ndst3* mutants show no delay in mineralization, Pallerla *et al.* suggested that the delayed osteoblastic mineralization may arise from impaired IHh signaling with limited IHh binding sites provided by the undersulfated HS produced by *Ndst1* mutants.

2. Importance of BMP

In the other publication on the *Ndst1* skeletal phenotype, Hu *et al.* also reported less mineralization in *Ndst1* mutants, with a 1–2 days delay in the osteogenic process compared to wild-type as judged from longbones and phalanges.[53] A closer assessment of the phalanges in the developing paws of *Ndst1*$^{-/-}$ mice revealed a delayed transformation of proliferative chondrocytes into hypertrophic stages and also a delay in the appearance of osteoblasts and vascular invasion into the perichondrium. The proliferative zone in the epiphyseal growth plate was wider in mutant bones and the chondrocyte proliferative rate was significantly increased at both E16.5 and E18.5. The delay in terminal differentiation of chondrocytes and osteoblasts was confirmed by immunostaining of several stage dependent chondrocyte and osteoblast markers. In an "overlay" assay to test BMP-binding to endogenous HS in phalanges, BMP-2, -4, and -6 all showed poor binding to mutant HS. It was found that PTHrP was highly upregulated in E18.5 mutant embryos while no difference was detected between mutant and wild type in Gli-1, -3, IHh, or Ptch protein levels. As both BMP-receptor (BMPR) IA, an important downstream mediator of IHh function, and phosphorylated Smad1 were strongly upregulated in *Ndst1*$^{-/-}$ phalanges, increased BMPR signaling was suggested as a molecular mechanism for the delay in maturation of mutant bones. This theory was also tested by blocking BMPR signaling with the BMP antagonist noggin. Treatment of E15.5 *Ndst1*$^{-/-}$ limb explants with noggin significantly downregulated PTHrP and phosphorylated Smad1, decreased the proliferation of chondrocytes, and accelerated the hypertrophic differentiation which further strengthened the hypothesis. These results are in line with the enhanced BMP-signaling suggested to cause the delayed maturation of *Ndst1*$^{-/-}$ lungs, recently presented in a publication from the same lab.[52] In contrast to the lung phenotype, the bone phenotype could, however, not be rescued by administration of exogenous heparin to explant cultures.

F. Reduced Pericyte Recruitment During Vascular Development

There are no reports on defective vasculature in Ndst1 deficient embryos at late developmental stages; however, a defect in the developing primary plexus of the hind brain at E11.5 was revealed by Abramsson *et al.*[54] This study showed that platelet derived growth factor (PDGF) BB interactions with sulfated HS regions are necessary for correct recruitment of pericytes during vascular development. A lack of pericytes in the vascular front of the newly formed plexus at E11.5 was revealed in *Ndst1*$^{-/-}$ embryos. This phenotype was, as many of the other phenotypes found in *Ndst1* knockouts, highly variable in severity. The extent of pericyte deficiency spanned from severely affected embryos to those with almost normal pericyte recruitment. The Ndst1 deficient hindbrains were compared to hindbrains from the *Pdgf-b*$^{ret/ret}$ mouse strain that produces PDGF-B that lacks the HS retention motif. Earlier studies of the *Pdgf-b*$^{ret/ret}$ mouse strain have shown that the retention of PDGF-B is crucial for normal recruitment and organization of pericytes in the microvasculature.[80] The phenotype seen in Ndst1 deficient embryos resembled the one seen in the *Pdgf-b*$^{ret/ret}$ embryos but was less severe. Comparisons were also made with C5-epimerase deficient embryos.[81] The pericyte phenotype in these embryos was much less prominent than in the Ndst1 deficient embryos and of a very transient nature. Taken together, sufficiently extended *N*- and *O*-sulfated HS domains were needed whereas the detailed sequence of sulfations seemed to be of less importance for the retention of PDGF-BB and correct activation of PDGFR-β.

G. Lack of Both Ndst1 and Ndst2 Causes Early Embryonic Lethality

While *Ndst1*$^{+/-}$*Ndst2*$^{-/-}$ mice are viable, *Ndst1/2*$^{-/-}$ embryos die before E6.5 and embryos dissected out at E5.5 lack parts of the embryo/extraembryonic tissue (Dagälv, Holmborn, Eriksson, Ringvall, Kjellén, unpublished data). Characterization of HS structure in embryonic stem cells isolated from blastocysts of different *Ndst* genotypes revealed big differences. Wild-type cells synthesize HS with the highest degree of sulfation closely followed by HS from *Ndst2*$^{-/-}$ cells, which in turn is more sulfated than HS produced by *Ndst1*$^{-/-}$ cells. HS from *Ndst1/2*$^{-/-}$ cells only contains 6-*O*-sulfate groups. Notably, in these cells, lack of one Ndst isoform does not result in increased expression of any of the others (unpublished data). *Ndst1/2*$^{-/-}$ embryonic stem cells have been used in several *in vitro* differentiation studies, some of which are ongoing. When vascular endothelial growth factor-receptor (VEGFR)-mediated angiogenesis was studied in chimeric embryoid bodies composed of *Ndst1/2*$^{-/-}$ and *Vegfr2*$^{-/-}$ embryonic stem cells, capillary endothelial structures were explicitly formed by *Ndst1/2*$^{-/-}$ cells which had their VEGFR signaling fully supported by HS produced in *trans* by perivascular smooth muscle cells/pericytes differentiated from VEGFR-2 deficient cells only.[82] This is in agreement with the results of Fuster *et al.* who demonstrated that physiological angiogenesis, studied as cutaneous wound healing, was normal in *Ndst1*$^{f/f}$*Tek*Cre^{+} mice where sulfation of endothelial HS was reduced. In contrast, tumor angiogenesis was affected in these mice.[83] The *Ndst1/2*$^{-/-}$ embryonic stem cells have also been used to show that *N*-sulfated HS is required for FGF4 signaling to maintain the cells primed for differentiation.[84]

Deleting the *Ndst1* gene from T lymphocytes in Ndst2 deficient mice should result in T cells devoid of HS as no Ndst3 or 4 is expressed by these cells.[85] Using this approach, Garner *et al.* concluded that lymphocyte surface HS does not play any crucial role in lymphocyte development but that the proteoglycans may function to modulate the sensitivity of T cells to activation.[85]

VII. Effects of Ndst1 Deficiency in Adult Mice

While HS proteoglycans are essential for embryonic development, they also have several functions in adult physiology.[86] The conditionally deletable *Ndst1*$^{f/f}$ allele[47] has made it possible to study effects of Ndst1 unmodified HS in adult animals, avoiding the perinatal death of the systemic knockouts (see Table II). Two chapters in this book (Chapters 9 and 10) where the roles of endothelial HS in angiogenesis, and hepatocyte HS in lipoprotein metabolism are described will cover several of these studies. Therefore, they will only briefly be mentioned here.

In a publication by Wang *et al.*[88] it was shown that the decreased sulfation of HS on endothelial cells in *Ndst1*$^{f/f}$*Tek*Cre^{+} mice reduced leukocyte extravasation in several inflammatory models. Both weaker binding of leukocytes to the endothelium through L-selectin reduced chemokine transcytosis across the endothelial cells and presentation of chemokines on the lumenal endothelial surface were caused by the decreased sulfation of HS. Additionally, ovalbumin induced allergic airway inflammation has been shown to be attenuated in *Ndst1*$^{f/f}$*Tek*Cre^{+} mice[89] with significantly lower responsiveness to methacolin in allergen challenged *Ndst1*$^{f/f}$*Tek*Cre^{+} mice than in wild-type allergen challenged mice. Allergen challenged *Ndst1*$^{f/f}$*Tek*Cre^{+} mice also seemed to have reduced levels of inflammatory cells both in the lung tissue and in broncho-alveolar lavage fluid. In another study with the *Ndst1*$^{f/f}$*Tek*Cre^{+} mice, tumor angiogenesis was shown to be largely dependent on Ndst1 modified HS produced by endothelial cells while

TABLE II
Ndst1 Cell Specific Knockout Phenotypes in Adult Mice

Affected region/process—cellular specificity	Effect	Reference
Digestive system—hepatocytes	Hypertriglyceridaemia	87
	Delayed clearance of hepatic and dietary triglyceride-rich lipoprotein particles	
Inflammation— endothelial cells leukocytes	Decreased leukocyte rolling and chemokine transcytosis	88
	Reduction in recruitment of inflammatory cells	88,89
Hemostasis— endothelial cells leukocytes	Mild decrease in bleeding time	88
Vascular system— endothelium leukocytes	Reduced tumor angiogenesis	83
	Attenuated tumor growth	
Vascular system—VSMCs	Decreased VSMC proliferation	90
	Reduced lesion formation after vascular injury	
Toxoplasma infection—mammary tumor cells	Reduced replication of intracellular tachyzoites	91,92

the HS status on leukocytes was of no major importance.[83] Although the Ndst1 deficient endothelial cells are surrounded by Ndst1 sufficient nonendothelial cells such as pericytes, fibroblasts, and tumor-cells, the production of wild-type HS in trans was not sufficient for persistent angiogenesis in the tumor.

A reduction of the clearance of intestinally derived and hepatic lipoprotein particles was observed in *Ndst1*$^{f/f}$*AlbCre*$^{+}$ mice, where liver HS sulfation was reduced to about 50%.[87] In later publications, it was demonstrated that syndecan-1 was responsible for the HS mediated uptake and that in addition to *N*-sulfate groups also 2-*O*-sulfate but not 6-*O*-sulfate groups were essential for clearance of the particles.[93,94]

Specific deletion of *Ndst1* in vascular smooth muscle cells (VSMCs) led to decreased *N*- and 2-*O*-sulfation of HS and it was shown that these HS modifications are important for vascular size, VSMC proliferation, and vascular remodeling in response to injury.[90] A reduced size of the femoral artery was observed in both neonatal and adult mice and VSMC proliferation was reduced by about 50%. Mice with *Ndst1* mutant VSMCs exhibited a significant reduction in lesion formation that was associated with reduced collagen content.

Finally, HS has long been considered to be used as a necessary receptor/docking device for some infecting parasites such as *Toxoplasma gondii*. However, no difference in infection rate was seen in livers of mice where *Ndst1* had been inactivated,[91] while replication of the organism was reduced *in vitro* in HS-deficient cells.[92]

VIII. Concluding Remarks

Many of the phenotypes described for the *Ndst1* knockout mice can be explained by a general delay in development whereas other phenotypes rather arise from patterning defects, proliferative dysregulation, or disturbed cell migration. This can be exemplified by the disturbed skeletal development where less skeletal mineralization seen in mutant embryos can be interpreted as a delayed maturation process while the fusion of vertebrae may be caused rather by defective embryo patterning. Further, the lack of neural crest derived craniofacial skeletal elements arises from problems for neural crest cells to migrate to their right location. Also the lung phenotype seen in $Ndst1^{-/-}$ pups can be referred to as delayed maturation with difficulties for precursor cells to differentiate into their final mature phenotype but there is also a proliferative upregulation leading to hyperplastic lungs.

Most, if not all, developmental processes involve signaling systems that are more or less dependent on HS for function or regulation. For soluble ligands, HS may support the building and maintenance of morphogen gradients, for storage or as a scaffold to escape degradation. For some signaling systems, HS is required as a device that facilitates receptor dimerization and receptor–ligand interactions; some events may be completely dysfunctional without HS whereas other signaling events may only have a slight change of kinetics. There is also a difference in the request of specific sulfation patterns, the degree of sulfation, and the spacing of sulfated regions between signaling systems.

Taken together, the studies of *Ndst1* knock-out mice and cells suggest that lack of Ndst1 affects several signaling pathways where FGF, BMP, and Hh seem to be the most prominent members. However, which pathways and to what degree these are affected seem to be highly dependent on the specific organ or tissue with specific cell types and specific requirement of a well-regulated HS production in both time and space.

References

1. Häcker U, Lin X, Perrimon N. The *Drosophila* sugarless gene modulates Wingless signaling and encodes an enzyme involved in polysaccharide biosynthesis. *Development* 1997;**124**:3565–73.
2. Haerry TE, Heslip TR, Marsh JL, O'Connor MB. Defects in glucuronate biosynthesis disrupt Wingless signaling in *Drosophila*. *Development* 1997;**124**:3055–64.

3. Binari RC, Staveley BE, Johnson WA, Godavarti R, Sasisekharan R, Manoukian AS. Genetic evidence that heparin-like glycosaminoglycans are involved in wingless signaling. *Development* 1997;**124**:2623–32.
4. Lin X, Buff EM, Perrimon N, Michelson AM. Heparan sulfate proteoglycans are essential for FGF receptor signaling during *Drosophila* embryonic development. *Development (Cambridge, England)* 1999;**126**:3715–23.
5. Lin X, Perrimon N. Dally cooperates with *Drosophila* Frizzled 2 to transduce Wingless signalling. *Nature* 1999;**400**:281–4.
6. The I, Bellaiche Y, Perrimon N. Hedgehog movement is regulated through tout velu-dependent synthesis of a heparan sulfate proteoglycan. *Mol Cell* 1999;**4**:633–9.
7. Grobe K, Ledin J, Ringvall M, Holmborn K, Forsberg E, Esko JD, et al. Heparan sulfate and development: differential roles of the *N*-acetylglucosamine *N*-deacetylase/*N*-sulfotransferase isozymes. *Biochim Biophys Acta* 2002;**1573**:209–15.
8. Lindahl U, Kusche-Gullberg M, Kjellen L. Regulated diversity of heparan sulfate. *J Biol Chem* 1998;**273**:24979–82.
9. Esko JD, Lindahl U. Molecular diversity of heparan sulfate. *J Clin Invest* 2001;**108**:169–73.
10. Ringvall M, Ledin J, Holmborn K, van Kuppevelt T, Ellin F, Eriksson I, et al. Defective heparan sulfate biosynthesis and neonatal lethality in mice lacking *N*-deacetylase/*N*-sulfotransferase-1. *J Biol Chem* 2000;**275**:25926–30.
11. Forsberg E, Pejler G, Ringvall M, Lunderius C, Tomasini-Johansson B, Kusche-Gullberg M, et al. Abnormal mast cells in mice deficient in a heparin-synthesizing enzyme. *Nature* 1999;**400**:773–6.
12. Humphries DE, Wong GW, Friend DS, Gurish MF, Qiu WT, Huang C, et al. Heparin is essential for the storage of specific granule proteases in mast cells. *Nature* 1999;**400**:769–72.
13. Ledin J, Staatz W, Li JP, Gotte M, Selleck S, Kjellen L, et al. Heparan sulfate structure in mice with genetically modified heparan sulfate production. *J Biol Chem* 2004;**279**:42732–41.
14. Brandan E, Hirschberg CB. Purification of rat liver *N*-heparan-sulfate sulfotransferase. *J Biol Chem* 1988;**263**:2417–22.
15. Hashimoto Y, Orellana A, Gil G, Hirschberg CB. Molecular cloning and expression of rat liver *N*-heparan sulfate sulfotransferase. *J Biol Chem* 1992;**267**:15744–50.
16. Dixon J, Loftus SK, Gladwin AJ, Scambler PJ, Wasmuth JJ, Dixon MJ. Cloning of the human heparan sulfate-*N*-deacetylase/*N*-sulfotransferase gene from the Treacher Collins syndrome candidate region at 5q32-q33.1. *Genomics* 1995;**26**:239–44.
17. Kusche-Gullberg M, Eriksson I, Pikas DS, Kjellen L. Identification and expression in mouse of two heparan sulfate glucosaminyl *N*-deacetylase/*N*-sulfotransferase genes. *J Biol Chem* 1998;**273**:11902–7.
18. Pettersson I, Kusche M, Unger E, Wlad H, Nylund L, Lindahl U, et al. Biosynthesis of heparin. Purification of a 110-kDa mouse mastocytoma protein required for both glucosaminyl N-deacetylation and *N*-sulfation. *J Biol Chem* 1991;**266**:8044–9.
19. Orellana A, Hirschberg CB, Wei Z, Swiedler SJ, Ishihara M. Molecular cloning and expression of a glycosaminoglycan *N*-acetylglucosaminyl *N*-deacetylase/*N*-sulfotransferase from a heparin-producing cell line. *J Biol Chem* 1994;**269**:2270–6.
20. Eriksson I, Sandback D, Ek B, Lindahl U, Kjellen L. cDNA cloning and sequencing of mouse mastocytoma glucosaminyl *N*-deacetylase/*N*-sulfotransferase, an enzyme involved in the biosynthesis of heparin. *J Biol Chem* 1994;**269**:10438–43.
21. Kjellen L, Pettersson I, Unger E, Lindahl U. Two enzymes in one: N-deacetylation and *N*-sulfation in heparin biosynthesis are catalyzed by the same protein. *Adv Exp Med Biol* 1992;**313**:107–11.
22. Wei Z, Swiedler SJ, Ishihara M, Orellana A, Hirschberg CB. A single protein catalyzes both N-deacetylation and *N*-sulfation during the biosynthesis of heparan sulfate. *Proc Natl Acad Sci USA* 1993;**90**:3885–8.
23. Kakuta Y, Sueyoshi T, Negishi M, Pedersen LC. Crystal structure of the sulfotransferase domain of human heparan sulfate *N*-deacetylase/*N*-sulfotransferase 1. *J Biol Chem* 1999;**274**:10673–6.

24. Duncan MB, Liu M, Fox C, Liu J. Characterization of the *N*-deacetylase domain from the heparan sulfate *N*-deacetylase/*N*-sulfotransferase 2. *Biochem Biophys Res Commun* 2006;**339**:1232–7.
25. Wei Z, Swiedler SJ. Functional analysis of conserved cysteines in heparan sulfate *N*-deacetylase-*N*-sulfotransferases. *J Biol Chem* 1999;**274**:1966–70.
26. Bengtsson J, Eriksson I, Kjellen L. Distinct effects on heparan sulfate structure by different active site mutations in NDST-1. *Biochemistry* 2003;**42**:2110–5.
27. Aikawa J, Esko JD. Molecular cloning and expression of a third member of the heparan sulfate/heparin GlcNAc *N*-deacetylase/*N*-sulfotransferase family. *J Biol Chem* 1999;**274**:2690–5.
28. Aikawa J, Grobe K, Tsujimoto M, Esko JD. Multiple isozymes of heparan sulfate/heparin GlcNAc *N*-deacetylase/GlcN *N*-sulfotransferase. Structure and activity of the fourth member, NDST4. *J Biol Chem* 2001;**276**:5876–82.
29. Pallerla SR, Lawrence R, Lewejohann L, Pan Y, Fischer T, Schlomann U, et al. Altered heparan sulfate structure in mice with deleted NDST3 gene function. *J Biol Chem* 2008;**283**:16885–94.
30. Grobe K, Esko JD. Regulated translation of heparan sulfate *N*-acetylglucosamine *N*-deacetylase/*N*-sulfotransferase isozymes by structured 5'-untranslated regions and internal ribosome entry sites. *J Biol Chem* 2002;**277**:30699–706.
31. Bornemann DJ, Park S, Phin S, Warrior R. A translational block to HSPG synthesis permits BMP signaling in the early *Drosophila* embryo. *Development (Cambridge, England)* 2008;**135**:1039–47.
32. Shworak NW, Liu J, Fritze LM, Schwartz JJ, Zhang L, Logeart D, et al. Molecular cloning and expression of mouse and human cDNAs encoding heparan sulfate D-glucosaminyl 3-*O*-sulfotransferase. *J Biol Chem* 1997;**272**:28008–19.
33. Kinnunen T, Huang Z, Townsend J, Gatdula MM, Brown JR, Esko JD, et al. Heparan 2-*O*-sulfotransferase, hst-2, is essential for normal cell migration in *Caenorhabditis elegans*. *Proc Natl Acad Sci USA* 2005;**102**:1507–12.
34. Harfouche R, Hentschel DM, Piecewicz S, Basu S, Print C, Eavarone D, et al. Glycome and transcriptome regulation of vasculogenesis. *Circulation* 2009;**120**:1883–92.
35. Pikas DS, Eriksson I, Kjellen L. Overexpression of different isoforms of glucosaminyl *N*-deacetylase/*N*-sulfotransferase results in distinct heparan sulfate *N*-sulfation patterns. *Biochemistry* 2000;**39**:4552–8.
36. Westling C, Lindahl U. Location of *N*-unsubstituted glucosamine residues in heparan sulfate. *J Biol Chem* 2002;**277**:49247–55.
37. Holmborn K, Ledin J, Smeds E, Eriksson I, Kusche-Gullberg M, Kjellen L. Heparan sulfate synthesized by mouse embryonic stem cells deficient in Ndst1 and Ndst2 is 6-O-sulfated but contains no *N*-sulfate groups. *J Biol Chem* 2004;**279**:42355–8.
38. Carlsson P, Presto J, Spillmann D, Lindahl U, Kjellen L. Heparin/heparan sulfate biosynthesis: processive formation of *N*-sulfated domains. *J Biol Chem* 2008;**283**:20008–14.
39. Esko JD, Selleck SB. Order out of chaos: assembly of ligand binding sites in heparan sulfate. *Annu Rev Biochem* 2002;**71**:435–71.
40. Kobayashi S, Morimoto K, Shimizu T, Takahashi M, Kurosawa H, Shirasawa T. Association of EXT1 and EXT2, hereditary multiple exostoses gene products, in Golgi apparatus. *Biochem Biophys Res Commun* 2000;**268**:860–7.
41. McCormick C, Duncan G, Goutsos KT, Tufaro F. The putative tumor suppressors EXT1 and EXT2 form a stable complex that accumulates in the Golgi apparatus and catalyzes the synthesis of heparan sulfate. *Proc Natl Acad Sci USA* 2000;**97**:668–73.
42. Pinhal MA, Smith B, Olson S, Aikawa J, Kimata K, Esko JD. Enzyme interactions in heparan sulfate biosynthesis: uronosyl 5-epimerase and 2-*O*-sulfotransferase interact *in vivo*. *Proc Natl Acad Sci USA* 2001;**98**:12984–9.
43. Presto J, Thuveson M, Carlsson P, Busse M, Wilen M, Eriksson I, et al. Heparan sulfate biosynthesis enzymes EXT1 and EXT2 affect Ndst1 expression and heparan sulfate sulfation. *Proc Natl Acad Sci USA* 2008;**105**:4751–6.

44. Schwartz NB, Roden L, Dorfman A. Biosynthesis of chondroitin sulfate: interaction between xylosyltransferase and galactosyltransferase. *Biochem Biophys Res Commun* 1974;**56**:717–24.
45. Ledin J, Ringvall M, Thuveson M, Eriksson I, Wilen M, Kusche-Gullberg M, et al. Enzymatically active *N*-deacetylase/*N*-sulfotransferase-2 is present in liver but does not contribute to heparan sulfate *N*-sulfation. *J Biol Chem* 2006;**281**:35727–34.
46. Jenniskens GJ, Ringvall M, Koopman WJ, Ledin J, Kjellen L, Willems PH, et al. Disturbed Ca2+ kinetics in *N*-deacetylase/*N*-sulfotransferase-1 defective myotubes. *J Cell Sci* 2003;**116**:2187–93.
47. Grobe K, Inatani M, Pallerla SR, Castagnola J, Yamaguchi Y, Esko JD. Cerebral hypoplasia and craniofacial defects in mice lacking heparan sulfate Ndst1 gene function. *Development (Cambridge, England)* 2005;**132**:3777–86.
48. Kreuger J, Spillmann D, Li JP, Lindahl U. Interactions between heparan sulfate and proteins: the concept of specificity. *J Cell Biol* 2006;**174**:323–7.
49. Fan G, Xiao L, Cheng L, Wang X, Sun B, Hu G. Targeted disruption of NDST-1 gene leads to pulmonary hypoplasia and neonatal respiratory distress in mice. *FEBS Lett* 2000;**467**:7–11.
50. Pallerla SR, Pan Y, Zhang X, Esko JD, Grobe K. Heparan sulfate Ndst1 gene function variably regulates multiple signaling pathways during mouse development. *Dev Dyn* 2007;**236**:556–63.
51. Pan Y, Woodbury A, Esko JD, Grobe K, Zhang X. Heparan sulfate biosynthetic gene Ndst1 is required for FGF signaling in early lens development. *Development (Cambridge, England)* 2006;**133**:4933–44.
52. Hu Z, Wang C, Xiao Y, Sheng N, Chen Y, Xu Y, et al. Ndst1-dependent heparan sulfate regulates BMP signaling and internalization in lung development. *J Cell Sci* 2009;**122**:1145–54.
53. Hu Z, Yu M, Hu G. NDST-1 modulates BMPR and PTHrP signaling during endochondral bone formation in a gene knockout model. *Bone* 2007;**40**:1462–74.
54. Abramsson A, Kurup S, Busse M, Yamada S, Lindblom P, Schallmeiner E, et al. Defective N-sulfation of heparan sulfate proteoglycans limits PDGF-BB binding and pericyte recruitment in vascular development. *Genes Dev* 2007;**21**:316–31.
55. Pan Y, Carbe C, Powers A, Zhang EE, Esko JD, Grobe K, et al. Bud specific *N*-sulfation of heparan sulfate regulates Shp2-dependent FGF signaling during lacrimal gland induction. *Development (Cambridge, England)* 2008;**135**:301–10.
56. Weaver M, Batts L, Hogan BL. Tissue interactions pattern the mesenchyme of the embryonic mouse lung. *Dev Biol* 2003;**258**:169–84.
57. Kaufman MH, Bard JBL. *The anatomical basis of mouse development*. San Diego, CA: Academic Press; 1999.
58. Bloom W, Fawcett DW. *A textbook of histology*. Philadelphia, PA: W. B. Saunders Company; 1975.
59. Adamson IY, Bowden DH. Origin of ciliated alveolar epithelial cells in bleomycin-induced lung injury. *Am J Pathol* 1977;**87**:569–80.
60. Uhal BD. Cell cycle kinetics in the alveolar epithelium. *Am J Physiol* 1997;**272**:L1031–45.
61. Shu W, Jiang YQ, Lu MM, Morrisey EE. Wnt7b regulates mesenchymal proliferation and vascular development in the lung. *Development* 2002;**129**:4831–42.
62. Colvin JS, White AC, Pratt SJ, Ornitz DM. Lung hypoplasia and neonatal death in Fgf9-null mice identify this gene as an essential regulator of lung mesenchyme. *Development* 2001;**128**:2095–106.
63. Warburton D, Schwarz M, Tefft D, Flores-Delgado G, Anderson KD, Cardoso WV. The molecular basis of lung morphogenesis. *Mech Dev* 2000;**92**:55–81.
64. Sannes PL. Differences in basement membrane-associated microdomains of type I and type II pneumocytes in the rat and rabbit lung. *J Histochem Cytochem* 1984;**32**:827–33.
65. van Kuppevelt TH, Cremers FP, Domen JG, Kuyper CM. Staining of proteoglycans in mouse lung alveoli. II. Characterization of the Cuprolinic blue-positive, anionic sites. *Histochem J* 1984;**16**:671–86.

66. Li Z-Y, Hirayoshi K, Suzuki Y. Expression of *N*-deacetylase/sulfotransferase and 3-*O*-sulfotransferase in rat alveolar type II cells. *Am J Physiol Lung Cell Mol Physiol* 2000;**279**:L292–301.
67. Newman DR, Walsh E, Apparao KB, Sannes PL. Fibroblast growth factor-binding protein and *N*-deacetylase/*N*-sulfotransferase-1 expression in type II cells is modulated by heparin and extracellular matrix. *Am J Physiol Lung Cell Mol Physiol* 2007;**293**:L1314–20.
68. Michos O, Panman L, Vintersten K, Beier K, Zeller R, Zuniga A. Gremlin-mediated BMP antagonism induces the epithelial-mesenchymal feedback signaling controlling metanephric kidney and limb organogenesis. *Development (Cambridge, England)* 2004;**131**:3401–10.
69. Nohe A, Hassel S, Ehrlich M, Neubauer F, Sebald W, Henis YI, et al. The mode of bone morphogenetic protein (BMP) receptor oligomerization determines different BMP-2 signaling pathways. *J Biol Chem* 2002;**277**:5330–8.
70. Fehrenbach H. Alveolar epithelial type II cell: defender of the alveolus revisited. *Respir Res* 2001;**2**:33–46.
71. Yoneda A, Couchman JR. Regulation of cytoskeletal organization by syndecan transmembrane proteoglycans. *Matrix Biol* 2003;**22**:25–33.
72. Dietl P, Haller T, Mair N, Frick M. Mechanisms of surfactant exocytosis in alveolar type II cells *in vitro* and *in vivo*. *News Physiol Sci* 2001;**16**:239–43.
73. Hogan B, Beddington R, Costantin F, Lacy E. *Manipulating the mouse embryo*. Cold Spring Harbour: Cold Spring Harbour Laboratory Press; 1994.
74. Jeong J, Mao J, Tenzen T, Kottmann AH, McMahon AP. Hedgehog signaling in the neural crest cells regulates the patterning and growth of facial primordia. *Genes Dev* 2004;**18**:937–51.
75. Trumpp A, Depew MJ, Rubenstein JL, Bishop JM, Martin GR. Cre-mediated gene inactivation demonstrates that FGF8 is required for cell survival and patterning of the first branchial arch. *Genes Dev* 1999;**13**:3136–48.
76. Schneider RA, Hu D, Rubenstein JL, Maden M, Helms JA. Local retinoid signaling coordinates forebrain and facial morphogenesis by maintaining FGF8 and SHH. *Development (Cambridge, England)* 2001;**128**:2755–67.
77. Stringer SE, Mayer-Proschel M, Kalyani A, Rao M, Gallagher JT. Heparin is a unique marker of progenitors in the glial cell lineage. *J Biol Chem* 1999;**274**:25455–60.
78. Kinto N, Iwamoto M, Enomoto-Iwamoto M, Noji S, Ohuchi H, Yoshioka H, et al. Fibroblasts expressing Sonic hedgehog induce osteoblast differentiation and ectopic bone formation. *FEBS Lett* 1997;**404**:319–23.
79. Koziel L, Kunath M, Kelly OG, Vortkamp A. Ext1-dependent heparan sulfate regulates the range of Ihh signaling during endochondral ossification. *Dev Cell* 2004;**6**:801–13.
80. Lindblom P, Gerhardt H, Liebner S, Abramsson A, Enge M, Hellstrom M, et al. Endothelial PDGF-B retention is required for proper investment of pericytes in the microvessel wall. *Genes Dev* 2003;**17**:1835–40.
81. Li JP, Gong F, Hagner-McWhirter A, Forsberg E, Abrink M, Kisilevsky R, et al. Targeted disruption of a murine glucuronyl C5-epimerase gene results in heparan sulfate lacking L-iduronic acid and in neonatal lethality. *J Biol Chem* 2003;**278**:28363–6.
82. Jakobsson L, Kreuger J, Holmborn K, Lundin L, Eriksson I, Kjellen L, et al. Heparan sulfate in trans potentiates VEGFR-mediated angiogenesis. *Dev Cell* 2006;**10**:625–34.
83. Fuster MM, Wang L, Castagnola J, Sikora L, Reddi K, Lee PH, et al. Genetic alteration of endothelial heparan sulfate selectively inhibits tumor angiogenesis. *J Cell Biol* 2007;**177**:539–49.
84. Lanner F, Lee KL, Sohl M, Holmborn K, Yang H, Wilbertz J, et al. Heparan sulfation-dependent fibroblast growth factor signaling maintains embryonic stem cells primed for differentiation in a heterogeneous state. *Stem Cells* 2010;**28**:191–200.
85. Garner OB, Yamaguchi Y, Esko JD, Videm V. Small changes in lymphocyte development and activation in mice through tissue-specific alteration of heparan sulphate. *Immunology* 2008;**125**:420–9.

86. Bishop JR, Schuksz M, Esko JD. Heparan sulphate proteoglycans fine-tune mammalian physiology. *Nature* 2007;**446**:1030–7.
87. MacArthur JM, Bishop JR, Stanford KI, Wang L, Bensadoun A, Witztum JL, et al. Liver heparan sulfate proteoglycans mediate clearance of triglyceride-rich lipoproteins independently of LDL receptor family members. *J Clin Invest* 2007;**117**:153–64.
88. Wang L, Fuster M, Sriramarao P, Esko JD. Endothelial heparan sulfate deficiency impairs L-selectin- and chemokine-mediated neutrophil trafficking during inflammatory responses. *Nat Immunol* 2005;**6**:902–10.
89. Zuberi RI, Ge XN, Jiang S, Bahaie NS, Kang BN, Hosseinkhani RM, et al. Deficiency of endothelial heparan sulfates attenuates allergic airway inflammation. *J Immunol* 2009;**183**:3971–9.
90. Adhikari N, Basi DL, Townsend D, Rusch M, Mariash A, Mullegama S, et al. Heparan sulfate Ndst1 regulates vascular smooth muscle cell proliferation, vessel size and vascular remodeling. *J Mol Cell Cardiol* 2010.
91. Bishop JR, Esko JD. The elusive role of heparan sulfate in *Toxoplasma gondii* infection. *Mol Biochem Parasitol* 2005;**139**:267–9.
92. Bishop JR, Crawford BE, Esko JD. Cell surface heparan sulfate promotes replication of *Toxoplasma gondii*. *Infect Immun* 2005;**73**:5395–401.
93. Stanford KI, Bishop JR, Foley EM, Gonzales JC, Niesman IR, Witztum JL, et al. Syndecan-1 is the primary heparan sulfate proteoglycan mediating hepatic clearance of triglyceride-rich lipoproteins in mice. *J Clin Invest* 2009;**119**:3236–45.
94. Stanford KI, Wang L, Castagnola J, Song D, Bishop JR, Brown JR, et al. Heparan sulfate 2-*O*-sulfotransferase is required for triglyceride-rich lipoprotein clearance. *J Biol Chem* 2010;**285**:286–94.

Glucuronyl C5-Epimerase: An Enzyme Converting Glucuronic Acid to Iduronic Acid in Heparan Sulfate/ Heparin Biosynthesis

JIN-PING LI

Department of Medical Biochemistry and Microbiology, Uppsala University, Uppsala, Sweden

The glucuronyl C5 epimerase (HSepi) is one of the modification enzymes involved in biosynthesis of heparan sulfate (HS) and heparin, catalyzing the epimerization of D-glucuronic acid (GlcA) to L-iduronic acid (IdoA) at polymer level. IdoA is critical for HS and heparin to interact with protein ligands, because of its flexible conformation. Although the enzyme recognizes both GlcA and IdoA as substrates catalyzing a reversible reaction of the hexuronic acids *in vitro*, the reaction appears irreversible *in vivo*. Targeted interruption of the gene, *Glce*, in mice resulted in neonatal lethality accompanied with kidney agenesis, premature lung, and skeletal malformations, demonstrating that the single gene coded enzyme is essential for animal development. Elimination of the enzyme resulted in abnormal HS and heparin structure that completely lacks IdoA residues. Loss of 2-*O*-sulfation due to lacking IdoA in HS chains appears compensated by increased *N*- and *O*-sulfation of the glucosamine residues. Recombinant HSepi is used to generate HS/heparin related compounds having potential to be used for therapeutic purposes.

Progress in Molecular Biology
and Translational Science, Vol. 93
DOI: 10.1016/S1877-1173(10)93004-4

I. Introduction

Today, it is natural to find that heparan sulfate (HS) and heparin are composed of three different sugar components, that is, glucosamine (GlcN), glucuronic acid (GlcA), and iduronic acid (IdoA). However, historically in the long march to dissect the fine compositional structure of heparin, it took half of a century to identify IdoA as one of the sugar units in heparin,[1] and took additional 10 years to confirm that IdoA residue is generated by epimerization of GlcA residues that are incorporated into the saccharide polymers.[2] The later discovery was on the basis of the observation that the monosugars, UDP-*N*-acetylated glucosamine (UDP-GlcNAc), and UDP-GlcA are detected, but not UDP-IdoA. An explanation for the observation was made possible by the report that D-mannuronic acid is epimerized to L-guluronic acid in the polymer chain of alginate biosynthesis.[3] This investigation led to discovery of the enzyme, glucuronyl C5-epimerase (HSepi; 5.1.3.12) that catalyzes the epimerization of the hexuronic acid (HexA) from GlcA to IdoA in biosynthesis of heparin.[4] Subsequent studies by concerted approaches have elucidated that heparin and HS share a common biosynthesis process,[5] resulting in the polysaccharide products with high similarity in molecular structure. Heparin is produced by connective-tissue type mast cells, whereas HS is expressed in most other mammalian (and many nonmammalian) cells.

The biosynthesis process of HS/heparin is initiated by glycosylation reactions that generate saccharide sequences composed of alternating D-glucuronic acid (GlcA) and *N*-acetyl-D-glucosamine (GlcNAc) units, covalently bound to the respective core proteins. The resulting polymer of (GlcAβ1,4-GlcNAcα1,4)n disaccharide repeats is modified through a series of enzyme catalyzed reactions, including *N*-deacetylation/*N*-sulfation of GlcNAc residues, C5-epimerization of the D-GlcA units to yield L-IdoA residues, and finally *O*-sulfation at various positions.[6] The modification reactions occur in a stepwise manner, such that early steps provide substrates for subsequent reactions. Thus, the HSepi only recognizes the polysaccharide that is *N*-sulfated at the GlcN units. Heparin is extensively sulfated and has a high content of IdoA (typically 50–90% of the total HexA), whereas HS has a more varied structure, generally less sulfated than heparin and with lower IdoA content (typically 30–55 % of the total hexuronic acid).[7]

II. Molecular Biology of HSepi

A. HSepi and the Gene

The enzymatic activity that converts GlcA to IdoA was first demonstrated in a mouse mastocytoma tissue[2,4] and thought to be committed only to heparin biosynthesis. Efforts were made to purify the enzyme from the mastocytoma tissue; however, because of the limited availability of the tumor tissue, only partial purification was reached.[8] Later, appreciable enzymatic activity was detected in a variety of mammalian tissues, including liver, lung, and spleen that provide ample quantity of materials for purification of proteins. By a combination of various chromatographic techniques, the enzyme was finally purified to homogeneity, ~ 1 million-fold, from bovine liver.[9] Peptide sequence information obtained from the purified protein was used to clone the enzyme.[10] A single cDNA species was isolated from a bovine lung cDNA library by hybridizing screening. The gene encompassing the entire open reading frame for the protein was then characterized, encoding for a type II transmembrane protein of 618 amino acid residues.[11,12] Aligning the deduced human, murine, and bovine amino acid sequences indicated 97% identity in these species. This enzyme, in most publications, has been described as glucuronyl C5-epimerase, but also as hexuronyl C5-epimerase. In fact, an enzyme with same function involved in dermatan sulfate (DS) biosynthesis is also a glucuronyl C5-epimerase. To distinguish these two epimerases, it is proposed to introduce abbreviations of HSepi for HS and DSepi for DS biosynthesis.[13]

The HSepi gene is assigned as *Glce*. Contrary to several other enzymes (see Other chapters in this book) involved in HS/heparin biosynthesis, *Glce* has only been found to occur as a single species without any isoforms in mammalians, although duplicated genes were isolated from zebrafish.[14] The gene composed of three exons (Fig. 1A) is localized at chromosome 9 in mouse and chromosome 15q23 in humans, spanning a ~ 14-kb nucleotide sequence. The gene is expressed in nonvertebrates, for example, *Drosophila melanogaster* and *Caenorhabditis elegans* (Fig. 1B).

B. Enzyme Activity Assay and Substrate Specificity

It is impossible to quantify the HSepi protein in tissues because all available antibodies tested failed to detect the protein by Western blot analysis. However, as a result of the sensitive analysis method, the enzymatic activity is detected in all tissues examined, for example, brain, lung, heart spleen, liver, and kidney. Hence, HSepi appears expressed in all tissues in mammalians, albeit to a very low level. The quantity of the enzyme activity seems not directly correlated to the production of IdoA residue in HS chain *in vivo*, as the heterozygous embryos expressing 50% of the enzymatic activity produced identical HS as their wild-type littermates.[15]

The native HSepi purified from bovine liver appeared to be composed of two subunits, as the product migrated into two bands on SDS-PAGE, one with an apparent M_r of ~ 52 kDa and the other with an apparent M_r of ~ 20 kDa.[9] Analysis of several purified batches confirmed the N-terminal truncation of the native HSepi, resulting in two peptides that apparently remain associated (Fig. 2). Both fragments are required for the epimerase activity, because recombinant expression of the N-terminal fragment failed to show the

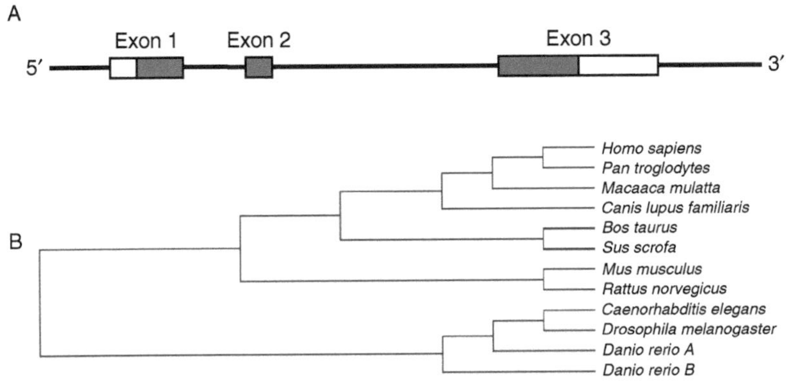

FIG. 1. Illustration of the HSepi (*Glce*) gene organization (A) and the phylogenetic tree of the gene in selected organisms (B).

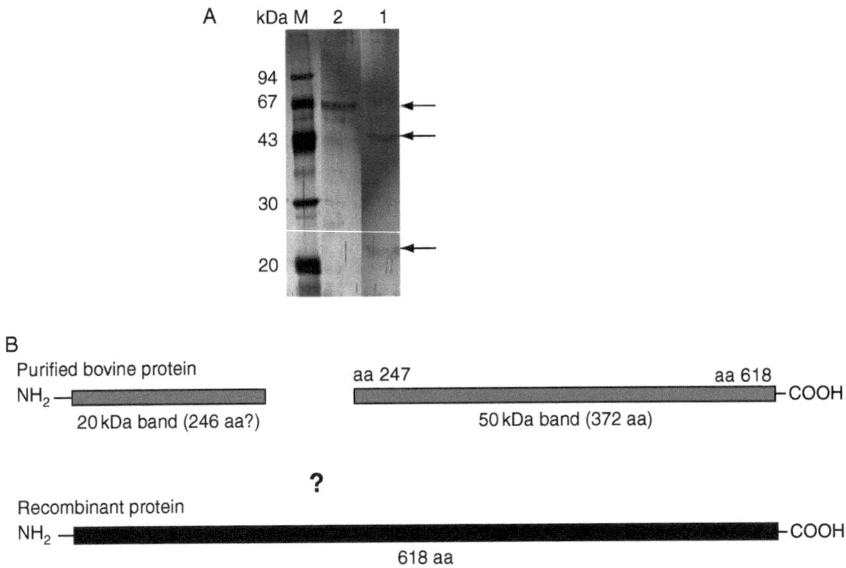

FIG. 2. Molecular structure of HSepi. (A) SDS-PAGE analysis of the purified bovine native HSepi (*lane 1*) and recombinant (*lane 2*) HSepi. The molecular marker is indicated as (*M*). The *arrows* indicate the respective bands around 20 and 50 kDa in the native protein, as well as the 70 kDa band of purified recombinant proteins. (B) The purified native protein is composed of two peptides; the one migrating close to 50 kDa is confirmed to be containing amino acids 247–618; the exact length of the smaller peptide that migrated close to the 20 kDa marker is unknown (*upper* part). The *question mark* indicates that whether an elimination of peptide occurs in connection with fragmentation of the HSepi nascent protein (*lower* part) is unclear.

epimerase activity (unpublished data). However, whether the cleavage of N-terminal is accompanied by elimination of peptide between the N- and C-terminal fragments and what is the mechanism of truncation are unknown.

Initially, recombinant HSepi was expressed as an N-termini truncated form of the protein composed of amino acids 174–618 in an insect expression system.[10] The product showed considerably lower activity in comparison to the purified native form. Later, the full-length mouse cDNA, lacking the signal sequence of 33 amino acids at N-termini, produced a substantial active enzyme in a yeast system, at the same level as the native purified protein (unpublished data). A recent report showed epimerase activity of a recombinant product encoded by full-length human *Glce* expressed in a bacterial system.[16] Notably no successful overexpression of the active enzyme in mammalian cells has been reported, though a coexpression of HSepi with 2-*O*-sulfotransferase (Hs2st) showed colocalization of the two enzymes in the Golgi of CHO cells.[17]

HSepi activity can be determined by measuring ^{3}H in the form of $^{3}H_2O$ upon incubation of the enzyme with a C5-[^{3}H]GlcA/IdoA labeled substrate, for example, modified heparin or K5 polysacccharide (*Escherichia coli* K5 capsular polysaccharide). This assay, the most convenient method available, was developed on the basis of the findings that [^{14}C]IdoA was detected after incubation of a [^{14}C] GlcA-labeled heparin precursor with mastocytoma microsome preparations.[4] Subsequently, it was found that incubation of ^{3}H and ^{14}C dual labeled UDP-GlcA with microsomal fraction prepared from mastocytoma, along with UDP-GlcNAc and PAPS (3′-phosphoadenosine 5′-phosphosulfate; a sulfate donor), resulted in loss of ^{3}H from C5, but not from C2 or C4.[18] These findings permitted the development of the assay for HSepi, in which ^{3}H released from the polysaccharides forms $^{3}H_2O$. The $^{3}H_2O$ generated after incubation of the enzyme with the C5-[^{3}H]GlcA-labeled heparin precursor substrate was measured by distillation and liquid scintillation counting.[19] By this way, HSepi substrate could be prepared; however, preparation of such a substrate is a time-consuming process, involving enzymatic conversion of C5-[^{3}H]glucose to UDP-C5-[^{3}H]GlcA, and subsequent formation of the substrate polysaccharide by incubation of mastocytoma microsomes in the presence of UDP-GlcNAc and PAPS. Thereafter, it was found that incubation of modified heparin with HSepi in the presence of $^{3}H_2O$ resulted in ^{3}H incorporation at C5 of both GlcA and IdoA residues. The reversible C5 hydrogen exchange mechanism led to development of a simplified procedure for preparation of HSepi substrate. Incubation of *O*-desulfated heparin with partially purified liver HSepi in the presence of $^{3}H_2O$ generated substrate that contained both C5-[^{3}H]GlcA and C5-[^{3}H]IdoA residues.[20] Further, the detection was simplified from counting the $^{3}H_2O$ after distillation of the product to measuring the $^{3}H_2O$ after a cocktail biphasic separation. In this procedure, $^{3}H_2O$ in the aqueous reaction mixture is extracted into an organic phase, while the polysaccharide substrate remains in the aqueous phase and does not generate scintillations.[21]

Moreover, a substrate with exclusively C5-[^{3}H]GlcA units can be prepared following metabolic labeling of the *E. coli* K5 capsular polysaccharide, using C5-[^{3}H]glucose. The metabolically labeled polysaccharide, identical to the precursor polysaccharide in HS/heparin biosynthesis, can be purified from the bacterial culture media. This backbone polysaccharide is modified by chemical *N*-deacetylation followed by *N*-sulfation, producing a substrate for HSepi.[22] After the modification, the polysaccharide is recovered and used directly as a substrate for HSepi assay. The ^3H_2O generated upon incubation of this substrate with HSepi is quantified by scintillation counting, after separation from the polysaccharide substrate either by partition in a biphasic scintillation system or by anion-exchange chromatography.[22]

Structural analysis of HS and heparin has revealed the occurrence of -GlcNSO$_3$-IdoA-GlcNSO$_3$- and -GlcNSO$_3$-IdoA-GlcNAc-, but not of -GlcNAc-IdoA-GlcNSO$_3$- sequences.[23] IdoA thus occurs in contiguous *N*-sulfated domains of the HS chain, as well as in regions composed of interspersed *N*-acetylated and *N*-sulfated disaccharide units.[24] Accordingly, enzymatic radioisotope labeling experiments indicated that substrate recognition requires a potential HexA target residue (4GlcNSO$_3$α1-4HexAβ1-4GlcNAc/SO$_3$α1-) to be linked at C4 to an *N*-sulfated GlcN neighbor, whereas the unit linked at C1 may be either *N*-acetylated or *N*-sulfated. *O*-Sulfation at C2 of a HexA unit, or at C6 of one or both of the neighboring GlcN residues, appears to preclude C5-epimerization. The C5-epimerase involved in heparin/HS formation does not attack chondroitin [4GlcAβ1-3GalNAcβ1-]*n*, nor does the C5-epimerase involved in DS biosynthesis cross-react with heparin/HS precursor polysaccharide.

HSepi is active over a broad pH range, with optimum at pH 7.0–7.5. Salt is required for maximal activity, 50 mM NaCl or 100 mM KCl; higher concentrations are inhibitory. The enzyme shows no requirement for divalent cations.[8] A range of K_m values, 4–70 μM expressed as substrate disaccharide units, was found for both the *E. coli* K5 and heparin based substrates.[22] The native as well as recombinant enzyme is fairly stable in purified forms when stored in 10% glycerol at 4 °C.

C. Reaction Mechanism

The action mechanism of HSepi is not fully understood. Epimerization of a HexA at polymer level is rare; so far only three hexuronyl C5-epimerization reactions have been demonstrated, involving conversion of β-D-mannuronic acid to α-L-guluronic acid in alginate biosynthesis by bacteria or seaweeds, and of D-GlcA to L-IdoA in HS/heparin as well as in DS biosynthesis.[25]

Incubation of HSepi with either substrates, that is (4GlcAβ1-4GlcNSO$_3$α1-)*n* that is obtained by chemical *N*-deacetylation/*N*-sulfation of *E. coli* K5 capsular polysaccharide or (4IdoAβ1-4GlcNSO$_3$α1-)*n* that is the major disaccharide unit

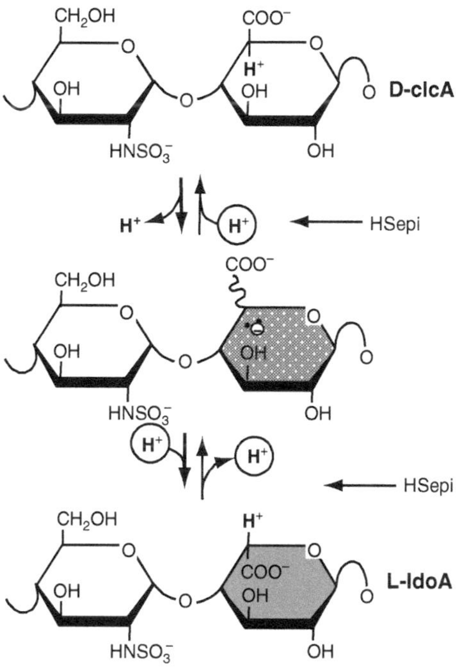

FIG. 3. Proposed reaction mechanism for epimerization of the HexA units in HS/heparin biosynthesis. The reaction involves reversible abstraction of a proton at C5 of target HexA, leading to the formation of a carbanion intermediate. The C5 proton is exchangeable with protons of the medium.

in chemically *O*-desulfated heparin, in the presence of ^3H_2O yields polysaccharide products with C5-^{3}H-labeled GlcA as well as IdoA residues.[26] These results indicated that the inversion of configuration at C5 is freely reversible between GlcA and IdoA. This reaction occurs by abstraction and readdition of the C5 hydrogen atom, presumably through formation of a carbanion intermediate (Fig. 3). However, this inversion reaction seems favoring retention of D-gluco configuration at equilibrium, as incubation of C5-[^{3}H]glucose-labeled K5-derived substrate (containing exclusively GlcA) and heparin-derived substrate (containing both GlcA and IdoA) with HSepi resulted in a distribution of ~65% GlcA and ~35% IdoA from both substrates.[26]

The mode of HSepi action in the membrane-bound Golgi biosynthetic machinery differs from that of the solubilized enzymes used for *in vitro* incubations. Heparin produced by connective tissue mast cells can contain as much as 90% IdoA residue of total HexA,[7] which is, so far, impossible to reach by *in vitro* incubations with any of the substrates.[22] In fact, dual isotope labeling (C5–^{3}H; ^{14}C) of GlcA revealed that equilibrium is never approached, as the GlcA residues recovered in the product retain their C5–^{3}H label[27] and is resistant to further epimerization upon incubation with purified enzymes *in vitro*. It is assumed that a strongly linked action of HSepi and 2-*O*-sulfotransferase in biosynthesis of HS/heparin blocks back-epimerization, pulling the equilibrium of the epimerization reaction for IdoA generation. Moreover, the modification reactions, encompassing *N*-deacetylation/*N*-sulfation, C5-epimerization, and *O*-sulfations, of an individual heparin chain *in vivo* are completed within a few minutes as revealed by pulse-chase labeling experiment,[28] apparently much rapid than the *in vitro* incubation for equilibrium to be established. This suggests that, in the intact cell, interaction between the HSepi and its GlcA target residue is restricted to a single encounter that leads to the formation of IdoA. However, the precise mechanism determining whether a given GlcA substrate residue will be converted to IdoA or escapes C5-epimerization remains unclear. Such a mechanism is in accord with a processive mode of the HS modification.[6,29,30]

Nevertheless, labeling kinetics during incubations for different periods of time in the presence of ^3H_2O resulted in approximately equal labeling of GlcA and IdoA units due to reversible abstraction and readdition of a hydrogen (or tritium) atom at C5. The distribution of label was not influenced by the HexA composition of the initial substrate, suggesting that the hydrogen exchange process is affected by two polyprotic bases, most likely lysine residues.[26] A proton is abstracted from C5 through action of a functional enzyme base, yielding a stabilized carbanion intermediate (Fig. 3). Another proton is introduced with change in configuration at carbon 5. Our ongoing effort in illustrating the molecular structure of HSepi is expected to shed light on the mechanism of the epimerization action.

III. Transformation-Related Phenotypes Attributable to HSepi

A. Gene Knockout of HSepi in Mouse

Targeted interruption of the *Glce* gene in mouse was approached by partial deletion of exon 3 corresponding to the 341 C-terminal amino acid residues of the 618-residue protein (Fig. 1A). The *Glce* mutant mouse was generated through the classic procedure by selecting ES cell clone with homologous recombination after electroporation, followed by microinjection of the positive clone into C57BL/6 blastocysts, yielding chimeric animals and germ line transmission.[15]

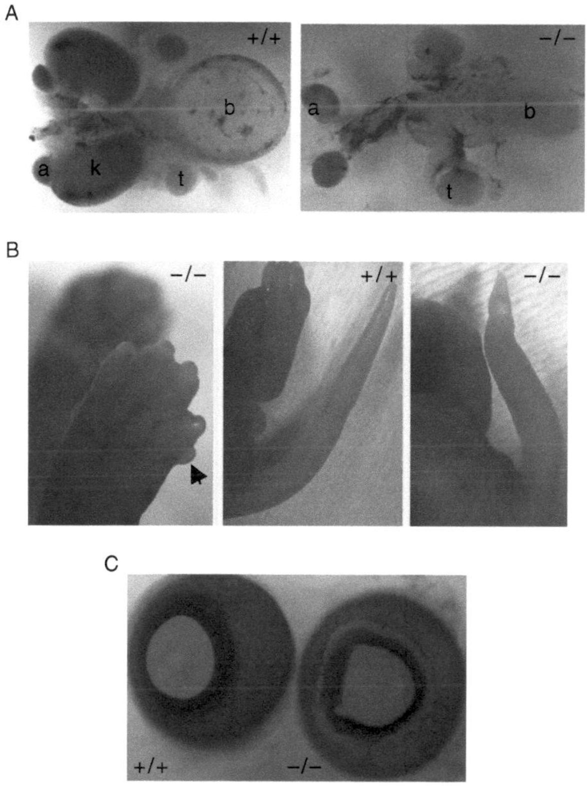

FIG. 4. Some of the developmental defects observed in the HSepi mutant mice. (A) Kidney agenesis; (B) extra digits in fore limbs (*arrowhead*) and curling tail; (C) iris coloboma. +/+, wild type; −/−, *Glce* mutant.

The offspring of heterozygous intercrosses was born on d.p.c. (days postcoitus) 18.5. The heterozygous littermates showed no aberrant phenotype and are fertile, with a normal lifespan. By contrast, $Glce^{-/-}$ pups are neonatal lethal accompanied by multiple defects (Fig. 4). All $Glce^{-/-}$ mice lacked kidneys, but showed no overt abnormalities in other abdominal organs. The lungs of mutant animals were poorly inflated and immature with thickened, cell-rich alveolar walls. Further, all $Glce^{-/-}$ animals showed bilateral iris coloboma. Abundant skeletal abnormalities included shorter body length (average 19 mm for $Glce^{-/-}$ animals, 26 mm for wild-type littermates) with generally excessive mineralization. Skeletal malformations include lack of proximal phalanges and tarsal bones, curling tail, postaxial polydactyly in one or both forelimbs, malformed ribcage and sternum, and often cleft palate. Other organs seem gross-anatomically normal, but further characterization is needed.

B. Structure Alteration of HS/Heparin in HSepi Mutant Mice

Analysis of metabolically labeled HS isolated from embryos showed no difference between wild-type and heterozygous littermates; however, the HS from mutant embryos completely lacked IdoA.[15] Gel chromatography of metabolically [^{3}H]GlcN-labeled HS indicated similar chain length for the samples of *Glce*$^{+/+}$, *Glce*$^{+/-}$, and *Glce*$^{-/-}$, whereas the *Glce*$^{-/-}$ polysaccharide emerged more retarded than the corresponding *Glce*$^{+/+}$ and *Glce*$^{+/-}$ products on anion-exchange chromatography. The wild-type and heterozygous patterns of *O*-sulfated disaccharides were indistinguishable and typical of normal HS, with a predominance of mono-*O*- and di-*O*-sulfated, IdoA-containing species. By contrast, these disaccharides were completely absent from the *Glce*$^{-/-}$ samples, which instead showed increased proportions of GlcA-containing species, including a disaccharide with 2-*O*-sulfated GlcA (Fig. 5A). The proportion of disaccharides, derived from contiguous *N*-sulfated domains in mutant HS was increased relative to wild-type HS, indicating increased *N*-sulfation of mutant HS.[15] In summary, elimination of HSepi in mice resulted in a change in HS structure, from a largely intermixed distribution of *N*-substituents to a pattern of extended *N*-sulfated domains, along with a switch from -IdoA(2-OSO_3)-GlcNSO_3- to -GlcA-GlcNSO_3(6-OSO_3)- as the predominant sulfated disaccharide unit.

As discussed in previous sections, HSepi activity was first detected in a mastocytome tissue.[4] Although comparative studies regarding substrate recognition and catalytic kinetic have been conducted using enzymes partially purified from bovine liver and mouse mastocytome, resulting in indistinguishable activities,[22] direct evidence showing that the same protein is involved in HS and heparin biosynthesis was not available until generation of the HSepi mutant mice. As the mutant animals are neonatal lethal, it is impossible to obtain mature mast cells from the animals. Instead, mast cells were isolated from fetal (E16.5) skin and matured by *in vitro* cultures. Metabolically labeling of the fetal skin derived mast cells (FSMC) with [^{35}S]sulfate resulted in radiolabeled heparin. Analysis of the disaccharides derived from the *Glce*$^{-/-}$ heparin chain revealed a pattern resembling that of the disaccharides from *Glce*$^{-/-}$ HS (Fig. 5B). Thus, heparin from *Glce*$^{-/-}$ FSMC consisted more than 90% of *N*- and mono-6-*O*-sulfated, GlcA-containing disaccharide units, and was completely devoid of IdoA residues.[31]

C. Importance of IdoA Residue in the Functions of HS

Given the recognized importance of HS in early embryonic patterning and morphogenesis,[32–34] the selective nature of the IdoA-deficient phenotype, albeit severe, was unexpected. The requirements for defined HS structure in several functionally important interactions thus would not seem to include the presence

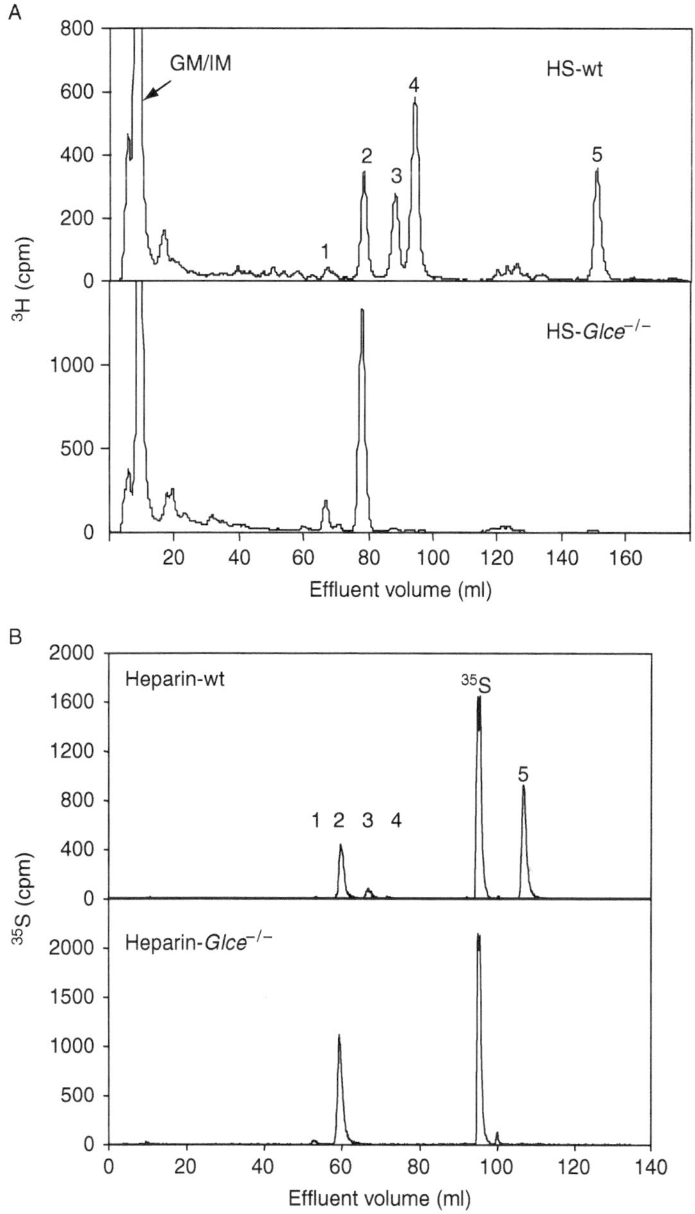

FIG. 5. Compositional analysis of disaccharides derived from HS (A) and heparin (B). The metabolically radioisotope-labeled samples were isolated from embryos (for HS) or fetal skin derived mast cells (FSMC) and treated with HNO_2 at pH 1.5 to convert the polysaccharide chains to oligosaccharides that were subsequently separated by gel chromatography. The disaccharide fractions were collected and analyzed by anion-exchange HPLC on a Partisil-10 SAX column. Peaks represent: 1, GlcA(2-OSO_3)-aMan$_R$; 2, GlcA-aMan$_R$(6-OSO_3); 3, IdoA-aMan$_R$(6-OSO_3); 4, IdoA(2-OSO_3)-aMan$_R$; 5, GlcA(2-OSO_3)-aMan$_R$6-OSO_3. The aMan$_R$ residues are derived from GlcNSO_3 units in intact polysaccharides. *GM/IM* indicates not sulfated GlcA-aMan$_R$ *and* IdoA-aMan$_R$; ^{35}S indicates free sulfate released from metabolically 35[S]-labeled heparin upon pH 1.5 cleavage.

of IdoA units. Comparison of the defects in the *Glce*$^{-/-}$ mice with phenotypes due to elimination of other enzymes involved in HS biosynthesis shows similarities as well as differences. HS synthesized by mice deficient in *N*-deacetylase/*N*-sulfotransferase isoform 1 gene (*Ndst1*) is low in *N*-sulfate, which consequently results in reduced IdoA residues[35]; however, *Hs2st* mutant embryos produced HS with normal pattern of epimerization, but the IdoA residues are completely nonsulfated.[36] Comparison of the phenotypes of *Ndst1*$^{-/-}$, *Glce*$^{-/-}$ and *Hs2st*$^{-/-}$ embryos revealed a differential dependence of development on HS fine structure, albeit all mutants are neonatal lethal (with an exception that ~30% of *Ndst1*$^{-/-}$ pups die earlier). Strikingly, *Hs2st*$^{-/-}$ and *Glce*$^{-/-}$ displayed penetration phenotype of kidney agenesis,[15,37] while *Ndst1*$^{-/-}$ embryos had normal kidneys[35]; on the other hand, contrary to the *Ndst1*$^{-/-}$ and *Glce*$^{-/-}$ mice that showed defect in lung maturation, the *Hs2st*$^{-/-}$ embryos had apparently normal lungs and remained alive for some hours after birth.

Accumulated information provides some clues to structure–function relations in HS biology.[6,38] The multiple functions ascribed to HS proteoglycans throughout embryonic development involve several signaling networks, including FGF, PDGF, VEGF, Wnt, TGF-β, and Hedgehog pathways. Some of the corresponding signaling mechanisms of these cytokines appear either unaffected or perturbed by the lack of IdoA in HS.[39,40] Reduction of *N*-sulfation of HS in *Ndst1*$^{-/-}$ impaired PDGF-BB signaling, resulting in pericyte detachment and delayed pericyte migration *in vivo*. In contrast, the HS in *Glce*$^{-/-}$, lacking IdoA and 2-*O*-sulfation, retained PDGF-BB *in vitro* and only transiently delayed pericyte recruitment *in vivo*.[39] It appears that appropriately spaced *N*-sulfated domains are critical to retain PDGF-BB and activate PDGF receptor signaling, whereas the HexA moiety does not appear to be important for this interaction. In contrast, IdoA with 2-*O*-sulfation is critical for FGF2 mediated signaling.[40] Studies using mouse embryonic fibroblast cells (MEF) isolated from the HSepi mutant and wild-type embryos revealed a defective FGF2-dependent intracellular signaling in the HSepi mutant MEF cell, exhibiting delayed proliferation and migration upon FGF2 stimulation. Binding studies revealed an aberrant mode of interaction between mutant HS chains and FGF2, indicating an important role of IdoA in the structural–functional correlation of FGF2 with HS. Further, it was found that the mutant HS lacking IdoA interacted poorly with GDNF (glia-derived neurotrophic factor), in a similar way as with FGF2. As GDNF is known to be important for kidney development, the defected interaction of GDNF with the mutant HS may have contributed to the kidney agenesis phenotype in HSepi mutant mice. In contrast, the IdoA residue seems not required for the interaction of HS with FGF10.

Although the HS sequences identified to interact with protein ligands invariably contain one or more IdoA residues, as IdoA units are believed to promote the interactions because of their marked conformational flexibility,[41]

TABLE I
Differential Phenotypes in Mice Lacking Different Enzymes

	Kidney defect	Lung defect
Ndst1$^{-/-}$	No	Yes
Hs2st$^{-/-}$	Yes	No
Glce$^{-/-}$	Yes	Yes

it seems the upregulation in *N*-sulfation and 6-*O*-sulfation observed in *Glce*$^{-/-}$ could conceivably "compensate" for the loss of IdoA and 2-*O*-sulfate residues, in some interactions crucial to development. On the other hand, certain phenotype traits have been identified to be critically dependent on the presence of IdoA. Some of these features can be differentially related to the sulfation state of this monosaccharide component, as revealed by the *Glce*$^{-/-}$ and *Hs2st*$^{-/-}$ phenotypes (Table I). Lung maturation thus requires a minimal proportion of IdoA units (or a structure associated with such units) that do not need to be sulfated. Other events, including kidney, iris, and skeletal development, depend on HS with 2-*O*-sulfated IdoA residues. Thus, to understand the mechanisms of HS in development, complete elucidation of the molecular interactions continues to provide challenges for future work.

IV. Regulation of the Gene Expression

The HSepi gene, *Glce*, is expressed in many species examined. Alignment of the selected peptide sequences shows high homology in different species (Fig. 6), especially in mammalians, indicating that the gene is generally conserved through evolution. So far, there are no reported human diseases associated with *Glce* mutations. A systemic study on *Glce* gene expression in mammals has not been reported; however, a spatial–temporal expression is observed during zebrafish development.[42] Both *Glce* genes were detected in fertilized zebrafish embryos prior to the onset of zygotic transcription. At later developmental stages, the genes are expressed widely throughout gastrulation and become restricted to the hindbrain at 24 h postfertilization.[14] Expression of the gene is found to be an important factor for dorsoventral axis formation and patterning in zebrafish, as knockdown of the gene by antisense morpholino oligonucleotides promoted axis dorsalization. Although the HS-2-*O*-sulfotransferase and HSepi are predicted to be interdependent for protein translocation from the endoplasmic reticulum to the Golgi of CHO cells,[17] their expression is

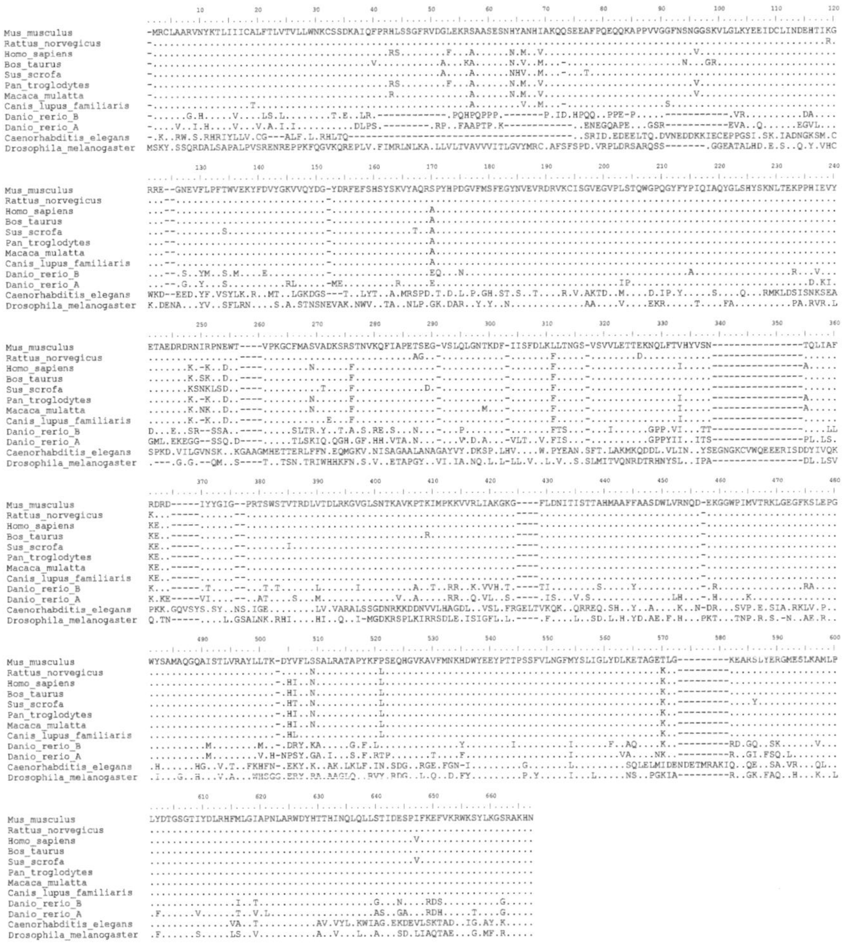

FIG. 6. Alignment of HSepi peptide sequences from selected animal species.

not coordinately regulated in zebrafish development.[42] Expression of the gene in *C. elegans* (termed *hse* 5) was found important to neuron development,[43] and was involved in regulating axon guidance and cell migration.[44]

Regulatory mechanism for *Glce* expression has not been fully elucidated. However, it has been reported that activation of beta-catenin-TCF4 transactivation complex is correlated with *Glce* expression.[45] Nonetheless, *Glce* expression is affected by a number of factors, including changed expression of other genes. Overexpression of catalase in human HaCaT keratinocytes was reported to result in upregulation of a number of genes involved in HS biosynthesis, including *Glce*.[46] However, hydrogen peroxide treatment resulted in down-regulation of the *Glce* mRNA level.[46] It is worth noting that a recent report suggested that *Glce* might be a tumor suppress gene, because analysis of 74 patients with breast tumor along with 21 control patients without malignancy revealed either downregulation or loss of *Glce* mRNA expression in 82–84% of human breast tumors.[47] In accordance, *Glce* expression was found suppressed in endothelial cells that overexpress the EJ-ras oncogene[48] and in the mouse liver that overexpresses human heparanase[49], while heparanase is known to be upregulated in most metastasis tumors.[50] Whether *Glce* is indeed involved in tumor development is an interesting subject to be investigated.

V. Industrial Applications of HSepi

The anticoagulant and antithrombotic properties of heparin are extensively exploited in the clinic to prevent and treat thromboembolic disease.[51] As heparin is produced in connective tissue mast cells, crude heparin for preparation of clinical drugs, that is, unfractionated heparin and low molecule weight heparin, is exclusively isolated from animal tissues, primarily pig intestinal mucosa. Although such a preparation has been applied for more than 80 years and regarded as a safe and relatively cost effective resource, because of the growing general aversion toward drugs of animal origin, for example, potential risks of zoonotic diseases, and the environmental and ecological hazards resulting from producing heparin, a nonanimal heparin source replacement should be more preferred, not only by consumers but also by pharmaceutical industries. *De novo* chemical synthesis of such compounds is rapidly developing but still cumbersome.[52,53] An alternative approach exploits the *E. coli* K5 capsular polysaccharide (K5-PS) that has the same structure (GlcAβ1,4GlcNAα1,4) as the precursor polymer in HS/heparin biosynthesis.[54]

The K5-PS can be purified from large scale fermented *E. coli* K5 and modified by the reactions involved in biosynthesis of HS/heparin. All reactions of the polymer modification process can be mimicked by chemical procedures, albeit with considerable loss of specificity, with one exception, viz. the GlcA C5-epimerization that so far is not attainable by nonenzymatic reactions. To satisfy this unique requirement for enzymatic catalysis, recombinant HSepi is applied. Controlled modification of the *E. coli* K5 polysaccharide by chemical and enzymatic modifications produced "neoheparin" that exhibited anticoagulant activity and antithrombotic properties similar to those of heparin.[54] Alternatively, a complete enzyme-based modification is also reported.[55–57]

Moreover, it is observed that clinical applications of heparin for treatment of thrombosis in patients with cancer or inflammation showed a beneficial effect for the patients, in addition to the antithrombotic effect.[58,59] However, it is obvious that application of heparin to patients with no thrombotic complications is improper, because of the potential to lead to bleeding. Therefore, attention has been drawn to heparin preparations without anticoagulant activities[60] or HS/heparin-mimetics. With the progress in understanding the structure–function relationship of HS/heparin in their anti-inflammatory and antimetastatic effects, the prospect of producing tailor-made HS/heparin-related oligo- or polysaccharides for a variety of pharmaceutical purposes has attracted increasing attention.[61] The process for generating "neoheparin" by application of the HSepi in combination with other modification enzymes or chemical reactions can be readily adapted to produce structures preferably for other purposes.

VI. Concluding Remarks

The sugar residues in most polysaccharides are incorporated as their corresponding monomers during polymerization. There are three known exceptions to this rule, involving biosynthesis of HS/heparin, dermatan sulfate, and alginate. In HS/heparin biosynthesis, incorporated GlcA residue is C5-epimerized, catalyzed by the glucuronyl C5-epimerase (HSepi) at the polymer level, generating IdoA units that are important for the biological functions of HS/heparin. Targeted interruption of the gene coding for HSepi in mice revealed essential role of the enzyme, most likely because of the loss or reduced capacity of the mutant HS (lacking IdoA residues) to interact with cytokines. Although it is established that the severe development defects in the mutant mice are associated with the gene mutation, the exact mechanisms underlying the defects are largely uncovered. Moreover, the current observations are mainly limited to the macroscopic examination, so additional development defects may exist and are yet to be found out. In fact, a very recent study discovered that lymphoid organ development was impaired in the HSepi mutant mice.[62] Thus, a deeper insight into the structure and function correlation in HS-related biology requires further refined genetic approach. In particular, tissue specific conditional knockout of the selected biosynthetic enzymes has led to assessment of subtle functional defects in selected organ systems and cells that remain seemingly unaffected within an otherwise lethal phenotype. Moreover, several intriguing aspects of HSepi remain to be clarified and developed. Crystallization and mutational analysis of the protein will be required to define the active site of the enzyme and the precise catalytic mechanism. Efforts should be made to identify other proteins, enzymes, or auxiliary components, in the ER/Golgi compartments with which HSepi interacts. Further studies of the gene and its regulatory elements will provide more detailed information on the role of the enzyme in the complex enzymatic machinery involved in the assembly of HS/heparin chains.

References

1. Cifonelli JA, Dorfman A. The uronic acid of heparin. *Biochem Biophys Res Commun* 1962;**7**:41–4.
2. Lindahl U, Bäckström G. Biosynthesis of L-iduronic acids in heparin: epimerization of D-glucuronic acid on the polymer level. *Biochem Biophys Res Commun* 1972;**46**:985–91.
3. Haug A, Larsen B. Biosynthesis of alginate. Epimerisation of D-mannuronic to L-guluronic acid residues in the polymer chain. *Biochim Biophys Acta* 1969;**192**:557–9.
4. Höök M, Lindahl U, Backstrom G, Malmstrom A, Fransson L. Biosynthesis of heparin. 3. Formation of iduronic acid residues. *J Biol Chem* 1974;**249**:3908–15.
5. Lindahl U, Kusche M, Lidholt K, Oscarsson LG. Biosynthesis of heparin and heparan sulfate. *Ann NY Acad Sci* 1989;**556**:36–50.
6. Lindahl U, Li J-P. Interactions between heparan sulfate and proteins—design and functional implications. *Int Rev Cell Mol Biol* 2009;105–59.
7. Taylor RL, Shively JE, Cifonelli JA. Uronic acid composition of heparins and heparan sulfates. *Biochemistry* 1973;**12**:3633–7.
8. Malmstrom A, Roden L, Feingold DS, Jacobsson I, Backstrom G, Lindahl U. Biosynthesis of heparin. Partial purification of the uronosyl C-5 epimerase. *J Biol Chem* 1980;**255**:3878–83.
9. Campbell P, Hannesson HH, Sandbäck D, Rodén L, Lindahl U, Li J-P. Biosynthesis of heparin/heparan sulfate. Purification of the D-glucuronyl C-5 epimerase from bovine liver. *J Biol Chem* 1994;**269**:26953–8.
10. Li J-P, Hagner-McWhirter A, Kjellen L, Palgi J, Jalkanen M, Lindahl U. Biosynthesis of heparin/heparan sulfate. cDNA cloning and expression of D-glucuronyl C5-epimerase from bovine lung. *J Biol Chem* 1997;**272**:28158–63.
11. Li JP, Gong F, El Darwish K, Jalkanen M, Lindahl U. Characterization of the D-glucuronyl C5-epimerase involved in the biosynthesis of heparin and heparan sulfate. *J Biol Chem* 2001;**276**:20069–77.
12. Crawford BE, Olson SK, Esko JD, Pinhal MA. Cloning, Golgi localization, and enzyme activity of the full-length heparin/heparan sulfate-glucuronic acid C5-epimerase. *J Biol Chem* 2001;**276**:21538–43.
13. Maccarana M, Olander B, Malmstrom J, Tiedemann K, Aebersold R, Lindahl U, Li JP, Malmstrom A. Biosynthesis of dermatan sulfate: chondroitin glucuronate C5-epimerase is identical to SART2. *J Biol Chem* 2006;**281**:11560–8.
14. Ghiselli G, Farber SA. D-glucuronyl C5-epimerase acts in dorso-ventral axis formation in zebrafish. *BMC Dev Biol* 2005;**5**:19.
15. Li J-P, Gong F, Hagner-McWhirter Å, Forsberg E, Åbrink M, Kisilevsky R, Zhang X, Lindahl U. Targeted disruption of a murine glucuronyl C5-epimerase gene results in heparan sulfate lacking L-iduronic acid and in neonatal lethality. *J Biol Chem* 2003;**278**:28363–6.
16. Li K, Bethea HN, Liu J. Using engineered 2-*O*-sulfotransferase to determine the activity of heparan sulfate C5-epimerase and its mutants. *J Biol Chem* 2010;**285**:11106–13.
17. Pinhal MA, Smith B, Olson S, Aikawa J, Kimata K, Esko JD. Enzyme interactions in heparan sulfate biosynthesis: uronosyl 5-epimerase and 2-*O*-sulfotransferase interact in vivo. *Proc Natl Acad Sci USA* 2001;**98**:12984–9.
18. Lindahl U, Jacobsson I, Höök M, Bäckström G, Feingold DS. Biosynthesis of heparin. Loss of C-5 hydrogen during conversion of D-glucuronic to L-iduronic acid residues. *Biochem Biophys Res Commun* 1976;**70**:492–9.
19. Jacobsson I, Bäckström G, Höök M, Lindahl U, Feingold DS, Malmström A, Rodén L. Biosynthesis of heparin. Assay and properties of the microsomal uronosyl C-5 epimerase. *J Biol Chem* 1979;**254**:2975–82.

20. Jensen JW, Rodén L, Jacobsson I, Lindahl U, Prihar H, Feingold DS. Biosynthesis of heparin. A new substrate for heparosan-*N*-sulfate-D-glucopyranosyl-uronate 5-epimerase. *Carbohydr Res* 1983;**117**:241–53.
21. Campbell P, Feingold DS, Jensen JW, Malmström A, Rodén L. New assay for uronosyl 5-epimerases. *Anal Biochem* 1983;**131**:146–52.
22. Hagner-McWhirter A, Hannesson HH, Campbell P, Westley J, Roden L, Lindahl U, Li J-P. Biosynthesis of heparin/heparan sulfate: kinetic studies of the glucuronyl C5-epimerase with *N*-sulfated derivatives of the *Escherichia coli* K5 capsular polysaccharide as substrates. *Glycobiology* 2000;**10**:159–71.
23. Jacobsson I, Lindahl U, Jensen JW, Rodén L, Prihar H, Feingold DS. Biosynthesis of heparin. Substrate specificity of heparosan *N*-sulfate D-glucuronosyl 5-epimerase. *J Biol Chem* 1984;**259**:1056–63.
24. Maccarana M, Sakura Y, Tawada A, Yoshida K, Lindahl U. Domain structure of heparan sulfates from bovine organs. *J Biol Chem* 1996;**271**:17804–10.
25. Valla S, Li J, Ertesvag H, Barbeyron T, Lindahl U. Hexuronyl C5-epimerases in alginate and glycosaminoglycan biosynthesis. *Biochimie* 2001;**83**:819–30.
26. Hagner-Mcwhirter A, Lindahl U, Li J-P. Biosynthesis of heparin/heparan sulphate: mechanism of epimerization of glucuronyl C-5. *Biochem J* 2000;**347**:69–75.
27. Hagner-McWhirter A, Li JP, Oscarson S, Lindahl U. Irreversible glucuronyl C5-epimerization in the biosynthesis of heparan sulfate. *J Biol Chem* 2004;**279**:14631–8.
28. Höök H, Lindahl U, Hallén A, Bäckström G. Biosynthesis of heparin. Studies on the microsomal sulfation process. *J Biol Chem* 1975;**250**:6065–71.
29. Salmivirta M, Lidholt K, Lindahl U. Heparan sulfate—a piece of information. *FASEB J* 1996;**10**:1270–9.
30. Carlsson P, Presto J, Spillmann D, Lindahl U, Kjellen L. Heparin/heparan sulfate biosynthesis: processive formation of *N*-sulfated domains. *J Biol Chem* 2008;**283**:20008–14.
31. Feyerabend TB, Li JP, Lindahl U, Rodewald HR. Heparan sulfate C5-epimerase is essential for heparin biosynthesis in mast cells. *Nat Chem Biol* 2006;**2**:195–6.
32. Lin X, Wei G, Shi Z, Dryer L, Esko JD, Wells DE, Matzuk MM. Disruption of gastrulation and heparan sulfate biosynthesis in EXT1-deficient mice. *Dev Biol* 2000;**224**:299–311.
33. Nybakken K, Perrimon N. Heparan sulfate proteoglycan modulation of developmental signaling in *Drosophila*. *Biochim Biophys Acta* 2002;**1573**:280–91.
34. Grobe K, Ledin J, Ringvall M, Holmborn K, Forsberg E, Esko JD, Kjellen L. Heparan sulfate and development: differential roles of the *N*-acetylglucosamine *N*-deacetylase/*N*-sulfotransferase isozymes. *Biochim Biophys Acta* 2002;**1573**:209–15.
35. Ringvall M, Ledin J, Holmborn K, van Kuppevelt T, Ellin F, Eriksson I, Olofsson AM, Kjellén L, Forsberg E. Defective heparan sulfate biosynthesis and neonatal lethality in mice lacking *N*-deacetylase/*N*-sulfotransferase-1. *J Biol Chem* 2000;**275**:25926–30.
36. Merry CL, Bullock SL, Swan DC, Backen AC, Lyon M, Beddington RS, Wilson VA, Gallagher JT. The molecular phenotype of heparan sulfate in the Hs2st−/− mutant mouse. *J Biol Chem* 2001;**276**:35429–34.
37. Bullock SL, Fletcher JM, Beddington RS, Wilson VA. Renal agenesis in mice homozygous for a gene trap mutation in the gene encoding heparan sulfate 2-sulfotransferase. *Genes Dev* 1998;**12**:1894–906.
38. Gallagher JT. Heparan sulfate: growth control with a restricted sequence menu. *J Clin Invest* 2001;**108**:357–61.
39. Abramsson A, Kurup S, Busse M, Yamada S, Lindblom P, Schallmeiner E, Stenzel D, Sauvaget D, Ledin J, Ringvall M, Landegren U, Kjellen L, Bondjers G, Li JP, Lindahl U, Spillmann D, Betsholtz C, Gerhardt H. Defective *N*-sulfation of heparan sulfate proteoglycans limits PDGF-BB binding and pericyte recruitment in vascular development. *Genes Dev* 2007;**21**:316–31.

40. Jia J, Maccarana M, Zhang X, Bespalov M, Lindahl U, Li JP. Lack of L-iduronic acid in heparan sulfate affects interaction with growth factors and cell signaling. *J Biol Chem* 2009;**284**:15942–50.
41. Casu B, Lindahl U. Structure and biological interactions of heparin and heparan sulfate. *Adv Carbohydr Chem Biochem* 2001;**57**:159–206.
42. Cadwallader AB, Yost HJ. Combinatorial expression patterns of heparan sulfate sulfotransferases in zebrafish: III. 2-*O*-sulfotransferase and C5-epimerases. *Dev Dyn* 2007;**236**:581–6.
43. Bulow HE, Hobert O. Differential sulfations and epimerization define heparan sulfate specificity in nervous system development. *Neuron* 2004;**41**:723–36.
44. Rhiner C, Gysi S, Frohli E, Hengartner MO, Hajnal A. Syndecan regulates cell migration and axon guidance in *C. elegans*. *Development* 2005;**132**:4621–33.
45. Ghiselli G, Agrawal A. The human D-glucuronyl C5-epimerase gene is transcriptionally activated through the beta-catenin-TCF4 pathway. *Biochem J* 2005;**390**:493–9.
46. Nakayama F, Hagiwara A, Yamamoto T, Akashi M. Hydrogen peroxide as a potential mediator of the transcriptional regulation of heparan sulphate biosynthesis in keratinocytes. *Cell Mol Biol Lett* 2008;**13**:475–92.
47. Grigorieva E, Eshchenko T, Rykova VI, Chernakov A, Zabarovsky E, Sidorov SV. Decreased expression of human D-glucuronyl C5-epimerase in breast cancer. *Int J Cancer* 2008;**122**:1172–6.
48. Lopes CC, Toma L, Pinhal MA, Porcionatto MA, Sogayar MC, Dietrich CP, Nader HB. EJ-ras oncogene transfection of endothelial cells upregulates the expression of syndecan-4 and downregulates heparan sulfate sulfotransferases and epimerase. *Biochimie* 2006;**88**:1493–504.
49. Escobar Galvis ML, Jia J, Zhang X, Jastrebova N, Spillmann D, Gottfridsson E, van Kuppevelt TH, Zcharia E, Vlodavsky I, Lindahl U, Li JP. Transgenic or tumor-induced expression of heparanase upregulates sulfation of heparan sulfate. *Nat Chem Biol* 2007;**3**:773–8.
50. Vlodavsky I, Elkin M, Abboud-Jarrous G, Levi-Adam F, Fuks L, Shafat I, Ilan N. Heparanase: one molecule with multiple functions in cancer progression. *Connect Tissue Res* 2008;**49**:207–10.
51. Gray E, Mulloy B, Barrowcliffe TW. Heparin and low-molecular-weight heparin. *Thromb Haemost* 2008;**99**:807–18.
52. Petitou M, Hérault J-P, Bernat A, Driguez P-A, Duchaussoy P, Lormeau J-C, Herbert J-M. Synthesis of thrombin-inhibiting heparin mimetics without side effects. *Nature* 1999;**398**:417–22.
53. Petitou M, Nancy-Portebois V, Dubreucq G, Motte V, Meuleman D, de Kort M, van Boeckel CA, Vogel GM, Wisse JA. From heparin to EP217609: the long way to a new pentasaccharide-based neutralisable anticoagulant with an unprecedented pharmacological profile. *Thromb Haemost* 2009;**102**:804–10.
54. Lindahl U, Li JP, Kusche-Gullberg M, Salmivirta M, Alaranta S, Veromaa T, Emeis J, Roberts I, Taylor C, Oreste P, Zoppetti G, Naggi A, Torri G, Casu B. Generation of "neoheparin" from *E. coli* K5 capsular polysaccharide. *J Med Chem* 2005;**48**:349–52.
55. Kuberan B, Lech MZ, Beeler DL, Wu ZL, Rosenberg RD. Enzymatic synthesis of antithrombin III-binding heparan sulfate pentasaccharide. *Nat Biotechnol* 2003;**21**:1343–6.
56. Xu D, Moon AF, Song D, Pedersen LC, Liu J. Engineering sulfotransferases to modify heparan sulfate. *Nat Chem Biol* 2008;**4**:200–2.
57. Peterson S, Frick A, Liu J. Design of biologically active heparan sulfate and heparin using an enzyme-based approach. *Nat Prod Rep* 2009;**26**:610–27.
58. Kakkar AK, Levine MN, Kadziola Z, Lemoine NR, Low V, Patel HK, Rustin G, Thomas M, Quigley M, Williamson RC. Low molecular weight heparin, therapy with dalteparin, and survival

in advanced cancer: the fragmin advanced malignancy outcome study (FAMOUS). *J Clin Oncol* 2004;**22**:1944–8.

59. Green D, Hull RD, Brant R, Pineo GF. Lower mortality in cancer patients treated with low-molecular-weight versus standard heparin. *Lancet* 1992;**339**:1476.
60. Ludwig RJ. Therapeutic use of heparin beyond anticoagulation. *Curr Drug Discov Technol* 2009;**6**:281–9.
61. Lindahl U. Heparan sulfate–protein interactions—a concept for drug design? *Thromb Haemost* 2007;**98**:109–15.
62. Reijmers RM, Vondenhoff MF, Roozendaal R, Kuil A, Li JP, Spaargaren M, Pals ST, Mebius RE. Impaired lymphoid organ development in mice lacking the heparan sulfate modifying enzyme glucuronyl C5-epimerase. *J Immunol* 2010;**184**:3656–64.

Mice Deficient in Heparan Sulfate 6-O-Sulfotransferase-1

Hiroko Habuchi and
Koji Kimata

Research Complex for the Medicine Frontiers, Aichi Medical University, Nagakute, Aichi, Japan

Heparan sulfate chains are initially synthesized on core proteins as linear polysaccharides composed of glucuronic acid-*N*-acetylglucosamine repeating units and subjected to marked structural modification by sulfation at various places and epimerization of hexuronic acid residues (C5-epimerase) at the Golgi lumen and further by 6-*O*-desulfation at the cell surface, which generates

Progress in Molecular Biology
and Translational Science, Vol. 93
DOI: 10.1016/S1877-1173(10)93005-6

1877-1173/10 $35.00

their characteristic divergent fine structures. This chapter focuses on the biological and physiological functions of 6-*O*-sulfation in HS and the characterization of the enzymes catalyzing 6-*O*-sulfation (HS6ST). HS6STs in mammals such as humans and mice comprise of three isoforms (HS6ST-1, -2, and -3) and one alternatively spliced form of HS6ST-2 (HS6ST-2S). Each of these isoforms has distinct substrate preferences, albeit overlapping each other. These HS6ST isoforms are expressed in a spatiotemporally regulated manner in most organs. HS6ST-1-deficient mice are lethal mostly at later embryonic stages and exhibit abnormal angiogenesis in labyrinthine zone of placenta and aberrant lung morphology similar to pulmonary emphysema. These knockout mice also exhibit retinal axon guidance abnormality at the optic chiasm. Other HS6ST-deficient animals reveal various malformations in muscle development and branching morphology of the caudal vein of zebrafish, in tracheal formation of *Drosophila*, and in axon guidance of ventral nerve cord interneurons of *Caenorhabditis elegans*. Mouse embryonic fibroblasts prepared from HS6ST-1/HS6ST-2 double knockout mice did produce HS lacking 6-*O*-sulfation and responded differently to various FGFs dependent signaling.

I. Introduction

Heparan sulfate proteoglycans are ubiquitously present on the cell surface and extracellular matrix including basement membrane and act as microenvironmental regulatory factor. HS chains attached to HSPG interact with a huge number of proteins including growth factors, morphogens, extracellular matrix components, proteases, and protease inhibitors, and play pivotal roles in the developmental and pathophysiological processes.[1–9] The interactions between HS and ligands are thought to be dependent not only upon the patterns of sulfate and hexuronic acid isomer but also on their total negative charge.[10–13] The polysaccharide backbone of HS is synthesized by the alternative transfer of glucuronic acid and *N*-acetylglucosamine residues. It is then partially modified through the coordinated action of various enzymes in the Golgi apparatus, namely, *N*-deacetylase/*N*-sulfotransferases (NDSTs), C5-epimerase (Hsepi), and 2-*O*-, 6-*O*-, and 3-*O*-sulfotransferases to generate HS with a variety of structures.[14,15] HS is further modified by 6-*O*-endosulfatase (Sulf) at the cell surface (Fig. 1).[16] Thus, the structures containing 6-*O*-sulfate groups in HS are regulated by two steps, one is in the process of biosynthesis and another is at the postbiosynthesis stage. Recent loss-of-function studies on enzymes involved in HS biosynthesis have demonstrated that HS has critical roles in developmental and pathological processes. In this chapter, on the basis of the observations of phenotypes of HS6ST-knockdown or -knockout organisms, the functional aspects of 6-*O*-sulfation in HS are described.

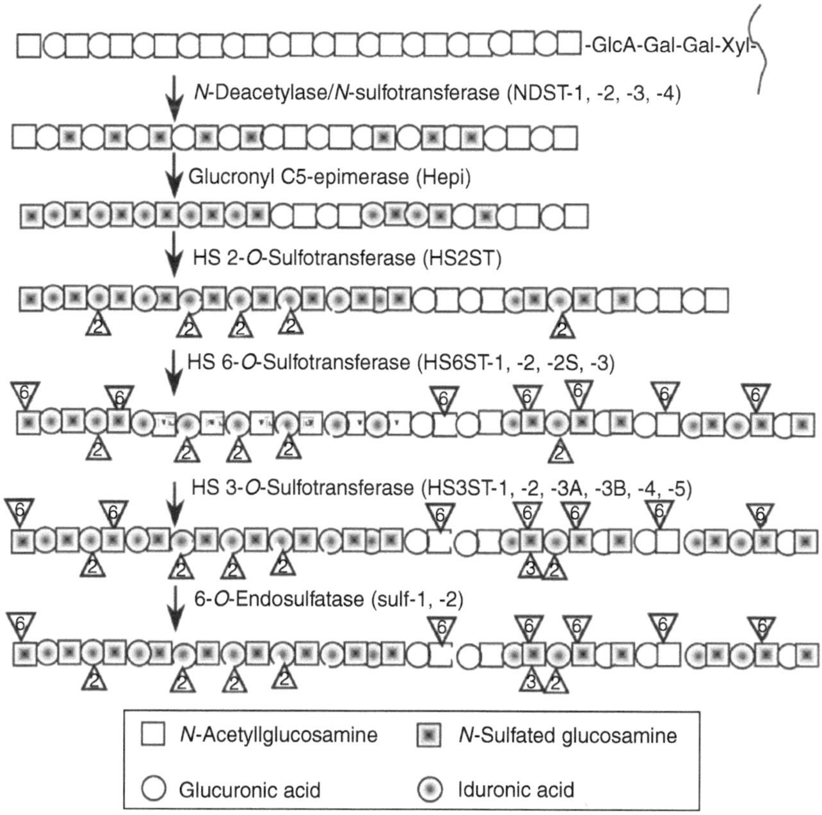

FIG. 1. Schema for the modification pathway of heparan sulfate biosynthesis. Some of the sulfotransferases are presumed to be involved on the basis of tissue and substrate specificities of sulfotransferases. (Redrawn, with permission, from Ref. 18.)

II. Characterization of Heparan Sulfate 6-O-Sulfotransferases

A. Isolation and Purification of HS6ST-1 and Cloning of HS6ST-1 and Its Isoforms

Heparan sulfate 6-*O*-sulfotransferases (HS6STs), which catalyze the transfer of sulfate from 3′-phospho adenosine 5′-phosphosulfate (PAPS) to position 6 of *N*-sulfoglucosamine and *N*-acetylglucosamine residues in HS, are derived from three different genes, Hs6st-1, -2, and -3 and one alternatively spliced form of Hs6st-2 (Hs6st-2S).[17–19] The first isolation and cloning of protein with the HS6ST activity were as follows.[20] The protein was purified to homogeneity from the culture medium of Chinese-hamster ovary (CHO)-cells. At 72-h culture, most of the HS6ST activity existed in the conditioned medium and only 4% of the total activity existed in the cell layer. The purified enzyme showed two protein bands with 52 and 42 kDa, both of which were glycoproteins with *N*-glycans. The optimum pH was around 6.3, and the activity was markedly stimulated by protamine and inhibited strongly by dithiothreitol. Interestingly, in contrast to the HS6ST activity, most of the activity of HS 2-*O*-sulfotransferase (HS2ST), which transfers sulfate to position 2 of hexuronic acid in HS, was found in the cell layer.[21,22]

On the basis of the amino acid sequence of the purified enzyme, hamster and human HS6ST-1 cDNAs were cloned[17] and subsequently two isoforms, HS6ST-2 and HS6ST-3 cDNAs from the mouse brain cDNA library were cloned by cross-hybridization with human HS6ST-1 cDNA,[18] and a HS6ST-2 spliced form, HS6ST-2S, was found.[19] The cDNAs of the three HS6ST isoforms, HS6ST-1, HS6ST-2, and HS6ST-3, contained single open reading frames that predicted type II transmembrane proteins composed of 401, 506, and 470 amino acid residues, respectively (Fig. 2). Amino acid sequence of HS6ST-1 was 51% and 57%, identical to those of HS6ST-2 and HS6ST-3, respectively. HS6ST-2 and HS6ST-3 had 50% identity. Although only a little identity was observed in both the amino-terminal and carboxyl-terminal sequences among these isoforms, amino acid sequences of the center regions were well conserved. Two putative PAPS binding sites, which were predicted from the X-ray crystallographic analysis of estrogen sulfotransferase[23] and later became the useful marker for computer-screening of various sulfotransferases,[24] were present in these center regions of the three isoforms. The sequence for the putative 5′-phosphosulfate binding site (KTGGTTF) and the putative 3′-phosphate binding site (RDPSVSRYLS) were well conserved among HS6ST-1, -2, and -3. In fact, when the lysine residue in the putative 3′-phosphate binding site was mutated to glycine, the activities of all three isoforms of HS6STs were completely abolished. HS6ST-2 had a unique cluster composed of six Glu-Ser repeating units at the C-terminal domain. The function of this domain remains to be determined although the sequence appears to match the *N*-myristoylation site. HS6ST-3 had another unique hydrophilic sequence composed of 10 glutamic acid residues. This domain may endow the sulfotransferase molecule with a negatively charged hydrophilic surface, but the functional role of this domain is not evident. Mammals such as humans and mouse have three genes for HS6ST, but *Drosophila* and *Caenorhabditis elegans* have only one gene for HS6ST. The substrate specificity of *Drosophila* HS6ST exhibits the intermediate property between those of mouse HS6ST-1 and HS6ST-2.[25] Zebrafish has four genes for HS6ST, which are thought to be generated by genome duplication.[26] In mice, Hs6st-1 gene locates on chromosome 1B, Hs6st-2 gene on X A3.3, and Hs6st-3 on 14 E4 (from NCBI database).

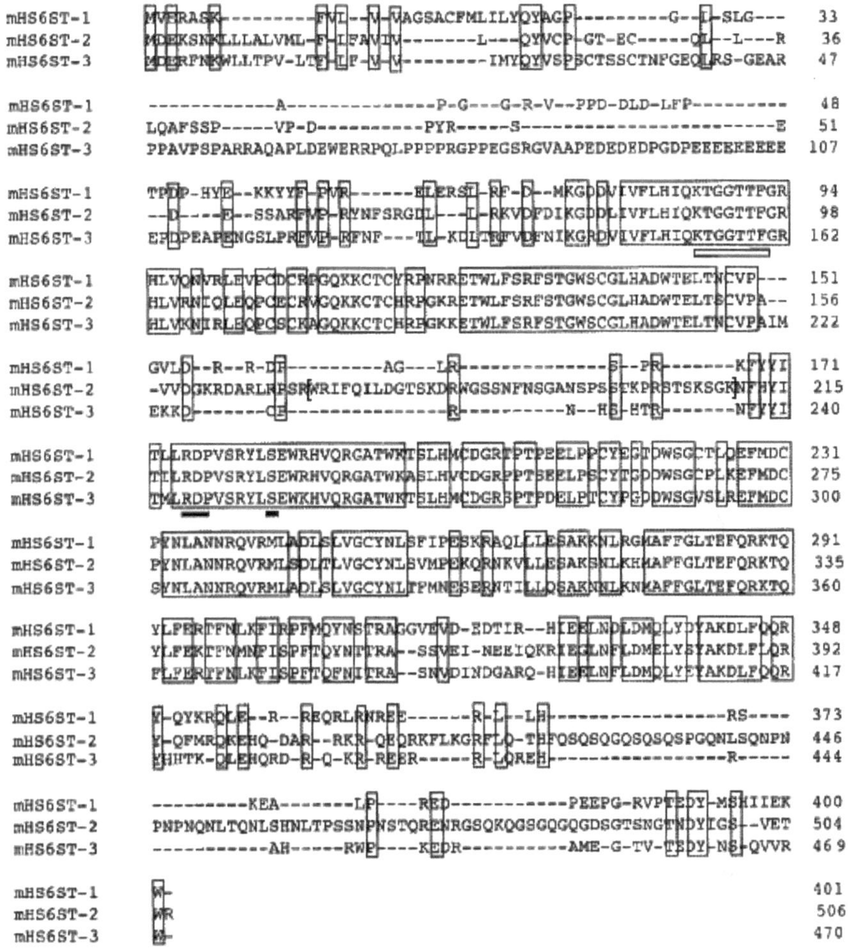

FIG. 2. Sequence comparison among mouse HS6ST-1, -2, and -3. The predicted amino acid sequences are aligned using Genetyx computer program. The dashes indicate the skipped positions for alignment. The boxes indicate that the predicted amino acid in the alignment is identical among the three sequences. The PAPS binding sites are shown with a white box for 5′-phosphosulfate and black boxes for 3′-phosphate in PAPS, respectively. The deleted portion in HS6ST-2S is shown within [ ] of HS6ST-2.

B. Substrate Specificities of HS6STs

When the open reading frames in HS6ST cDNAs were inserted into expression vectors with N-terminal tag sequence and transfected into COS-7 cells, HS6ST activities in the extracts obtained from the cells transfected with these cDNAs were 10-fold higher than the activity of the host cells.[18] The activities found in the culture medium at 48-h culture were 62%, 50%, and 64% of the total HS6ST activities for HS6ST-1, HS6ST-2, and HS6ST-3, respectively. The HS6ST proteins secreted into the culture medium lost the N-terminal tag, while the recombinant HS6STs proteins in the cell extracts still retained the tag, suggesting that there are some specific mechanisms for the expressed HS6STs to be cleaved at the N-terminal sites during the secretion into the culture medium.

All the purified recombinant HS6ST isoforms transferred sulfate to position 6 of *N*-sulfoglucosamine residues located not only in the internal regions but also at the nonreducing terminal. However, they showed little activity toward *N*-acetylglucosamine residues *in vitro*. Interestingly, all the HS6ST isoforms were shown to sulfate these residues *in vivo* to produce $GlcNAc6SO_4$ residues, because CHO cells transfected with any of the Hs6st isoforms all produced much more $GlcNAc(6SO_4)$ residues in HS than nontransfected wild-type CHO cells did. Such apparently discrepant results observed between *in vitro* and *in vivo* may reflect the biosynthetic pathway of heparan sulfate in which 6-*O*-sulfation of the GlcNAc residue may occur as a reaction coordinating with the elongation of HS backbone. The individual isoforms showed characteristic preference for the uronic acid residue neighboring the *N*-sulfoglucosamine although the substrate specificities of these isoforms largely overlap.[18] Figure 3 shows the activities of these isoforms toward completely desulfated *N*-sulfated (CDSNS)-heparin (major repeating unit: $IdoA\text{-}GlcNSO_3$) and *N*-sulfated heparosan (major repeating unit: $GlcA\text{-}GlcNSO_3$). HS6ST-1 prefers the $IdoA\text{-}GlcNSO_3$ while HS6ST-2 has a different preference, depending upon the substrate concentration, and HS6ST-3 acts on either substrate. When HS was used as the substrate for the HS6ST isoforms, digestion of the products with a mixture of heparinase and haparanases yielded $HexA\text{-}GlcNSO_3(6SO_4)$, $HexA(2SO_4)\text{-}GlcNSO_3(6SO_4)$, and $GlcNSO_3(6SO_4)$ derived from nonreducing end in the proportion of 48.5%, 27.3%, and 24.2%, respectively, for HS6ST-1 and 18.2%, 57.6%, and 24.2%, respectively, for HS6ST-2.[19] Furthermore, enzymatic reactions using octasaccahrides $(IdoA\text{-}GlcNSO_3)3\text{-}IdoA\text{-}AMan$ with different number of 2-*O*-sulfate groups as the substrates revealed that both HS6ST-2 and HS6ST-3 preferred

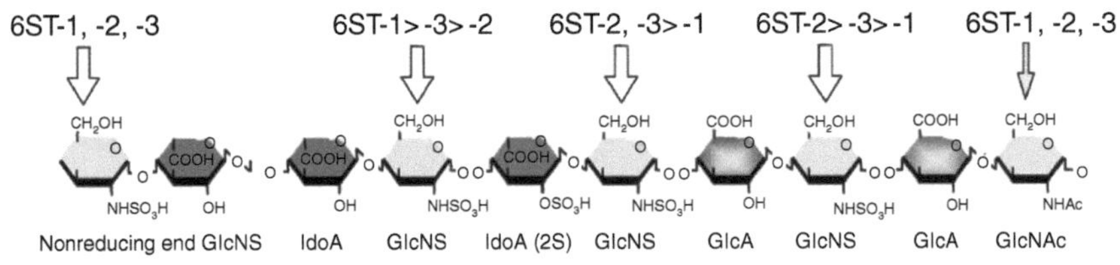

FIG. 3. Substrate preference of HS6ST isoforms.

octasaccharides with two to three 2-*O*-sulfate groups to those with no or one 2-*O*-sulfate group.[27] Taken together, it is most probable that HS6ST-1 sulfates predominantly the position 6 of $GlcNSO_3$ residues in HexA-$GlcNSO_3$ unit, whereas HS6ST-2 and HS6ST-3 transfer sulfate preferentially to position 6 of $GlcNSO_3$ residues in IdoA($2SO_4$)-$GlcNSO_3$ unit.

As for biosynthesis of trisulfated disaccharide units (IdoA($2SO_4$)-$GlcNSO_3$($6SO_4$)) in HS/heparin, 6-*O*-sulfation of $GlcNSO_3$ should be preceded by 2-*O*-sulfation of IdoA, because HS2ST transfers sulfate to position 2 of the iduronic acid residues in IdoA-$GlcNSO_3$ unit of HS but not in IdoA-$GlcNSO_3$($6SO_4$) unit.[21] The substrate specificities of HS6ST isoforms are summarized in Fig. 3.

C. Localization of HS6STs in the Golgi Complexes and Secretion into the Extracellular Space

The most enzymes involved in the biosynthesis of sulfated glycosaminoglycans are Golgi and rough endoplasmic reticulum (ER) resident proteins and may form multienzyme complexes; EXT1/EXT2,[28,29] HS2ST/C-5 epimerase,[30] and EXT2/NDST1.[31] HS6ST-1, -2, and -3 colocalized with a Golgi marker (GM130), but not with an ER marker protein (disulfide isomerase) under the forced expression conditions.[32] These HS6ST isoform proteins were further clarified to localize in the *trans*-side of the Golgi stack because of the following observation: (1) When the GFP-tagged HS6ST transfectant cells were treated with brefeldin A (BFA), which is known to cause the redistribution of the most Golgi proteins into the ER without affecting the proteins in the *cis*-Golgi network,[33–35] the GFP protein was redistributed to the ER, while *cis*-Golgi protein GM130 was found in dispersed and punctate structures. (2) On the treatment with nocodazole, which disrupts the microtubule network, the scattering of the *cis*-Golgi protein lagged behind the scattering of the *trans*-side protein.[31] The scattering of the HS6ST-GFP upon nocodazole treatment preceded the scattering of the *cis*-Golgi marker GM130, suggesting that HS6ST-GFP is localized at the more trans-side region than GM130.

Which domains in the HS6ST proteins are responsible for the localization in the Golgi apparatus? By using mutated HS6STs lacking various domains, the Golgi localization signal of the HS6STs was found mainly within the stem regions, which are present between transmembrane domain and catalytic domain. The cytoplasmic and/or transmembrane regions also assist Golgi localization.[32]

Where and how are HS6STs proteins secreted? To address these issues, HS6STs having GFP-tags at the C-terminals were expressed stably in CHO-K1 cells and monitored their GFP fluorescence. The treatment of HS6ST-3-GFP-expressing cells with brefeldin A or monensin, which blocks different steps of the intracellular transport process, exhibited that the N-terminal hydrophobic domain of HS6ST-3 was processed in the ER or *cis/medial* Golgi. Furthermore, the addition of the various protease inhibitors into the cells exhibited that only cell-permeable β-secretase inhibitor[36] specifically inhibited HS6ST-3 secretion but not the secretion of HS6ST-1 and HS6ST-2. By the cotransfection of β-secretase into the stable transfectants, the secretion of HS6ST-3 was much more enhanced with a concomitant decrease in the intracellular HS6ST-3-GFP. These observations suggest that β-secretase regulates at least partly the secretion of HS6ST-3. However, *in vitro* incubation of HS6ST-3 with β-secretase did not cause any cleavage, and even when the stable HS6ST-3-GFP-expressing cells were treated with the β-secretase inhibitor which blocks the secretion, intracellular HS6ST-3-GFP showed the N-terminally processed short form. Therefore, the stimulation of the secretion of HS6ST-3 by β-secretase appears to be due to an indirect mechanism involving unknown factors other than β-secretase. As for HS6ST-1 and HS6ST-2, the mechanism of secretion still remains unclear.

D. Expression Patterns of Three Isoforms of HS6ST and Fine Structure of HS

HS6ST transcripts are differentially expressed not only in various tissues but also at different developmental stages, and therefore it would be possible that distinctive structures of HS produced in a spatiotemporally regulated manner control the function of heparin-binding proteins such as growth factors, morphogens, and extracellular proteins. In the adult mice, Northern blot analyses exhibited that these three isoforms were expressed quite differently from each other.[18] HS6ST-1 transcript was expressed mainly in the liver and moderately in the kidney, lung, heart, and brain. On the other hand, HS6ST-2 was expressed strongly in the brain and spleen. In contrast to the HS6ST-1 and -2, HS6ST-3 transcripts were expressed rather ubiquitously in most tissues. Using *in situ* hybridization, expression patterns of HS6ST isoforms were investigated in detail during embryonic day 8.5 (E8.5)–E14.[37] The expression patterns of three HS6ST isoforms in most organs were regulated spatially and temporally. On the whole, HS6ST-1 transcript is expressed predominantly in the epithelial and neural-derived tissues, whereas HS6ST-2 is expressed in the more mesenchymal tissues. HS6ST-3 appears at later stages in a more restricted manner. From the different expression patterns and different substrate preferences among three HS6ST isoforms, it is expected that these sulfotransferases might play different functions *in vivo*. For example, studies on the formation of limb bud described below highlight the function of the HS6ST isoforms.

In developing limb bud, various growth factors and morphogens such as FGFs, hedgehog, Wnt, and bone morphogenic proteins are expressed in a spatiotemporally specific manner. At present, it is known that chick has two HS6STs (HS6ST-1, -2). In the formation of chick limb bud during stage 21 to stage 24, HS6ST-1 transcripts are preferentially localized in the anterior proximal regions, while HS6ST-2 transcripts are localized mainly in the posterior proximal regions and weakly in the anterior region (Fig. 4A). In a transverse section of the wing bud region, both transcripts are expressed strongly in the mesenchyme and

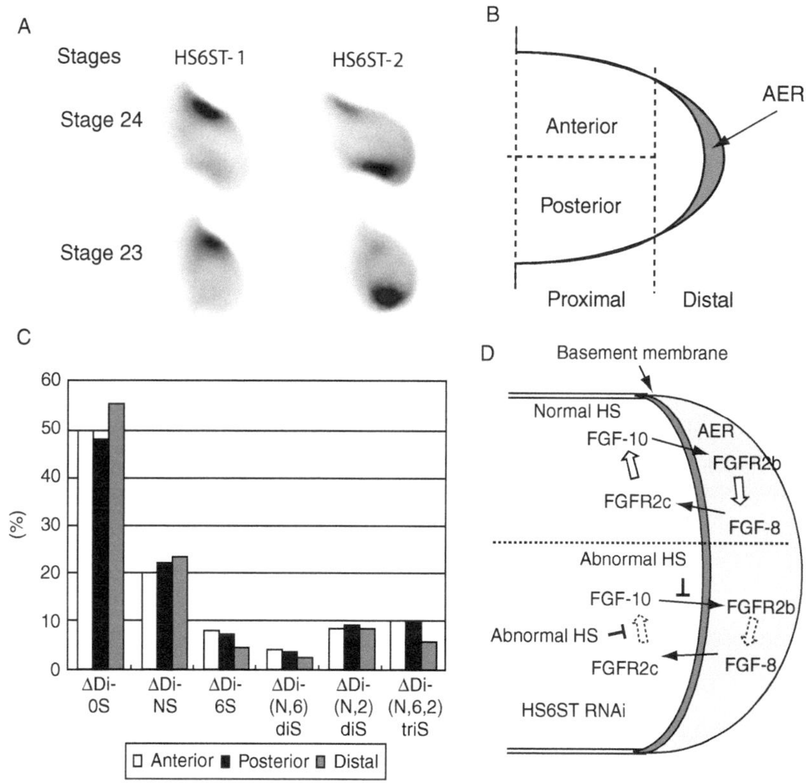

FIG. 4. Expression of HS6ST-1 and HS6ST-2 in chick limb bud, and HS structures. (A) Whole mount *in situ hybridization* in chick limb bud at stages 23 and 24. (B) Each wing bud dissected from chick embryos from stage 24 was separated into three regions. The distal region contained the AER and mesenchyme underlying the AER. (C) HS was isolated from three regions, and was digested with a heparitinase-I, -II, and heparinase mixture. The histograms show the percentage compositions of unsaturated disaccharide. (D) Hypothetical roles of HS in FGF signaling loops between the AER and the mesoderm in developing limb bud. (Redrawn, with permission, from Refs. 38,40.)

weakly in the overlying ectoderm. In contrast, HS2ST transcripts are expressed uniformly throughout the limb bud during early developmental stages.[38–40] A structural analysis of HS from different regions of the wing bud showed that 6-*O*-sulfated *N*-sulfoglucosamine/*N*-acetylglucosamine residues are more abundant in the proximal region than in the distal region (Fig. 4C), and that iduronic acid 6-*O*-sulfated *N*-sulfoglucosmine units tend to be concentrated in the anterior proximal region, while glucuronic acid 6-*O*-sulfated *N*-sulfoglucosamine units are abundant in the posterior proximal region. Such structural changes of HS in the different regions are well consistent with the substrate preferences of HS6ST isoforms that are present in each region. These observations suggest that HS with different sulfation patterns generated by the multiple sulfotransferases could provide appropriate extracellular environments in which growth factors and morphogens could transduce their proper signals. In fact, the inhibition of Hs6st-1 or Hs6st-2 in the prospective limb region using RNA interference (RNAi) resulted in the truncation of limb buds and reduced the expression of *fgf-8* and *fgf-10* in the apical ectodermal ridge and underlying mesenchyme, respectively (Fig. 4D). Hs6st-2 RNAi resulted in a higher frequency of limb bud truncation and a more marked change in both *fgf-8* and *fgf-10* expression than Hs6st-1 RNAi. Each of Hs6st-1 RNAi and Hs6st-2 RNAi caused a significant reduction of 6-*O*-sulfate groups in HS differently; Hs6st-2 RNAi caused a marked reduction of IdoA($2SO_4$)-GlcNSO_3($6SO_4$) units, while Hs6st-1 RNAi brought about significant decrease of HexA-GlcNSO_3($6SO_4$) units and increase of IdoA($2SO_4$)-GlcNSO_3 units. These isoform-specific structural changes are likely to be attributed to their different substrate specificities. These observations support the idea that proper sulfation pattern of HS is important in developmental processes via the regulation of growth factor binding.

III. HS6ST-1-Deficient Mice Exhibit Defective HS Biosynthesis, Abnormal Placentation and Organ Morphogenesis, and Late Embryonic Lethality[41]

A. Targeted Disruption of HS6ST-1 Gene

Construction of the targeting vectors and strategy for screening mice for the homologous recombinations are shown in Fig. 5A. In the recombinant allele, two 3′-phosphoadenosine 5′-phosphosulfate binding sites were deleted. Chimeric mice were generated from ES clones that had undergone homologous recombination at the HS6ST-1 locus and had been confirmed not to contain any additional site of the vector integration. Complete destruction of the HS6ST-1 gene was confirmed by Southern and Northern blot analyses. The HS6ST-1 transcript was not detected in the homozygous mutant mice, whereas

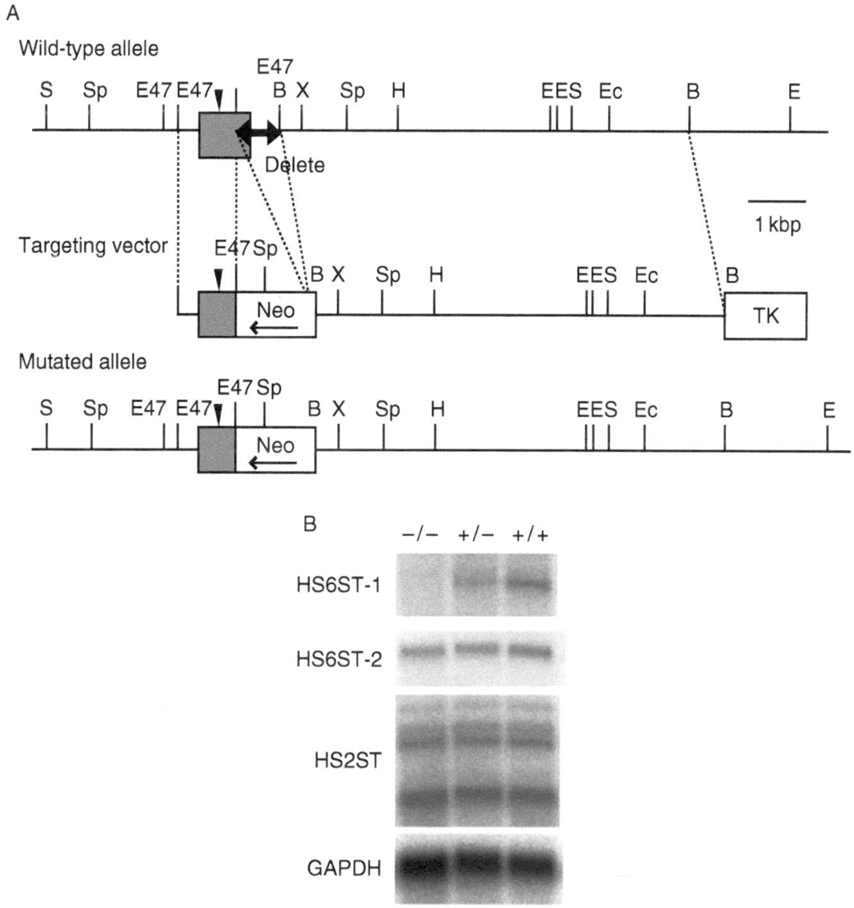

FIG. 5. Targeted disruption of *HS6ST-1* and the generation of *HS6ST-1*-deficient mice. (A) Depiction of the targeting vector for *HS6ST-1* disruption. The restriction maps of the HS6ST-1 gene (*top*), the targeting vector (*middle*), and the mutant allele (*bottom*) are shown. A 0.7 kb-fragment containing two PAPS binding sites (indicated by arrows in the *top* panel) was deleted and replaced with a neo gene cassette. The translation start site is indicated by triangles. Ec, *Eco*RV; E, *Eco*RI; B, *Bam*HI; H, *Hin*dIII; Sp, *Sph*I; S, *Sac*I; X, *Xba*I; E47, *Eco*47III. (B) Northern blot analysis. RNA was prepared from the trunks of the E15.5 wild-type, HS6ST-1$^{+/-}$, and HS6ST-1$^{-/-}$ mice. (Redrawn, with permission, from Ref. 41.)

levels of the HS6ST-2 and HS2ST transcripts in the homozygous mice were similar to those observed in wild-type mice (Fig. 5B). Additionally, any upregulation of HS6ST-2, HS6ST-3, and HS2ST transcripts in the homozygous mice as a compensation for the loss of HS6ST-1 transcript was hardly detected by real-time reverse transcription-PCR.

TABLE I
LETHALITY IN THE HS6ST-$1^{-/-}$ MICE FROM E15.5 TO THE PERINATAL STAGE

		HS6ST-1 genotype			
Age	No. of mice	+/+	+/−	−/−	R°
P1	133	45 (34%)	82 (62%)	6 (4%)	
E17.5–E18.5	27	9 (33%)	16 (59%)	2 (8%)	
E15.5	23	6 (26%)	12 (52%)	4 (17%)	1 (4%)
E14.5	18	7 (39%)	6 (33%)	5 (28%)	
E12.5	23	7 (30%)	10 (44%)	6 (26%)	

R°, resorbed embryo was homozygous. Genotypes of the offspring of the intercrosses between C57BL/6 heterozygotes (3F–4F) were determined.

B. Lethality of HS6ST-1-Deficient Mice at Later Embryonic Stages

Genotype analysis of the progeny derived from intercrosses of the heterozygous mice (C57BL/6 strains) indicated that most of the HS6ST-1 null mice died between E15.5 and the perinatal stage (Table I). The genotypes of pups at E12.5 and E14.5 agreed well with the Mendelian rule, indicating that HS6ST-1 null mice at the early embryonic stages appeared to be fully viable. From E17.5 to E18.5, the number of homozygous embryos was ~30% of that expected from the Mendelian rule. On the 21st postnatal day, less than 4% of the offspring from the heterozygotes were knockout mice, that is, ~15% of the homozygotes survive. These surviving knockout mice were grown up to adults and both the adult male and female mice were fertile, although their mean litter sizes were not investigated in detail. In the C3H/He strain, 7% of the offspring from the heterozygous intercrosses were homozygotes on the 21st postnatal day. In the BALB/c strain, there were no homozygous offspring from the heterozygous intercrosses at birth. Thus, the survival rate appears to depend on the genetic backgrounds.

C. Reduction of 6-*O*-Sulfate Content in the HS Isolated from HS6ST-1KO-Mice

HS structures from various tissues of newborn mice were compared among the wild-type, heterozygous, and homozygous mice (Fig. 6). There was marked decrease in the relative contents of GlcNAc($6SO_4$) and HexA-GlcNSO_3($6SO_4$) residues in the HS isolated from the kidney and lung; these tissues normally express relatively high levels of HS6T-1 but rather low levels of HS6ST-2. In contrast, the decrease in 6-*O*-sulfation was modest enough in the HS isolated

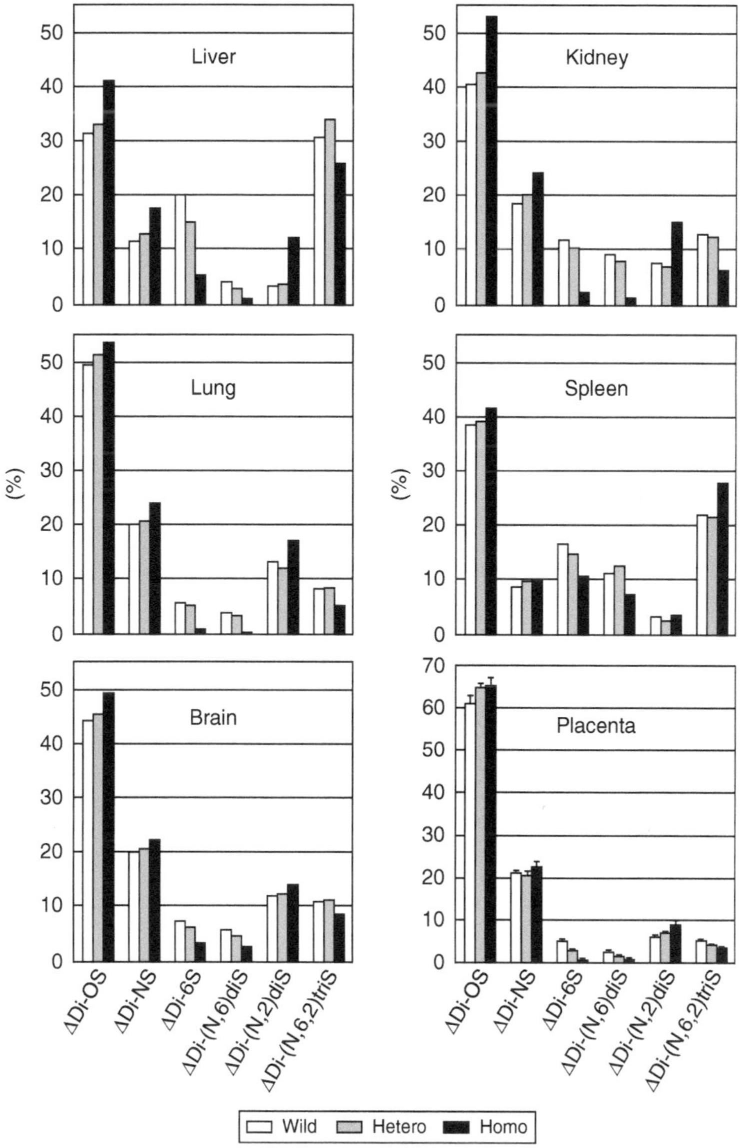

FIG. 6. Reduction of 6-*O*-sulfate in the HS of various tissues taken from mutant mice. HS was isolated from various tissues of neonatal mice and from the placentas of E15.5 embryos. HS was digested with a heparitinase-I, -II, and heparinase mixture. The histograms show the percentage compositions of unsaturated disaccharide (ΔDi-OS, ΔDi-6S, ΔDi-NS, ΔDi-(N,6)diS, ΔDi-(N,2)diS, and ΔDi-(N,6,2) triS are the disaccharide units derived from -HexA-GlcNAc-, -HexA-GlcNAc($6SO_4$)-, -HexA-GlcNSO$_3$-, -HexA-GlcNSO$_3$($6SO_4$)-, -HexA($2SO_4$)-GlcNSO$_3$-, and -HexA($2SO_4$)-GlcNSO$_3$($6SO_4$)-, respectively) in the HS isolated from the organs of wild-type, heterozygous, and homozygous mutant mice. (Redrawn, with permission, from Ref. 41.)

from the spleens and brains where HS6ST-2 was expressed at high levels but HS6ST-1 at low levels. Structural alterations of heparan sulfate from the homozygous tissues are considerably consistent with the expected ones from the expression levels of HS6ST-1 in wild-type tissues by Northern blot analysis.[18] In addition, the decrease in trisulfated disaccharide unit, IdoA($2SO_4$)-GlcNSO_3($6SO_4$), was less in comparison with the decrease in GlcNAc($6SO_4$) and HexA-GlcNSO_3($6SO_4$) units, even in liver, kidney, and lung. Such alterations of the HS structure appeared to be consistent with the substrate specificities of HS6ST-1.

D. Growth Retardation and Tissue Abnormality of HS6ST-1-Deficient Mice

The HS6ST-1-deficient mice tended to be smaller than the wild-type mice during embryonic stages and until at the latest P34 days. Comparison of the skeletons at birth revealed systemic growth retardation in the HS6ST-1 null mice (Fig. 7). The KO-mice missed the hind limb phalange and either had

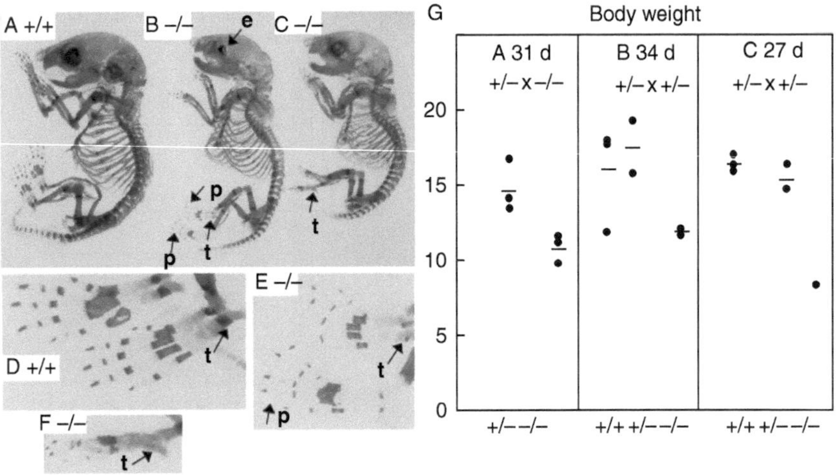

FIG. 7. Growth retardation of the HS6ST-1-deficient mice. The skeletal samples of newborn mice were stained with alcian blue and arizarin red. On the whole, the skeletons of the HS6ST-1-deficient mice (B, C) were smaller than those of wild-type mice (A). They often lacked a hind limb phalange, and either possessed a smaller tarsus in the hind limb compared to the tarsus in wild-type mice, or were completely lacking a tarsus. (D–F) Higher magnification of the respective hind limbs shown in (A)–(C). *Arrows* indicate the tarsus (t), phalanges (p), and eyes (e). Right panel (G) shows the body weight of about 4–5-week-old mice. (Redrawn, with permission, from Ref. 41.) (See Color Insert.)

smaller hind limb tarsus or missed it completely, which was expected from the above-described results from the knockdown of Hs6sts in chick limb buds. As the ossification of these bones occurs at a late stage of development, these phenotypes more likely result from growth retardation rather than impaired ossification. HS6ST-1-deficient mice sometimes exhibited an aberrant eye morphology that was rarely observed in wild-type mice.

Another group also found that body weight of homozygotes was reduced at least until 20 weeks compared to that of wild type.[42] Sulf 2 KO mice, which produce an excess of 6-O-sulfated HS, also showed the decreased body weight.[43,44] These observations suggest that an optimal degree of 6-O-sulfation in HS may be necessary for proper regulation of body weight by some unknown mechanisms.

E. Aberrant Angiogenesis in the Placenta

It is thought that impaired placental function might be implicated in homozygous embryo growth retardation and/or perinatal lethality. HS of placenta from 15.5-day wild-type embryo had the higher content of nonsulfated disaccharide unit (HexA-GlcNAc) and the lower contents of 6-O-sulfated disaccharides units, compared with those of other tissues from neonatal mice (Fig. 6). In the placentas from the homozygous embryos, 6-O-sulfated disaccharide units almost disappeared with the exception of trisulfated disaccharide unit. Therefore, HS6ST-1 plays a particularly important role in HS biosynthesis in the placentas.

In comparison with the placental development of the homozygous and wild-type embryos by staining with hematoxylin/eosin, embryo-derived nucleated red cells that were abundant in the placentas of wild-type embryos were reduced in the placentas of homozygous embryos (Fig. 8). These observations suggest that angiogenesis should be impaired in the placentas of the HS6ST-1-deficient embryos. In fact, immunohistochemistry using anti-CD31 antibody, an antibody specific for vascular endothelial cells, supported such assumption. In the wild-type placentas, intense staining of numerous microvessels was observed in the labyrinthine zone, where the exchange of the nutrients and gases occurs (Fig. 9A). In contrast, a weak staining of microvessels was observed in the labyrinthine zone of the HS6ST-1-deficient placentas. Analysis of these tissues by confocal microscopy showed that the number of microvessels in the homozygous placenta was reduced to about 60% of that in the wild-type placentas (Fig. 9B and C). Such an aberrant angiogenesis might jeopardize the nutritional supply and gas exchange in the HS6ST-1-deficient placentas.

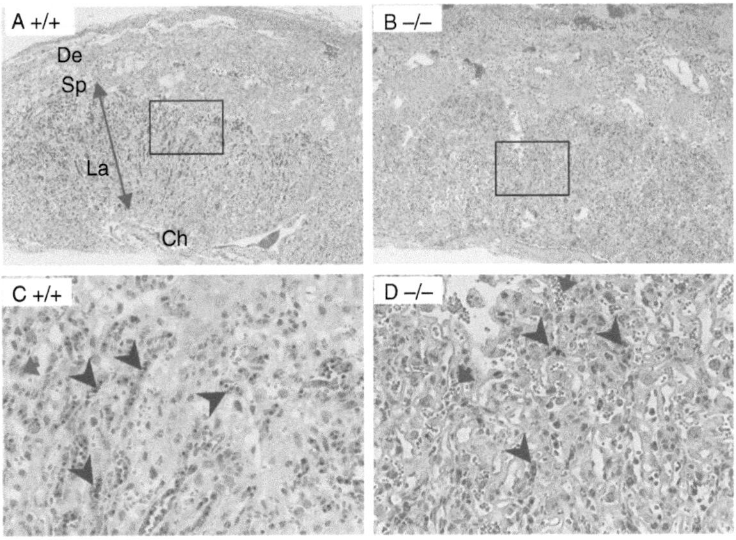

FIG. 8. Comparison of placenta morphology in HS6ST-$1^{+/+}$ (A, C) and HS6ST-$1^{-/-}$ (B, D) mice. The E15.5 placentas were stained with H & E. The lower panels show a higher magnification of the rectangle in the upper panels. The orange arrowheads indicate nucleated fetal red blood cells, and the green arrows indicate maternal red blood cells. In the labyrinthine zone of HS6ST-$1^{-/-}$ embryo placentas, there were a reduced number of fetal red blood cells. The yellow arrow indicates the labyrinthine zone. De, maternal deciduas; Ch, chorionic plate; La, labyrinthine zone; Sp, spongiotrophoblast layer. (Redrawn, with permission, from Ref. 41.) (For interpretation of the references to color in this figure legend, the reader is referred to the Web version of this chapter.)

F. Reduction of VEGF-A mRNA and Its Protein in the Placenta

How does this abnormal angiogenesis occur in the homozygous embryos? Wnt2 knockout mice[45] have been reported to exhibit abnormal placentation, growth retardation, and perinatal lethality, grossly similar to that of the phenotype observed in HS6ST-1-deficient mice. Wnt is known to induce the expression of VEGF-A, a strong inducer of angiogenesis.[46] Members of the Wnt family and some spliced forms of VEGF-A have an affinity for heparin, and evidence is accumulating that HS regulates signaling through these factors. In addition, such upregulation of VEGF-A signaling was reported to require 6-*O*-sulfate in HS.[47] Expression level of VEGF-A mRNA in HS6ST-1-deficient placentas was about 60% of that in the wild-type placentas (Fig. 10A). However, no significant difference in the expression level of Wnt2 mRNA was observed between the wild-type and the deficient placentas. Immunohistochemical staining of

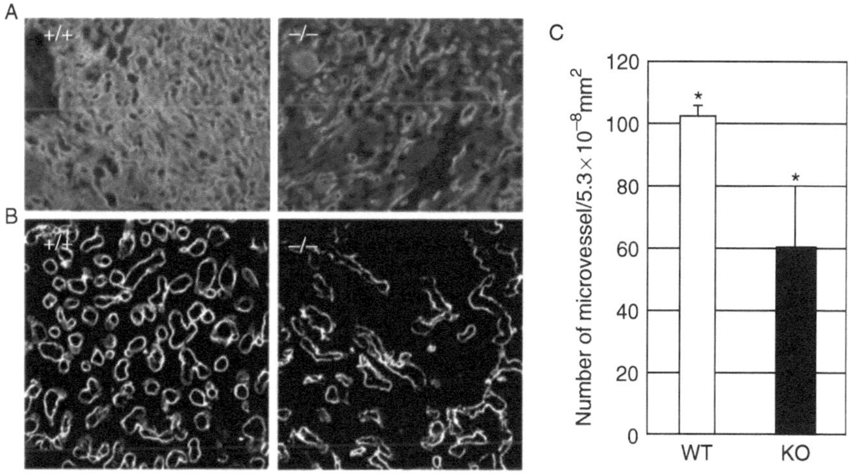

FIG. 9. Reduced microvessels in the placenta of a homozygous embryo from a heterozygous mother. (A) Frozen sections from the E15.5 placentas were stained with an anti-CD31 antibody (green). (B) The left panel shows the photographs of a wild-type embryo, and the right panel shows the photographs of a homozygote embryo analyzed by confocal microscopy. (C) Bars indicate the mean number of microvessels in the labyrinthine zone per 5.3×10^{-8} mm^2 counted from the photographs obtained from the confocal microscopic analysis. $^{*}p < 0.03$. (Redrawn, with permission, from Ref. 41.) (See Color Insert.)

placentas with an antibody against VEGF-A revealed that VEGF-A protein was expressed in the labyrinthine zones of both the wild-type and HS6ST-1-deficient placentas; however, intensity of the staining was apparently reduced in the HS6ST-1-deficient placentas compared with that in the wild-type placentas (Fig. 10B). The lower expression of VEGF-A in HS6ST-1-deficient-placentas was confirmed by Western blot analysis (Fig. 10C, D). These observations suggest that the decreased VEGF-A expression in HS6ST-1-deficient placentas result in impaired vascular development and may thereby lead to growth retardation and lethality.

G. Interaction of Wnt2 with Mutant HS

The expression of VEGF-A is known to be induced by Wnt, which is bound to heparin and regulated by heparin and HS. Therefore, the affinity of Wnt2 to wild-type HS and homozygous HS from placentas was analyzed by surface plasmon resonance. The K_d values for Wnt2/wild-type placentas-HS and Wnt2/homozygous placentas-HS were 172 and 330 nM, respectively. Although HSs from placentas showed relatively weak affinity for Wnt2, the interaction of

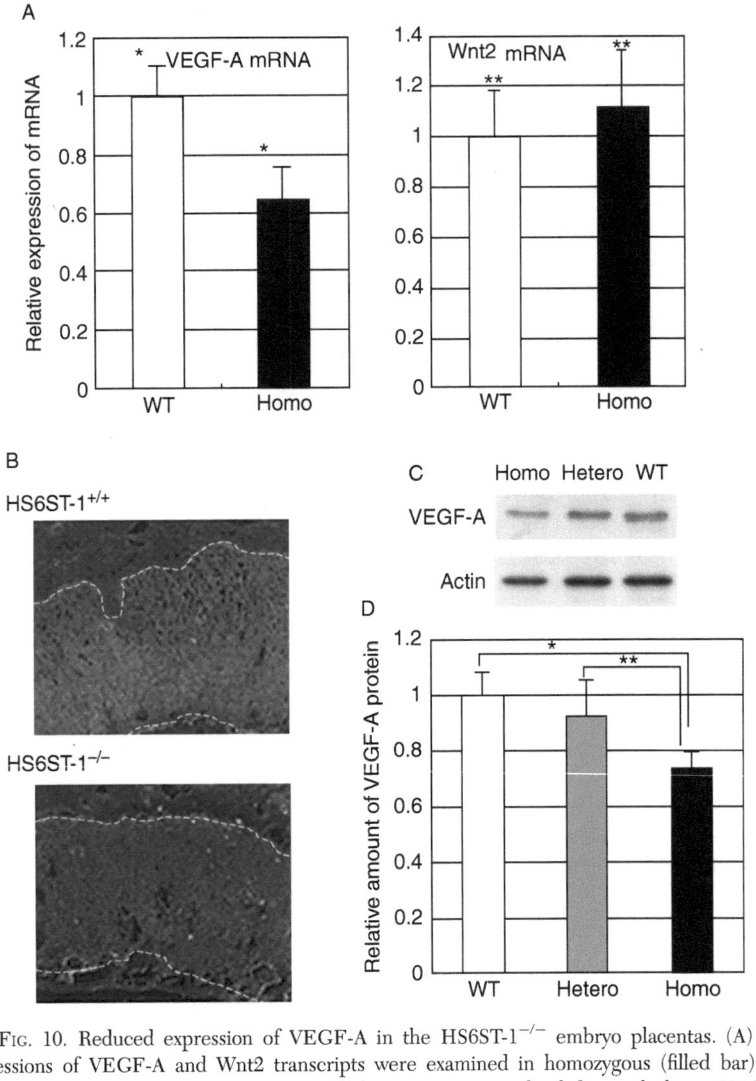

FIG. 10. Reduced expression of VEGF-A in the HS6ST-1$^{-/-}$ embryo placentas. (A) The expressions of VEGF-A and Wnt2 transcripts were examined in homozygous (filled bar) and wild-type (WT) placentas (open bar) by real-time RT-PCR. The left panel shows VEGF-A mRNA ($^{*}p < 0.01$), and the right panel shows Wnt2 mRNA ($^{**}p = 0.23$). (B) Frozen sections from the E15.5 placentas were stained with an anti-VEGF antibody. (C) Proteins extracts from the E15.5 placentas were subjected to 10% SDS-PAGE. Western blots were probed using an antibody against VEGF-A or actin. (D) The relative amounts of the VEGF-A protein in the wild-type (open bar), heterozygote (gray bar), or knockout (filled bar) mice placentas were measured from the Western blots and normalized against the actin protein bands ($^{*}p < 0.01$; $^{**}p < 0.01$). (Redrawn, with permission, from Ref. 41.)

Wnt2 with wild-type HS occurred significantly and the magnitude of interaction was about two times higher than with Wnt2/homozygous HS. Thus, 6-O-sulfate residues in HS are likely to participate in its recognition of Wnt2.

H. Aberrant Lung Morphology

The formation of lungs alveoli is known to be regulated by Wnt5a, FGF10, and BMP4,[48,49] which bind to heparin/HS. Further, HS6ST-1 is expressed strongly at the tips of branching tubules in the developing lung,[50] and FGF10 is preferentially bound to 6-O-sulfate residues in HS.[11] Thus, a HS6ST-1 deficiency is potentially relevant to impaired lung development. The lung of 7-day-old-mice and that of adult mice (10-month-old) were stained with hematoxylin/eosin. The lung of homozygous mice exhibited the alveolar enlargement (Fig. 11), and the intra-alveolar distance by measuring the mean linear intercept (MLI) indicated that, in 7-day-old mice, MLI of the homozygous mice was significantly increased compared with that of the heterozygous mice (40 ± 5 vs. 30 ± 4 μm). In adult mice, the more marked difference was observed between the homozygous mice (42 ± 6 μm) and the wild-type mice (29 ± 4 μm). This phenotype is characteristic of emphysema. As an increase in MLI is inversely proportional to the internal lung surface area,[51] these results suggest that HS6ST-1-deficient mice have impaired alveolarization. These observations raise the possibility that decreased 6-O-sulfation adversely affects the signaling of undefined growth factors and possibly extracellular matrix proteins in the lung, which adversely impact alveolarization. Later studies suggested that such alveolar enlargement appears to be due to abnormal conformation of elastin, which was fragmented and irregularly deposited in the alveoli of the adult lung of Hs6st-1 null mice.[42] In addition, alveolar macrophages were present more frequently in the alveolar spaces of HS6ST-1-deficient adult mice compared with that of wild-type. These observations are consistent with the previous studies showing that alveolar macrophages accumulate in the lung in pulmonary emphysema and that elastin fragments generated at the diseased sites are potent chemoattractants for monocytes, the precursors of the macrophages.[52] As the model mouse of pulmonary emphysema, HS6ST-1-deficient mice might contribute to developing prevention strategies and elucidating mechanisms for pulmonary emphysema.

HS6ST-1-deficient embryonic mice also have revealed that more retinal ganglion cell (RGC) axons entered the contralateral optic neuron and growth toward the opposite eye.[53] These axon navigation defects colocalized with regions of high Hs6st-1 expression at the optic chiasm. Hs6st-1 expression also coincided with Slit expression domains at the region of axon guidance error in the mutant embryos. Such a navigation error observed in HS6ST-$1^{-/-}$ mice was similar to that observed in Slit$1^{-/-}$/Slit$2^{-/-}$ double knockout mice.[54] When retinal explants were cultured on a bed of collagen gel containing COS-7 cells transfected with Slit2 (COS-7/Slit-2) or mock-vector (COS-7/mock), the retinal axon navigation toward

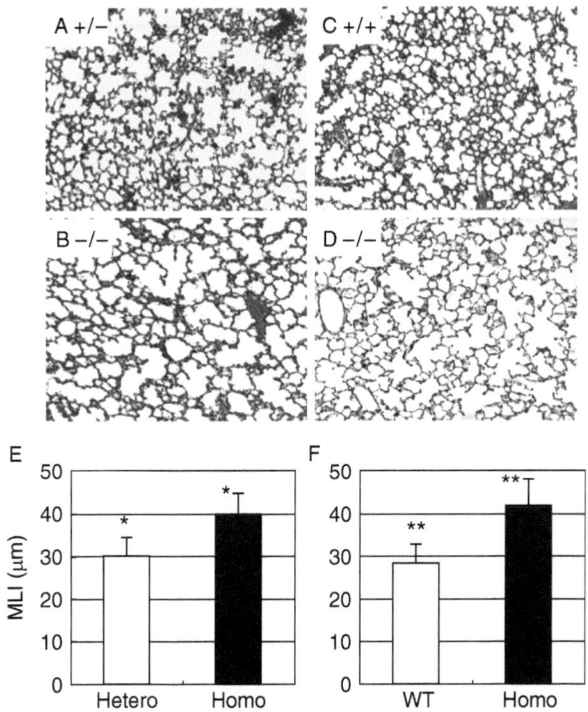

FIG. 11. Aberrant morphology of lungs in the HS6ST-1-deficient mice. Paraffin sections from 7-day-old mice (A, B) and adult mice (C, D) were stained with hematoxylin/eosin. (A) Heterozygous mice, (C) wild-type mice, (B, D) homozygous mice. The lungs of the HS6ST-1-deficient mice had apparently enlarged alveoli. (E, F) The intra-alveolar distance of the lungs from 7-days-old (E) and adult mice (F) was measured as the mean linear intercept (MLI). MLI was increased in the homozygous mice compared with the wild-type (WT) mice ($^{*}p < 0.001$; $^{**}p < 0.001$). (Redrawn, with permission, from Ref. 41)

COS-7/Slit2 in wild-type explants was inhibited strongly, compared to that toward COS-7/mock, while the retinal axon navigation toward COS-7/Slit2 HS6ST-1$^{-/-}$ explants was suppressed only weakly. Thus, the ability of RGC growth cone to repel Slit2 is dependent on the HS structure modified by HS6ST-1. Such a navigation error appears to be caused by the less sensitivity to Slit2 repulsion.

IV. Function of HS6ST in Other Organisms

Zebrafish Hs6st (zHs6st) was originally found as one of the genes that participate in neural plate regionalization, and later named Hs6st2. At present, four Zebrafish Hs6sts have been described, 6OST-1a, -1b, -2, and -3.[26] zHs6st2 is ubiquitously expressed in the blastoderm at the gastrulation and strongly expressed in the brain and eyes at 24 h postfertilization (hpf). At 48 hpf, a high level of Hs6st2 transcripts was detected in the fin buds and the head. Morpholino (MO)-mediated knockdown of HS6ST-2 resulted in pectoral fin bud defects and the disruption of somatic muscle development. Moreover, the size of the head was reduced; both the white matter composed of nerve tracts and the gray matter composed of somata were significantly reduced.[55] These phenotypes are reminiscent of *knypec/glypican* zebrafish mutants. Such impaired muscle differentiation appears to be caused by lingering expression of *myoD* that was induced by Wnt, suggesting that Hs6st2 regulates Wnt signaling. In these morphant embryos, the branching morphogenesis of the caudal vein was also defected and the expression domains of tie-1 and tie-2, the vascular markers, were reduced in the developing caudal vein.[56] Coinjection of the Hs6st2 MO and VEGF-A MO into embryos resulted in the synergistic increase of caudal vein branching defects, suggesting that Hs6st2 interacts with VEGF-A during angiogenesis *in vivo* and modulates the bioavailability of VEGF in the regions of the venous plexus. In contrast, Hs6st1 morphants did not affect vascular development. Interestingly, both HSs from Hs6st1 MO- and Hs6st2 MO-injected embryos contained moderately reduced 6-sulfated residues in HexA-GlcNSO$_3$(6SO$_4$) unit and HexA-GlcNAc6SO$_4$ unit, and slightly reduced 6-sulfated residues in HexA(2SO$_4$)-GlcNSO$_3$(6SO$_4$) unit. These observations suggest that the different arrangement of disaccharide unit in HS resulting from the tissue-specific expression of the Hs6sts may contribute to the biological functions of HSs.

Drosophila has only one Hs6st. The expression of Hs6st is highly specific in tracheal cells and resembles that of the *Drosophila* FGF receptor, *breathless* (*btl*), suggesting that Hs6st functions in tracheal development are regulated by FGF. In fact, RNAi-mediated knockdown of *Hs6st* resulted in abnormal formation of trachea.[25] These phenotypes are reminiscent of the defects observed in mutants of FGF signaling components, and actually MAPK activities, a downstream transducer of FGF signaling, were also reduced in the RNAi-injected embryos. These observations indicate that 6-*O*-sulfation of HS is critical for FGF signaling during development. Further, Hs6st null embryos showed partial lethality and a fraction (36%) of Hs6st null embryos exhibited tracheal defects.[57] However, the mutants did not completely inhibit *btl*-mediated tracheal formation. These mutants produced altered HS in which the loss of 6-*O*-sulfate groups was nearly completely compensated by the increase in 2-*O*-sulfate groups. The defect in formation of the tracheoblast was caused by the double destruction of Hs2st/Hs6st in *btl*-expressing (tracheal) cells but not in *branchless* (*bnl*)-expressing (nontracheal) cells. These observations demonstrate redundant roles of 2-*O*- and 6-*O*-sulfate groups of HS in FGF signaling during tracheal development of *Drosophila*.

C. elegans has also only one *Hs6st* orthologue, which is one of genes found in a suppressor screen of a HS-dependent axon branching phenotype.[58] Hs6st expresses specifically in neuronal tissues but not in hypodermal tissues. Hs6st mutants have revealed midline crossover defects of PVQL/PVQR and PVPL/PVPR, which are ventral cord interneurons and send axons along the entire length of the left and right ventral cord fascicles. In such PVQ axon guidance, 6-*O*-sulfated HS is required for the activation of sax-3/Robo.

V. Hs6st-1-/Hs6st-2-Deficient Fibroblasts Regulate Differentially the Signaling Induced with Various FGFs[59]

As described above, a line of evidence suggests that the 6-*O*-sulfate in HS plays a crucial role in development and pathophysiology and regulates the function of heparin-binding proteins negatively or positively in species-, tissue-, and developmental stage-dependent manner. As HS6ST-2 but not HS6ST-1 appeared to be mainly involved in the synthesis of 6-*O*-sulfation of the HS trisulfated units, we have also generated HS6ST-2 null mice.[59] The HS6ST-$2^{-/-}$ mice survive and are fertile without apparent abnormal phenotypes. The HS6ST-$1^{-/-}$/HS6ST-$2^{-/-}$ mice die at a slightly earlier stage than HS6ST-$1^{-/-}$ mice. Those differences suggest important roles of the 6-*O*-sulfation preference in the development of various tissues, which might result in the different survival rate. Thus, using mouse embryonic fibroblast (MEF) prepared from HS6ST-1- and/or HS6ST-2-deficient mouse, the function of 6-*O*-sulfation of the HS chain in the FGF response, especially in the FGF-inducible signaling, as a representative example was examined.

A. Characterization of HS Isolated from MEF

MEFs were prepared from WT, HS6ST-$1^{-/-}$, HS6ST-$2^{-/-}$, and HS6ST-$1^{-/-}$/HS6ST-$2^{-/-}$ mice at E14.5 (designated WT-MEF, 6ST1KO-MEFs, 6ST2KO-MEFs, and dKO-MEF, respectively). Expression levels of three isoforms of HS6ST were examined in WT-MEF by semiquantitative RT-PCR (Fig. 12). Transcription level of Hs6st-2 RNA was the highest and the expression level of Hs6st-3 mRNA ($\sim$40% of Hs6st-2 mRNA) was the lowest. These data support the results of *in situ hybridization* studies conducted previously,[37] in which Hs6st-2 mRNA is predominantly distributed in the mesenchyme, whereas Hs6st-1 mRNA is mainly expressed in the epithelium- and neuron-derived tissues during organogenesis. The HS isolated from WT-MEF (WT-HS) contains 20.6 mol of 6-*O*-sulfate residues per 100 mol of disaccharide. When compared with the WT-HS, the HS isolated from 6ST1KO-MEF (6ST1KO-HS) contained moderately decreased amount of 6-*O*-sulfate groups; GlcNAc6SO_4 residue,

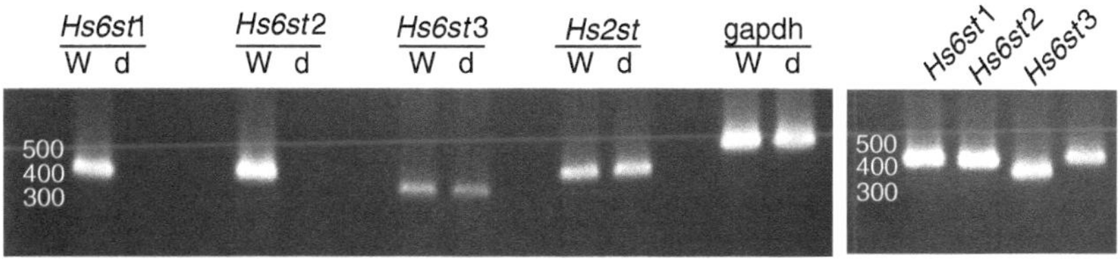

FIG. 12. Expression levels of the *Hs6st-1*, *Hs6st-2*, *Hs6st-3*, and *Hs2st* transcripts in WT-MEFs and dKO-MEFs. The expression levels of *Hs6st-1*, *Hs6st-2*, *Hs6st-3*, and *Hs2st* transcripts in the WT- and dKO-MEFs were semiquantitatively measured using RT-PCR. *GAPDH* was used as the control. To compare the relative expression levels of these transcripts, PCR was carried out using equal molar concentration of each plasmid containing their cDNA as the templates (right hand, 4 lanes). W, WT-MEFs; d, dKO-MEFs. (Redrawn, with permission, from Ref. 59.)

HexA-GlcNSO_3(6SO_4) unit, and HexA(2SO_4)-GlcNSO_3(6SO_4) unit were reduced to 70%, 74%, and 82%, respectively, of WT-HS (Fig. 13). In contrast, 6-*O*-sulfation in the HS isolated from 6ST2KO-MEF (6ST2KO-HS) was markedly decreased; the amounts of GlcNAc(6SO_4) residue, HexA-GlcNSO_3(6SO_4) unit, and HexA(2SO_4)-GlcNSO_3(6SO_4) unit were decreased to 33%, 21%, and 64% of WT-HS. Thus, the gene destruction of each HS6ST isoform alone did not result in the complete loss of 6-*O*-sulfation in HS from embryonic fibroblast. As expected, 6-*O*-sulfate residues in the dKO-HS were less than the detectable level regardless of the expression of Hs6st-3 mRNA. The activity of HS6ST in the dKO-MEFs was less than 4% of that in the WT-MEFs. The residual activity is due probably to the action of HS6ST-3. These data are consistent with almost complete loss of 6-*O*-sulfate residues observed in the dKO-HS.

Is the lack of 6-*O*-sulfation compensated by 2-*O*-sulfation? Expression of Hs2st in dKO-MEF was 1.2-fold higher than that in WT-MEFs. The content of 2-*O*-sulfate in the HS isolated from all the mutant cells was increased by 1.2–1.9-fold, compared to WT-HS. WT-HS contained 9.8 mol of 2-*O*-sulfate residues per 100 mol of disaccharide, and the amount of these residues in dKO-HS increased to 14.3/100 mol of disaccharide. Thus, the 2-*O*-sulfation partially compensates for the lack of 6-*O*-sulfation in dKO-HS. However, extent of the compensation increase in 2-*O*-sulfation varies in a species-specific manner.[57] The different extent of compensation among animals and/or organs may be due to differences in the substrate specificity and expression levels of modification enzymes.

B. Control of 6-*O*-Sulfated HS on FGF-Dependent Signaling

There are many *in vitro* studies on the specific binding structure between FGFs and heparin/modified heparin or its oligosaccharides, and the specific formation of the ternary complex comprised of FGFs, FGFRs, heparin/

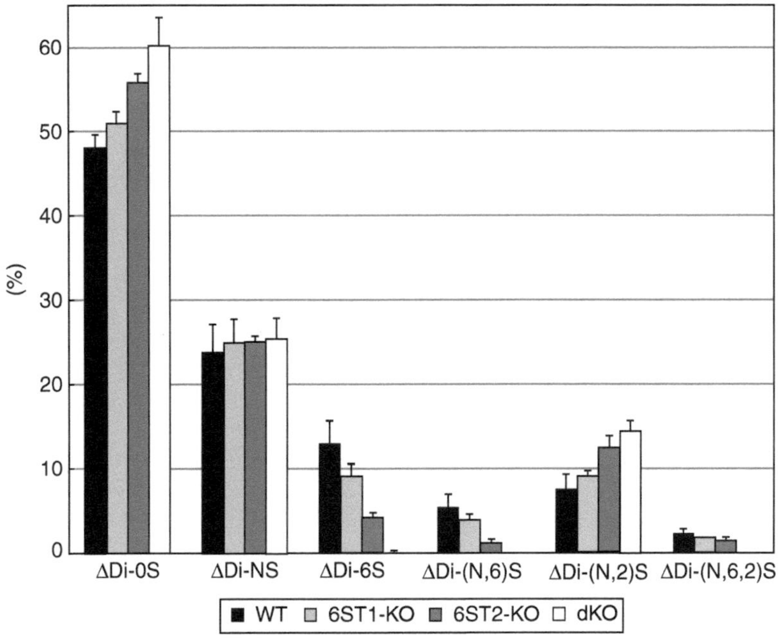

FIG. 13. Comparison among disaccharide compositions of HS from the MEFs of WT, HS6ST-$1^{-/-}$, HS6ST-$2^{-/-}$, and dKO mice. HS was prepared from the cell layers of MEFs. HS was digested with a heparitinase-I, -II, and heparinase mixture. The histogram shows the percentage of unsaturated disaccharide compositions (ΔDi-OS, ΔDi-6S, ΔDi-NS, ΔDi-(N,6)diS, ΔDi-(N,2)diS, and ΔDi-(N,6,2)triS are the disaccharide units derived from -HexA-GlcNAc-, -HexA-GlcNAc(6SO_4)-, -HexA-GlcNSO_3-, -HexA-GlcNSO_3(6SO_4)-, -HexA(2SO_4)-GlcNSO_3-, and -HexA(2SO_4)-GlcNSO_3(6SO_4)-, respectively) in the HS isolated from the MEFs of WT-, HS6ST1-KO, HS6ST2-KO, and dKO mice. (Redrawn, with permission, from Ref. 59.)

modified heparin, or its oligosaccharides. To assess the effects of 6-*O*-sulfation in HS on the response to heparin-binding growth factors, expression and signaling of FGFs in the HS6ST-deficient cells and binding of FGFs to HSs synthesized by these cells were examined. The expression levels of each of *fgfrs* in WT-MEFs and dKO-MEFs were as follows: the WT-MEFs and dKO-MEFs strongly expressed *fgfr1b*, *fgfr1c*, *fgfr2c*, and *fgfr3* and moderately *fgfr2b*, whereas these cells weakly expressed *fgfr4*. The expression profile of *fgfrs* in dKO-MEF was thus very similar to that observed in WT-MEF. These data suggest that the 6-*O*-sulfation defect in HS has a negligible effect on the expression of *fgfr* mRNA. Therefore, the effects of 6-*O*-sulfation on the signaling of FGF-1, -2, and -4 mediated mainly through FGFRs that expressed strongly in MEFs were analyzed by the phosphorylation of FGFR1 and ERK. In order to ascertain whether the cell surface HS in the MEFs contributed to the regulation of FGF signaling, HS and/or CS in WT-MEFs were removed by digestion with a heparitinase mixture and/or chondroitinase ABC. Digestion with a heparitinase mixture but not with chondroitinase ABC resulted in a marked reduction in FGF-dependent phosphorylation of FGFR1 with regard to all three FGFs. These results indicate that cell surface HS but not CS promotes the activities of FGF-1, -2, and -4. The phosphorylation of ERK1/2 induced by FGF-2 and FGF-4 was remarkably decreased when HS was removed, indicating that the signal transduction of both FGF-2 and FGF-4 required the HS chain at least in the MEFs. However, FGF-1-dependent phosphorylation of ERK was moderately decreased by the removal of HS. The double digestion with a heparitinase mixture and chondroitinase ABC did not alter the extent of decrease. These results suggest that the signaling of FGF-1 mediated through FGFR1 depends on HS, but that mediated through other FGFRs, such as FGFR2 and/or FGFR3, appears to depend only partially on HS. The effect of 6-*O*-sulfate in HS on the response to FGFs was compared between WT-MEFs and dKO-MEFs.

The FGF-1-dependent phosphorylation of FGFR1 was clearly reduced in dKO-MEFs, indicating that 6-*O*-sulfate residues in HS stimulate FGF-1 signaling mediated through FGFR1 (Fig. 14). At lower concentration of FGF-1 (0.2–1 ng/ml), the phosphorylation of ERK1/2 in dKO-MEFs was decreased to $\sim$70% of the WT-MEFs, and the degree of reduction in the phosphorylation of ERK1/2 in KO-MEFs was slightly lower than that in WT-MEFs from which HS was removed. Thus, lower phosphorylation of ERK1/2 in dKO-MEF induced by FGF-1 is likely to be caused, at least in part, by the lower affinity between FGF-1/dKO-HS compared with WT-MEF; the K_d value for FGF-1/dKO-HS is about three times higher than that for FGF-1/WT-HS.

With respect to heparin/HS binding, FGF-2 exhibits a stronger preference for heparin/HS containing 2-*O*-sulfate residues than for heparin/HS containing 6-*O*-sulfate residues.[60–63] Therefore, FGF-2 is expected to interact with dKO-HS rather strongly, because the content of the 2-*O*-sulfate residues in dKO-HS was 1.9-fold higher than that in the dKO-MEFs. Indeed, the affinity of FGF-2 for dKO-HS was three times higher than that for WT-HS. But, nevertheless, FGF-2-inducible signaling appeared to be reduced in dKO-MEFs, because the phosphorylation of ERK in dKO-MEFs at the lowest concentration of FGF-2 (0.5 ng/ml) was decreased to $\sim$60% of that in the WT-MEFs. The level of FGFR1 phosphorylation in the dKO-MEFs was apparently decreased compared with that in WT-MEFs although the level of FGFR1 phosphorylation was very low in both WT- and dKO-MEFs. Taken together, 6-*O*-sulfate residues would be important for formation of the FGF-2/FGFR/HS ternary complex which is indispensable for the signaling. In fact, as described below, the ability of dKO-HS to form the ternary complex with FGF-2/FGFR1c was lower than that of WT-HS.

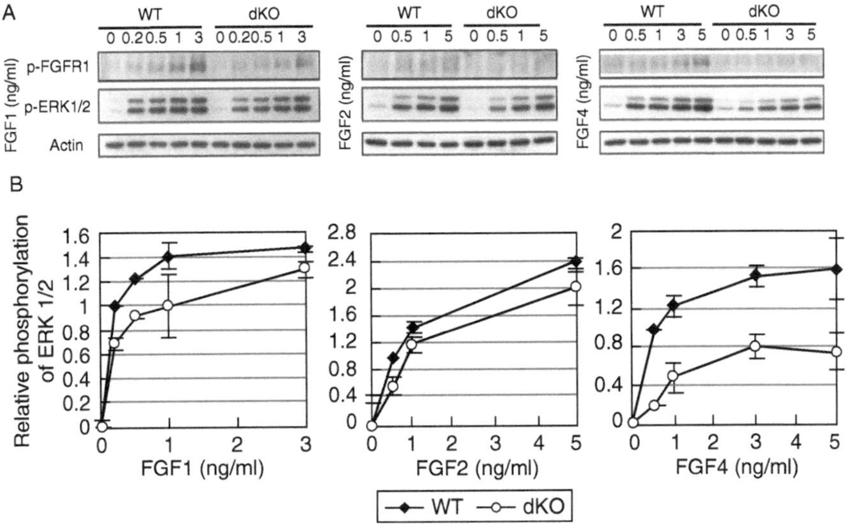

FIG. 14. FGF-1-, FGF-2-, and FGF-4-induced phosphorylation of ERK1/2 and FGFR1 in WT- and dKO-MEFs. The WT- and dKO-MEFs were incubated with different concentrations of FGF-1, FGF-2, and FGF-4 for 5 min at 37 °C. (A) Phosphorylation of FGFR-1 and ERK1/2 in the cells was analyzed by Western blotting. (B) Amount of pERK1/2 was quantified from (A), and was normalized by actin bands. (Redrawn, with permission, from Ref. 59.)

Among three FGFs, FGF-4 signaling was most distinctly affected by the decreased 6-*O*-sulfation in HS. The affinity of FGF-4 for dKO-HS was markedly decreased to approximately one-fifth of that of FGF-4 for WT-HS. As expected from such a marked decrease in the affinity of FGF-4, the level of ERK phosphorylation in the dKO-MEF at the concentrations of 0.5–5 ng/ml FGF-4 was reduced to 15–40% of the level in the WT-MEF. FGFR1 phosphorylation in dKO-MEF was apparently decreased compared with that in the WT-MEFs, and the extent of decrease in FGFR1 phosphorylation in dKO-MEFs was higher than that in ERK phosphorylation in dKO-MEF. These data are consistent with the results obtained from biochemical and X-ray crystallography studies in which both 2-*O*- and 6-*O*-sulfate groups in heparin are demonstrated to be crucial for the interaction of FGF-4 with heparin. As FGF-4 signaling has been reported to be mediated mainly through FGFR1c and FGFR2c, the FGF-4 signaling also appears to be promoted by HS having 6-*O*-sulfate residues.

To summarize, FGF-1-, FGF-2-, and FGF-4-dependent signaling through FGFR1 requires strong binding to the HS chain and is modified by 6-*O*-sulfate content in HS. These results support the previous structural analysis by X-ray crystallography showing that the Arg 209 in FGFR1 forms a hydrogen bond with 6-*O*-sulfate residues in heparin and thereby induces FGF/FGFR1 dimer formation.[64]

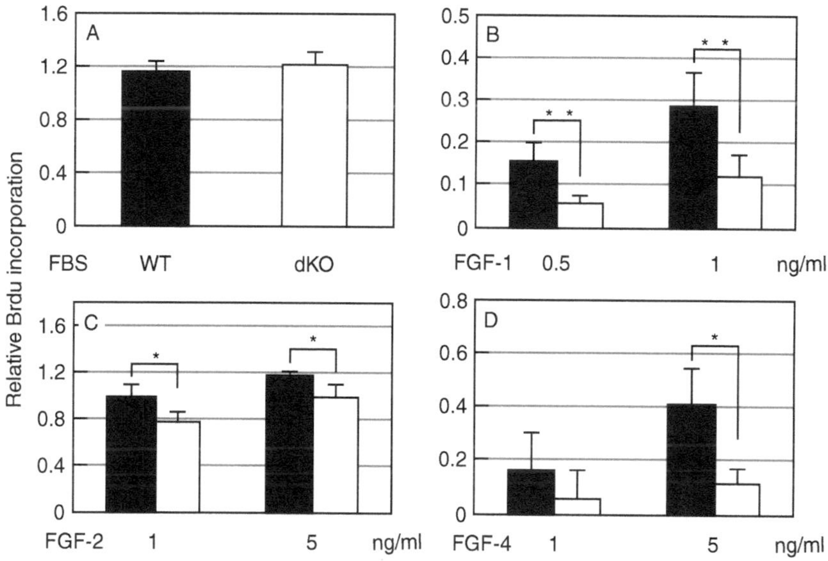

FIG. 15. FGF-dependent dKO-MEF proliferation was apparently decreased in comparison to that of WT-MEF. MEFs were incubated in serum-free DMEM containing 0.1% BSA for 48 h. (A) 10% of FBS; (B) different concentrations of FGF-1 (0.5, 1 ng/ml); (C) FGF-2 (1, 5 ng/ml); (D) FGF-4 (1, 5 ng/ml), and BrdU labeling solution were added into the wells. After further incubation for 24 h, BrdU incorporated to DNA was measured. Each bar (filled bars, WT-MEF; open bars, dKO-MEF) represents the mean ± S.D. $^{*}p < 0.01$; $^{**}p < 0.005$. (Redrawn, with permission, from Ref. 59.)

C. Reduction of FGF-Dependent Cell Growth in dKO-MEFs

As described above, the deletion of 6-*O*-sulfate in HS caused the reduction of FGF-induced signaling. Furthermore, all three FGFs-dependent DNA syntheses in dKO-MEF were decreased compared with that in the WT-MEF (Fig. 15). Thus, 6-*O*-sulfation in HS plays the critical role on the cell proliferation as well as the phosphorylation of ERK1/2.

D. Effects of 6-*O*-Sulfation on the Ternary Complex Formation of FGF-FGFR1c-HS

Although FGF-2 had the higher affinity for dKO-HS rather than for WT-HS, FGF-2-dependent signaling in dKO-MEF was lower than that in WT-MEFs. One mechanism is likely that the formation of ternary complex, FGF-2/FGFR/HS, might be affected by deficient 6-*O*-sulfation of HS. Therefore, ternary complex formation of FGF/FGFR1c was examined in the absence or presence of WT-HS or dKO-HS. The FGF-2/FGFR1c complexes were only slightly formed in the

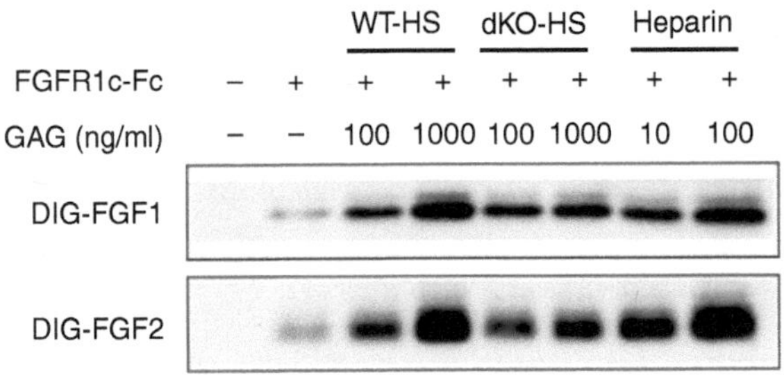

FIG. 16. Effects of 6-*O*-sulfation on the ternary complex formation of FGF/FGFR1/HS. Twenty nanograms of Dig-labeled FGF-1 or FGF-2 and 50 ng of FGFR1c-Fc were added into buffer containing WT-HS, dKO-HS, or heparin of the indicated concentration. After the incubation at 4 °C for 1 h, protein G-Sepharose bead was added into the mixtures and incubated at 4 °C for 2 h. The beads were washed with 2 M NaCl in 20 mM HEPES buffer, pH 7.5 and then with PBS and subjected to Western blotting. The coprecipitated Dig-FGF was determined by an antidigoxigenin antibody. (Redrawn, with permission, from Ref. 59.)

absence of the glycosaminoglycans (Figs. 16 and 17). In contrast, the addition of the glycosaminoglycans extremely promoted the complex formation, and the stimulation was detected at 100 ng/ml WT-HS and dKO-HS. However, the ability of WT-HS on the ternary complex formation was apparently higher than that of dKO-HS at 1 μg/ml. These results indicate that 6-*O*-sulfate residues in HS promote the ternary complex formation of FGF-2/FGFR1c/HS and support the reduction of FGF-2-dependent signaling in dKO-MEF. Likewise, FGF-1/FGFR1c/HS ternary complex with dKO-HS was apparently decreased at 1 μg/ml in comparison with that of WT-HS. It is expected from these results that FGFs-dependent signaling in dKO-MEF might be rescued by exogenously added heparin or WT-HS but not by dKO-HS. In fact, heparin did recover the signaling in dKO-MEF to the level in WT-MEF, suggesting that the responses of the dKO-MEFs to FGFs become lower principally because of the altered structure of HS (Fig. 6). However, addition of WT-HS or dKO-HS in dKO-MEF could not rescue FGF-4-induced signaling, suggesting that HSPG rather than free HS chains might be effective for the activation of FGFs. Heparin having multivalent binding sites may be anchored into extracellular matrix through heparin-binding proteins such as fibronectin, and forms the active ternary complex without core protein. On the other hand, the free HS chains prepared from WT-MEFs probably have only a few binding sites. Therefore, these HS chains might be less effective in forming the active ternary complex than HS-proteoglycan bearing multiple HS chains. As a model of 2:2:2 FGF-1/FGFR1/nonreducing site in HS was shown previously,[65] one FGF-4 and one FGFR1 might form a half-complex between each nonreducing end of HS. However, the free HS chain might only form a half-ternary complex, which is not the active ternary complex.[66]

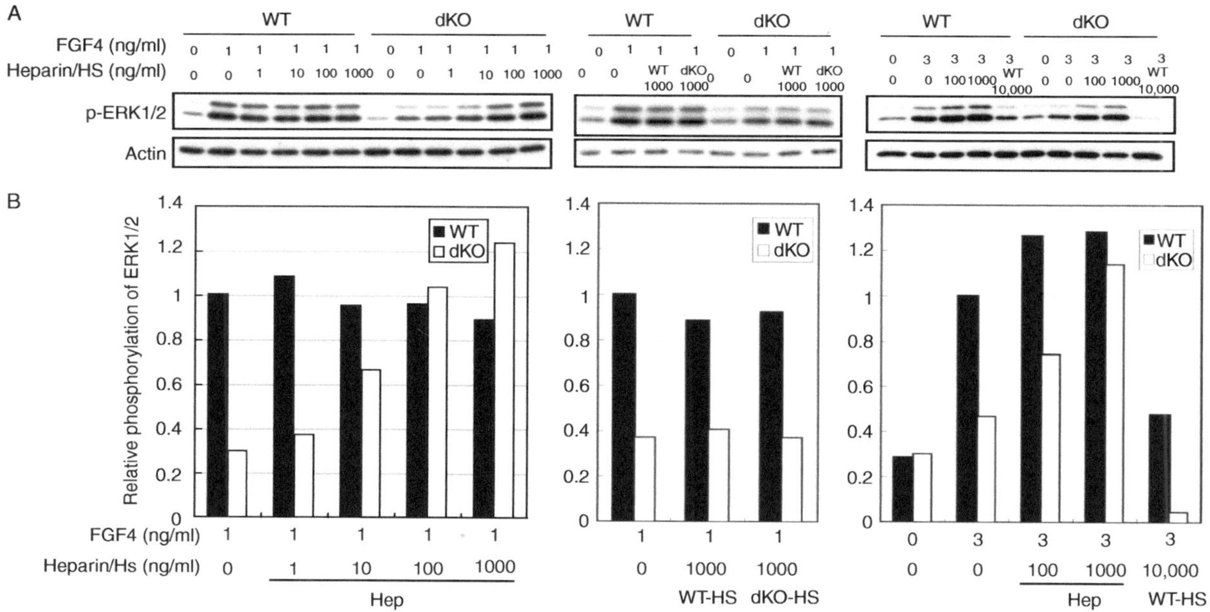

FIG. 17. FGF signaling in dKO-MEFs was recovered by exogenous heparin. MEFs were incubated for 5 min at 37 °C with 1 or 3 ng/ml FGF-4 in the absence or presence of heparin (1–1000 ng/ml), WT-HS (1, 10 μg/ml), and dKO-HS (1, 10 μg/ml). (A) Phosphorylation of ERK1/2 in these cell lysates was analyzed by Western blotting. (B) Amount of pERK1/2 was quantified from (A) and was normalized by actin bands. (Redrawn, with permission, from Ref. 59.)

In any case, 6-*O*-sulfation appears to distinctly regulate the formation of the FGF/FGFR/HS ternary complex. Further studies will investigate the 6-*O*-sulfate requirement of various FGF/FGFR combinations of the ~ 23 members in the FGF family. HS also interacts with many HB proteins other than growth factors and regulates cellular behaviors such as adhesion and invasion.[1,8,67,68] The function and mechanism of the regulation of HS 6-*O*-sulfation during such phenomena could be clarified by using the cells isolated from various organs of 6ST1-KO, 6ST2-KO, and dKO mice.

References

1. Bernfield M, Götte M, Park PW, Reizes O, Fitzgerald ML, Lincecum J, Zako M. Functions of cell surface heparan sulfate proteoglycans. *Annu Rev Biochem* 1999;**68**:729–77.
2. Esko JD, Selleck SB. Order out of chaos: assembly of ligand binding sites in heparan sulfate. *Annu Rev Biochem* 2002;**71**:435–71.
3. Habuchi H, Habuchi O, Kimata K. Sulfation pattern in glycosaminoglycan: does it have a code? *Glycoconj J* 2004;**21**:47–52.
4. Bishop JR, Schuksz M, Esko JD. Heparan sulphate proteoglycans fine-tune mammalian physiology. *Nature* 2007;**446**:1030–7.
5. Gallagher JT. Heparan sulfate: growth control with a restricted sequence menu. *J Clin Invest* 2001;**108**:357–61.
6. Bülow HE, Hobert O. The molecular diversity of glycosaminoglycans shapes animal development. *Annu Rev Cell Dev Biol* 2006;**22**:375–407.
7. Kimata K, Habuchi O, Habuchi H, Watanabe H. Knockout mice and proteoglycans. In: Kamerling JP, Boons G-J, Lee YC, Suzuki A, Taniguchi N, Voragen AGJ, editors. *Comprehensive glycoscience, Vol. 4*. Amsterdam: Elsevier Science Publishers; 2007. p. 159–91.
8. Parish CR. The role of heparan sulphate in inflammation. *Nat Rev Immunol* 2006;**6**:633–43.
9. Nakato H, Kimata K. Heparan sulfate fine structure and specificity of proteoglycan functions. *Biochim Biophys Acta* 2002;**1573**:312–8.
10. Allen BL, Rapraeger AC. Spatial and temporal expression of heparan sulfate in mouse development regulates FGF and FGF receptor assembly. *J Cell Biol* 2003;**163**:637–48.
11. Ashikari-Hada S, Habuchi H, Kariya Y, Itoh N, Reddi AH, Kimata K. Characterization of growth factor-binding structures in heparin/heparan sulfate using an octasaccharide library. *J Biol Chem* 2004;**279**:12346–54.
12. Powell AK, Yates EA, Fernig DG, Turnbull JE. Interactions of heparin/heparan sulfate with proteins: appraisal of structural factors and experimental approaches. *Glycobiology* 2004;**14**: 17R–30R.
13. Kreuger J, Spillmann D, Li JP, Lindahl U. Interactions between heparan sulfate and proteins: the concept of specificity. *J Cell Biol* 2006;**174**:323–7.
14. Esko JD, Lindahl U. Molecular diversity of heparan sulfate. *J Clin Invest* 2001;**108**:169–73.
15. Sugahara K, Kitagawa H. Heparin and heparan sulfate biosynthesis. *IUBMB Life* 2002;**54**:163–75.
16. Ai X, Do AT, Kusche-Gullberg M, Lindahl U, Lu K, Emerson Jr. CP. Substrate specificity and domain functions of extracellular heparan sulfate 6-*O*-endosulfatases, QSulf1 and QSulf2. *J Biol Chem* 2006;**281**:4969–76.

17. Habuchi H, Kobayashi M, Kimata K. Molecular characterization and expression of heparan-sulfate 6-sulfotransferase. Complete cDNA cloning in human and partial cloning in Chinese hamster ovary cells. *J Biol Chem* 1998;**273**:9208–13.
18. Habuchi H, Tanaka M, Habuchi O, Yoshida K, Suzuki H, Ban K, Kimata K. The occurrence of three isoforms of heparan sulfate 6-*O*-sulfotransferase having different specificities for hexuronic acid adjacent to the targeted *N*-sulfoglucosamine. *J Biol Chem* 2000;**275**:2859–68.
19. Habuchi H, Miyake G, Nogami K, Kuroiwa A, Matsuda Y, Kusche-Gullberg M, Habuchi O, Tanaka M, Kimata K. Biosynthesis of heparan sulphate with diverse structures and functions: two alternatively spliced forms of human heparan sulphate 6-*O*-sulphotransferase-2 having different expression patterns and properties. *Biochem J* 2003;**371**:131–42.
20. Habuchi H, Habuchi O, Kimata K. Purification and characterization of heparan sulfate 6-sulfotransferase from the culture medium of Chinese hamster ovary cells. *J Biol Chem* 1995;**270**:4172–9.
21. Kobayashi M, Habuchi H, Habuchi O, Saito M, Kimata K. Purification and characterization of heparan sulfate 2-sulfotransferase from cultured Chinese hamster ovary cells. *J Biol Chem* 1996;**271**:7645–53.
22. Kobayashi M, Habuchi H, Yoneda M, Habuchi O, Kimata K. Molecular cloning and expression of Chinese hamster ovary cell heparan-sulfate 2-sulfotransferase. *J Biol Chem* 1997;**272**:13980–5.
23. Kakuta Y, Pedersen LG, Carter CW, Negishi M, Pedersen LC. Crystal structure of estrogen sulphotransferase. *Nat Struct Biol* 1997;**4**:904–8.
24. Habuchi O. Diversity and functions of glycosaminoglycan sulfotransferases. *Biochim Biophys Acta* 2000;**1474**:115–27.
25. Kamimura K, Fujise M, Villa F, Izumi S, Habuchi H, Kimata K, Nakato H. *Drosophila* heparan sulfate 6-*O*-sulfotransferase (dHS6ST) gene. Structure, expression, and function in the formation of the tracheal system. *J Biol Chem* 2001;**276**:17014–21.
26. Cadwallader AB, Yost HJ. Combinatorial expression patterns of heparan sulfate sulfotransferases in zebrafish: III. 2-*O*-sulfotransferase and C5-epimerases. *Dev Dyn* 2007;**236**:581–6.
27. Jemth P, Smeds E, Do AT, Habuchi H, Kimata K, Lindahl U, Kusche-Gullberg M. Oligosaccharide library-based assessment of heparan sulfate 6-*O*-sulfotransferase substrate specificity. *J Biol Chem* 2003;**278**:24371–6.
28. Kobayashi S, Morimoto K, Shimizu T, Takahashi M, Kurosawa H, Shirasawa T. Association of EXT1 and EXT2, hereditary multiple exostoses gene products, in Golgi apparatus. *Biochem Biophys Res Commun* 2000;**268**:860–7.
29. McCormick C, Duncan G, Goutsos KT, Tufaro F. The putative tumor suppressors EXT1 and EXT2 form a stable complex that accumulates in the Golgi apparatus and catalyzes the synthesis of heparan sulfate. *Proc Natl Acad Sci USA* 2000;**97**:668–73.
30. Pinhal MA, Smith B, Olson S, Aikawa J, Kimata K, Esko JD. Enzyme interactions in heparan sulfate biosynthesis: uronosyl 5-epimerase and 2-*O*-sulfotransferase interact in vivo. *Proc Natl Acad Sci USA* 2001;**98**:12984–9.
31. Presto J, Thuveson M, Carlsson P, Busse M, Wilén M, Eriksson I, Kusche-Gullberg M, Kjellén L. Heparan sulfate biosynthesis enzymes EXT1 and EXT2 affect NDST1 expression and heparan sulfate sulfation. *Proc Natl Acad Sci USA* 2008;**105**:4751–6.
32. Nagai N, Habuchi H, Esko JD, Kimata K. Stem domains of heparan sulfate 6-*O*-sulfotransferase are required for Golgi localization, oligomer formation and enzyme activity. *J Cell Sci* 2004;**117**:3331–41.
33. Lippincott-Schwartz J, Donaldson JG, Schweizer A, Berger EG, Hauri HP, Yuan LC, Klausner RD. Microtubule-dependent retrograde transport of proteins into the ER in the presence of brefeldin A suggests an ER recycling pathway. *Cell* 1990;**60**:821–36.

34. Lippincott-Schwartz J, Yuan LC, Bonifacino JS, Klausner RD. Rapid redistribution of Golgi proteins into the ER in cells treated with brefeldin A: evidence for membrane cycling from Golgi to ER. *Cell* 1989;**56**:801–13.
35. Saraste J, Svensson K. Distribution of the intermediate elements operating in ER to Golgi transport. *J Cell Sci* 1991;**100**:415–30.
36. Nagai N, Habuchi H, Kitazume S, Toyoda H, Hashimoto Y, Kimata K. Regulation of heparan sulfate 6-*O*-sulfation by beta-secretase activity. *J Biol Chem* 2007;**282**:4942–51.
37. Sedita J, Izvolsky K, Cardoso WV. Differential expression of heparan sulfate 6-*O*-sulfotransferase isoforms in the mouse embryo suggests distinctive roles during organogenesis. *Dev Dyn* 2004;**231**:782–94.
38. Nogami K, Suzuki H, Habuchi H, Ishiguro N, Iwata H, Kimata K. Distinctive expression patterns of heparan sulfate *O*-sulfotransferases and regional differences in heparan sulfate structure in chick limb buds. *J Biol Chem* 2004;**279**:8219–29.
39. Kobayashi T, Habuchi H, Tamura K, Ide H, Kimata K. Essential role of heparan sulfate 2-*O*-sulfotransferase in chick limb bud patterning and development. *J Biol Chem* 2007;**282**:19589–97.
40. Kobayashi T, Habuchi H, Nogami K, Ashikari-Hada S, Tamura K, Ide H, Kimata K. Functional analysis of chick heparan sulfate 6-*O*-sulfotransferases in limb bud development. *Dev Growth Differ* 2010;**52**:146–56.
41. Habuchi H, Nagai N, Sugaya N, Atsumi F, Stevens RL, Kimata K. Mice deficient in heparan sulfate 6-*O*-sulfotransferase-1 exhibit defective heparan sulfate biosynthesis, abnormal placentation, and late embryonic lethality. *J Biol Chem* 2007;**282**:15578–88.
42. Izvolsky KI, Lu J, Martin G, Albrecht KH, Cardoso WV. Systemic inactivation of Hs6st1 in mice is associated with late postnatal mortality without major defects in organogenesis. *Genesis* 2008;**46**:8–18.
43. Lamanna WC, Baldwin RJ, Padva M, Kalus I, Ten DamG, van Kuppevelt TH, Gallagher JT, von Figura K, Dierks T, Merry CL. Heparan sulfate 6-*O*-endosulfatases: discrete in vivo activities and functional co-operativity. *Biochem J* 2006;**400**:63–73.
44. Lum DH, Tan J, Rosen SD, Werb Z. Gene trap disruption of the mouse heparan sulfate 6-*O*-endosulfatase gene, Sulf2. *Mol Cell Biol* 2007;**27**:678–88.
45. Monkley SJ, Delaney SJ, Pennisi DJ, Christiansen JH, Wainwright BJ. Targeted disruption of the Wnt2 gene results in placentation defects. *Development* 1996;**122**:3343–53.
46. Zhang X, Gaspard JP, Chung DC. Regulation of vascular endothelial growth factor by the Wnt and K-ras pathways in colonic neoplasia. *Cancer Res* 2001;**61**:6050–4.
47. Ashikari-Hada S, Habuchi H, Kariya Y, Kimata K. Heparin regulates vascular endothelial growth factor165-dependent mitogenic activity, tube formation, and its receptor phosphorylation of human endothelial cells. Comparison of the effects of heparin and modified heparins. *J Biol Chem* 2005;**280**:31508–15.
48. Bellusci S, Grindley J, Emoto H, Itoh N, Hogan BL. Fibroblast growth factor 10 (FGF10) and branching morphogenesis in the embryonic mouse lung. *Development* 1997;**124**:4867–78.
49. Izvolsky KI, Zhong L, Wei L, Yu Q, Nugent MA, Cardoso WV. Heparan sulfates expressed in the distal lung are required for Fgf10 binding to the epithelium and for airway branching. *Am J Physiol Lung Cell Mol Physiol* 2003;**285**:L838–46.
50. Izvolsky KI, Shoykhet D, Yang Y, Yu Q, Nugent MA, Cardoso WV. Heparan sulfate-FGF10 interactions during lung morphogenesis. *Dev Biol* 2003;**258**:185–200.
51. Balasubramaniam V, Tang JR, Maxey A, Plopper CG, Abman SH. Mild hypoxia impairs alveolarization in the endothelial nitric oxide synthase-deficient mouse. *Am J Physiol Lung Cell Mol Physiol* 2003;**284**:L964–71.
52. Hunninghake GW, Davidson JM, Rennard S, Szapiel S, Gadek JE, Crystal RG. Elastin fragments attract macrophage precursors to diseased sites in pulmonary emphysema. *Science* 1981;**212**:925–7.

53. Pratt T, Conway CD, Tian NM, Price DJ, Mason JO. Heparan sulphation patterns generated by specific heparan sulfotransferase enzymes direct distinct aspects of retinal axon guidance at the optic chiasm. *J Neurosci* 2006;**26**:6911–23.
54. Plump AS, Erskine L, Sabatier C, Brose K, Epstein CJ, Goodman CS, Mason CA, Tessier-Lavigne M. Slit1 and Slit2 cooperate to prevent premature midline crossing of retinal axons in the mouse visual system. *Neuron* 2002;**33**:219–32.
55. Bink RJ, Habuchi H, Lele Z, Dolk E, Joore J, Rauch GJ, Geisler R, Wilson SW, den Hertog J, Kimata K, Zivkovic D. Heparan sulfate 6-*O*-sulfotransferase is essential for muscle development in zebrafish. *J Biol Chem* 2003;**278**:31118–27.
56. Chen E, Stringer SE, Rusch MA, Selleck SB, Ekker SC. A unique role for 6-*O* sulfation modification in zebrafish vascular development. *Dev Biol* 2005;**284**:364–76.
57. Kamimura K, Koyama T, Habuchi H, Ueda R, Masu M, Kimata K, Nakato H. Specific and flexible roles of heparan sulfate modifications in *Drosophila* FGF signaling. *J Cell Biol* 2006;**174**:773–8.
58. Bülow HE, Hobert O. Differential sulfations and epimerization define heparan sulfate specificity in nervous system development. *Neuron* 2004;**41**:723–36.
59. Sugaya N, Habuchi H, Nagai N, Ashikari-Hada S, Kimata K. 6-*O*-sulfation of heparan sulfate differentially regulates various fibroblast growth factor-dependent signalings in culture. *J Biol Chem* 2008;**283**:10366–76.
60. Habuchi H, Suzuki S, Saito T, Tamura T, Harada T, Yoshida K, Kimata K. Structure of a heparan sulphate oligosaccharide that binds to basic fibroblast growth factor. *Biochem J* 1992;**285**:805–13.
61. Turnbull JE, Fernig DG, Ke Y, Wilkinson MC, Gallagher JT. Identification of the basic fibroblast growth factor binding sequence in fibroblast heparan sulfate. *J Biol Chem* 1992;**267**:10337–41.
62. Guimond S, Maccarana M, Olwin BB, Lindahl U, Rapraeger AC. Activating and inhibitory heparin sequences for FGF-2 (basic FGF). Distinct requirements for FGF-1, FGF-2, and FGF-4. *J Biol Chem* 1993;**268**:23906–14.
63. Ishihara M. Structural requirements in heparin for binding and activation of FGF-1 and FGF-4 are different from that for FGF-2. *Glycobiology* 1994;**4**:817–24.
64. Schlessinger J, Plotnikov AN, Ibrahimi OA, Eliseenkova AV, Yeh BK, Yayon A, Linhardt RJ, Mohammadi M. Crystal structure of a ternary FGF-FGFR-heparin complex reveals a dual role for heparin in FGFR binding and dimerization. *Mol Cell* 2000;**6**:743–50.
65. Wu ZL, Zhang L, Yabe T, Kuberan B, Beeler DL, Love A, Rosenberg RD. The involvement of heparan sulfate (HS) in FGF1/HS/FGFR1 signaling complex. *J Biol Chem* 2003;**278**:17121–9.
66. Robinson CJ, Harmer NJ, Goodger SJ, Blundell TL, Gallagher JT. Cooperative dimerization of fibroblast growth factor 1 (FGF1) upon a single heparin saccharide may drive the formation of 2:2:1 FGF1.FGFR2c.heparin ternary complexes. *J Biol Chem* 2005;**280**:42274–82.
67. Bishop JR, Stanford KI, Esko JD. Heparan sulfate proteoglycans and triglyceride-rich lipoprotein metabolism. *Curr Opin Lipidol* 2008;**19**:307–13.
68. Ilan N, Elkin M, Vlodavsky I. Regulation, function and clinical significance of heparanase in cancer metastasis and angiogenesis. *Int J Biochem Cell Biol* 2006;**38**:2018–39.

The Roles of Chondroitin-4-Sulfotransferase-1 in Development and Disease

Michael Klüppel[*,†,‡]

[*]Human Molecular Genetics Program, Children's Memorial Research Center, Northwestern University, Feinberg School of Medicine, Chicago, Illinois, USA

[†]Department of Pediatrics, Northwestern University, Feinberg School of Medicine, Chicago, Illinois, USA

[‡]Robert H. Lurie Comprehensive Cancer Center, Northwestern University, Feinberg School of Medicine, Chicago, Illinois, USA

The glycosaminoglycan chondroitin sulfate (CS) consists of long linear chains of repeating disaccharide units, which are covalently attached to core proteins to form CS-proteoglycans. These molecules have been shown to fulfill important biological functions in development, disease, and signaling. Biosynthesis of CS takes place in the Golgi apparatus. Concomitant to chondroitin chain elongation, sulfation of specific carbon residues by chondroitin sulfotransferase enzymes takes place. The sulfation balance and pattern of CS on

Progress in Molecular Biology
and Translational Science, Vol. 93
DOI: 10.1016/S1877-1173(10)93006-8

specific carbon residues are tightly regulated during development, injury, and disease, with the temporal and spatial expression of chondroitin sulfotransferase genes believed to be a crucial determinant of this fine balance of chondroitin sulfation. Chondroitin-4-sulfotransferase-1 (C4ST-1)/carbohydrate sulfotransferase 11 (CHST11) is one of the enzymes involved in the sulfation of chondroitin by catalyzing the transfer of sulfate groups from a sulfate donor to the carbon-4 position of the *N*-acetylgalactosamine sugar of the repeating disaccharide units.Here, I summarize the significant recent advances in our understanding of the roles of C4ST-1 in vertebrate development, disease, and signaling pathways, and the transcriptional regulation of the *C4ST-1* gene. Proper 4-sulfation of chondroitin by C4ST-1 plays a crucial role in the skeletal development and signaling events, and new evidence is suggestive of a potential role for C4ST-1 in human disease, including cancer.

I. Introduction

Glycosaminoglycans (GAGs) such as chondroitin sulfate (CS) are long chains of repeating disaccharide units, which are covalently linked to core proteins to form proteoglycans.[1–4] Mature proteoglycans can be cell membrane-bound or are part of the extracellular matrix (ECM).[5,6] Proteoglycans are important for a wide range of biological processes, including cell migration, proliferation and survival, as well as modulation of growth factor signaling.[6–9] During the synthesis of CS in the Golgi apparatus, a number of chondroitin sulfotransferase enzymes modify the elongating sugar chains through transfer of sulfate groups from a donor to specific positions on the sugar moieties.[1–4,10] The sulfation balance and pattern of CS on specific carbon residues is tightly regulated during development, injury and disease, with the temporal and spatial expression of chondroitin sulfotransferase genes believed to be a crucial determinant of this fine balance of chondroitin sulfation.[3,4] Chondroitin-4-sulfotransferase-1 (C4ST-1)/carbohydrate sulfotransferase 11 (CHST11) is one of the enzymes involved in the sulfation of chondroitin by catalyzing the transfer of sulfate groups from a sulfate donor to the carbon-4 position of the *N*-acetylgalactosamine (GalNAc) sugar of the repeating disaccharide units.[11–14]

It has become increasingly clear over the last years that the sulfation balance of GAGs plays a critical role in the biological roles of the proteoglycans carrying the GAG chains.[3,6,15–17] For example, loss-of-function mutations in mouse C4ST-1 have severe embryonic defects,[18] a phenotype not recapitulated by knock out studies in mouse C6ST-1,[19] thus demonstrating distinct roles of these chondroitin sulfotransferases, and their CS products, in mammalian development. Interestingly, the sulfation balance of chondroitin sulfation chains can facilitate GAG–protein interactions with growth factors and other proteins,[15,20–33] thus establishing specific chondroitin sulfation patterns as important contributors to the overall function of CS-proteoglycans. Moreover, specific CS forms also have been shown to affect cancer cell behavior, as well as signaling pathways involved in cancer progression,[6,16,22,26,34–42] suggesting important roles of chondroitin sulfotransferases in tumor biology as well. This chapter will discuss the significant recent advances in our understanding of the roles of C4ST-1 in vertebrate development, disease, and signaling pathways, and the transcriptional regulation of the *C4ST-1* gene. Appropriate 4-sulfation of chondroitin by C4ST-1 plays a crucial role in embryonic skeletal development and signaling events, and new evidence is suggestive of a potential role for C4ST-1 in human disease, including cancer.

II. Gene Structure

The human *C4ST-1* gene is located on chromosome 12q23.3,[12] and covers approximately 301 kilo bases (kb) of genomic sequence (Fig. 1; NCBI Genome Reference Consortium—GRCh37). The transcript is encoded by four exons, the first two of which are located in close proximity to each other, and separated by a small 121 bp first intron (NCBI Genome Reference Consortium—GRCh37). The second exon contains the putative main start codon. The third exon is small, while the fourth intron is large and contains most of the coding

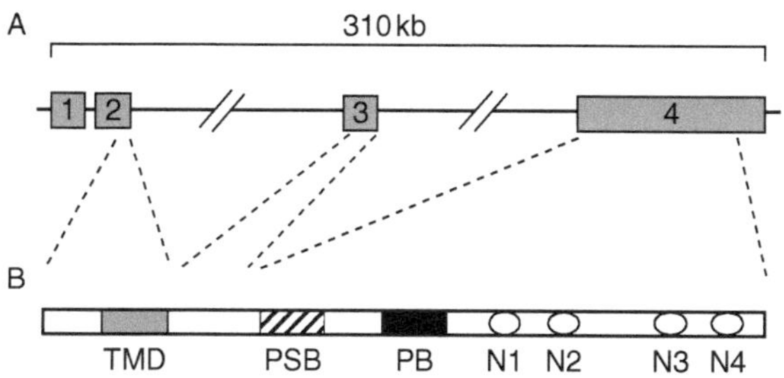

FIG. 1. Gene and protein structure of C4ST-1. (A) Schematic illustration of the human *C4ST-1* locus on chromosome 12q.23.3. The C4ST-1 gene consists of four exons and spans approximately 310 kb of genomic sequence. The ATG start codon resides in exon 2, the STOP codon is located in exon 4. (B) Schematic illustration of the protein structure of C4ST-1 with known motifs indicated. C4ST-1 contains a transmembrane domain (TMD) for anchorage in the Golgi membrane, and a large luminal catalytic domain harboring a 5′-phosphosulfate site (PSB), a 3′ phosphate binding site (PB), as well as four C-terminal *N*-glycosylation sites (N1–N4). DNA sequences encoding various parts of the protein are indicated by dashed lines.

sequence. Introns 2 and 3 are large and approximately 150 kb in size. The mouse *C4ST-1* locus is located on chromosome 10-C1 and is organized in a very similar fashion. However, it spans only approximately 210 kb of genomic sequence, due to smaller intron size. Moreover, mouse *C4ST-1* exon 1 contains sequences homologous to both human exons 1 and 2, thus effectively fusing these two exons (NCBI m37 mouse assembly).

So far, there have been 41 putative orthologues identified, from organisms ranging from the zebrafish to the humans (NCBI Genome Reference Consortium—GRCh37). Amino acid (aa) identity between human and zebrafish C4ST-1 is 79%,[43] suggesting important conserved functions in different organisms.

III. cDNA and Protein Structure of C4ST-1

A number of different approaches were employed to clone a C4ST-1 cDNA. One study identified human C4ST-1 through the screening of expressed sequence tag (EST) databases with conserved sequences of human HNK-1 sulfotransferase.[11] Another approach was the identification of a mouse *C4ST-1* cDNA through the sequence of partial peptides of a purified rat C4ST-1 protein.[14] Subsequently, the same group cloned a human *C4ST-1* cDNA.[12] Northern blot analysis showed the tissue-specific expression of several mRNA species, including 1.3, 1.9, 5.0, and 6.0 kb messages.[11,12,14] It is not clear whether these different mRNA species represent different C4ST-1 splice forms, or whether the utilization of cell-type specific promoters can result in transcription of different mRNA forms. Human *C4ST-1* has an open reading frame of 1059 bp and encodes a protein of 352 aa with the characteristics of a type II transmembrane protein, with a short N-terminal cytoplasmic domain, a Golgi transmembrane domain, and a luminal catalytic domain (Fig. 1; Refs. 11,12,14). The catalytic domain contains binding sites for the sulfate donor 3′-phosphoadenosine 5′-phosphosulfate (PAPS): a K^{125}VACT 5′-phosphosulfate binding site, and an R186EPFERLVS 3′-phosphate binding site, which function as catalytic domains for the transfer of sulfate groups.[11,12,14] Four N-linked glycosylation sites are present in the C-terminal part of the protein (Fig. 1).[11,12,14,44] The predicted size of the protein is approximately 42 kDa. However, gel analysis of either purified or Flag-tagged protein showed multiple product bands, ranging from 50 to 64 kDa,[13,44] suggesting extensive posttranslational modification of C4ST-1. Indeed, removal of *N*-glycans from the C4ST-1 protein by enzymatic digestion reduced the bands to the expected size,[44] demonstrating that C4ST-1 undergoes extensive posttranslational *N*-glycosylation.

IV. Enzymatic Activities

CS-proteoglycans have been shown to fulfill important biological functions in development and disease.[6,9,45–48] The CS side chains are linear GAGs comprised of glucoronic acid (GlnA)—GalNAc disaccharide units.[3,16,17] Biosynthesis of the CS chains takes place in the Golgi apparatus and involves three major steps: the formation of a linker tetrasaccharide to anchor the GAG chain to the core protein, polymerization of disaccharide units, and chemical modifications, including sulfation, in various positions of the sugar backbone (Fig. 2).[3,16,17] It has become increasingly clear that the sulfation balance of the chondroitin side chains plays an important role in the overall function of these proteoglycans.[3,16,17] The sulfation balance and pattern of chondroitin is tightly regulated during development,[49] and has been shown to be disturbed during malignant transformation.[6,45,50] For example, the ratio of 4- to 6-sulfated GAGs, including CS, changes from 4:1 in normal human colon tissue to 1:3 in colon adenocarcinoma. This sulfation balance is thought to be mainly determined by the temporal and spatial expression of chondroitin sulfotransferase genes.[3,16]

C4ST-1 belongs to a group of Golgi-associated chondroitin sulfotransferases, and has been shown to be able to transfer sulfate groups from the sulfate donor PAPS to several acceptors, including chondroitin and desulfated dermatan sulfate.[11–14,51] Only slight activity was observed toward CS-A and CS-C. Dermatan sulfate, keratan sulfate, CS-E, and heparan sulfate did not serve as acceptors for C4ST-1.[11–14,44,52] C4ST-1 functions by facilitating the addition of a sulfate group exclusively to the 4-position of the GalNAc sugar of chondroitin and dermatan, leading to the production of GalNAc $(4SO_4)$.[11–14,44,52] Thus, C4ST-1 is one of the key enzymes in the production of CS-A, and contributes to the production of dermatan sulfate (Fig. 2B and C). CS-A units can subsequently be utilized by another sulfotransferase, GalNAc4S-6ST, for 6-*O*-sulfation of the 4-sulfated GalNAc$(4SO_4)$ sugars, to produce disulfated CS-E units containing GalNAc$(4,6\text{-}SO_4)$ sugars (Fig. 2C).[2,53,54] Both CS-A and CS-E products of C4ST-1 have been shown to have important and distinct biological functions in vertebrates (see below). Interestingly, loss-of-function mutations in C4ST-1 not only lead to a reduction in CS-A and CS-E level but to an overall reduction in CS levels and CS chain length, indicating that elongation of CS chains is partly dependent on 4-sulfation.[18,27] CS-A units have been shown to significantly enhance the activity of GalNAc transferase-II, an enzyme involved in the elongation of the CS backbone.[55] Thus, C4ST-1 activity plays important roles not only in the biosynthesis of 4-sulfated chondroitin, but also in CS chain elongation.

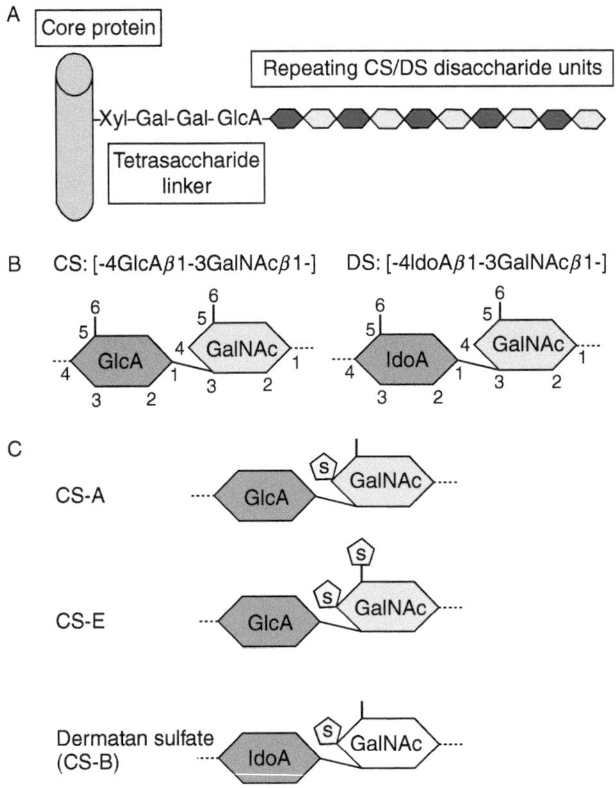

FIG. 2. Chondroitin and dermatan sulfate structures dependent on 4-sulfation. (A) Schematic illustration of a CS/DS-proteoglycan. The repeating CS/DS disaccharide units are attached to a core protein via a tetrasaccharide linker region (Xyl, Xylose; Gal, Galactose; GlcA, Glucoronic acid). (B) Schematic illustration of CS [-4GlcAβ1-3GalNAcβ1-] and DS [-4IdoAβ1-3GalNAcβ1-] disaccharide units. CS consists of repeating GlcA—GalNAc units, whereas DS consists of IdoA and GalNAc sugars; numbers of individual carbon atoms are indicated (GalNAc, *N*-acetylgalactosamine; IdoA, Iduronic acid). (C) CS-A, CS-E, and dermatan sulfate structures require sulfation on carbon-4 of GalNAc. CS-E requires the subsequent sulfation of carbon 6 of GalNAc-4S by GalNAc4S-6ST.

C4ST-1 has been shown to be *N*-glycosylated on four C-terminal amino acids; this posttranslational modification appears to be crucial for proper function of the enzyme, as mutational analysis demonstrated a nearly complete loss of enzyme activity when two or more of these *N*-glycosylation sites were eliminated.[44]

C4ST-1 is thought to be located in the membrane of the Golgi apparatus through its hydrophobic transmembrane domain. However, it was initially purified as a secreted protein from the culture medium of a rat chondrosarcoma cell line.[13] While heparan sulfatases have been shown to be secreted and function in the ECM,[56] a putative extracellular role of sulfotransferases, including C4ST-1 has not been established to date.

V. Homologous Proteins

C4ST-1 belongs to a group of Golgi-associated chondroitin sulfotransferases whose founding member was the human HNK-1 sulfotransferase.[11] Phylogenetic analysis showed that C4ST-1 is more closely related to the homologous proteins C4ST-2/CHST12 and C4ST-3/CHST13.[11,57] Amino acid identity between C4ST-1 and C4ST-2 is 41.8%, between C4ST-1 and C4ST-3 45%, and between C4ST-2 and C4ST-3 27%.[11,57] However, extensive sequence conservation is observed in the C terminal part of the proteins, which contains the catalytic domains. The *C4ST-3* gene is located on human chromosome 3q21.3 and displays a very similar structure to *C4ST-1*.[11,57] The *C4ST-2* gene, on the other hand, is located on human chromosome 7p22.3 and displays a quite distinct gene structure with only one coding exon (NCBI Genome Reference Consortium—GRCh37). Substrate specificities also revealed a closer relationship between C4ST-1 and C4ST-3. Both sulfotransferases transfer sulfate preferentially to chondroitin, although activity toward dermatan was also observed, while C4ST-2 appeared to be able to utilize both chondroitin and dermatan as substrates to approximately equal degrees.[11,51,57] In contrast to human *C4ST-1* and *C4ST-2*, which have been shown to have widespread, but not ubiquitous expression in adult tissues,[11,12,14] expression of human *C4ST-3* was restricted to the adult liver, kidney, and lymph nodes. Expression was also observed in the fetal liver.[57]

The existence of multiple chondroitin-4-sulfotransferase genes with distinct expression domains might reflect spatial requirements for each gene in order to maintain proper 4-sulfation of chondroitin and dermatan.

VI. Expression Pattern of C4ST-1

The expression pattern of *C4ST-1* showed specific as well as dynamic domains of expression. *C4ST-1* were shown to be expressed in embryonic stem (ES) cells differentiated into embryoid bodies (EB) in a flat culture system.[58] To establish the expression domains of *C4ST-1* during mouse embryogenesis, tetraploid aggregation chimeras utilizing ES cells heterozygous for a gene trap mutation in the *C4ST-1* gene were generated and analyzed for the expression of lacZ, which is under the control of the endogenous *C4ST-1* locus in these cells. Normal diploid embryos were also analyzed by whole mount- and section-*in situ* hybridization.[58] Expression of *C4ST-1* was observed at embryonic day 9.75

(e9.75) in the apical ectodermal ridge (AER), an important signaling center of the developing limbs. *C4ST-1* was also expressed in the epithelium of the branchial arches, notochord, the ventral neural tube, sympathetic ganglia, and the mesonephric tubules of the developing kidneys. In the heart at that stage, expression was observed in the myocardium of the atrium and the endocardial cushions of both inflow tract and atrio-ventricular canal. At e15.5 of embryonic development, *C4ST-1* was prominently expressed in the cartilage component of developing bones, but was absent from joints and bone that had initiated mineralization. No expression was observed in the ventral neural tube at that stage anymore, suggesting that expression of *C4ST-1* might be regulated in a temporal manner in this structure. In the heart at this stage, *C4ST-1* continued to be expressed in the atrial myocardium, as well as atrial and atrio-ventricular valves. In the kidney, *C4ST-1* expression was restricted to collecting tubules and Bowman's capsules. A punctate staining was observed in the liver at e15.5, suggestive of a potential expression in hematopoietic cells. Expression was also seen in tooth primordia, hair follicles, and mammary glands. These results demonstrated a highly specific temporal and spatial expression pattern of C4ST-1 during embryogenesis.[58]

The roles of specific CSs as regulators of neurite outgrowth *in vitro* have been well established. Thus, several mRNA expression studies have analyzed in more detail the expression domains of chondroitin sulfotransferases, including *C4ST-1*, in mouse embryonic and postnatal brain.[59] In e13.5 brain, expression was found in the ventricular zone of the telencephalon, an embryonic neurogenic region.[59] Moreover, neural stem/progenitor cells isolated from e13.5 mouse embryonic cerebral cortex and ganglionic eminence, and grown as neurospheres, also expressed *C4ST-1*, as well as other chondroitin sulfotransferases, suggestive of a potential role of chondroitin sulfotransferases in general, and C4ST-1 and its CS products in particular, in neural stem cell biology.

In postnatal brain, *C4ST-1* expression has been observed to be widespread from postnatal days 7 to 21, with lower expression levels observed at 7 weeks of age.[60] Areas of the brain with high *C4ST-1* mRNA levels include the olfactory bulb, caudate putamen, cerebral cortex, hippocampus, thalamus, midbrain, and cerebellum. Disaccharide composition of CS chains from postnatal brains showed a high percentage of 4-sulfated chains, suggestive of an important role of C4ST-1 and its CS products in brain development.[60]

In summary, *C4ST-1* displayed a highly specific temporal and spatial expression pattern during mouse embryogenesis, suggestive of important functions during development.

In contrast to the specific spatial and temporal expression patterns during mouse embryogenesis, expression analysis of human *C4ST-1* suggested a more widespread expression in adult tissues, with highest levels seen in lung, placenta, spleen, thymus, and peripheral leukocytes. It remains to be seen whether the same widespread expression is observed in tissues of adult mice.

VII. C4ST-1 in Cellular Signaling Pathways

A. C4ST-1 as a Target of TGFβ Signaling

Relatively few studies have investigated a potential involvement of C4ST-1 in cellular signaling events. One of the first approaches utilized a genetic induction gene trap approach to identify targets of transforming growth factor beta (TGFβ) signaling in undifferentiated and differentiated mouse ES cells.[58] This screen identified the C4ST-1 gene as a direct target of several members of the TGFβ family of growth factors, including TGFβ, bone morphogenetic protein-2 (BMP2), and activin. Mouse embryonic explant and whole embryo cultures revealed that treatment with BMP2 could induce C4ST-1 expression in liver and lung explants, but not in kidney explants or whole embryo cultures, suggesting a tissue- and cell-type specific regulation of C4ST-1 gene expression by signaling through TGFβ-like growth factors.[58]

A recent study shed more light on the regulation of *C4ST-1* expression by TGFβ signaling events.[61] This study utilized a bioinformatical approach to identify potential *cis*-regulatory modules conserved between human and mouse *C4ST-1* loci. The comparative analysis of 120 kb of genomic DNA surrounding the C4ST-1 translational start site (from -100 to $+20$ kb) identified several conserved putative regulatory modules. Functional tests of these elements using luciferase reporter assays in human embryonic kidney (HEK293T) and mouse normal murine mammary gland (NmuMG) cells revealed a functional *C4ST-1* promoter just upstream of exon 1, as well positive regulatory modules at -3.5, $+2$, and $+14$ kb, and negative regulatory modules at -55 and $+0.1$ kb (Fig. 3). Treatment with TGFβ ligand was able to significantly enhance transcriptional activity of the $+14$ kb element (in HEK293T cells), and the -55, -3.5, and $+2$ kb elements in NmuMG cells (Fig. 3). These data suggested that activation of the TGFβ signaling cascade could regulate *C4ST-1* transcription in a cell-type specific manner via multiple long-range *cis*-regulatory modules.[61]

B. C4ST-1 and CS-E in Canonical Wnt Signaling

C4ST-1 and its product CS-E have recently been shown to play a role in the canonical Wnt signaling cascade. Canonical Wnt signaling involves the activation, accumulation and nuclear translocation of intracellular β-catenin, and is involved in numerous biological processes, including embryonic development, tissue homeostasis, and disease.[62] Initial observations in articular chondrocytes indicated that a decrease in chondroitin sulfation, or in CS levels, both hallmarks of osteoarthritis (OA), could diminish β-catenin accumulation in response to exogenously administered canonical Wnt3a ligand.[25] In a subsequent study, investigators utilized mutant sog9 L-cell fibroblasts, which are deficient in expression of both *EXT1* (*Extosis-1*, required for heparan

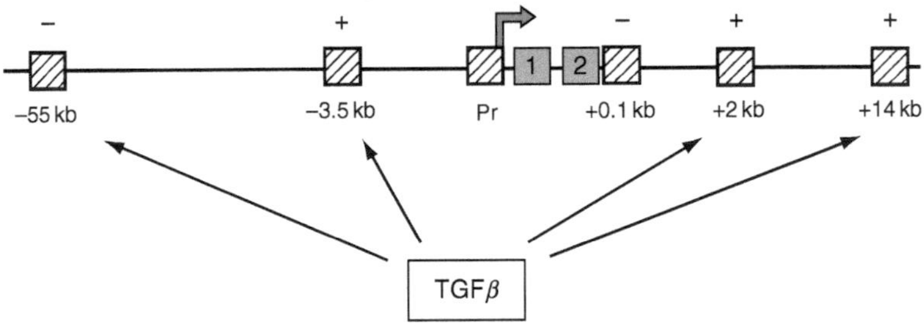

FIG. 3. Overview of long-range transcriptional regulatory elements in the *C4ST-1* locus. Transcriptional repressors are located at − 55 and + 0.1 kb (relative to *C4ST-1* exon 1), whereas transcriptional activators are located at − 3.5, + 2, and + 14 kb. A functional *C4ST-1* promoter (pr) is located upstream of exon 1. TGFβ signaling can positively influence activity of elements at − 55, − 3.5, + 2, and + 14 kb. Transcriptional regulatory elements are shown as hatched boxes, *C4ST-1* exons 1 and 2 are demarked by solid boxes.

sulfate biosynthesis) and *C4ST-1* genes.[23] When compared to normal L-cells, sog9 cells display a diminished β-catenin accumulation in response to Wnt3a stimulation, which can be rescued by stable reintroduction of C4ST-1, but not EXT1. This suggested that C4ST-1, most likely through CS-proteoglycans modified by C4ST-1, might play a role in canonical Wnt signaling. Analysis of various CS isoforms showed that CS-E, but not other CS isoforms, including CS-A, was able to bind Wnt3a ligand with high affinity. Indeed, treatment of cells with exogenously added CS-E, but not other CS isoforms, could reduce β-catenin levels.[23] The authors suggested a model in which C4ST-1 expression leads to an increase in so far unknown cell-surface CS-E-proteoglycan(s), which can bind Wnt3a via CS-E side chains and facilitate presentation of the ligand to the core Wnt receptor complex, thus increasing the efficiency of ligand–receptor interactions.[23] In this model, the core protein of this hypothetical proteoglycan appeared to have an important function in localizing the proteoglycan in the cell membrane, since addition of exogenous CS-E could not mimic the enhancing effect of C4ST-1, but instead interfered with accumulation of β-catenin.[23] These data might suggest that treatment with exogenous CS-E could be utilized to interfere with the tumor-promoting effects of canonical Wnt signaling during cancer progression.[62]

VIII. Functional Roles of C4ST-1 in Development

A. C4ST-1 in Mouse Embryonic Cartilage Morphogenesis

In order to analyze potential roles of C4ST-1 in mammalian development, ES cells carrying a gene trap mutation in the *C4ST-1* locus were used to generate heterozygous mice ($C4ST\text{-}1^{gt/+}$).[18] These heterozygous mice were viable and fertile. *C4ST-1*$^{gt/+}$ mice were mated in order to generate mice homozygous for the C4ST-1 gene trap mutation (*C4ST-1*$^{gt/gt}$). In contrast to their littermates, homozygous offspring displayed a perinatal lethality, most likely due to severe respiratory distress, and did not survive the first day after birth. Moreover, when compared to their wild type and heterozygous littermates, C4ST-1$^{gt/gt}$ animals and late gestation embryos appeared severely dwarfed. This dwarfism related to a severe growth defect of bones generated through endochondral ossification, which requires a cartilage intermediate. Affected bones included limb bones (femur, tibia, fibula, metatarsals), rib cage, scapula, as well as bones of the skull derived through endochondral ossification, including maxilla, mandible, and nasal bones. In contrast, the growth of skull bones that develop without cartilage intermediate through a process called intramembraneous ossification, including frontal, parietal, interparietal and supraoccipital bones, were not affected by the loss of C4ST-1. Moreover, these animals also displayed an impaired segmentation of digits, and altered shapes of vertebrae. These data demonstrated that C4ST-1 is required for cartilage morphogenesis during embryogenesis.[18]

The process of cartilage morphogenesis is a highly regulated multistage process that involves the aggregation of mesenchymal cells, differentiation into chondrocytes, and the subsequent coordinated proliferation, differentiation and apoptosis of these chondrocytes in the cartilage growth plate in order to generate a cartilage scaffold for the developing bone.[63,64] The cartilage growth plate is a transient structure comprised of three main compartments containing proliferating, columnar and hypertrophic chondrocyte subpopulations.[63,64] Detailed analysis of embryonic cartilage development in *C4ST-1*$^{gt/gt}$ mutant embryos revealed that early processes endochondral ossification, including the aggregation of mesencymal cells and differentiation into chondrocytes were not affected by the loss of C4ST-1.[18] First alterations in cartilage elements in *C4ST-1*$^{gt/gt}$ embryos became apparent at e15.5, with defects in the cartilage segmentation in digits, and an overall shortening of the cartilage growth plate. This growth plate defect increased at later stages of embryogenesis. Cartilage growth plates appeared disorganized, with a severe reduction in size of all three chondrocyte compartments as well as cell numbers (Fig. 4). Columnar chondrocytes were almost completely absent, and the few remaining columns were misoriented. Hypertrophic chondrocytes appeared smaller and less differentiated (Fig. 4).

Molecular analysis of mutant growth plates showed that the loss of C4ST-1 leads to an overall loss of CS, as well as a reduction of the CS proteoglycan Aggrecan. However, expression of other ECM markers, including Collagen-II, was not affected. Interestingly, the loss of C4S also led to an expansion of the spatial expression domain of C6S in the growth plate. However, patterning of mutant growth plates appeared not to be disturbed, consistent with the notion that early phases of cartilage development were not affected by the loss of C4ST-1.

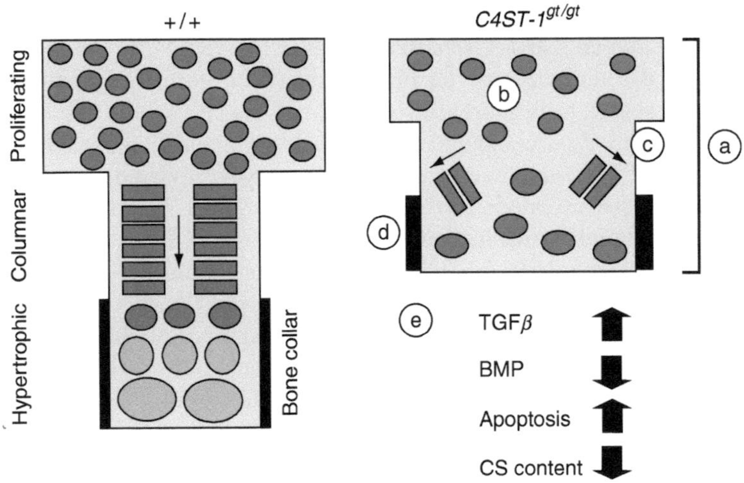

FIG. 4. Schematic illustration of cartilage growth plate defects in e18.5 *C4ST-1*$^{gt/gt}$ embryos. Wild-type cartilage growth plates (+/+) have clearly defined zones of proliferating, columnar, and hypertrophic chondrocytes. Columnar chondrocytes are oriented longitudinally to the axis of the developing bone (arrow). A bone collar is surrounding the hypertrophic zone. In *C4ST-1*$^{gt/gt}$ cartilage growth plates, a number of defects were observed: (a) severely shortened growth plates, (b) hypocellularity, (c) relatively small numbers of short and misoriented (arrows) chondrocyte columns, (d) thickened bone collar, and (e) increased TGFβ signaling, decreased BMP signaling, increased apoptosis, and decreased CS content.

Mechanistically, the loss of C4ST-1 caused an imbalance of apoptotic signals, leading to a dramatic increase in the growth plate area prone to undergo apoptosis (Fig. 4). Paradoxically, mutant growth plates also showed an increase in proliferating cells; the net effect of these two opposing phenomena, however, was a loss of chondrocytes.

Analysis of signaling events affected in mutant growth plates showed that the absence of C4ST-1 lead to an increase in endogenous TGFβ signaling, a negative regulator of chondrocyte maturation, but a decrease in BMP signaling, which functions as a positive regulator. Thus, C4ST-1 and the endogenous chondroitin sulfation balance affected signaling pathways controlling chondrocytes development (Fig. 4).[18]

In summary, C4ST-1 plays a crucial role in mammalian chondrocyte development and growth factor signaling during embryonic cartilage morphogenesis.[18]

B. C4ST-1 in Zebrafish Development

Additional evidence for *in vivo* functions of C4ST-1 came from a study in zebrafish in which antisense morpholino oligonucleotides were injected into two- or four-cell stage embryos.[43] Resulting C4ST-1 morphants displayed multiple embryonic morphological defects, including a ventrally bent trunk and a twisted tail, due to perturbations in somatic muscle development. Also observed were aberrant projections of spinal motor axons. It is of interest to note that the defects in zebrafish C4ST-1 morphants appeared distinct from the phenotype observed in *C4ST-1*$^{gt/gt}$ embryos. While *C4ST-1*$^{gt/gt}$ embryos also displayed a kinky tail, spinal motor axonal projections were not analyzed in these embryos. Importantly, the zebrafish C4ST-1 morphants did not recapitulate the severe skeletal abnormalities observed in *C4ST-1*$^{gt/gt}$ embryos. This was most likely due to the inability of mopholinos to completely eliminate *C4ST-1* expression; indeed, the levels of 4-sulfated chondroitins in C4ST-1 morphants were approximately 37% of wild type levels, suggesting that a further reduction in C4ST-1 levels might be necessary to observe the additional defects seen in mouse *C4ST-1*$^{gt/gt}$ embryos.[43]

IX. C4ST-1 in Disease

A. C4ST-1 in Cancer

The roles of CS-proteoglycans in cancer have been studied in great detail.[6] Moreover, changes in CS levels and chondroitin sulfation balance have also been described during tumor progression. For example, colon adenocarcinoma was associated with a reduction in 4-sulfated chondroitin, whereas malignant melanoma showed an increase in disulfated CS-E units.[6] However, the function of enzymes involved in the biosynthesis of CS in tumor biology are not well understood. Indeed, relatively few studies have attempted a functional analysis of C4ST-1 in cancer progression thus far. In one case report of a patient with B-cell chronic lymphocytic leukemia (B-CLL), a chromosomal translocation with breakpoints in the IGH locus on chromosome 14, and the C4ST-1 locus on chromosome 12 [t(12;14)(q23;q32)] was identified.[65] The translocation breakpoint mapped to intron 2 of the *C4ST-1* locus, and resulted in the expression of two truncated forms of C4ST-1. However, a potential functional involvement of disturbed C4ST-1 expression in the pathogenesis of B-CLL remains to be investigated.

Another study compared the expression levels of genes involved in chondroitin and heparan sulfate biosynthesis in normal bone marrow plasma cells (BMPC) versus multiple myeloma cells (MMC).[66] This study demonstrated increased expression of a group of genes, including C4ST-1, involved in HS and CS biosynthesis, in malignant cells, suggesting a potential role in the biology of multiple myeloma. Increased expression of *C4ST-1* was also observed in microarray studies of human breast carcinoma samples.[67] However, *C4ST-1* expression appeared to decrease with the increased stage of colorectal cancer.[68]

Together, these studies suggest that misregulation of *C4ST-1* expression might play a role in progression in a number of different tumors; however, a functional analysis of any involvement of C4ST-1 in tumorigenesis is not available at present.

However, several studies have identified potential roles of specific CS forms in tumor progression, including CS-A and CS-E. Chondroitin-4-sulfotransferases, which include C4ST-1, have been shown to be involved in the biosynthesis of these CS forms, thus suggesting a role for these enzymes in tumorigenesis as well. For example, higher amounts of CS-E units were found on a highly metastatic mouse osteosarcoma cell line, when compared to the nonmetatstatic parental tumor line.[35] Tissue colonization experiments demonstrated that preincubation of these metastatic tumor cells with an antibody against endogenous CS-E, or administration of exogenous CS-E together with tumor cells, could interfere with colonization of the liver.[35] Similar results were obtained with mouse lung carcinoma cells in a different study.[69] CS-E has also been shown to bind P-selectin on mouse breast cancer cells *in vitro*, suggestive of a potential role of CS-proteoglycans in tumor cell adhesion to the endothelium during extravasation of the circulating tumor cells to target organs.[39] However, this putative function of CS-E has not been demonstrated *in vivo*. High levels of CS-E were also observed in ovarian adenocarcinomas.[70]

In addition, CS-A has been shown to regulate fibrosarcoma cell adhesion, motility and migration through JNK and tyrosine kinase signaling pathways.[37]

Work in melanoma cells has provided insight into potential mechanisms of CS action. One study investigated a potential role of CS in matrix metalloproteinase (MMP)-mediated invasion of melanoma cells.[22] C4S-, but not C6S-, chains on melanoma-specific CS proteoglycan (MCSP) facilitate the formation of a tripartite complex of MCSP, membrane-type 3 (MT3)-MMP, and pro-MMP-2, and the subsequent activation of MMP-2. Activation of MMP-2 can facilitate invasion of tumor cells through basement membranes and connective tissue by degrading type IV collagen.

Other studies have shown an involvement of C4S chains of cell-surface proteoglycans in FGF ligand–receptor interactions, thus potentially enhancing a tumor-promoting signaling pathway.[20,21]

In summary, misregulation of C4ST-1 expression has been observed in a number of cancers. Moreover, specific CS forms, including CS-A and CS-E, have been shown to regulate biological events that are associated with tumor progression. This might suggest a putative functional role of C4ST-1 in the regulation of tumor cell biology through its role in the biosynthesis of CS-A and CS-E.

B. C4ST-1 and CS-E in Herpes Simplex Virus Infectivity

Herpes simplex virus (HSV) envelope glycoproteins have been shown to utilize cell-surface GAGs to efficiently bind to and infect host cells.[71] The gC HSV envelope protein has been suggested to bind cell-surface CS-E-proteoglycans with high affinity, and treatment with exogenous CS-E could potently inhibit HSV infectivity, thus identifying CS-E chains of cell-surface proteoglycans as key receptors for HSV entry into a host cell.[72]

A recent study demonstrated a drastically reduced susceptibility to HSV infection in L-cell fibroblasts deficient in C4ST-1 expression and CS-E production (sog9 cells).[27] Stable expression of C4ST-1 in sog9 cells led to increased production of CS-E chains, and rendered these cells susceptible to HSV infection again. Thus, C4ST-1 expression is a critical factor in cell sensitivity to HSV infection.[72]

C. Altered Expression of C4ST-1 in Osteoarthritis

OA is a degenerative disease characterized by loss of matrix GAGs and cartilage integrity.[73] A genome-wide gene expression analysis, comparing human OA cartilage to normal donor cartilage, aimed to identify genes that are misregulated during disease progression.[74] This study identified increased levels of *C4ST-1* expression in OA.[74] Interestingly, a loss-of-function mutation in the mouse *C4ST-1* gene leads to severe cartilage defects, and is characterized by disorganized and hypocellular cartilage growth plates with fibrillated ECM and an overall loss of CS.[18] Many of these cartilage deficiencies are characteristic of the degenerative alterations observed in OA.[74] Combined, these data suggest a requirement for tightly controlled regulation of *C4ST-1* expression in the development and maintenance of healthy cartilage.

D. CS-A is a Cell-Surface Receptor for *Plasmodium falciparum*-Infected Erythrocytes

Malaria is an infectious disease that affects approximately 250 million people each year, leading to 1 million deaths (www.cdc.gov/malaria). Malaria is caused by the parasites of the species *Plasmodium* and is transmitted through infected mosquitos. It has been shown that high affinity adherence of *P. falciparum*-infected erythrocytes to endothelial cells is mediated by CS-A, but not CS-B or CS-C, units, on endothelial cell-surface proteoglycans, thus implicating a specific form of CS in the pathogenesis of Malaria.[24] Although C4ST-1 is one of the principal enzymes in the biosynthesis of CS-A, no studies so far have addressed a potential role of the *C4ST-1* gene in *Plasmodium* infections and Malaria.

X. Concluding Remarks

Unlike heparan sulfate and heparan sulfotransferases, which have been the subject of intense investigations, CS and chondroitin sulfotransferases have not received as much attention until recently. However, several observations suggest important biological roles for specific CS forms, and the chondroitin sulfotransferases involved in their biosynthesis. Specifically, C4ST-1 has been shown to be important for mouse embryonic skeletal development, and for motor axon guidance in zebrafish. More thorough analysis of mouse and zebrafish mutants might reveal additional biological processes that require functional C4ST-1. In addition, several lines of evidence might also suggest a role of C4ST-1 in tumor biology, OA, and other diseases. For example, CS-A and CS-E, two CS forms created by C4ST-1 enzyme function, have been shown to affect tumor cell behavior. In addition, misregulation of C4ST-1 gene expression has been demonstrated for a number of different tumors. These lines of evidence suggest that C4ST-1 is not only important for mammalian development, but might also play a role in tumor progression. The next years will undoubtedly give novel insights into the roles of C4ST-1 and the balance of chondroitin sulfation in development, disease and signaling pathways.

References

1. Habuchi O. Biological function of chondroitin sulfate and sulfotransferases. *Seikagaku* 2000;**72**:427–36.
2. Habuchi O. Diversity and functions of glycosaminoglycan sulfotransferases. *Biochim Biophys Acta* 2000;**1474**:115–27.
3. Honke K, Taniguchi N. Sulfotransferases and sulfated oligosaccharides. *Med Res Rev* 2002;**22**:637–54.
4. Kusche-Gullberg M, Kjellen L. Sulfotransferases in glycosaminoglycan biosynthesis. *Curr Opin Struct Biol* 2003;**13**:605–11.
5. Selleck SB. Proteoglycans and pattern formation: sugar biochemistry meets developmental genetics. *Trends Genet* 2000;**16**:206–12.
6. Wegrowski Y, Maquart FX. Chondroitin sulfate proteoglycans in tumor progression. *Adv Pharmacol* 2006;**53**:297–321.
7. Mythreye K, Blobe GC. Proteoglycan signaling co-receptors: roles in cell adhesion, migration and invasion. *Cell Signal* 2009;**21**:1548–58.
8. Okina E, Manon-Jensen T, Whiteford JR, Couchman JR. Syndecan proteoglycan contributions to cytoskeletal organization and contractility. *Scand J Med Sci Sports* 2009;**19**:479–89.
9. Heinegard D. Proteoglycans and more—from molecules to biology. *Int J Exp Pathol* 2009;**90**:575–86.
10. Habuchi O, Miyashita N. Separation and characterization of chondroitin 6-sulfotransferase and chondroitin 4-sulfotransferase from chick embryo cartilage. *Biochim Biophys Acta* 1982;**717**:414–21.

11. Hiraoka N, Nakagawa H, Ong E, Akama TO, Fukuda MN, Fukuda M. Molecular cloning and expression of two distinct human chondroitin 4-*O*-sulfotransferases that belong to the HNK-1 sulfotransferase gene family. *J Biol Chem* 2000;**275**:20188–96.
12. Okuda T, Mita S, Yamauchi S, Matsubara T, Yagi F, Yamamori D, et al. Molecular cloning, expression, and chromosomal mapping of human chondroitin 4-sulfotransferase, whose expression pattern in human tissues is different from that of chondroitin 6-sulfotransferase. *J Biochem* 2000;**128**:763–70.
13. Yamauchi S, Hirahara Y, Usui H, Takeda Y, Hoshino M, Fukuta M, et al. Purification and characterization of chondroitin 4-sulfotransferase from the culture medium of a rat chondrosarcoma cell line. *J Biol Chem* 1999;**274**:2456–63.
14. Yamauchi S, Mita S, Matsubara T, Fukuta M, Habuchi H, Kimata K, et al. Molecular cloning and expression of chondroitin 4-sulfotransferase. *J Biol Chem* 2000;**275**:8975–81.
15. Gama CI, Tully SE, Sotogaku N, Clark PM, Rawat M, Vaidehi N, et al. Sulfation patterns of glycosaminoglycans encode molecular recognition and activity. *Nat Chem Biol* 2006;**2**:467–73.
16. Malavaki C, Mizumoto S, Karamanos N, Sugahara K. Recent advances in the structural study of functional chondroitin sulfate and dermatan sulfate in health and disease. *Connect Tissue Res* 2008;**49**:133–9.
17. Yamada S, Sugahara K. Potential therapeutic application of chondroitin sulfate/dermatan sulfate. *Curr Drug Discov Technol* 2008;**5**:289–301.
18. Klüppel M, Wight TN, Chan C, Hinek A, Wrana JL. Maintenance of chondroitin sulfation balance by chondroitin-4-sulfotransferase 1 is required for chondrocyte development and growth factor signaling during cartilage morphogenesis. *Development* 2005;**132**:3989–4003.
19. Uchimura K, Kadomatsu K, Nishimura H, Muramatsu H, Nakamura E, Kurosawa N, et al. Functional analysis of the chondroitin 6-sulfotransferase gene in relation to lymphocyte subpopulations, brain development, and oversulfated chondroitin sulfates. *J Biol Chem* 2002;**277**:1443–50.
20. Asada M, Shinomiya M, Suzuki M, Honda E, Sugimoto R, Ikekita M, et al. Glycosaminoglycan affinity of the complete fibroblast growth factor family. *Biochim Biophys Acta* 2009;**1790**:40–8.
21. Deepa SS, Umehara Y, Higashiyama S, Itoh N, Sugahara K. Specific molecular interactions of oversulfated chondroitin sulfate E with various heparin-binding growth factors. Implications as a physiological binding partner in the brain and other tissues. *J Biol Chem* 2002;**277**:43707–16.
22. Iida J, Wilhelmson KL, Ng J, Lee P, Morrison C, Tam E, et al. Cell surface chondroitin sulfate glycosaminoglycan in melanoma: role in the activation of pro-MMP-2 (pro-gelatinase A). *Biochem J* 2007;**403**:553–63.
23. Nadanaka S, Ishida M, Ikegami M, Kitagawa H. Chondroitin 4-*O*-sulfotransferase-1 modulates Wnt-3a signaling through control of E disaccharide expression of chondroitin sulfate. *J Biol Chem* 2008;**283**:27333–43.
24. Rogerson SJ, Chaiyaroj SC, Ng K, Reeder JC, Brown GV. Chondroitin sulfate A is a cell surface receptor for *Plasmodium falciparum*-infected erythrocytes. *J Exp Med* 1995;**182**:15–20.
25. Shortkroff S, Yates KE. Alteration of matrix glycosaminoglycans diminishes articular chondrocytes' response to a canonical Wnt signal. *Osteoarthr Cartil* 2007;**15**:147–54.
26. Sugahara KN, Hirata T, Tanaka T, Ogino S, Takeda M, Terasawa H, et al. Chondroitin sulfate E fragments enhance CD44 cleavage and CD44-dependent motility in tumor cells. *Cancer Res* 2008;**68**:7191–9.
27. Uyama T, Ishida M, Izumikawa T, Trybala E, Tufaro F, Bergstrom T, et al. Chondroitin 4-*O*-sulfotransferase-1 regulates E disaccharide expression of chondroitin sulfate required for herpes simplex virus infectivity. *J Biol Chem* 2006;**281**:38668–74.
28. Zou P, Zou K, Muramatsu H, Ichihara-Tanaka K, Habuchi O, Ohtake S, et al. Glycosaminoglycan structures required for strong binding to midkine, a heparin-binding growth factor. *Glycobiology* 2003;**13**:35–42.

29. Bao X, Mikami T, Yamada S, Faissner A, Muramatsu T, Sugahara K. Heparin-binding growth factor, pleiotrophin, mediates neuritogenic activity of embryonic pig brain-derived chondroitin sulfate/dermatan sulfate hybrid chains. *J Biol Chem* 2005;**280**:9180–91.
30. Hayashi K, Kadomatsu K, Muramatsu T. Requirement of chondroitin sulfate/dermatan sulfate recognition in midkine-dependent migration of macrophages. *Glycoconj J* 2001;**18**:401–6.
31. Ueoka C, Kaneda N, Okazaki I, Nadanaka S, Muramatsu T, Sugahara K. Neuronal cell adhesion, mediated by the heparin-binding neuroregulatory factor midkine, is specifically inhibited by chondroitin sulfate E. Structural and functional implications of the over-sulfated chondroitin sulfate. *J Biol Chem* 2000;**275**:37407–13.
32. Yamamoto H, Muramatsu H, Nakanishi T, Natori Y, Sakuma S, Ishiguro N, et al. Midkine as a molecular target: comparison of effects of chondroitin sulfate E and siRNA. *Biochem Biophys Res Commun* 2006;**351**:915–9.
33. Zou K, Muramatsu H, Ikematsu S, Sakuma S, Salama RH, Shinomura T, et al. A heparin-binding growth factor, midkine, binds to a chondroitin sulfate proteoglycan, PG-M/versican. *Eur J Biochem* 2000;**267**:4046–53.
34. Asimakopoulou AP, Theocharis AD, Tzanakakis GN, Karamanos NK. The biological role of chondroitin sulfate in cancer and chondroitin-based anticancer agents. *In Vivo* 2008;**22**:385–9.
35. Basappa, Murugan S, Sugahara KN, Lee CM, ten Dam GB, van Kuppevelt TH, et al. Involvement of chondroitin sulfate E in the liver tumor focal formation of murine osteosarcoma cells*Glycobiology* 2009;**19**:735–42.
36. Borsig L, Wang L, Cavalcante MC, Cardilo-Reis L, Ferreira PL, Mourao PA, et al. Selectin blocking activity of a fucosylated chondroitin sulfate glycosaminoglycan from sea cucumber. Effect on tumor metastasis and neutrophil recruitment. *J Biol Chem* 2007;**282**:14984–91.
37. Fthenou E, Zong F, Zafiropoulos A, Dobra K, Hjerpe A, Tzanakakis GN. Chondroitin sulfate A regulates fibrosarcoma cell adhesion, motility and migration through JNK and tyrosine kinase signaling pathways. *In Vivo* 2009;**23**:69–76.
38. Luo J, Kato M, Wang H, Bernfield M, Bischoff J. Heparan sulfate and chondroitin sulfate proteoglycans inhibit E-selectin binding to endothelial cells. *J Cell Biochem* 2001;**80**:522–31.
39. Monzavi-Karbassi B, Stanley JS, Hennings L, Jousheghany F, Artaud C, Shaaf S, et al. Chondroitin sulfate glycosaminoglycans as major P-selectin ligands on metastatic breast cancer cell lines. *Int J Cancer* 2007;**120**:1179–91.
40. Nikitovic D, Assouti M, Sifaki M, Katonis P, Krasagakis K, Karamanos NK, et al. Chondroitin sulfate and heparan sulfate-containing proteoglycans are both partners and targets of basic fibroblast growth factor-mediated proliferation in human metastatic melanoma cell lines. *Int J Biochem Cell Biol* 2008;**40**:72–83.
41. Sakko AJ, Butler MS, Byers S, Reinboth BJ, Stahl J, Kench JG, et al. Immunohistochemical level of unsulfated chondroitin disaccharides in the cancer stroma is an independent predictor of prostate cancer relapse. *Cancer Epidemiol Biomarkers Prev* 2008;**17**:2488–97.
42. Theocharis AD, Tsolakis I, Tzanakakis GN, Karamanos NK. Chondroitin sulfate as a key molecule in the development of atherosclerosis and cancer progression. *Adv Pharmacol* 2006;**53**:281–95.
43. Mizumoto S, Mikami T, Yasunaga D, Kobayashi N, Yamauchi H, Miyake A, et al. Chondroitin 4-*O*-sulfotransferase-1 is required for somitic muscle development and motor axon guidance in zebrafish. *Biochem J* 2009;**419**:387–99.
44. Yusa A, Kitajima K, Habuchi O. N-linked oligosaccharides are required to produce and stabilize the active form of chondroitin 4-sulphotransferase-1. *Biochem J* 2005;**388**:115–21.
45. Wegrowski Y, Maquart FX. Involvement of stromal proteoglycans in tumour progression. *Crit Rev Oncol Hematol* 2004;**49**:259–68.
46. Lambaerts K, Wilcox-Adelman SA, Zimmermann P. The signaling mechanisms of syndecan heparan sulfate proteoglycans. *Curr Opin Cell Biol* 2009;**21**:662–9.

47. Schaefer L, Schaefer RM. Proteoglycans: from structural compounds to signaling molecules. *Cell Tissue Res* 2010;**339**:237–46.
48. Yan D, Lin X. Shaping morphogen gradients by proteoglycans. *Cold Spring Harb Perspect Biol* 2009;**1**:a002493.
49. Kitagawa H, Tsutsumi K, Tone Y, Sugahara K. Developmental regulation of the sulfation profile of chondroitin sulfate chains in the chicken embryo brain. *J Biol Chem* 1997;**272**:31377–81.
50. Kovalszky I, Pogany G, Molnar G, Jeney A, Lapis K, Karacsonyi S, et al. Altered glycosaminoglycan composition in reactive and neoplastic human liver. *Biochem Biophys Res Commun* 1990;**167**:883–90.
51. Mikami T, Mizumoto S, Kago N, Kitagawa H, Sugahara K. Specificities of three distinct human chondroitin/dermatan *N*-acetylgalactosamine 4-*O*-sulfotransferases demonstrated using partially desulfated dermatan sulfate as an acceptor: implication of differential roles in dermatan sulfate biosynthesis. *J Biol Chem* 2003;**278**:36115–27.
52. Yamada T, Ohtake S, Sato M, Habuchi O. Chondroitin 4-sulphotransferase-1 and chondroitin 6-sulphotransferase-1 are affected differently by uronic acid residues neighbouring the acceptor GalNAc residues. *Biochem J* 2004;**384**:567–75.
53. Habuchi O, Moroi R, Ohtake S. Enzymatic synthesis of chondroitin sulfate E by *N*-acetylgalactosamine 4-sulfate 6-*O*-sulfotransferase purified from squid cartilage. *Anal Biochem* 2002;**310**:129–36.
54. Ohtake S, Kondo S, Morisaki T, Matsumura K, Kimata K, Habuchi O. Expression of sulfotransferases involved in the biosynthesis of chondroitin sulfate E in the bone marrow derived mast cells. *Biochim Biophys Acta* 2008;**1780**:687–95.
55. Kitagawa H, Tsutsumi K, Ujikawa M, Goto F, Tamura J, Neumann KW, et al. Regulation of chondroitin sulfate biosynthesis by specific sulfation: acceptor specificity of serum beta-GalNAc transferase revealed by structurally defined oligosaccharides. *Glycobiology* 1997;**7**:531–7.
56. Lamanna WC, Frese MA, Balleininger M, Dierks T. Sulf loss influences *N*-, 2-*O*-, and 6-*O*-sulfation of multiple heparan sulfate proteoglycans and modulates fibroblast growth factor signaling. *J Biol Chem* 2008;**283**:27724–35.
57. Kang HG, Evers MR, Xia G, Baenziger JU, Schachner M. Molecular cloning and characterization of chondroitin-4-*O*-sulfotransferase-3. A novel member of the HNK-1 family of sulfotransferases. *J Biol Chem* 2002;**277**:34766–72.
58. Klüppel M, Vallis KA, Wrana JL. A high-throughput induction gene trap approach defines C4ST as a target of BMP signaling. *Mech Dev* 2002;**118**:77–89.
59. Akita K, von Holst A, Furukawa Y, Mikami T, Sugahara K, Faissner A. Expression of multiple chondroitin/dermatan sulfotransferases in the neurogenic regions of the embryonic and adult central nervous system implies that complex chondroitin sulfates have a role in neural stem cell maintenance. *Stem Cells* 2008;**26**:798–809.
60. Mitsunaga C, Mikami T, Mizumoto S, Fukuda J, Sugahara K. Chondroitin sulfate/dermatan sulfate hybrid chains in the development of cerebellum. Spatiotemporal regulation of the expression of critical disulfated disaccharides by specific sulfotransferases. *J Biol Chem* 2006;**281**:18942–52.
61. Willis CM, Wrana JL, Klüppel M. Identification and characterization of TGFbeta-dependent and -independent *cis*-regulatory modules in the C4ST-1/CHST11 locus. *Genet Mol Res* 2009;**8**:1331–43.
62. Nusse R. Wnt signaling in disease and in development. *Cell Res* 2005;**15**:28–32.
63. Aigner T, Gerwin N. Growth plate cartilage as developmental model in osteoarthritis research-potentials and limitations. *Curr Drug Targets* 2007;**8**:377–85.
64. Orth MW. The regulation of growth plate cartilage turnover. *J Anim Sci* 1999;**77** (Suppl. 2):183–9.

65. Schmidt HH, Dyomin VG, Palanisamy N, Itoyama T, Nanjangud G, Pirc-Danoewinata H, et al. Deregulation of the carbohydrate (chondroitin 4) sulfotransferase 11 (CHST11) gene in a B-cell chronic lymphocytic leukemia with a t(12;14)(q23;q32). *Oncogene* 2004;**23**:6991–6.
66. Bret C, Hose D, Reme T, Sprynski AC, Mahtouk K, Schved JF, et al. Expression of genes encoding for proteins involved in heparan sulphate and chondroitin sulphate chain synthesis and modification in normal and malignant plasma cells. *Br J Haematol* 2009;**145**:350–68.
67. Potapenko IO, Haakensen VD, Luders T, Aslaug H, Bukholm I, Sorlie T, et al. Glycan gene expression signatures in normal and malignant breast tissue; possible role in diagnosis and progression. *Mol Oncol* 2010;**4**(2):98–118.
68. Kalathas D, Theocharis DA, Bounias D, Kyriakopoulou D, Papageorgakopoulou N, Stavropoulos MS, et al. Alterations of glycosaminoglycan disaccharide content and composition in colorectal cancer: structural and expressional studies. *Oncol Rep* 2009;**22**:369–75.
69. Li F, Ten Dam GB, Murugan S, Yamada S, Hashiguchi T, Mizumoto S, et al. Involvement of highly sulfated chondroitin sulfate in the metastasis of the Lewis lung carcinoma cells. *J Biol Chem* 2008;**283**:34294–304.
70. ten Dam GB, van de Westerlo EM, Purushothaman A, Stan RV, Bulten J, Sweep FC, et al. Antibody GD3G7 selected against embryonic glycosaminoglycans defines chondroitin sulfate-E domains highly up-regulated in ovarian cancer and involved in vascular endothelial growth factor binding. *Am J Pathol* 2007;**171**:1324–33.
71. Shukla D, Liu J, Blaiklock P, Shworak NW, Bai X, Esko JD, et al. A novel role for 3-*O*-sulfated heparan sulfate in herpes simplex virus 1 entry. *Cell* 1999;**99**:13–22.
72. Bergefall K, Trybala E, Johansson M, Uyama T, Naito S, Yamada S, et al. Chondroitin sulfate characterized by the E-disaccharide unit is a potent inhibitor of herpes simplex virus infectivity and provides the virus binding sites on gro2C cells. *J Biol Chem* 2005;**280**:32193–9.
73. Krasnokutsky S, Samuels J, Abramson SB. Osteoarthritis in 2007. *Bull NYU Hosp Jt Dis* 2007;**65**:222–8.
74. Karlsson C, Dehne T, Lindahl A, Brittberg M, Pruss A, Sittinger M, et al. Genome-wide expression profiling reveals new candidate genes associated with osteoarthritis. *Osteoarthr Cartilage* 2010;**18**(4):581–92.

Roles of Heparan Sulfate in Mammalian Brain Development: Current Views Based on the Findings from *Ext1* Conditional Knockout Studies

Yu Yamaguchi,[*] Masaru Inatani,[*,†] Yoshihiro Matsumoto,[*,‡] Junko Ogawa,[*] and Fumitoshi Irie[*]

[*]*Sanford Children's Health Research Center, Sanford-Burnham Medical Research Institute, La Jolla, California, USA*

[†]*Department of Ophthalmology and Visual Science, Kumamoto University Graduate School of Medical Sciences, Kumamoto, Japan*

[‡]*Department of Orthopaedic Surgery, Graduate School of Medical Sciences, Kyushu University, Fukuoka, Japan*

Development of the mammalian central nervous system proceeds roughly in four major steps, namely the patterning of the neural tube, generation of neurons from neural stem cells and their migration to genetically predetermined destinations, extension of axons and dendrites toward target neurons to form neural circuits, and formation of synaptic contacts. Earlier studies on spatiotemporal expression patterns and *in vitro* function of heparan sulfate (HS) suggested that HS is functionally involved in various aspects of neural development. Recent studies using knockout of genes involved in HS

DOI: 10.1016/S1877-1173(10)93007-X

1877-1173/10 $35.00

biosynthesis have provided more physiologically relevant information as to the role of HS in mammalian neural development. This chapter reviews the current understanding of the *in vivo* function of HS deduced from the phenotypes of conditional *Ext1* knockout mice.

I. Introduction

Following neurulation, development of the vertebrate central nervous system (CNS) proceeds in four major steps: (i) regionalization of the neural tube into distinct subdivisions, which later become major domains of the CNS, such as the cerebrum, midbrain, and cerebellum (patterning); (ii) generation of neurons from neural stem cells within each CNS subdivisions and migration of newly generated neurons toward the sites of final settlement (neurogenesis); (iii) extension of axons and dendrites toward target neurons to form neural circuits (axon growth and guidance); and (iv) formation and maturation of synaptic contacts (synaptogenesis). These developmental processes are regulated by a number of morphogens, growth factors, axon guidance molecules, and adhesion molecules.

Heparan sulfate (HS) is one of the four classes of glycosaminoglycans.[1,2] There are three major family of heparan sulfate proteoglycans (HSPGs), including syndecans, glypicans, and perlecan. There are a large amount of data indicating that HSPGs are involved in various biological processes, such as growth factor/morphogen signaling, morphogen gradient formation, cell adhesion, and extracellular matrix assembly.[3,4] It was therefore anticipated that knockout of individual HSPGs would lead to substantial developmental phenotypes. However, the attempts to define the physiological roles of HSPGs in brain development by knocking out HSPG core protein genes have been met with relatively disappointing outcomes.[5] Most illuminating is the case of glypicans. All six mammalian glypicans are expressed in the developing CNS with some very interesting expression patterns.[6] Yet none of the glypican mutant mice showed gross brain defects. (Recent detailed analyses have begun to uncover interesting phenotypes in glypican knockout mice, though. See, e.g., Ref. 6) Obviously, this is in large part due to the functional redundancy of various HSPGs. Although shared core protein functions may at least partly be responsible for such redundancy, it is more likely that many important functions of HSPGs are endowed by their HS chains. Knockout studies of enzymes that are involved in HS biosynthesis have revealed much greater and informative phenotypes, supporting this notion. In this chapter, we review the recent progress in our understanding of the role of HS in mammalian brain development, focusing mainly on our observation with *Nestin-Cre;Ext1*[7] and *Wnt1-Cre;Ext1*[8] conditional knockout mice.

II. Studying Heparan Sulfate Function *In Vivo* by Genetic Manipulation of *Ext1*

HS biosynthesis is governed by a series of enzymes. Following the attachment of a xylose residue to a serine residue on the core protein, the tetrasaccharide linkage region shared by HS and chondroitin sulfate is synthesized by the actions of a series of transferases. Then, a linear polymer of alternating glucuronic acid (GlcA) and *N*-acetylglucosamine (GlcNAc) residues, which forms the "core" of HS, is synthesized by the action of GlcA/GlcNAc polymerase. Once a linear unsulfated HS chain is assembled, the individual sugar residues undergo sequential modification by *N*-deacetylase/*N*-sulfotransferases (NDSTs), glucuronyl C-5-epimerase, 2-*O*-, 6-*O*-, and 3-*O*-sulfotransferases.[9] Studies during last several years have solidly established the Ext1 protein, which was originally identified as one of the causative genes for multiple hereditary exostosis (MHE), a genetic bone disorder, as the GlcNAc/GlcA copolymerase.[10] The human *EXT1* gene, which resides in 8q24 and contains 11 exons, encodes a 746 amino acid glycoprotein with a type II transmembrane topology, which is typical for glycosyltransferases. The Ext1 protein is localized in the endoplasmic reticulum and the Golgi apparatus. Ext1 and its structurally related protein, Ext2, form hetero-oligomers within the Golgi apparatus,[11] and this Ext1/Ext2 association is essential for the expression of biologically relevant GlcA/GlcNAc copolymerase activity.[12,13] This biochemical model is consistent with human genetic observations that the mutations in either of the two *EXT* genes could cause MHE.[13] Moreover, a mouse genetic study demonstrated that HS biosynthesis is completely abolished in *Ext1*$^{-/-}$ ES cells, and that constitutive *Ext1* knockout results in very early embryonic lethality due to gastrulation defects.[14] Unlike other enzymes involved in HS biosynthesis, such as NDSTs and 3-*O*-, and 6-*O*-sulfotransferases, there are no other isoforms or related molecules that have the GlcA/GlcNAc copolymerase activity. These properties make *Ext1* as an excellent target for genetic disruption of HS biosynthesis in model animals. Unlike the disruption of HS modifying enzymes (e.g., NDSTs, C5-epimerase, *O*-sulfotransferases), HS chains will not be produced in *Ext1* null cells, thereby allowing more direct interpretation of the causal relationship between HS and resultant phenotypes. On the other hand, *Ext1* knockout does not provide information as to the function of specific modifications of HS or "fine structure" of HS. Still, *Ext1* null phenotypes provide important baseline data to understand the function of fine modification of HS (see Section VI).

III. *Cre* Systems for Studies of Brain Development

There are a number of *Cre* driver mice, both transgenic and knock-in, available for studies on brain development. The brain contains a number of different cell types. The three major neural cell types are neurons, astrocytes, and oligodendrocytes. *Nestin-Cre* is a useful and reliable *Cre* driver to ablate the targeted gene in all these neural cells from embryonic day (E) 9.5–10.5 onward[15] (see Section IV). Each neural cell type contains subpopulations of further differentiated cells. It is now possible to target many of these subpopulations of neural cells. For example, there are *Cre* drivers that selectively target different types of neurons, such as glutamatergic, dopaminergic, GABAergic, catecholaminergic, cerebellar Purkinje and granule neurons.[16] Moreover, there are nearly 10 independent *CaMKII-Cre* drivers that selectively target glutamatergic neurons in different brain regions; some of them have a broad specificity toward the entire forebrain, while the others target almost solely the hippocampus.[16] It should be kept in mind that none of these Cre drivers are entirely specific for a certain cell population, and none of them are 100% efficient in inducing recombination in a given cell type. The specificity and efficiency of *Cre* driver are limited by the specificity of promoters used and the efficiency of *Cre* recombinase. For example, glial fibrillary acidic protein (GFAP) is a well-known marker for astrocytes. However, GFAP is also expressed in embryonic and adult neural stem cells.[17] As a result, the first generation *GFAP-Cre* driver induces pan-neural recombination, not only in astrocytes but also in neurons and oligodendrocytes, a pattern essentially the same as that by *Nestin-Cre*. In the case of conditional knockout of dystroglycan, this led to ablation of the gene in the entire embryonic brain.[18] *GFAP-Cre* drivers that do not act in the embryonic brain are available,[19] although these late-onset *GFAP-Cre* drivers still act in adult neural stem cells as well as in astrocytes. It is thus important to carefully consider the specificity of *Cre* in designing conditional knockout studies.

Certain aspects of brain development can be studied much more readily and economically using nonmammalian model animals, such as *C. elegans*, *Drosophila*, and zebrafish. Axon pathfinding is one of such developmental events, in which nonmammalian model animals have a great advantage over mice. On the other hand, mice are indispensible to study physiological and behavioral aspects of the neural function. Another advantage of the mouse system is that the dissection of gene function in different cell types can be done by using multiple *Cre* systems. This capability is especially useful in studies of HS and other glycosaminoglycans, because they tend to be expressed broadly in different cell types within a single tissue. For example, the use of different *Cre* systems enables the dissection of cell autonomous and nonautonomous roles of HS in axon guidance. We used *Wnt1-Cre* to ablate *Ext1* only in the dorsal part of the developing spinal cord to demonstrate cell autonomous role of HS in the pathfinding of spinal cord commissural axons[8] (see Section V).

IV. *Nestin-Cre*-Mediated *Ext1* Knockout Mice

Nestin is an intermediate filament protein expressed in neural stem cells. Although it is also expressed in some nonneural cell types, *Cre* transgenes utilizing the nestin promoter are one of the most reliable and widely used *Cre* to ablate genes in the brain. *Nestin-Cre* has been shown to delete loxP-modified genes in neural stem cells and their progeny as early as E9.5.[20] As it acts on neural stem cells, a targeted gene will be ablated in all three lineages originated from neural stem cells, namely neurons, astrocytes, and oligodendrocytes. Microglia, which are mononuclear phagocytes derived from bone marrow precursor cells, are not affected by *Nestin-Cre*. Thus *Nestin-Cre* is useful to conduct "pan-neural" conditional knockout studies. *Nestin-Cre*-mediated *Ext1* knockout has demonstrated that HS plays multiple functional roles in brain development.[7] Three aspects of brain development are prominently disrupted in these mice, namely the patterning and regionalization of the brain, cortical neurogenesis, and axon guidance.

A. Role of Heparan Sulfate in Brain Patterning

Homozygous *Nestin-Cre;Ext1* conditional knockout mice (i.e., *Nestin-Cre; Ext1$^{flox/flox}$*; called *Nestin-Cre;Ext1* cKO mice, thereafter) invariably die at birth. This is likely due to respiratory failure, considering severe brain defects involving the hindbrain (see below). Despite the grossly normal appearance, *Nestin-Cre;Ext1* cKO mice exhibit multiple brain patterning defects (Fig. 1). Most notably, these mice lack a discrete cerebellum.[7] Separation of the caudal midbrain and the cerebellum is actually initiated, but fails to complete. As a result, *Nestin-Cre;Ext1* cKO mice show a truncation in the region. However, this is not due to the total agenesis of the cerebellum. Analysis with a marker for Purkinje cells (calbindin) shows a disorganized collection of calbindin-positive cells in the truncated end of the midbrain, indicating that Purkinje cells are generated.[7] Thus, the specification of the cerebellum does occur in these mice but the subsequent developmental processes necessary for complete separation of the midbrain and the cerebellum do not progress in the absence of HS.

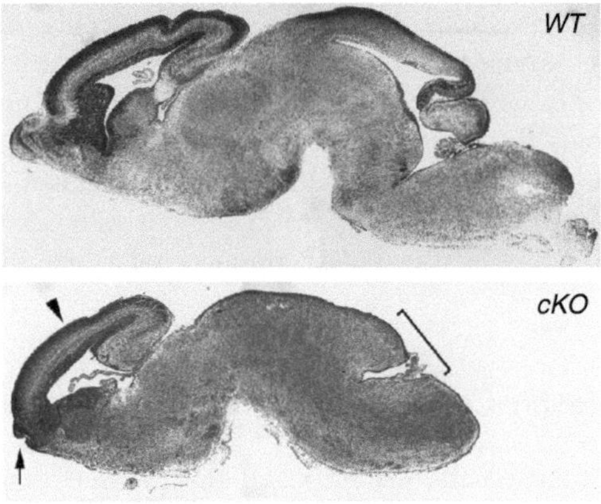

FIG. 1. Gross brain phenotype of *Nestin-Cre;Ext1* cKO mice. Thionin-stained sagittal sections of the entire brain of wild-type (*WT*) and *Nestin-Cre;Ext1* cKO (*cKO*) mice at E18.5. Note the lack of discrete cerebellum and the truncation of the midbrain in the midbrain–hindbrain region (*bracket*), small cerebral cortex (*arrowhead*), and the absence of the olfactory bulbs (*arrow*).

This phenotype of the *Nestin-Cre;Ext1* cKO mouse is similar to the phenotypes of hypomorphic *Fgf8*,[21] natural *Wnt1* (*swaying*),[22] and *Lrp6* null[23] mutants. FGF8 is a key mediator of the isthmic organizer, which induces the expression of *Wnt1*, *Engrailed 1* (*En1*), *Engrailed 2* (*En2*), and *Fgf8* itself during midbrain–hindbrain patterning.[24] LRP6, a member of the low-density lipoprotein receptor family, mediates Wnt signaling as a coreceptor for Wnt proteins.[25] FGFs, including FGF8, and Wnts bind to and are functionally modulated by HS/heparin.[26,27] It is thus speculated that the lack of HS perturbs normal functioning of these morphogens in the midbrain–hindbrain boundary region ("isthmus"), causing the defective separation of the midbrain and cerebellum.

In the isthmus of E9.5 wild-type embryos, expression of *Wnt1*, *En1*, and *En2* are detected as an intense band by *in situ* hybridization.[28] In *Nestin-Cre; Ext1* cKO mice, expression domains of *Wnt1*, *En1*, and *En2* in the isthmus are expanded and diffuse, whereas expression domain of *Fgf8* is unaltered.[7] Together with the cascade of the developmental gene expression mentioned above, these results are consistent with the hypothesis that the lack of HS affects the pathway downstream of *Fgf8*. The expanded pattern of expression of downstream genes suggests that the distribution of FGF8 protein in the midbrain–hindbrain boundary is abnormal in *Nestin-Cre;Ext1* cKO mice.

HS is thought to modulate various morphogens in several different, not mutually exclusive, modes of action. HS can act as an obligatory coreceptor for a morphogen. In this case, the presence of HS on the surface of signal-receiving cells is essential for the cells to transduce the signal of the morphogen. On the other hand, HS can modulate the diffusion of a morphogen in developing tissues. The interaction with HS may also protect morphogens from degradation. It is not easy to definitively conclude which mode of action applies to the case of midbrain–hindbrain patterning. Nevertheless, it is likely that HS in the isthmus acts in the second mode of action mentioned above. The expanded expression domains of downstream genes suggest that FGF8 has acted in wider areas, presumably due to more diffuse distribution in the region. At the same time, the fact that HS-deficient cells in the region surrounding the isthmus can induce FGF8-dependent downstream genes suggests that HS is not essential for these cells to transduce FGF8 signals, indicating that HS does not act as an obligatory coreceptor. There is more direct evidence supporting this notion. Immunostaining of FGF8 proteins in the isthmus revealed diffuse, weakened immunoreactivity of FGF8.[7] Thus HS probably acts to control the distribution of FGF8, which is secreted by cells of the isthmus organizer, in the region, so that appropriate levels of FGF8 signal are provided to surrounding cells. The weakened FGF8 immunoreactivity may also suggest that in the absence of HS, FGF8 protein may be more susceptible to proteolytic degradation. More recently, Chen *et al.*[29] demonstrated using a ligand and carbohydrate engagement assay (LICE)[30] combined with heparitinase treatment that FGF forms a posterior-to-anterior gradient spanning the isthmus and the midbrain, and that HS is required for the formation of this FGF gradient in the region.[29]

Another prominent patterning defect observed in *Nestin-Cre;Ext1* cKO mice is the lack of olfactory bulbs. Molecular mechanisms underlying this phenotype have not been elucidated yet. It is interesting, however, that both *Fgf8* hypomorphic mutants and brain-specific *Fgfr1* null mice show a lack of olfactory bulb formation.[21,31] Thus, it is possible that the lack of olfactory bulb formation in the absence of HS is also due to defective FGF signaling, as in the midbrain–hindbrain patterning defect.

B. Role of Heparan Sulfate in Cortical Neurogenesis

Neurogenesis is the process of generation and differentiation of neurons from neural stem cells. In the cerebral cortex, neural stem cells residing in the ventricular zone, the deepest layer of the developing cerebrum facing the ventricular lumen, proliferate to produce newly born neurons, which then migrate toward the surface of the brain. Neurogenesis occurs predominantly during embryonic stages and complete before birth in most parts of the mammalian brain. In rodents, low levels of neurogenesis continue in a few regions of the adult brain, such as the subventricular zone and the dentate gyrus. Neurogenesis is regulated by several growth factors, including FGFs. Earlier studies suggested that the functions of FGF1 and FGF2 in neurogenesis are modulated by HSPGs expressed in the ventricular zone.[32] In the developing brain, several HSPGs are highly expressed. Spatiotemporal expression patterns suggest that glypicans play important roles in neurogenesis. Glypican-4 is expressed in the ventricular zone and in neural stem cells in culture.[33] Glypican-4 expression is restricted to cells that retain stem cell properties: once cells commit neuronal differentiation, glypican-4 expression ceases. Glypican-4 expression is also found in the adult dentate gyrus,[33] where neurogenesis continues to occur during adult life. Glypican-1 (*Gpc1*) is expressed in the ventricular zone, although unlike glypican-4, it is also expressed in postmitotic and differentiated neurons during the late stages of embryonic development.[34] Glypican-2 is expressed mainly in postmitotic neurons and along axon tracts during the phase of active axon elongation.[35] Glypican-5 is also expressed in postmitotic neurons.[36] Other than glypicans, perlecan is strongly expressed in the ventricular zone, predominantly associated with basement membranes.[37]

Consistent with these pieces of circumstantial evidence indicating the role of HSPGs in neurogenesis, *Nestin-Cre;Ext1* cKO mice display microcephaly and impaired cortical neurogenesis.[7] In these mutant mice, the forebrains are abnormally small and the thickness of the cortical wall is greatly reduced (Fig. 1). Replication of neural stem cells in the ventricular zone is reduced by ~30% in *Nestin-Cre;Ext1* cKO mice compared with wild-type mice. Suggestive of the role of perturbed FGF signaling in these *in vivo* and *in vitro* phenotypes, *Ext1*-deficient neural stem cells show greatly reduced proliferative responses to FGF2 and FGF8.[7] Neither enhancement in apoptotic cell death nor abnormalities in cortical lamination patterns occurs in *Nestin-Cre;Ext1* cKO mutants. Together, it is likely that reduction in proliferation of neural stem cells is the primary cause of the microcephaly phenotype in *Nestin-Cre;Ext1* cKO mice. Defective FGF signaling in HS-deficient neural stem cells may be the key defect underlying the microcephalic phenotype. In contrast, no abnormalities in the lamination pattern of the cerebral cortex are found in *Nestin-Cre;Ext1* cKO mice, suggesting that HS is not functionally involved in the migration of newly generated neurons. The absence of lamination abnormalities in knockout mice of perlecan and *Gpc1*, (see below) seems consistent with this observation. It remains to be determined whether the initial specification and the patterning of the subregions within the cerebral cortex are affected in the absence of HS. The microcephaly of *Nestin-Cre;Ext1* cKO mice may be due to a combined effect of abnormal specification of the early telencephalon and impaired neural stem cell proliferation following the patterning of the cortex.

Reduced neurogenesis and microcephaly have also been reported in perlecan and *Gpc1* knockout mice. *Perlecan*$^{-/-}$ mice exhibit variable expressivity in terms of the brain phenotype, with ~55% embryos develop exencephaly.[38] Subpopulation (~45%) of *Perlecan*$^{-/-}$ embryos without exencephaly show hypoplastic forebrains and thin cerebral wall.[39] In the neocortical ventricular zone of these mice, the BrdU labeling index is normal at E13.5 but decreased to a 60% of that in wild-type mice at E16.5, indicating that cell proliferation is normal at the beginning of cortical neurogenesis but decreases at later stages. On the other hand, the cortex of *Perlecan*$^{-/-}$ mice shows a normal lamination pattern, suggesting normal radial migration of neurons. *Gpc1*$^{-/-}$ mice also show a reduction in the size of the brain, which appears less pronounced than in *Nestin-Cre;Ext1* cKO mice.[6] Again, no obvious changes in the lamination pattern of the cortex are observed. Only patterning defect observed in *Gpc1*$^{-/-}$ mice is a mild abnormality in the foliation pattern of the cerebellar cortex. The analysis of signaling pathways and genetic interaction experiments strongly suggest that the neurogenesis phenotype in *Gpc1*$^{-/-}$ mice results from a reduction in FGF17 signaling during the early phase of brain development.[6]

In view of the role of HS in neurogenesis in the embryonic brain, it is of great interest whether HS also plays a role in adult neurogenesis. Again, replication and differentiation of adult neural stem cells are regulated by growth factors and morphogens that are thought to be modulated by HS, such as FGFs, Sonic hedgehog (Shh), and Wnts.[40] Initial answers to this issue would be obtained from conditional *Ext1* knockout mice targeted to adult neural stem cells. Late onset *GFAP-Cre* transgene[19] may be useful for this purpose.

C. Role of Heparan Sulfate in Axon Guidance

Neuronal progenies generated from neural stem cells in the ventricular zone migrate toward specific locations in the developing brain, and extends axons and dendrites to form stereotypic neuronal connections. Physiological role of HS in axon guidance has been suggested since the 1990s by experiments using lower model animals.[41,42] Phenotypes of *Nestin-Cre;Ext1* cKO mice revealed that the role of HS in axon guidance in mice is not only physiological but also unexpectedly pervasive.

Nestin-Cre;Ext1 cKO mice show severe defects in commissural fiber tract development. Three major commissures in the forebrain, namely the corpus callosum, hippocampal commissure, and anterior commissure, are all absent in *Nestin-Cre;Ext1* cKO mice.[7] In the corpus callosum, axons from the cortex fail to approach the midline and instead extend ventrally, while axons in wild-type mice turn to the midline and cross toward the contralateral side. In the anterior commissure, axons in *Nestin-Cre;Ext1* cKO mice extend straight toward the ventral surface of the forebrain without turning toward the midline. These defects are somewhat similar to that of mutant mice in which the guidance molecule Slit1 and Slit2 are ablated.[43] Slit proteins have been shown to bind and be modulated by HS,[44] suggesting the possibility that the lack of HS disrupts Slit signaling necessary for the formation of forebrain commissures. On the other hand, perturbation of FGF signaling also causes the lack of forebrain commissures.[45–47] Thus, although HS is indeed involved in guidance signaling by the Slit/Robo system (see below), the lack of commissures in *Nestin-Cre;Ext1* cKO mice may be due to defective FGF signaling. At present, it has not been determined whether the loss of commissures in *Nestin-Cre;Ext1* mice is due to the disruption of Slit function or FGF function. In FGF mutants, the formation of midline glial structures is defective.[46,47] Thus it is not entirely clear whether the forebrain commissure phenotype in *Nestin-Cre;Ext1* cKO (as well as *Ndst1*$^{-/-}$ mice; Ref. 48) is due to defective guidance signaling or is secondary to the loss of critical midline glial structures.

The defect in retinal axon trajectory at the optic chiasm is one of the examples demonstrating that HS plays a direct role in axon guidance. In the optic chiasm, retinal axons cross the midline and project into the contralateral tectum. In *Nestin-Cre;Ext1* cKO mice, these axons project ectopically into the contralateral optic nerve.[7] This phenotype is remarkably similar to that of *Slit1*$^{-/-}$;*Slit2*$^{-/-}$ double mutant mice.[49] Slit proteins are a family of secreted repulsive guidance molecules.[50] These molecules are expressed in (but not limited to) midline structures and signal through Robo receptors on the surface of incoming growth cones. In the optic chiasm, Slit1 and Slit2 act cooperatively to guide retinal axons to contralateral sides.[49] Slit proteins bind HS. Moreover, the presence of HS promotes Slit–Robo interaction, and axon repulsion activity of Slit requires HS in explant assays.[44] The physiological role of HS in Slit-mediated axon guidance has been demonstrated by genetic interaction experiments using *Slit2* and *Ext1* null alleles.[7] Although little guidance defects were found in *Slit2*$^{-/-}$ mice (due to the intact Slit1 function), a reduction of one *Ext1* allele in *Slit2*$^{-/-}$ background causes profound retinal axon misguidance, as observed in *Slit1*$^{-/-}$;*Slit2*$^{-/-}$ double mutant mice and *Nestin;Cre;Ext1* cKO mice.[7] Such a strong dosage-sensitive genetic interaction between *Slit* and *Ext1* indicates that HS plays a physiologically essential role in Slit-mediated retinal axon guidance. It remains to be determined whether the requirement of HS in this axon guidance event is either cell autonomous or environmental. At present, no genetic experiments have been performed to determine whether HS needs to be expressed on the extending axons (cell autonomous mode of action) or in the cells forming guidance cues (environmental or cell nonautonomous mode of action) (Fig. 2). This issue can be addressed by using a conditional knockout system specifically targeted to retinal axons.

It is likely that detailed analysis of individual axon tracts would reveal more guidance defects in *Nestin-Cre;Ext1* cKO mice. Axons of the fasciculus retroflexus, a fiber tract connecting the limbic forebrain and the midbrain, have been shown to be defasciculated in *Nestin-Cre;Ext1* cKO mice.[51] The pathfinding of these axons is regulated by semaphorin 5A (Sema5A), which binds HS. Extension of sensory axons from dorsal root ganglia into the spinal cord is also disrupted in *Nestin-Cre;Ext1* cKO mice (J. Ogawa *et al.*, unpublished data).

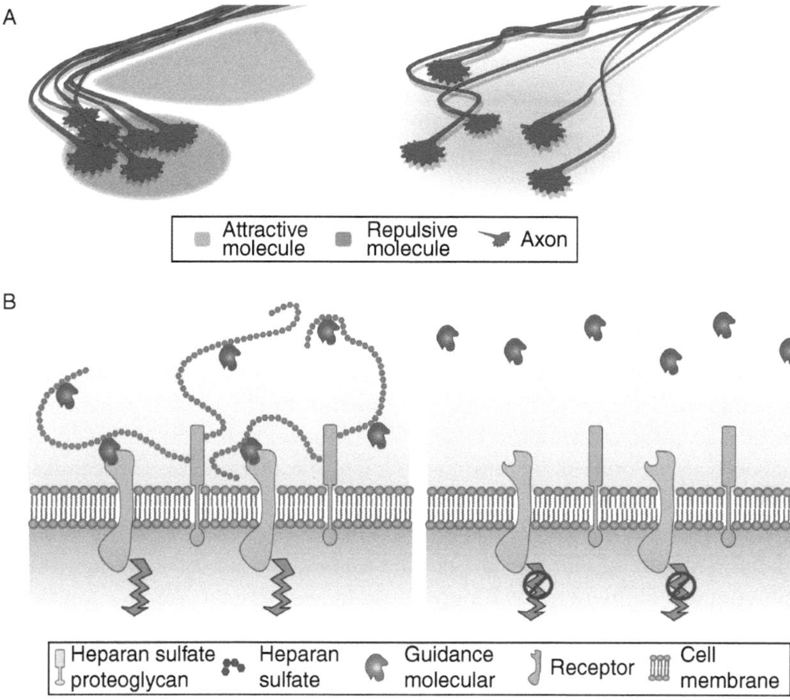

FIG. 2. Hypothetical modes of action of HS in axon guidance. (A) HS can act as an environmental (cell nonautonomous) factor that controls the distribution of guidance molecules in the critical guidepost areas. In the absence of HS, such topologically controlled distribution of guidance molecules is disrupted, resulting in loss or weakening of guidance signals and misguidance of axons (*right panel*). (B) HS can act as an obligatory coreceptor on the surface of extending growth cones. In the absence of HS, the interaction of guidance molecules to their cognate receptors is abolished or greatly reduced, resulting in loss of response to the guidance signal (*right panel*). This mode of action of HS has been genetically demonstrated in the case of netrin-1-dependent guidance of commissural axons, using a *Wnt1-Cre*-driven, dorsal spinal cord-specific conditional knockout approach.[8] These two hypothetical modes of action are not mutually exclusive and can act simultaneously in a guidance process.

As such, the role of HS in axon guidance seems to be quite pervasive. Other than the aforementioned guidance molecules for which the requirement of HS has been demonstrated *in vivo*, there are probably other molecules and guidance events that require HS. For instance, ephrin-A3 binds HS and the presence of cell surface HS is required for the full biological activity of ephrin-A3.[52] Ephrin-A3 has been shown to function in the guidance of various types of axons in the developing nervous system.[53–55] It is interesting to examine whether these guidance events controlled by ephrin-A3/Eph receptor signaling are disturbed in *Ext1* mutant mice.

V. *Wnt1-Cre*-Mediated *Ext1* Knockout Mice

As discussed above, there is now ample evidence that HS plays an essential role in various axon guidance processes. However, the mechanistic basis for these functions has not been clarified yet in most of the cases. It is important to note that HS is expressed rather ubiquitously in the developing nervous system. Although genes encoding *O*-sulfotransferases show somewhat more restricted spatiotemporal expression patterns than *Ext1* and *Ndst* genes, they have never been found to be expressed in a highly restricted manner as seen in morphogens and guidance molecules. In other words, in essentially all axon guidance events, HS is expressed both in extending axons and in cells forming the terrain where these growing axons navigate, including cells forming critical guidance cues. Thus, to understand the function of HS in axon guidance, it is important to address whether HS is the required cell autonomously as an axonal coreceptor or as an environmental factor that modulates the localization of guidance molecules (see Fig. 2). One of the methods to address this question *in vivo* is to eliminate *Ext1* in extending axons and in the terrain surrounding the axons separately. The use of *Cre* that targets specific types of neurons or specific regions of the nervous system can achieve this goal. We took advantage of regional specificity of *Wnt1-Cre* in the spinal cord (Fig. 3) to test this cell autonomous versus environmental question in the context of netrin-1-dependent commissural axon pathfinding.[8]

In the spinal cord, cell bodies of commissural neurons reside in the most dorsal part of the spinal cord. Axons from commissural neurons project ventrally toward the ventral midline, and then cross the midline and turn rostrally to project longitudinally in the ventral funiculus.[56] The initial trajectory of commissural axons toward the ventral midline is controlled primarily by the long-range chemoattractant netrin-1, which is expressed in the floor plate.[57] Netrin-1 secreted from the floor plate binds to its receptor DCC (*deleted in colorectal cancer*) expressed on the extending growth cones of commissural axons, triggering intracellular signaling that leads to attraction of the growth cone along the netrin-1 gradient.[58] Netrin-1 binds to heparin with a high affinity,[59] suggesting the physiological significance of HS in netrin-1 function.

With *Nestin-Cre*, recombination of floxed alleles occurs in the entire spinal cord.[8] The *Wnt1-Cre* transgene[60] has a different specificity (see Fig. 3). In the mouse spinal cord, *Wnt1-Cre*-mediated recombination begins at E9.5 and occurs selectively in the most dorsal part of the spinal cord, where cell bodies of commissural neurons are located.[61] In *Nestin-Cre;Ext1* cKO mice, the thickness of the ventral commissure is reduced,[8] as seen in *Netrin-1*$^{-/-}$ and *Dcc*$^{-/-}$ mice.[62,63] Remarkably, *Wnt1-Cre*-mediated ablation of *Ext1* also causes commissural axon path finding defects that share similarities with those of *Netrin-1*$^{-/-}$ and *Dcc*$^{-/-}$ mice.[8] As HS is not eliminated in the ventral

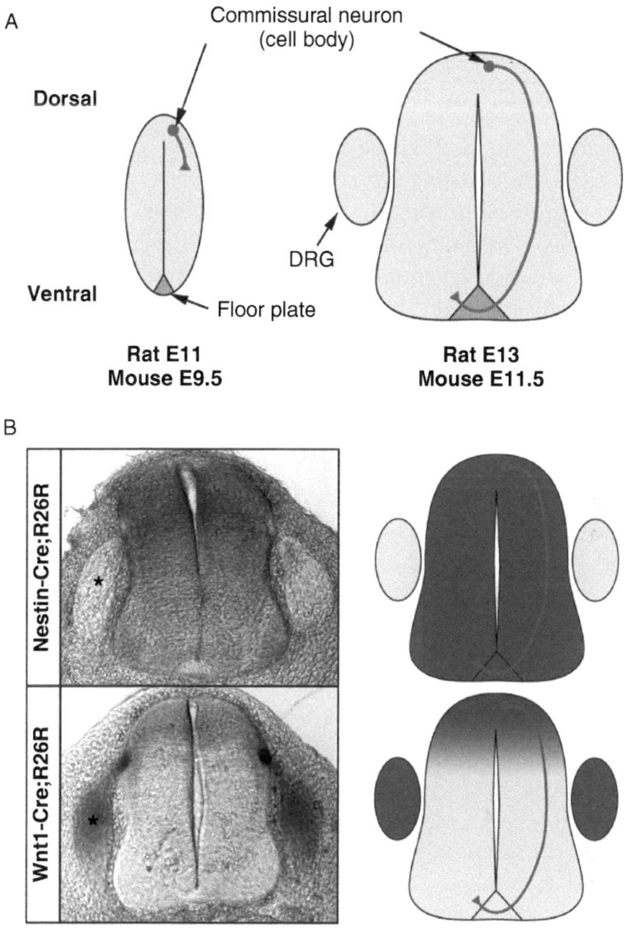

FIG. 3. *Wnt1-Cre* can be used to dissect the mode of action of HS in the guidance of spinal cord commissural axons. (A) Development of commissural axons in the spinal cord. Cell bodies of commissural neurons reside in the very dorsal part of the developing spinal cord. Commissural axons are attracted toward the ventral midline by the action of netrin-1 secreted from the floor plate. (B) Recombination patterns of *Nestin-Cre* and *Wnt1-Cre* in the spinal cord. With *Wnt1-Cre*, recombination occurs only in the very dorsal part of the spinal cord. As a result, HS expression is ablated in commissural neurons and their axons, whereas expression in the cells consisting of the ventral part of the spinal cord in which commissural axons navigate is not. Also, note the difference in the activity of *Nestin-Cre* and *Wnt1-Cre* in the dorsal root ganglia, which can potentially be useful to dissect the function of HS in the pathfinding of sensory axons from dorsal root ganglia into the spinal cord.

three-quarters of the spinal cord, this indicates that HS needs to be autonomously expressed by commissural neurons in order for their axons to respond to netrin-1 chemoattractive signals. Consistent with this conclusion, various *in vitro* experiments show that cells lacking cell surface HS cannot transduce netrin-1 signals via the MAP kinase pathway.[8]

These observations strongly suggest that in netrin-1/DCC-mediated axon guidance, HS expressed in commissural axons acts as an obligatory coreceptor. This notion seems to resonate with the earlier suggestion that there should be a coreceptor for netrin-1.[64] It remains to be determined how HS is involved in netrin-1/DCC signaling at the cell surface. It is interesting to note that not only netrin-1 but also DCC has affinity to heparin.[65] Therefore, it is possible that netrin-1, DCC, and HS may form a ternary complex, like the case with FGF/FGF receptor signaling.[66]

Cell autonomous requirements for HS have also been shown in the Sema5A-mediated guidance of the fasciculus retroflexus axons mentioned above. In this case, elimination of HS from axons by heparitinase treatment or by chemical inhibition of HS synthesis greatly diminishes Sema5A-mediated attractive effects *in vitro*.[51] Although there is no genetic evidence for cell autonomous requirement for HS in this case, these *in vitro* data strongly suggest the importance for cell surface HS in the transduction of Sema5A signals.

VI. Phenotypic Comparison with Mutant Mice of HS-Modifying Enzymes

Phenotypic comparison between *Ext1* mutants and mutants of HS modifying enzymes that catalyze steps downstream of *Ext1* can provide genetic evidence for the role of fine modification of HS *in vivo*. One problem with this approach is that for the steps in which multiple enzyme isoforms are present, the absence of phenotype in a given knockout mouse model may be due to the compensatory effect by other isoforms. This is a potential issue with NDST, 3-*O*-, and 6-*O*-sulfotransferase genes. (This is less of an issue with NDSTs, because *Ndst1* is the predominant gene among the four *Ndst* genes in many tissues.) On the other hand, C5-epimerase and 2-*O*-sulfotransferase are encoded by single genes (*Glce* and *Hs2st*, respectively), which makes comparison of phenotypes more straightforward than NDSTs, 3-*O*-, and 6-*O*-sulfotransferases. Another potential problem is the difference in the nature of gene inactivation between conventional and conditional knockout systems. Cells in conventional knockout mice have been adapted to the null environment from the beginning of embryogenesis, whereas cells in conditional knockout mice may represent an acute phase of adaptation. Thus, one should be cautious in comparing phenotypes of conventional and conditional knockout mice. Keeping in mind these limitations, phenotypic comparison between different HS mutant mice could still be very informative to understand *in vivo* significance of HS fine structure.

Brain phenotype has been fairly well characterized in *Ndst1*$^{-/-}$ mice.[48] In comparison with *Nestin-Cre;Ext1* cKO mice, penetrance of phenotypes in *Ndst1*$^{-/-}$ mice is lower and their expressivity variable. Also, craniofacial defects due to systemic ablation of the gene complicate the interpretation of brain phenotypes. Yet it is interesting to note that the brain phenotype of strongly affected *Ndst1*$^{-/-}$ mice substantially overlaps with that of *Nestin-Cre;Ext1* cKO mice. These *Ndst1*$^{-/-}$ mice display the agenesis of olfactory bulbs, the lack of the hippocampal and anterior commissures, and microcephaly.[48] These results support the notion that HSPGs with unsulfated HS chains are essentially nonfunctional, equivalent to the same core protein with no HS chains (only with linkage tetrasaccharides).

Comparison of brain phenotypes with *Glce*$^{-/-}$ mice should provide insight into the physiological function of epimerization and *O*-sulfation of HS chains. Thus far, only cursory characterization has been performed on the brain phenotype of *Glce*$^{-/-}$ mice. Still it is interesting to note that no gross abnormalities in brain morphology have been reported in these mice.[67] Since C5-epimerase is encoded by a single gene (i.e., *Glce*), this lack of significant phenotype is not due to the compensating function by an isoform. The lack of gross brain phenotype does not rule out minor and/or quantitative abnormalities. Indeed, the size of forebrain commissures is moderately reduced in *Glce*$^{-/-}$ mice (J. Ogawa *et al.*, unpublished data). Nevertheless, stark difference in the severity of brain phenotypes between *Glce*$^{-/-}$ and *Ndst*$^{-/-}$ mice suggests that *N*-sulfated HS can functionally substitute fully modified HS in most of the gross neural developmental processes.

Another interesting observation is the brain phenotype of 2-*O*-sulfotransferase (*Hs2st*) and 6-*O*-sulfotransferase 1 (*Hs6st1*) knockout mice.[68] Like *Glce*$^{-/-}$ mice, the gross brain structure of these mutant mice is normal. Yet detailed analysis of axon tracts revealed subtle but significant abnormalities in retinal axon guidance at the optic chiasm.[68] Thus for retinal axons to correctly navigate, HS chains modified by *O*-sulfation seems to be required. The mode of action of HS, whether it acts cell autonomously or environmentally, has not been defined in these cases, because *Hs2st* and *Hs6st1* are expressed both at the chiasm and by retinal ganglion cells.[68] Again, this issue can be addressed by retina-specific conditional knockout, when conditional alleles for these genes become available. In addition to the retinal axon guidance defect, proliferation of cortical precursor cells is reduced in *Hs2st*$^{-/-}$ mice, suggesting that fine modification of HS is necessary for robust replication of cortical neural stem cells.[69]

VII. Concluding Remarks

In this chapter, we have summarized the current knowledge regarding the role of HS in brain development, focusing mainly on the phenotypes of *Nestin-Cre-* and *Wnt1-Cre*-based *Ext1* conditional knockout mice. The observations made on these knockout mice confirmed the long-term speculation that HS plays important roles in brain development. It turned out that the extent of the role of HS in brain development is much greater and diverse than originally speculated. Perhaps, the most unexpected finding is the very pervasive effect of HS on axon guidance. In the four major families of axon guidance molecules, namely netrins, Slit proteins, semaphorins, and ephrins, at least one member of each family has been shown to bind and be functionally modulated by HS. Undoubtedly, further and more detailed analysis should reveal additional guidance errors in *Ext1* and other HS mutant mice.

On the other hand, it is surprising that there are no obvious phenotypes in these mutant mice that suggest abnormal cell adhesion or migration. The lamination pattern of the cortex is unaltered not only in *Nestin-Cre;Ext1* cKO[8] but also in *Gpc1*$^{-/-}$[6] and *Perlecan*$^{-/-}$[39] mutant mice. This is in contrast to the observation from knockout mice of integrins; in these mice, abnormal neuronal migration is one of the most prominent neural phenotypes.[70] In a similar vein, it is also interesting to note that, despite numerous guidance errors, *Ext1*-deficient neurons do not seem to have problems in extending axons. This suggests that HS does not play a critical role in the adhesion and motility of growth cones.

Overall, the phenotypes of *Ext1* and other HS mutant mice indicate that the major role of HS *in vivo* is the modulation of signaling by morphogens and guidance molecules. In what mode of action, either cell autonomously or nonautonomously, HS acts in these signaling processes seems to be dependent on the type of morphogen/guidance molecule and the developmental context. The elucidation of this issue in individual developmental events would be one of the major challenges in the field. Meanwhile, phenotypic comparison among different HS mutant mice should allow us to define the physiological requirement of specific HS fine structures. Data obtained thus far suggest that physiological requirements for fine structure are not as stringent as that inferred from biochemical binding experiments. In any event, the next several years should witness a profound improvement in our understanding in the physiological significance of HS fine structure. Finally, *Ext1* and other HS mutant mice should help us to uncover and understand the role of HS in brain physiology and pathology.

References

1. Lindahl U, Kusche-Gullberg M, Kjellen L. Regulated diversity of heparan sulfate. *J Biol Chem* 1998;**273**:24979–82.
2. Esko JD, Lindahl U. Molecular diversity of heparan sulfate. *J Clin Invest* 2001;**108**:169–73.
3. Lander AD, Selleck SB. The elusive functions of proteoglycans: in vivo veritas. *J Cell Biol* 2000;**148**:227–32.
4. Iozzo RV. Heparan sulfate proteoglycans: intricate molecules with intriguing functions. *J Clin Invest* 2001;**108**:165–7.
5. Forsberg E, Kjellen L. Heparan sulfate: lessons from knockout mice. *J Clin Invest* 2001;**108**:175–80.
6. Jen YH, Musacchio M, Lander AD. Glypican-1 controls brain size through regulation of fibroblast growth factor signaling in early neurogenesis. *Neural Dev* 2009;**4**:33.
7. Inatani M, Irie F, Plump AS, Tessier-Lavigne M, Yamaguchi Y. Mammalian brain morphogenesis and midline axon guidance require heparan sulfate. *Science* 2003;**302**:1044–6.
8. Matsumoto Y, Irie F, Inatani M, Tessier-Lavigne M, Yamaguchi Y. Netrin-1/DCC signaling in commissural axon guidance requires cell-autonomous expression of heparan sulfate. *J Neurosci* 2007;**27**:4342–50.
9. Perrimon N, Bernfield M. Specificities of heparan sulphate proteoglycans in developmental processes. *Nature* 2000;**404**:725–8.
10. Zak BM, Crawford BE, Esko JD. Hereditary multiple exostoses and heparan sulfate polymerization. *Biochim Biophys Acta* 2002;**1573**:346–55.
11. McCormick C, Duncan G, Goutsos KT, Tufaro F. The putative tumor suppressors EXT1 and EXT2 form a stable complex that accumulates in the Golgi apparatus and catalyzes the synthesis of heparan sulfate. *Proc Natl Acad Sci USA* 2000;**97**:668–73.
12. Senay C, Lind T, Muguruma K, Tone Y, Kitagawa H, Sugahara K, Lidholt K, Lindahl U, Kusche-Gullberg M. The EXT1/EXT2 tumor suppressors: catalytic activities and role in heparan sulfate biosynthesis. *EMBO Rep* 2000;**1**:282–6.
13. Duncan G, McCormick C, Tufaro F. The link between heparan sulfate and hereditary bone disease: finding a function for the EXT family of putative tumor suppressor proteins. *J Clin Invest* 2001;**108**:511–6.
14. Lin X, Wei G, Shi Z, Dryer L, Esko JD, Wells DE, Matzuk MM. Disruption of gastrulation and heparan sulfate biosynthesis in EXT1-deficient mice. *Dev Biol* 2000;**224**:299–311.
15. Tronche F, Kellendonk C, Kretz O, Gass P, Anlag K, Orban PC, Bock R, Klein R, Schutz G. Disruption of the glucocorticoid receptor gene in the nervous system results in reduced anxiety. *Nat Genet* 1999;**23**:99–103.
16. Gaveriaux-Ruff C, Kieffer BL. Conditional gene targeting in the mouse nervous system: insights into brain function and diseases. *Pharmacol Ther* 2007;**113**:619–34.
17. Casper KB, McCarthy KD. GFAP-positive progenitor cells produce neurons and oligodendrocytes throughout the CNS. *Mol Cell Neurosci* 2006;**31**:676–84.
18. Moore SA, Saito F, Chen J, Michele DE, Henry MD, Messing A, Cohn RD, Ross-Barta SE, Westra S, Williamson RA, Hoshi T, Campbell KP. Deletion of brain dystroglycan recapitulates aspects of congenital muscular dystrophy. *Nature* 2002;**418**:422–5.
19. Bajenaru ML, Zhu Y, Hedrick NM, Donahoe J, Parada LF, Gutmann DH. Astrocyte-specific inactivation of the neurofibromatosis 1 gene (NF1) is insufficient for astrocytoma formation. *Mol Cell Biol* 2002;**22**:5100–13.
20. Knoepfler PS, Cheng PF, Eisenman RN. N-myc is essential during neurogenesis for the rapid expansion of progenitor cell populations and the inhibition of neuronal differentiation. *Genes Dev* 2002;**16**:2699–712.

21. Meyers EN, Lewandoski M, Martin GR. An Fgf8 mutant allelic series generated by Cre- and Flp-mediated recombination. *Nat Genet* 1998;**18**:136–41.
22. Thomas KR, Musci TS, Neumann PE, Capecchi MR. Swaying is a mutant allele of the proto-oncogene Wnt-1. *Cell* 1991;**67**:969–76.
23. Pinson KI, Brennan J, Monkley S, Avery BJ, Skarnes WC. An LDL-receptor-related protein mediates Wnt signalling in mice. *Nature* 2000;**407**:535–8.
24. Liu A, Joyner AL. Early anterior/posterior patterning of the midbrain and cerebellum. *Annu Rev Neurosci* 2001;**24**:869–96.
25. Tamai K, Semenov M, Kato Y, Spokony R, Liu C, Katsuyama Y, Hess F, Saint-Jeannet JP, He X. LDL-receptor-related proteins in Wnt signal transduction. *Nature* 2000;**407**:530–5.
26. Loo BM, Salmivirta M. Heparin/Heparan sulfate domains in binding and signaling of fibroblast growth factor 8b. *J Biol Chem* 2002;**277**:32616–23.
27. Bradley RS, Brown AM. The proto-oncogene int-1 encodes a secreted protein associated with the extracellular matrix. *EMBO J* 1990;**9**:1569–75.
28. Liu A, Losos K, Joyner AL. FGF8 can activate Gbx2 and transform regions of the rostral mouse brain into a hindbrain fate. *Development* 1999;**126**:4827–38.
29. Chen Y, Mohammadi M, Flanagan JG. Graded levels of FGF protein span the midbrain and can instruct graded induction and repression of neural mapping labels. *Neuron* 2009;**62**:773–80.
30. Allen BL, Rapraeger AC. Spatial and temporal expression of heparan sulfate in mouse development regulates FGF and FGF receptor assembly. *J Cell Biol* 2003;**163**:637–48.
31. Hebert JM, Lin M, Partanen J, Rossant J, McConnell SK. FGF signaling through FGFR1 is required for olfactory bulb morphogenesis. *Development* 2003;**130**:1101–11.
32. Nurcombe V, Ford MD, Wildschut JA, Bartlett PF. Developmental regulation of neural response to FGF-1 and FGF-2 by heparan sulfate proteoglycan. *Science* 1993;**260**:103–6.
33. Hagihara K, Watanabe K, Chun J, Yamaguchi Y. Glypican-4 is an FGF2-binding heparan sulfate proteoglycan expressed in neural precursor cells. *Dev Dyn* 2000;**219**:353–67.
34. Litwack ED, Ivins JK, Kumbasar A, Paine-Saunders S, Stipp CS, Lander AD. Expression of the heparan sulfate proteoglycan glypican-1 in the developing rodent. *Dev Dyn* 1998;**211**:72–87.
35. Ivins JK, Litwack ED, Kumbasar A, Stipp CS, Lander AD. Cerebroglycan, a developmentally regulated cell-surface heparan sulfate proteoglycan, is expressed on developing axons and growth cones. *Dev Biol* 1997;**184**:320–32.
36. Saunders S, Paine-Saunders S, Lander AD. Expression of the cell surface proteoglycan glypican-5 is developmentally regulated in kidney, limb, and brain. *Dev Biol* 1997;**190**:78–93.
37. Ford-Perriss M, Turner K, Guimond S, Apedaile A, Haubeck HD, Turnbull J, Murphy M. Localisation of specific heparan sulfate proteoglycans during the proliferative phase of brain development. *Dev Dyn* 2003;**227**:170–84.
38. Costell M, Gustafsson E, Aszodi A, Morgelin M, Bloch W, Hunziker E, Addicks K, Timpl R, Fassler R. Perlecan maintains the integrity of cartilage and some basement membranes. *J Cell Biol* 1999;**147**:1109–22.
39. Girós A, Morante J, Gil-Sanz C, Fairen A, Costell M. Perlecan controls neurogenesis in the developing telencephalon. *BMC Dev Biol* 2007;**7**:29.
40. Suh H, Deng W, Gage FH. Signaling in adult neurogenesis. *Annu Rev Cell Dev Biol* 2009;**25**:253–75.
41. Wang L, Denburg JL. A role for proteoglycans in the guidance of a subset of pioneer axons in cultured embryos of the cockroach. *Neuron* 1992;**8**:701–14.
42. Walz A, McFarlane S, Brickman YG, Nurcombe V, Bartlett PF, Holt CE. Essential role of heparan sulfates in axon navigation and targeting in the developing visual system. *Development* 1997;**124**:2421–30.

43. Bagri A, Marin O, Plump AS, Mak J, Pleasure SJ, Rubenstein JL, Tessier-Lavigne M. Slit proteins prevent midline crossing and determine the dorsoventral position of major axonal pathways in the mammalian forebrain. *Neuron* 2002;**33**:233–48.
44. Hu H. Cell-surface heparan sulfate is involved in the repulsive guidance activities of Slit2 protein. *Nat Neurosci* 2001;**4**:695–701.
45. Shanmugalingam S, Houart C, Picker A, Reifers F, Macdonald R, Barth A, Griffin K, Brand M, Wilson SW. Ace/Fgf8 is required for forebrain commissure formation and patterning of the telencephalon. *Development* 2000;**127**:2549–61.
46. Smith KM, Ohkubo Y, Maragnoli ME, Rasin MR, Schwartz ML, Sestan N, Vaccarino FM. Midline radial glia translocation and corpus callosum formation require FGF signaling. *Nat Neurosci* 2006;**9**:787–97.
47. Tole S, Gutin G, Bhatnagar L, Remedios R, Hebert JM. Development of midline cell types and commissural axon tracts requires Fgfr1 in the cerebrum. *Dev Biol* 2006;**289**:141–51.
48. Grobe K, Inatani M, Pallerla SR, Castagnola J, Yamaguchi Y, Esko JD. Cerebral hypoplasia and craniofacial defects in mice lacking heparan sulfate Ndst1 gene function. *Development* 2005;**132**:3777–86.
49. Plump AS, Erskine L, Sabatier C, Brose K, Epstein CJ, Goodman CS, Mason CA, Tessier-Lavigne M. Slit1 and Slit2 cooperate to prevent premature midline crossing of retinal axons in the mouse visual system. *Neuron* 2002;**33**:219–32.
50. Brose K, Tessier-Lavigne M. Slit proteins: key regulators of axon guidance, axonal branching, and cell migration. *Curr Opin Neurobiol* 2000;**10**:95–102.
51. Kantor DB, Chivatakarn O, Peer KL, Oster SF, Inatani M, Hansen MJ, Flanagan JG, Yamaguchi Y, Sretavan DW, Giger RJ, Kolodkin AL. Semaphorin 5A is a bifunctional axon guidance cue regulated by heparan and chondroitin sulfate proteoglycans. *Neuron* 2004;**44**:961–75.
52. Irie F, Okuno M, Matsumoto K, Pasquale EB, Yamaguchi Y. Heparan sulfate regulates ephrin-A3/EphA receptor signaling. *Proc Natl Acad Sci USA* 2008;**105**:12307–12.
53. Kullander K, Mather NK, Diella F, Dottori M, Boyd AW, Klein R. Kinase-dependent and kinase-independent functions of EphA4 receptors in major axon tract formation in vivo. *Neuron* 2001;**29**:73–84.
54. Cutforth T, Moring L, Mendelsohn M, Nemes A, Shah NM, Kim MM, Frisen J, Axel R. Axonal ephrin-As and odorant receptors: coordinate determination of the olfactory sensory map. *Cell* 2003;**114**:311–22.
55. Cang J, Kaneko M, Yamada J, Woods G, Stryker MP, Feldheim DA. Ephrin-as guide the formation of functional maps in the visual cortex. *Neuron* 2005;**48**:577–89.
56. Bovolenta P, Dodd J. Guidance of commissural growth cones at the floor plate in embryonic rat spinal cord. *Development* 1990;**109**:435–47.
57. Kennedy TE, Serafini T, de la Torre JR, Tessier-Lavigne M. Netrins are diffusible chemotropic factors for commissural axons in the embryonic spinal cord. *Cell* 1994;**78**:425–35.
58. Keino-Masu K, Masu M, Hinck L, Leonardo ED, Chan SS, Culotti JG, Tessier-Lavigne M. Deleted in Colorectal Cancer (DCC) encodes a netrin receptor. *Cell* 1996;**87**:175–85.
59. Serafini T, Kennedy TE, Galko MJ, Mirzayan C, Jessell TM, Tessier-Lavigne M. The netrins define a family of axon outgrowth-promoting proteins homologous to *C. elegans* UNC-6. *Cell* 1994;**78**:409–24.
60. Danielian PS, Muccino D, Rowitch DH, Michael SK, McMahon AP. Modification of gene activity in mouse embryos in utero by a tamoxifen-inducible form of Cre recombinase. *Curr Biol* 1998;**8**:1323–6.
61. Charron F, Stein E, Jeong J, McMahon AP, Tessier-Lavigne M. The morphogen sonic hedgehog is an axonal chemoattractant that collaborates with netrin-1 in midline axon guidance. *Cell* 2003;**113**:11–23.

62. Serafini T, Colamarino SA, Leonardo ED, Wang H, Beddington R, Skarnes WC, Tessier-Lavigne M. Netrin-1 is required for commissural axon guidance in the developing vertebrate nervous system. *Cell* 1996;**87**:1001–14.
63. Fazeli A, Dickinson SL, Hermiston ML, Tighe RV, Steen RG, Small CG, Stoeckli ET, Keino-Masu K, Masu M, Rayburn H, Simons J, Bronson RT, Gordon JI, Tessier-Lavigne M, Weinberg RA. Phenotype of mice lacking functional Deleted in colorectal cancer (Dcc) gene. *Nature* 1997;**386**:796–804.
64. Stein E, Zou Y, Poo M, Tessier-Lavigne M. Binding of DCC by netrin-1 to mediate axon guidance independent of adenosine A2B receptor activation. *Science* 2001;**291**:1976–82.
65. Bennett KL, Bradshaw J, Youngman T, Rodgers J, Greenfield B, Aruffo A, Linsley PS. Deleted in colorectal carcinoma (DCC) binds heparin via its fifth fibronectin type III domain. *J Biol Chem* 1997;**272**:26940–6.
66. Ornitz DM. FGFs, heparan sulfate and FGFRs: complex interactions essential for development. *Bioessays* 2000;**22**:108–12.
67. Li JP, Gong F, Hagner-McWhirter A, Forsberg E, Abrink M, Kisilevsky R, Zhang X, Lindahl U. Targeted disruption of a murine glucuronyl C5-epimerase gene results in heparan sulfate lacking L-iduronic acid and in neonatal lethality. *J Biol Chem* 2003;**278**:28363–6.
68. Pratt T, Conway CD, Tian NM, Price DJ, Mason JO. Heparan sulphation patterns generated by specific heparan sulfotransferase enzymes direct distinct aspects of retinal axon guidance at the optic chiasm. *J Neurosci* 2006;**26**:6911–23.
69. McLaughlin D, Karlsson F, Tian N, Pratt T, Bullock SL, Wilson VA, Price DJ, Mason JO. Specific modification of heparan sulphate is required for normal cerebral cortical development. *Mech Dev* 2003;**120**:1481–8.
70. Schmid RS, Anton ES. Role of integrins in the development of the cerebral cortex. *Cereb Cortex* 2003;**13**:219–24.

Anticoagulant Heparan Sulfate: To Not Clot—Or Not?

Nicholas W. Shworak,[*,†] Takashi Kobayashi,[*] Ariane de Agostini,[‡,§] and Nicole C. Smits[*]

[*]*Department of Medicine, Dartmouth Medical School, Hanover, New Hampshire, USA*

[†]*Department of Pharmacology and Toxicology, Dartmouth Medical School, Hanover, New Hampshire, USA*

[‡]*Department of Gynaecology and Obstetrics, Laboratory of Reproductive Biology, Service of Clinical Pathology, Geneva University Hospitals and University of Geneva, Geneva, Switzerland*

[§]*Department of Genetic and Laboratory Medicine, Service of Clinical Pathology, Geneva University Hospitals and University of Geneva, Geneva, Switzerland*

Progress in Molecular Biology
and Translational Science, Vol. 93
DOI: 10.1016/S1877-1173(10)93008-1

Vascular endothelial cells (ECs) produce anticoagulant heparan sulfate (HS^{AT+})—a small subpopulation of heparan sulfate (HS) containing a specific pentasaccharide motif with high affinity for plasma antithrombin (AT). This pentasaccharide is responsible for the anticoagulant action of therapeutic heparin, which dramatically catalyzes AT neutralization of coagulation proteases. Consequently, HS^{AT+} has been designated as "anticoagulant HS," and has long been thought to convey antithrombotic properties to the blood vessel wall. The *Hs3st1* gene encodes HS 3-*O*-sulfotransferase-1, whose rate limiting action regulates EC production of HS^{AT+}. To elucidate the biologic role of HS^{AT+}, we generated *Hs3st1*$^{-/-}$ knock-out mice that have undetectable EC HS^{AT+}. Despite long held historic expectations, hemostasis was unaffected in *Hs3st1*$^{-/-}$ mice. In light of this surprising finding, herein we evaluate historic, biochemical, kinetic, physiologic, and molecular genetic studies of AT, heparin, and HS^{AT+}. We find that a hemostatic role for HS^{AT+} cannot presently be excluded; however, HS^{AT+} may well not be essential for AT's anticoagulant function. Specifically, in the absence of glycosaminoglycans, physiologic levels of AT can neutralize coagulation proteases at a sufficiently high throughput to account for most of AT's anticoagulant function. Moreover, at the vessel wall surface, glycosaminoglycans distinct from HS^{AT+} may be the predominant catalysts of AT's anticoagulant activity. We then explore the possibility that HS^{AT+} regulates a less well known function of AT, anti-inflammatory activity. We find that *Hs3st1*$^{-/-}$ mice exhibit a strong proinflammatory phenotype that is unresponsive to AT's anti-inflammatory activity. We conclude that the predominant function of HS^{AT+} is to mediate AT's anti-inflammatory activity.

Abbreviation: 3-OST, HS 3-*O*-sulfotransferase protein; AT, antithrombin; EC, endothelial cell; hFF, human follicular fluid; HS, heparan sulfate; HS^{AT+}, anticoagulant HS; HS^{AT-}, HS lacking AT-binding sites; HS^{gD+}, HS with gD-binding sites; Hs3st1, the gene encoding 3-OST-1; HSPG, HS proteoglycan; HSPG^{AT+}, HSPG with HS^{AT+}; HSPG^{AT-}, HSPG with HS^{AT-}; iNOS, inducible nitric oxide synthase; LPS, lipopolysaccharide; PAPS, adenosine 3′-phosphate 5′-phosphosulphate; PGI_2, prostacyclin; SIRS, systemic inflammatory response syndrome; SPLAG, rich in Ser, Pro, Leu, Ala, and Gly; T·AT, thrombin·antithrombin complex; TNF-α, tumor necrosis factor-α; type II_{HBS}, an AT mutant unable to bind HS^{AT+}

I. Introduction

Anticoagulant heparan sulfate (HS^{AT+}) is a subform of heparan sulfate (HS) that contains a specific pentasaccharide motif (Fig. 1) that has high affinity for antithrombin (AT), which is the most important natural anticoagulant protein found in plasma. The HS^{AT+} pentasaccharide motif was first

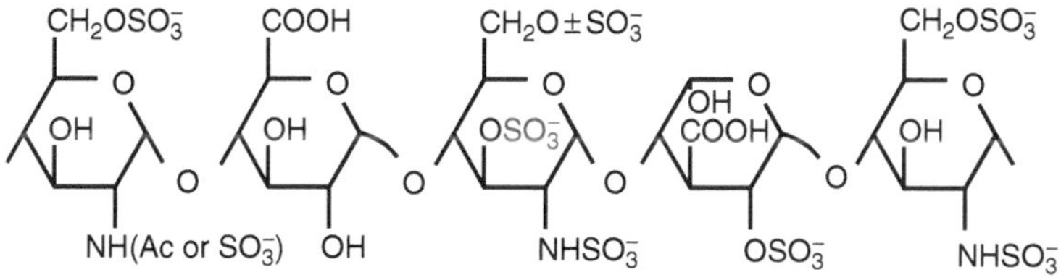

FIG. 1. The antithrombin-binding site of heparan sulfate and heparin. Although several moieties make contact with antithrombin, the central 3-*O*-sulfate (in grey) is absolutely essential for binding and catalysis.

identified as the critical structure that accounts for the anticoagulant activity of pharmaceutical heparin. Specifically, the binding of AT to heparin dramatically catalyzes the ability of AT to irreversibly neutralize thrombin, factor Xa, and other activated proteases of the coagulation cascade. Of all HS·protein complexes, structure–function relationships are best understood for the pentasaccharide·AT interaction. Despite intimate biochemical knowledge, the biological role of HS^{AT+} is unclear. HS^{AT+} is produced by endothelial cells (ECs), which line the inner surface of blood vessels. *In vitro*, this EC glycosaminoglycan functions like heparin and can also catalyze AT neutralization of coagulation proteases; consequently, HS^{AT+} has been designated as anticoagulant HS. On the basis of heparin's pharmacological mechanism of action, it has long been proposed that, *in vivo*, endothelial HS^{AT+} would activate AT and thereby regulate antithrombotic tone of the blood vessel wall. The objective of this chapter is to evaluate whether the endothelium deploys HS^{AT+} to regulate coagulation. It will be apparent that this issue has yet to be conclusively resolved. However, HS^{AT+} may participate in an alternative or additional function: the anti-inflammatory signaling activity of AT. Thus, data are emerging that the endothelium deploys HS^{AT+} as a component that mediates cell signaling.

II. Historical Perspective

Early investigations into the structure and function of HS evolved from the study of a specific HS subform—heparin. The historical timeline is dominated by almost a century of focus on this pharmaceutical anticoagulant. A review of this chronology reveals that the intense scrutiny on heparin placed heavy expectations that HS^{AT+} should function as a physiologic anticoagulant.

AT was first identified at the end of the nineteenth century as an activity that specifically inhibited thrombin.[1] In 1916, McLean found a distinct anticoagulant activity in extracts from ox heart and liver.[2] McLean's mentor, Howell, purified a potent form of this anticoagulant from dog liver, which was designated as heparin to denote its source of isolation.[3] Industrial quantities of heparin were first generated in the early 1930s by Connaught Laboratories.[4] The initial tissue source was beef liver, soon followed by beef lung; however, in the 1950s there was a shift to using pig intestinal mucus/mucosa.[5] Large scale production enabled the therapeutic deployment of heparin in the mid 1930s, and provided the material necessary to characterize its mechanism of action.[5] In 1939, Brinkhous and colleagues found that heparin's anticoagulant action required a plasma component termed "heparin cofactor".[6] Kinetic studies conducted by several groups during the 1950s, indicated that "plasma AT" and "plasma heparin cofactor" activities were intimately related.[5] Abildgaard purified AT in 1968, and demonstrated it embodied both activities.[7] Rosenberg and coworkers shortly afterward (1973) demonstrated that heparin catalyzed the inherent ability of AT to form covalent 1:1 inhibitory complexes with coagulation proteases.[8,9]

In 1965, Egeberg highlighted the importance of AT as a natural anticoagulant by identifying that its heritable deficiency was associated with thrombophilia (excessive clotting).[10] Prochownik (1983) isolated the gene and demonstrated that multiple mutations can lead to AT deficiency.[11] Today, it is appreciated that there are more than 120 distinct mutations of this gene that reduce AT levels or activity and thereby convey an increased risk of thrombosis.[12]

Solving heparin's structure was essential to ultimately elucidating heparin's molecular interactions. Howell[13] determined that heparin was a polysaccharide. Soon after, it was appreciated that heparin was a sulfated glycosaminoglycan—a linear polymer of alternating acid and amino sugars. However, heparin is a complex mixture of structures; consequently, almost 60 years of study were required to determine the key structural features that conveyed anticoagulant activity.[5] During this period, the prevailing view was that heparin's effects were mediated by nonspecific ionic interactions. However, in 1976, structural specificity was implicated when three separate investigations, from the groups of Lindahl, Rosenberg, and Sims, showed that only one-third of heparin molecules bound AT and only this population of molecules exhibited anticoagulant activity.[14–16] In the early 1980s, the groups of Casu, Choay, Lindahl, and Rosenberg demonstrated that a specific pentasaccharide motif (Fig. 1) was the active component of heparin. This motif contained a 3-*O*-sulfate group, a very rare substituent that was essential for both high-affinity AT-binding and enhancement of AT's anticoagulant activity.[5]

Heparin is not a normal constituent of blood, so the relevance of heparin's pharmacological action to endogenous physiology was initially unclear. As early as 1937, it was known that heparin was a minor body component, being present only in mast cell basophilic granules. (Subsequently, it has been shown that the liver and other sources of commercial heparin are indeed rich in mast cells.[5])

It became apparent in 1948 that the body also produces a structurally related compound with a lower sulfate content—HS.[17] Since the late 1960s, accumulating data have revealed that heparin is simply a minor form of the much more ubiquitous HS, which is produced by virtually all cell types as HS proteoglycans (HSPGs). In the 1980s, it was found that ECs produce a subpopulation of HSPGs that bear HS^{AT+}, heparin-like chains exhibiting the pentasaccharide motif that binds and enhances AT activity. On the basis of the pharmacologic mechanism of heparin, it was proposed that endothelial HS^{AT+} may be an endogenous activator of AT thereby regulating the anticoagulant tone of the blood vessel wall.[8,18] In the 1990s, Rosenberg, Shworak, and coworkers showed that EC synthesis of HS^{AT+} was regulated by addition of the very rare 3-*O*-sulfate group, which was performed by the rate-limiting action of HS 3-*O*-sulfotransferase-1 (3-OST-1).[19–22] They additionally cloned the corresponding mouse gene (*Hs3st1*) and generated *Hs3st1*$^{-/-}$ knockout mice. *Hs3st1*$^{-/-}$ mice finally provided a critical tool for evaluating the influence of endogenous HS^{AT+} on activities of AT.[23]

III. The Kinetic Mechanisms of AT's Anticoagulant Activity and Heparin Catalysis

Evaluating the *in vivo* function of HS^{AT+} first requires an understanding of how heparin affects AT's anticoagulant activity. AT is the most important natural anticoagulant found in plasma. AT is a serine protease inhibitor that can neutralize virtually all activated proteases of the coagulation cascade by forming a stable, inactivated, 1:1 AT·protease complex. In particular, AT is the major inhibitor of the terminal effectors factor Xa and thrombin. In the absence of heparin, AT is able to slowly neutralize activated clotting proteases. This heparin-independent activity is designated as the "progressive reaction." In contrast, heparin functions as a catalyst to radically increase the rate of protease neutralization. Heparin catalysis occurs by either of two mechanisms, depending on the specific coagulation protease.

For factors IXa and Xa, heparin catalysis simply requires the pentasaccharide (Fig. 1), which is present in only about one-third of heparin molecules that bind to AT with high affinity and exhibit anticoagulant activity (high-affinity heparin).[14–16] High-affinity heparin binds to AT with a K_d of 10–20 nM at pH 7.4 and $I = 0.15$.[24,25] Binding is a two-step process, involving formation of an initial complex in rapid equilibrium with a K_d of ~40 μM. Binding is followed by a conformational change with forward and reverse rate constants of ~500 and ~0.2 S^{-1}, respectively,[25] which enhances AT's affinity for the pentasaccharide by ~2000-fold. The conformational change not only enhances AT binding but also accelerates the docking of coagulation factors towards forming enzyme·AT complexes. Through this conformational change mechanism, heparin enhances AT neutralization of factors IXa and Xa by ~300-fold.[26,27]

In contrast, the pentasaccharide-induced conformational change only accelerates AT neutralization of thrombin by ~1.6-fold. Efficient neutralization of thrombin instead requires the pentasaccharide plus a minimal extension of 11 saccharides that contain a highly sulfated trisaccharide for binding to thrombin. In this mechanism, heparin functions as a bridge that facilitates thrombin docking to AT. Compared to the pentasaccharide, a bridging 16-mer enhances AT neutralization of thrombin by ~2000-fold.[28,29]

In vitro, HS^{AT+} catalyzes the neutralization of both factor Xa and thrombin, so it functions like heparin in performing both mechanisms (conformational change and bridging). Thus, HS^{AT+} has the biochemical potential to function as an endogenous anticoagulant. But is this biochemical potential actually fulfilled *in vivo*? It is important to note that ECs also produce nonanticoagulant HS (HS^{AT-}). This HS form is analogous to low-affinity heparin; both molecules lack the pentasaccharide for high-affinity AT binding. Low-affinity heparin binds AT about 1000-fold more weakly and does not induce the full conformational change of high-affinity heparin.[30] However at saturating concentrations, low-affinity heparin can still accelerate AT neutralization of factor Xa and thrombin by 140- and 60-fold respectively.[30] Thus, HS^{AT-} also has the biochemical potential to function as an endogenous anticoagulant. Clearly biologic studies are required to determine if either HS^{AT+} or HS^{AT-} functions as a natural anticoagulant.

IV. Key Features of HS^{AT+} Biosynthesis

An early step towards uncovering the biologic function of HS^{AT+} was to understand how its synthesis is regulated by ECs.

A. Identification of the Rate-Limiting Activity—3-OST-1

Vascular ECs produce two distinct populations of HSPGs. The bulk of HSPGs (90–99.5%) are $HSPG^{AT-}$, which contain HS^{AT-} chains that lack the high-affinity pentasaccharide. In contrast, only a minor subpopulation (0.5–10%) is $HSPG^{AT+}$, which bear HS^{AT+} chains that have the pentasaccharide. Although HS^{AT+} chains are extremely rare, Kojima *et al.* demonstrated that $HSPG^{AT+}$ bear almost exclusively HS^{AT+} chains (Fig. 2). Thus, biosynthesis must involve a high-fidelity mechanism, which ensures that every HS chain of $HSPG^{AT+}$ contains an AT-binding site. Moreover, this mechanism must be highly processive as HS^{AT+} chains contain multiple AT-binding sites.[31] As HS^{AT+} and HS^{AT-} chains are tightly segregated on two different HSPG subpopulations, it was initially

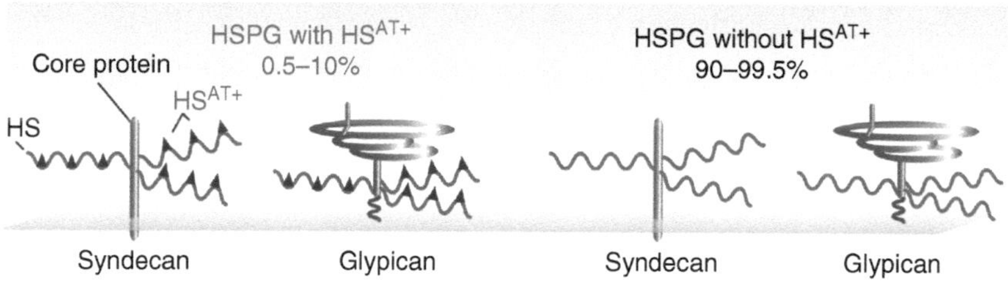

FIG. 2. Multiple core proteins are generated as both $HSPG^{AT+}$ and $HSPG^{AT-}$. Representative cell surface HSPGs. ECs use multiple core proteins to generate $HSPG^{AT+}$ (with virtually all HS^{AT+} chains) and $HSPG^{AT-}$ (with all HS^{AT-} chains).

thought that addition of HS^{AT+} was controlled by a unique core protein or a unique core protein structure. The former possibility was discounted by detecting HS^{AT+} on multiple cell surfaces and extracellular matrix proteoglycans, including syndecan-1, glypican-1, and perlecan.[31,32] Shworak *et al.* discounted the later possibility by expressing epitope-tagged syndecan-4 and demonstrating that this unique amino acid sequence was generated as both $HSPG^{AT+}$ and $HSPG^{AT-}$. Moreover, these experiments revealed that ECs have a fixed capacity to produce HS^{AT+}, which is synthesized by a unique biosynthetic pathway. Since multiple core proteins can create $HSPG^{AT+}$, it was immediately apparent that the biologic function of HS^{AT+} could not be solved by modulating core protein expression.

Instead, unique features of the biosynthetic pathway needed to be identified. Such identification was initiated by Rosenberg and colleagues, who developed an assay for the cell-free synthesis of HS^{AT+}.[21] Analysis of extracts from cells with different levels of HS^{AT+} synthesis revealed that the level of cellular HS^{AT+} is controlled by a rate-limiting activity. This activity was thought to be an HS 3-*O*-sulfotranserase (3-OST) because (1) the activity required a cofactor for sulfate donation (adenosine 3′-phosphate 5′-phosphosulphate; PAPS) and (2) HS^{AT+} only differs from its precursor by the presence of 3-*O*-sulfates.[19,21,33] The availability of this sensitive assay enabled the purification of the critical rate-limiting factor, which was indeed determined to be 3-OST-1.[20] Teleologically, regulating HS^{AT+} production by limiting 3-*O*-sulfation is an extremely elegant mechanism. Of course the AT-binding pentasaccharide contains a unique 3-*O*-sulfated glucosamine residue (Fig. 1).[34–38] This central 3-*O*-sulfate group is absolutely essential for both high-affinity AT-binding and inducing the conformational change that enhances AT anticoagulant activity. Moreover, of all HS modifications, the 3-*O*-sulfates are the rarest, typically comprising less than 0.5% of total sulfate moieties.[19,33] Thus, the rate limiting action of 3-OST-1 adds a very rare 3-*O*-sulfate group to a specific precursor structure so as to create the functional AT-binding sites within HS^{AT+}.

B. The 3-OST Multigene Family

Determining the amino acid sequence of 3-OST-1 enabled the isolation of the corresponding human and mouse cDNAs.[22] The encoded 3-OST-1 protein exhibits the unusual structure of an intraluminal resident protein. The precursor protein gains entry into the endoplasmic reticulum via a leader sequence of 20 amino acids. Signal peptidase activity removes these N-terminal residues, which produce a ~36,000 Da protein that lacks direct membrane attachment. The mature protein is comprised of a ~30 amino acid N-terminal SPLAG-domain (rich in Ser, *P*ro, *L*eu, *A*la, and *G*ly) and a 259 amino acid C-terminal sulfotransferase-domain and is modified by four N-linked glycans. It is presumed that 3-OST-1 is localized to the *trans*-Golgi compartment (where HS synthesis occurs) via binding of the SPLAG-domain to a hypothetical membrane-bound retention partner.[22,39] Most importantly, the sulfotransferase-domain encompasses determinants for (1) sequence specific recognition of appropriate HS precursors, (2) PAPS binding, and (3) catalysis.[22,39]

Bioinformatic analyses revealed that the 3-OST-1 sulfotransferase-domain exhibited high amino acid similarity to several unidentified mammalian genes. This homology approach initially enabled the isolation of cDNAs for 3-OST-2, -3A, -3B, and -4[40,41] and subsequently for 3-OST-5 and 3-OST-6 [42,43,97]. The existence of seven distinct 3-OST genes required extensive characterization to determine which genes might be responsible for endogenous HS^{AT+}. Characterization of substrate preferences and functions revealed that the large 3-OST multigene family was comprised of two major types of enzymes (Fig. 3).[40]

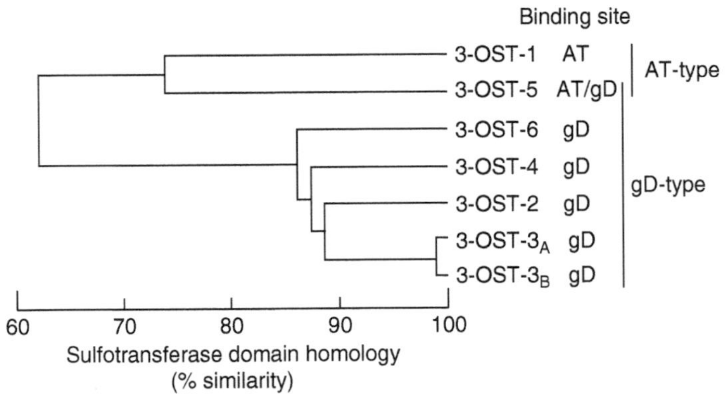

FIG. 3. The 3-OST multigene family is comprised of two structurally and functionally defined groups. The dendrogram of the human 3-OST multigene family shows the degree of sequence homology (% similarity) within the previously defined sulfotransferase domain.[39,41] Indicated are the preferred HS motifs generated by each enzyme, gD- or antithrombin (AT)-binding sites,[21,22,42–44] and the resulting structural/functional categories. Reprinted with permission from Ref. 40.

3-OST-1 preferentially generates HS^{AT+} and so it is an AT-type enzyme.[22,40,44] In contrast, 3-OST-2, -3A, -3B, -4, and -6 are designated as gD-type enzymes because they preferentially recognize a distinct precursor structure to create a different 3-*O*-sulfated motif (HS^{gD+}), known to bind glycoprotein gD of herpes simplex virus 1.[40,43,44] Cellular expression of gD-type enzymes enables cellular entry of this herpes virus.[44] Recent evidence suggests that the endogenous role of HS^{gD+} may be to bind and mediate signaling of cyclophilin B.[45] 3-OST-5 exhibits both specificities and is considered a member of both classes.[42] Considering these substrate specificities, and the fact that 3-OST-5 is not expressed in ECs (unpublished data), 3-OST-1 is likely to be the dominant isoform that regulates HS^{AT+} synthesis. Thus, to study the biologic function of HS^{AT+} we generated *Hs3st1*$^{-/-}$ mice, which lack 3-OST-1.[23] Characterization of *Hs3st1*$^{-/-}$ mice confirmed that 3-OST-1 is indeed the dominant isozyme that produces endogenous HS^{AT+} (described below).

V. Physiological Role of Endothelial HS^{AT+}: To Not Clot—Or Not?

Several factors place constraints on how HS^{AT+} could potentially function as an inhibitor of blood clotting. We shall first consider these factors, so as to provide appropriate background for evaluating experiments that use *Hs3st1*$^{-/-}$ mice to test the potential anticoagulant function of HS^{AT+}.

A. Solution Phase HS^{AT+} is An Anticoagulant

HS^{AT+} was designated as "anticoagulant HS" on the basis of the potent anticoagulant activity of heparin. Heparin is administered intravenously; consequently, its therapeutic activity involves solution phase heparin that can readily complex with AT. HS^{AT+} is known to occur in solution phase in only two situations. The first case is in the pathologic condition of hepatocellular carcinoma, which can occasionally secrete high levels of HS^{AT+} into the plasma. Such patients can exhibit fatal bleeding events,[46] which suggests that solution phase HS^{AT+} can indeed function as anticoagulant.

The second case occurs in the physiologic process of ovulation. During growth of the ovarian follicle, follicular fluid accumulates and contains numerous pro- and anticoagulant components of the coagulation cascade, including AT. Initially a fluid state is maintained, which is essential for the oocyte to be successfully delivered to the oviduct. After ovulation, clotting occurs within the ruptured follicle. Thus, mechanisms exist to prevent clotting of follicular fluid until after ovulation. Recent data indicate that this process is controlled by granulosa cells, which line the follicle and secrete select molecules into the follicular fluid.

Cultured granulosa cells are able to synthesize *bona fide* HS^{AT+}. Gonadotrophin stimulation enhances both synthesis and secretion of $HSPG^{AT+}$.[47] *In vivo*, $HSPG^{AT+}$ production by granulosa cells is also upregulated during gonadotrophin driven follicular growth and then decreases after ovulation.[48,49] These changes clearly correlate with the coagulation state of follicular fluid.

de Agostini *et al.* revealed the relevance of this correlation by characterizing HS^{AT+} purified from human follicular fluid (hFF) obtained from women undergoing oocyte pickup for *in vitro* fertilization.[50] AT-dependent clotting indices of preovulatory hFF were found to be very prolonged due to extremely high HS^{AT+} levels; antifactor Xa activity averaged 2.5 IU/ml, which is well above the therapeutic range for heparin (0.5–1 IU/ml). The relative abundance of hFF HS^{AT+} was also extremely high ($>50\%$ of total HS) and greatly exceeded that of EC HS ($<10\%$ HS^{AT+}). Purified hFF HS^{AT+} had an anticoagulant activity of ~ 167 IU/mg, which was comparable to that of unfractionated heparin (133 IU/mg) and constitutes the highest specific anticoagulant activity reported for any HS source.[50] Thus, development of the preovulatory follicle represents a physiologic scenario where HS^{AT+} occurs in the solution phase and exerts a strong anticoagulant activity.

B. Potential Anticoagulant Mechanisms for Surface Bound HS^{AT+}

The above two examples show that HS^{AT+} can function as an anticoagulant when it is present at high levels in solution; however, blood normally does not contain free HS^{AT+} nor heparin.[51] Instead, HS^{AT+} occurs on the endothelial lumenal surface and within the subendothelial matrix. The constraints of such localization suggest two potential ways in which HS^{AT+} might exert an anticoagulant effect. On one hand, the vast majority of HS^{AT+} (95%) exists ablumenally, in the subendothelial matrix, so is not in direct contact with the blood (Fig. 4).[52] However, this material would make contact with the blood after endothelial denudation; thus, such a reservoir of HS^{AT+} may limit clotting in the context of endothelial injury. On the other hand, the small amount of HS^{AT+} on the endothelial lumenal surface would potentially function in a different fashion. Lumenal HS^{AT+} is predicted to be saturated with bound AT because the plasma concentration of AT (~ 3.5 μM) exceeds the dissociation constant of AT for HS^{AT+} (~ 15 nM). Such saturation suggests that lumenal HS^{AT+} may continuously contribute to the anticoagulant tone of the endothelial surface. Cellular HS^{AT+} levels can be selectively altered without influencing the majority of HS,[53] which raises the further possibility of active modulation of HS^{AT+} in response to dynamic changes in hemostatic need.

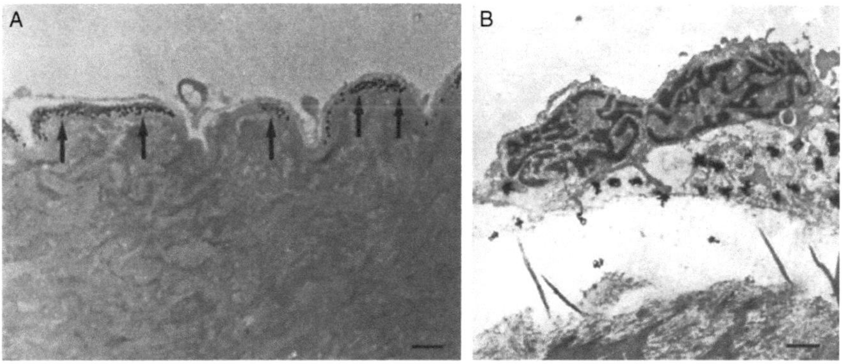

FIG. 4. Light and electron micrographs of ^{125}I-AT perfused normal rat aorta. (A) Light microscopic autoradiograph. Intense labeling is noted in the basement membrane beneath the ECs (*arrows*). *Bar*, 10 μm. (B) EM level autoradiograph. The labeling is clearly evident beneath the ECs. *Bar*, 1.0 μm. Reprinted with permission from Ref. 52.

C. Data Favoring Surface HS^{AT+} as an Anticoagulant

Three lines of evidence suggest HS^{AT+} has the potential to act as an anticoagulant. Marcum *et al.* were the first to examine this question by an elegant approach in which they perfused purified thrombin and AT through the vasculature of the rodent hind limb. Such perfusion led to elevated levels of thrombin·AT (T·AT) complexes.[54] HS^{AT+} and not endogenous heparin was implicated, as T·AT complex levels were reduced by enzymatic degradation of HS, but were normal when mast cell (heparin) deficient animals were used.[55] However, these experiments used AT concentrations well below physiological levels, so the physiological relevance is unclear. Second, in an analogous fashion, purified thrombin and AT were incubated with a cell line that does not express 3-OST-1 or with a derivative line that overexpresses the 3-OST-1 cDNA. 3-OST-1 expression drove high cellular synthesis of HS^{AT+} and elevated cell surface catalysis of T·AT complex formation.[23] Like the hindlimb approach, this experiment used subphysiologic AT levels, so physiologic conclusions cannot be drawn. In contrast, the third approach had clear physiologic relevance. Knock-in mice were generated to be homozygous for AT containing a point mutation that prevents AT from binding to HS^{AT+}. Such mice exhibit spontaneous clotting and have activated proteases of the poorly understood intrinsic pathway.[56] However, this mutant AT form has not been purified, so it is unknown if this point mutation also affects protease neutralization via the heparin-independent "progressive reaction." Although none of the above three experiments are conclusive, the results indicate that surface bound HS^{AT+} has the potential to act as an anticoagulant.

VI. Normal HS^{AT+} Levels are Not Required for Normal Hemostasis

A. Evaluation of *Hs3st1*$^{-/-}$ Mice

To examine the function of HS^{AT+} under physiologic conditions, we generated *Hs3st1*$^{-/-}$ mice that do not express 3-OST-1.[23] Even though we only eliminated one of the seven distinct 3-OST isoforms, HS isolated from tissues of *Hs3st1*$^{-/-}$ mice exhibited 75–98% reductions of anti-Xa activity compared to that of *Hs3st1*$^{+/+}$ mice (Fig. 5). Tissue proteoglycans also showed a comparable reduction in AT-binding sites.[23] Within the vasculature, endothelial HS^{AT+} was virtually undetectable, as illustrated for the aorta in Fig. 6. Combined, these results demonstrate that 3-OST-1 is the major source of HS^{AT+}, especially endothelial HS^{AT+}. Levels of total HS were comparable between genotypes, which confirms that *Hs3st1*$^{-/-}$ mice are ideal for selectively examining the function of endothelial HS^{AT+}.

If HS^{AT+} indeed contributes to hemostatic tone, then *Hs3st1*$^{-/-}$ mice should exhibit a procoagulant phenotype. We initially examined basal tissue fibrin levels, which provide an extremely sensitive index of microvascular hemostatic balance.[57–59] Unexpectedly, *Hs3st1*$^{-/-}$ mice and their wild-type

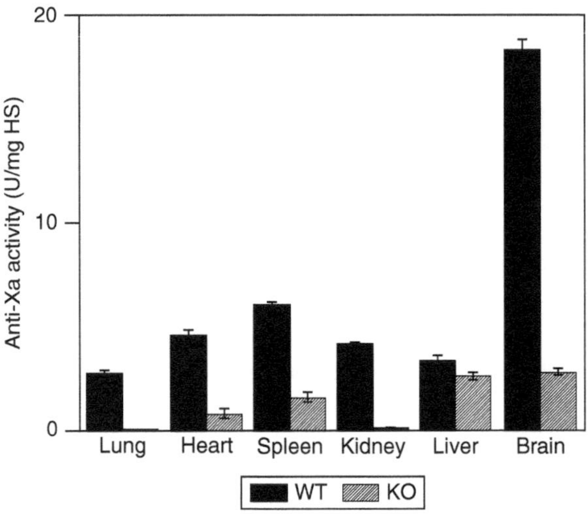

FIG. 5. *Hs3st1* disruption reduces tissue HS^{AT+} levels. HS^{AT+} content of HS purified from *Hs3st1*$^{+/+}$ (*WT*) and *Hs3st1*$^{-/-}$ (*KO*) littermates was measured as *in vitro* anti-Xa activity; $n = 3$ littermates per group. Modified from Ref. 23.

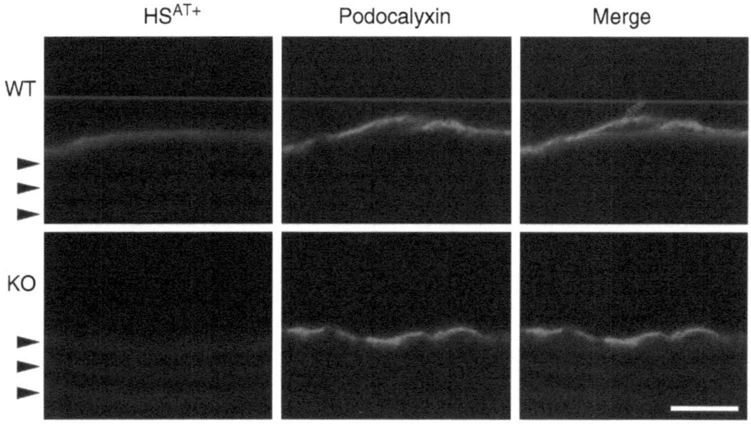

FIG. 6. Expression of HS^{AT+} on *Hs3st1*$^{-/-}$ aortic ECs is reduced to undetectable levels. Aorta cryosections (~5 μm) from *Hs3st1*$^{+/+}$ (*WT*) and *Hs3st1*$^{-/-}$ (*KO*) mice were costained with AT-647 to detect HS^{AT+} (*green*), and antipodocalyxin to detect ECs (*red*). The yellow color indicates coexpression of podocalyxin and HS^{AT+}. The *bar* indicates 10 μm. *Arrow heads*, elastic lamina; autofluorescence of elastin. (See Color Insert.)

littermates (*Hs3st1*$^{+/+}$) had indistinguishable levels of basal tissue fibrin. This was true even in *Hs3st1*$^{-/-}$ lungs and kidneys, organs that exhibited almost undetectable levels of HS^{AT+}.[23]

To test for a possible latent procoagulant state, mice were subjected to a thrombotic challenge; overnight hypoxia (8% O_2). Prolonged hypoxia enhances tissue factor expression in monocytes, macrophages, and pulmonary vascular ECs, which ultimately leads to elevated fibrin accumulation in the lung.[57–60] Despite the large reductions in lung HS^{AT+}, *Hs3st1*$^{-/-}$ mice, compared to *Hs3st1*$^{+/+}$ littermates, exhibited comparable elevations in lung fibrin. Thus, a strong procoagulant challenge failed to reveal a latent procoagulant state in *Hs3st1*$^{-/-}$ mice.[23]

As the majority of HS^{AT+} is localized in the subendothelial matrix, we tested for an anticoagulant role following endothelial denudation, which would allow direct contact of the blood with the subendothelial matrix. In response to chemical endothelial denudation, *Hs3st1*$^{-/-}$ and *Hs3st1*$^{+/+}$ mice had indistinguishable kinetics of occlusive thrombosis and showed comparable levels of T·AT complexes.[23] These observations reveal that physiologic hemostatic tone is not tightly linked to HS^{AT+} levels. This assertion was also supported by overexpression of 3-OST-1 in ECs, which elevated HS^{AT+} levels but failed to augment the preexisting high cell surface catalysis of T·AT complex formation.[23]

B. Study Limitations

The above findings demonstrate that normal levels of HS^{AT+} are not essential for normal hemostasis. However, the results cannot exclude some anticoagulant role for HS^{AT+} in the vascular system. There are a number of reasons why a procoagulant phenotype was not observed. (1) *Hs3st1*$^{-/-}$ mice still maintain some residual HS^{AT+}. Low HS^{AT+} levels might be sufficient for normal hemostatic balance and only a specific subpopulation of HS^{AT+} may be involved in hemostasis. For example, HS^{AT+} on the endothelial luminal surface is relatively minimal, but such localization might still be essential for anticoagulant activity. Moreover in *Hs3st1*$^{-/-}$ mice, residual levels of HS^{AT+} are derived from isoforms other than *Hs3st1*; therefore multiple genes may need to be eliminated to affect hemostasis. (2) The coagulation system exhibits tissue specific regulation. We may not have analyzed a particular tissue that selectively requires HS^{AT+} for hemostasis. (3) Reduced HS^{AT+} levels may trigger compensation. There are several natural anticoagulant pathways. One or more of these might be elevated in *Hs3st1*$^{-/-}$ mice, or procoagulant mechanisms might be reduced. However, it should be noted that such compensation has never been observed for other endogenous anticoagulants, where even a 50% deficiency affects hemostatic balance.[57–60]

VII. Data Against HS^{AT+} as an Anticoagulant

A. AT is Activated by Other Glycosaminoglycans

As we did not observe a procoagulant phenotype in *Hs3st1*$^{-/-}$ mice, we must also consider the possibility that HS^{AT+} does not function as an inhibitor of blood clotting. On the one hand, anticoagulant activity might result from AT engagement with glycosaminoglycans other than HS^{AT+}. Indeed, HS^{AT-} can catalyze AT neutralization of factor Xa. Although AT has a much lower affinity for HS^{AT-}, than for HS^{AT+}, substantial binding to HS^{AT-} can occur under physiologic AT concentrations.[61] Moreover, HS^{AT-} is 10- to 200-times more abundant than HS^{AT+}, so HS^{AT-} may be the predominant glycosaminoglycan that activates AT.[19] Alternatively, AT neutralization of thrombin can also be catalyzed by a specific form of thrombomodulin harboring a chondroitin sulfate chain.[62] Indeed, it appears that the EC surface must have molecules distinct from HS^{AT+} that enhance AT's anticoagulant activity because overexpression of 3-OST-1 in primary ECs elevated cell surface levels of HS^{AT+} but failed to increase AT neutralization of factor Xa.[23]

B. Physiologic Anticoagulation by AT is Largely Independent of Glycosaminoglycans

On the other hand, AT's anticoagulant activity may not even require a glycosaminoglycan cofactor. The dramatic catalysis of AT's anticoagulant activity by heparin, has led some to propose that AT's progressive (HS-independent) activity is biologically irrelevant. However, a large volume of evidence indicates that the progressive reaction is responsible for most, if not all, of AT's physiological anticoagulant effects.

Although the progressive reaction is about 1000-fold slower than heparin catalysis, physiologic concentrations of AT are about 1000-fold higher than AT levels used in biochemical assays. Both *in vitro* analyses and kinetic modeling show that plasma concentrations of AT completely compensate for the slow heparin-independent rate. Under physiologic conditions, the rate of the HS^{AT+}-free progressive reaction is sufficient to account for virtually all of AT's regulation of the coagulation cascade.[63,64] Moreover, AT's physiologic action occurs almost exclusively on platelet membranes,[65,66] which lack HS^{AT+} (unpublished data).

The lack of a need for a glycosaminoglycan cofactor is also supported by genetics. Human heterozygous mutations that reduce plasma AT activity by 50% cause increased risk for thrombosis. Under physiologic conditions, the progressive reaction occurs by pseudo first order kinetics with respect to AT. Thus, a 50% decrease in plasma AT will produce a 50% reduction in neutralization by the progressive reaction. Conversely, under physiologic conditions, endothelial HS^{AT+} is limiting and is saturated by AT. Thus, a 50% decrease in plasma AT would have no effect on neutralization catalyzed by HS^{AT+}.

Human mutations also occur that have at most a small effect on the progressive reaction but instead prevent AT binding to HS^{AT+} (type II_{HBS}). Hemostasis is normal in patients heterozygous for such mutations, which have negligible reduction in the progressive reaction. Hypercoagulability only occurs in homozygous patients and the phenotype is much less severe than seen in a 50% decrease in AT activity.[12] Similarly, embryonic death from a severe hypercoagulable state occurs in mice completely lacking AT; whereas mice homozygous for a type II_{HBS} AT form are viable (but eventually succumb to sudden onset thrombosis).[56,67] In this later case, it is unclear if thrombosis is predominantly due to the loss of AT's anticoagulant activity versus AT's anti-inflammatory activity (described below). Indeed, type II_{HBS} mice appeared as wild types for plasma T·AT complex levels and venous stasis induced thrombosis.[56] Combined kinetic and genetic data indicate that the majority, if not the entirety, of AT's anticoagulant activity occurs by the progressive reaction and is independent of HS^{AT+}.

VIII. Does HS^{AT+} Serve Alternative Functions?

A. *Hs3st1*$^{-/-}$ Mice Exhibit Unanticipated Phenotypes

That AT's anticoagulant function may be completely independent of HS^{AT+} introduces a critical question. What is the biologic function of HS^{AT+}? Although ECs are the major source of HS^{AT+}, expression is not limited to ECs. HS^{AT+} is also produced by numerous epithelial cells such as alveolar type II cells, glomerular epithelial cells, and ovarian granulosa cells that largely deposit HS^{AT+} in their basement membranes.[47,68,69] This variety of cell types suggests that HS^{AT+} may function in processes other than the coagulation cascade. This assumption is supported by further *in vivo* studies of *Hs3st1*$^{-/-}$ mice. As shown above, *Hs3st1*$^{-/-}$ mice have low levels of HS^{AT+} but lack an obvious procoagulant phenotype. However, these mice do have unanticipated phenotypes including intrauterine growth retardation and postnatal lethality when generated on an inbred strain.[53] The reasons for these consequences are unclear. Yet the existence of these phenotypes in the absence of an obvious procoagulant state suggests that HS^{AT+} regulates processes that are clearly distinct from coagulation.

B. HS^{AT+} Evolved Before AT

That HS^{AT+} controls processes distinct from coagulation is supported by evolutionary studies demonstrating that HS^{AT+} arose long before AT. AT is a member of the exceedingly large and diverse serine protease inhibitor (serpin) superfamily. Serpins are extremely ancient and are ubiquitous to all phylogenetic kingdoms.[70] However, serpins have undergone extremely rapid divergent evolution so as to fulfill highly specialized functions. Consequently, there is extremely little primary sequence homology between major phylogenic groupings; for example, serpins are very distinct between plants and animals. Yet among vertebrates, there is a common set of serpin orthologs.[71] The entire family of vertebrate serpins appears to have arisen from a common invertebrate serpin that was intracellular.[71] Thus, invertebrates do not have an ortholog of extracellular AT. Even among vertebrates, AT shows little homology to other serpins and has no paralog. AT's unique structure likely reflects its necessity to interact with many different activated coagulation proteases. Indeed, AT and the majority of components involved in the coagulation cascade appear to have coevolved with the divergence of vertebrates from cephalochordates.[72,73] Thus, a human-like coagulation cascade and AT are unique to vertebrates.

In contrast, the entire molecular machinery for generating HSPGs arose with the emergence of the earliest true multicellular animals, the Eumetazoans. The AT-binding site of heparin/HS^{AT+} arose at least as early as the invertebrates.[74] In particular, *Drosophila* exhibits a gene that is clearly orthologous to *Hs3st1*.[75] This indicates that HS^{AT+} emerged ~ 250 million years before AT/the coagulation cascade; thus, it is extremely unlikely that HS^{AT+} initially served as an anticoagulant. It is more likely that invertebrate HS^{AT+} plays a role in cell signaling, as this is a major function of ancient HSPGs. Moreover, *Drosophila* has a paralogous *Hs3st* gene that regulates the Notch signaling pathway.[75] The discrepancy in evolutionary appearance of HS^{AT+} and AT raises an important question. Did HS^{AT+} recruit AT to function in cell signaling or/and did AT recruit HS^{AT+} to regulate vertebrate coagulation?

IX. The Anti-Inflammatory Activity of AT

Although invertebrate HS^{AT+} must operate without AT, it is possible that vertebrate HS^{AT+} operates with AT, potentially controlling less appreciated functions of AT. In particular, several lines of evidence reveal that AT exhibits a potent anti-inflammatory activity that is clearly discrete from its well appreciated anticoagulant activity; that is, AT can protect against the deleterious effects of systemic inflammatory response syndromes (SIRS). This acute process can lead to multiple organ dysfunction, acute respiratory distress syndrome and shock. SIRS can result from many clinical conditions, including trauma, disseminated intravascular coagulation, severe burns, acute pancreatitis, hemorrhagic shock, and sepsis. All SIRS involve a systemic overactivation of leukocytes and ECs that leads to overwhelming inflammation and coagulation. A major initiator of SIRS is tumor necrosis factor-α (TNF-α; Fig. 7). TNF-α activates the transcription factor NF-κB and thereby induces expression of proinflammatory and procoagulant mediators such as interleukins and tissue factor, respectively. NF-κB also induces the expression of inducible nitric oxide synthase (iNOS). In turn, iNOS causes a massive increase in levels of nitric oxide that can lead to shock. Expression of TNF-α itself is NF-κB-dependent; thus positive feedback can drive the onset of SIRS (reviewed by Refs. 85,98).

A large volume of data from animal and clinical trials show that intravenous therapy with AT can reduce morbidity and/or mortality in SIRS such as sepsis, organ failure, and ischemia/reperfusion injury.[76–81] A minor contribution may derive from AT-mediated neutralization of thrombin and factor Xa, which should limit further thrombin and fibrin generation, as well as proinflammatory signaling through protease-activated receptors. However, there is substantial evidence that AT exerts its beneficial effects through direct effects on cell signaling. For example, in cultured ECs and in animals, AT stimulates synthesis of prostacyclin (PGI_2),[82,83] a potent inhibitor of leukocyte activation, leukocyte–endothelial interactions, platelet aggregation, and NF-κB activation (Fig. 7). Indeed, AT treatment of cultured ECs also abrogates TNF-α-mediated activation of NF-κB,[84] potentially through PGI_2 induction. Consequently,

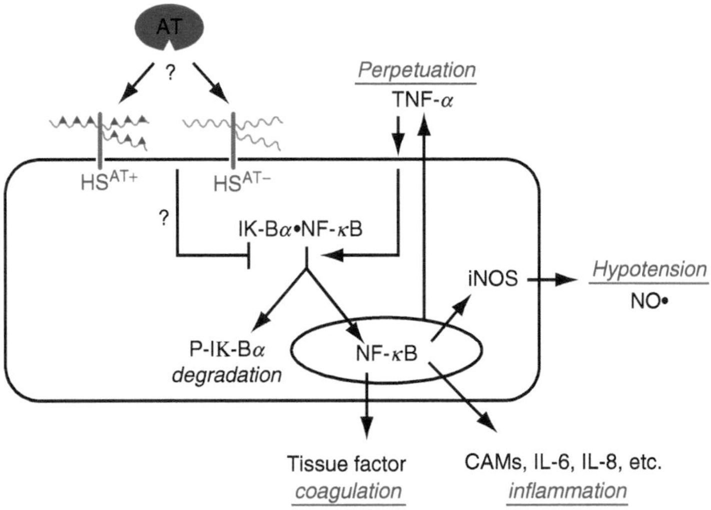

FIG. 7. Anti-inflammatory role of antithrombin. Signaling cascade described in text.

AT treatment prevents downstream expression of interleukin-6, interleukin-8, TNF-α, tissue factor, and iNOS (reviewed by Refs. 85,98). Indeed, in a rat sepsis model, AT blockade of NF-κB prevented iNOS-dependent shock.[76] AT also exerts signaling activities in leukocytes, which serve to inhibit leukocyte–endothelial interactions. These anti-inflammatory properties should not be surprising as there are many molecular interconnections between coagulation and inflammation. Moreover, it is increasingly apparent that several natural anticoagulants also have anti-inflammatory activity.[85]

In vitro and *in vivo* experiments show that the anti-inflammatory effects of AT (SIRS amelioration, PGI_2 production, and NF-κB inhibition) are clearly separate from its anticoagulant action. First, agents that exhibit only a pure anticoagulant activity do not induce these effects. Conversely, these effects require an AT concentration far in excess of that sufficient to resolve coagulation abnormalities. Finally, these effects are abolished by heparin, in striking contrast to heparin's enhancement of AT's anticoagulant activity.[76,84,88–93] However, these experiments cannot rule out that AT's anticoagulant activity at the cell surface might partially contribute to the anti-inflammatory effect.

Inhibition by heparin suggests the anti-inflammatory activity requires the binding of AT to cell surface HSPGs (Fig. 7). In support of this mechanism, the above effects are not induced by a chemically modified form of AT that lacks heparin binding but retains thrombin-neutralizing activity. Moreover, AT-induced PGI_2 synthesis by cultured ECs is prevented when cell surface HS chains are eliminated by enzymatic degradation or biosynthetic blockade.

Additionally, the β-AT glycosylation variant exhibits enhanced affinity for HS and enhanced anti-inflammatory signaling strength. Thus, it appears that AT binding to HS chains of cell surface HSPGs mediates a signaling event.[76,84,88–93]

It is well appreciated that HSPGs regulate the signaling activity of numerous HS-binding growth factors and chemokines.[86] The regulation of a given signaling pathway is conferred by distinct HS motifs that bind a particular ligand by a defined arrangement of sulfate groups along the HS chain.[86] However, it is presently unknown if AT's anti-inflammatory activity requires HS^{AT+}.

X. *Hs3st1*$^{-/-}$ Mice Show a Proinflammatory Phenotype

Evaluating the potential involvement of HS^{AT+} in AT's anti-inflammatory activity has clinical implications. The possible clinical utility of AT as a therapeutic for inflammatory conditions was called into question by the Phase III KyberSept trial of AT treatment in severe sepsis. The trial failed to achieve its primary end point of an improvement in mortality at 28 days posttherapy. In contrast, subgroup analysis revealed that AT treatment reduced 90 day mortality, but only in individuals who did not receive concomitant heparin.[81,87] This outcome should have been anticipated, as it was known that heparin negates AT's anti-inflammatory activity. Potentially, heparin acts as a competitive inhibitor that prevents AT from engaging with cell surface HSPGs that mediate AT signaling.[76,84,88–93] Thus, a clear understanding of AT's anti-inflammatory mechanism of action is required to ensure that subsequent clinical trials are designed with constraints that are ideal for detecting efficacy from AT treatment.

It is not immediately obvious that HS^{AT+} should be involved in AT's anti-inflammatory activity. AT's anti-inflammatory effects require supraphysiologic concentrations. The plasma AT concentration must be elevated from ~3.5 to ~7 μM. However, K_d of AT for HS^{AT+} is ~15 nM. Thus, HS^{AT+} should already be saturated at physiologic AT levels and site occupancy should not increase at higher AT levels. Such an argument suggests that HS^{AT+} could not function as a receptor that senses the concentration of AT. Thus, AT's anti-inflammatory effects might involve HS motifs with a lower affinity for AT. Alternatively, HS^{AT+} may not be a receptor but might function as an obligatory component for AT signaling.

If HS^{AT+} is critically required for AT's anti-inflammatory activity, then *Hs3st1*$^{-/-}$ mice (with low HS^{AT+}) should exhibit a proinflammatory phenotype. This possibility was tested by comparing the response of *Hs3st1*$^{-/-}$ mice and their wild-type littermates (*Hs3st1*$^{+/+}$) to the bacterial cell wall component, lipopolysaccharide (LPS). LPS killed ~60% of *Hs3st1*$^{-/-}$ mice compared to only ~30% of *Hs3st1*$^{+/+}$ mice. Kaplan-Meier survival analysis revealed the genotype effect was statistically significant ($P < 0.05$, manuscript in preparation). Thus, *Hs3st1*$^{-/-}$ mice exhibit a proinflammatory phenotype characterized by enhanced sensitivity to LPS. As shown above, *Hs3st1*$^{-/-}$ mice exhibited > 80% reduction in HS^{AT+} levels in most organs (Fig. 5). Moreover, HS^{AT+} on *Hs3st1*$^{-/-}$ aortic ECs is reduced to undetectable levels (Fig. 6). The proinflammatory phenotype of *Hs3st1*$^{-/-}$ mice, coupled with observations showing that AT treatment can reduce morbidity and/or mortality in SIRS,[76–80] suggests that HS^{AT+} mediates an anti-inflammatory activity of endogenous AT. To determine if HS^{AT+} mediates AT's anti-inflammatory activity, we examined the ability of exogenous AT's to ameliorate LPS-induced lethality in *Hs3st1*$^{+/+}$ and *Hs3st1*$^{-/-}$ mice. With AT pretreatment of LPS injected mice, *Hs3st1*$^{+/+}$ mice exhibit a dramatically improved survival compared to *Hs3st1*$^{-/-}$ mice (not shown). Thus, our results suggest that HS^{AT+} mediates an anti-inflammatory activity of AT. We are presently elucidating the involved molecular mechanisms.

Intriguingly, HS^{AT+} anti-inflammatory action might be disrupted in vascular inflammation. Levels of both total HS and HS^{AT+} decrease with the progression of vessel inflammation and atherosclerosis.[94] Moreover homocysteine, a proinflammatory molecule, selectively reduces HS^{AT+} levels.[95] Conversely, HS^{AT+} levels are selectively enhanced by the anti-atherogenic apolipoprotein E.[96] These alterations in HS expression are consistent with the possibility of HS^{AT+} mediating anti-inflammatory properties of AT. Furthermore, such studies suggest that HS^{AT+} might be a natural inhibitor of atherosclerosis. We are presently using *Hs3st1*$^{-/-}$ mice to explore this tantalizing prospect.

XI. Concluding Remarks

Historical biases propose that HS^{AT+} should be a critical regulator of hemostasis. Indeed, solution phase HS^{AT+} can catalyze AT neutralization of coagulation proteases. However, HS^{AT+} in the vascular system is not in solution but instead is on the vessel wall lumenal surface. Evaluation of *Hs3st1*$^{-/-}$ mice with greatly reduced vascular HS^{AT+} does not support a role for endothelial surface HS^{AT+} in hemostasis; however, such a function cannot presently be excluded. Yet other glycosaminoglycans may be more important in mediating surface catalysis of AT's anticoagulant activity. Moreover, kinetic and genetic studies indicate that the vast majority of AT's anticoagulant effects occur by the glycosaminoglycan-independent progressive reaction. The evolutionary emergence of HS^{AT+} long before AT/the coagulation cascade strongly suggests that HS^{AT+} participates in processes discrete from hemostasis. Our recent preliminary studies indicate that HS^{AT+} is actually an anti-inflammatory molecule and we are elucidating the molecular mechanisms by which endothelial HS^{AT+} mediates AT's anti-inflammatory activity. Such studies should provide inroads towards novel treatments for diseases involving vascular inflammation, potentially including atherosclerosis.

Acknowledgments

We thank members of the Heart and Vascular Research Center for their insight comments. This work was in part supported by grants to NWS from the National Institute of Aging (R01 AG023590), the National Heart, Lung and Blood Institute (R01 HL079104), and the Mitzutani Foundation (090105), and to AdA from the Swiss National Foundation (No. 3200B0-102148/1).

References

1. Contejean C. Recherches sur les injections intraveineuses de peptone et leur influence sur la coagulabilite du sang chez le chien. *Arch Psychol Norm Pathol* 1895;**7**:45–53.
2. McLean J. The thromboplastic action of cephalin. *Am J Physiol* 1916;**41**:250–7.
3. Howell WH, Holt E. Two new factors in blood coagulation—heparin and pro-antithrombin. *Am J Physiol* 1918;**47**:328–41.
4. Scott DA, Charles AF. Studies on heparin III. The purification of heparin. *J Biol Chem* 1933;**102**:431–48.
5. Röden L. Highlights in the history of heparin. In: Lane DA, Lindahl U, editors. *Heparin*. London: Edward Arnold; 1989. p. 1–24.
6. Brinkhous KM, Smith HP, Warner ED, Seegers WH. The inhibition of blood clotting: an unidentified substance which acts in conjunction with heparin to prevent the conversion of prothrombin to thrombin. *Am J Physiol* 1939;**125**:683–7.
7. Physiol AJ, Abildgaard U. Highly purified antithrombin III with heparin cofactor activity prepared by disc gel electrophoresis. *Scand J Clin Lab Invest* 1968;**21**:89–91.
8. Damus PS, Hicks M, Rosenberg RD. Anticoagulant action of heparin. *Nature* 1973;**246**:355–7.
9. Rosenberg RD, Damus PS. The purification and mechanism of action of human antithrombin-heparin cofactor. *J Biol Chem* 1973;**248**:6490–505.
10. Egeberg O. Inherited antithrombin deficiency causing thrombophilia. *Thromb Diath Haemorrh* 1965;**13**:516–30.
11. Prochownik EV, Antonarakis S, Bauer KA, Rosenberg RD, Fearon ER, Orkin SH. Molecular heterogeneity of inherited antithrombin III deficiency. *N Engl J Med* 1983;**308**:1549–52.
12. Lane DA, Bayston T, Olds RJ, Fitches AC, Cooper DN, Millar DS, et al. Antithrombin mutation database: 2nd (1997) update. For the plasma coagulation inhibitors subcommittee of the scientific and standardization committee of the international society on thrombosis and haemostasis. *Thromb Haemost* 1997;**77**:197–211.
13. Howell WH. The purification of heparin and its presence in blood. *Am J Physiol* 1925;**71**:553–62.
14. Andersson LO, Barrowcliffe TW, Holmer E, Johnson EA, Sims GE. Anticoagulant properties of heparin fractionated by affinity chromatography on matrix-bound antithrombin iii and by gel filtration. *Thromb Res* 1976;**9**:575–83.
15. Hook M, Bjork I, Hopwood J, Lindahl U. Anticoagulant activity of heparin: separation of high-activity and low-activity heparin species by affinity chromatography on immobilized antithrombin. *FEBS Lett* 1976;**66**:90–3.

16. Lam LH, Silbert JE, Rosenberg RD. The separation of active and inactive forms of heparin. *Biochem Biophys Res Commun* 1976;**69**:570–7.
17. Jorpes JE, Gardell S. On heparin monosulfuric acid. *J Biol Chem* 1948;**176**:267–76.
18. Marcum JA, Fritze L, Galli SJ, Karp G, Rosenberg RD. Microvasculature heparin-like species with anticoagulant activity. *Am J Physiol* 1983;**245**:H725–33.
19. Shworak NW, Shirakawa M, Colliec-Jouault S, Liu J, Mulligan RC, Birinyi LK, et al. Pathway-specific regulation of the synthesis of anticoagulantly active heparan sulfate. *J Biol Chem* 1994;**269**:24941–52.
20. Liu J, Shworak NW, Fritze LM, Edelberg JM, Rosenberg RD. Purification of heparan sulfate D-glucosaminyl 3-*O*-sulfotransferase. *J Biol Chem* 1996;**271**:27072–82.
21. Shworak NW, Fritze LM, Liu J, Butler LD, Rosenberg RD. Cell-free synthesis of anticoagulant heparan sulfate reveals a limiting converting activity that modifies an excess precursor pool. *J Biol Chem* 1996;**271**:27063–71.
22. Shworak NW, Liu J, Fritze LM, Schwartz JJ, Zhang L, Logeart D, et al. Molecular cloning and expression of mouse and human cDNAs encoding heparan sulfate D-glucosaminyl 3-*O*-sulfotransferase. *J Biol Chem* 1997;**272**:28008–19.
23. HajMohammadi S, Enjyoji K, Princivalle M, Christi P, Lech M, Beeler D, et al. Normal levels of anticoagulant heparan sulfate are not essential for normal hemostasis. *J Clin Invest* 2003;**111**:989–99.
24. Beeler D, Rosenberg R, Jordan R. Fractionation of low molecular weight heparin species and their interaction with antithrombin. *J Biol Chem* 1979;**254**:2902–13.
25. Olson ST, Srinivasan KR, Bjork I, Shore JD. Binding of high affinity heparin to antithrombin III. Stopped flow kinetic studies of the binding interaction. *J Biol Chem* 1981;**256**:11073–9.
26. Jordan RE, Oosta GM, Gardner WT, Rosenberg RD. The kinetics of hemostatic enzyme–antithrombin interactions in the presence of low molecular weight heparin. *J Biol Chem* 1980;**255**:10081–90.
27. Olson ST, Bjork I, Sheffer R, Craig PA, Shore JD, Choay J. Role of the antithrombin-binding pentasaccharide in heparin acceleration of antithrombin–proteinase reactions. Resolution of the antithrombin conformational change contribution to heparin rate enhancement. *J Biol Chem* 1992;**267**:12528–38.
28. Dementiev A, Petitou M, Herbert JM, Gettins PG. The ternary complex of antithrombin–anhydrothrombin–heparin reveals the basis of inhibitor specificity. *Nat Struct Mol Biol* 2004;**11**:863–7.
29. Langdown J, Johnson DJ, Baglin TP, Huntington JA. Allosteric activation of antithrombin critically depends upon hinge region extension. *J Biol Chem* 2004;**279**:47288–97.
30. Streusand VJ, Bjork I, Gettins PG, Petitou M, Olson ST. Mechanism of acceleration of antithrombin-proteinase reactions by low affinity heparin. Role of the antithrombin binding pentasaccharide in heparin rate enhancement. *J Biol Chem* 1995;**270**:9043–51.
31. Kojima T, Leone CW, Marchildon GA, Marcum JA, Rosenberg RD. Isolation and characterization of heparan sulfate proteoglycans produced by cloned rat microvascular endothelial cells. *J Biol Chem* 1992;**267**:4859–69.
32. Mertens G, Cassiman JJ, Van den Berghe H, Vermylen J, David G. Cell surface heparan sulfate proteoglycans from human vascular endothelial cells. Core protein characterization and antithrombin III binding properties. *J Biol Chem* 1992;**267**:20435–43.
33. Colliec-Jouault S, Shworak NW, Liu J, de Agostini AI, Rosenberg RD. Characterization of a cell mutant specifically defective in the synthesis of anticoagulantly active heparan sulfate. *J Biol Chem* 1994;**269**:24953–8.
34. Rosenberg RD, Lam L. Correlation between structure and function of heparin. *Proc Natl Acad Sci USA* 1979;**76**:1218–22.

35. Lindahl U, Backstrom G, Thunberg L, Leder IG. Evidence for a 3-*O*-sulfated D-glucosamine residue in the antithrombin-binding sequence of heparin. *Proc Natl Acad Sci USA* 1980;**77**:6551–5.
36. Casu B, Oreste P, Torri G, Zoppetti G, Choay J, Lormeau JC, et al. The structure of heparin oligosaccharide fragments with high anti-(factor Xa) activity containing the minimal antithrombin III-binding sequence. Chemical and 13C nuclear-magnetic-resonance studies. *Biochem J* 1981;**197**:599–609.
37. Choay J, Petitou M, Lormeau JC, Sinay P, Casu B, Gatti G. Structure–activity relationship in heparin: a synthetic pentasaccharide with high affinity for antithrombin III and eliciting high anti-factor Xa activity. *Biochem Biophys Res Commun* 1983;**116**:492–9.
38. Zhang L, Yoshida K, Liu J, Rosenberg RD. Anticoagulant heparan sulfate precursor structures in F9 embryonal carcinoma cells. *J Biol Chem* 1999;**274**:5681–91.
39. Yabe T, Shukla D, Spear PG, Rosenberg RD, Seeberger PH, Shworak NW. Portable sulphotransferase domain determines sequence specificity of heparan sulphate 3-*O*-sulphotransferases. *Biochem J* 2001;**359**:235–41.
40. Lawrence R, Yabe T, Hajmohammadi S, Rhodes J, McNeely M, Liu J, et al. The principal neuronal gD-type 3-*O*-sulfotransferases and their products in central and peripheral nervous system tissues. *Matrix Biol* 2007;**26**:442–55.
41. Shworak NW, Liu J, Petros LM, Zhang L, Kobayashi M, Copeland NG, et al. Multiple isoforms of heparan sulfate D-glucosaminyl 3-*O*-sulfotransferase. Isolation, characterization, and expression of human cdnas and identification of distinct genomic loci. *J Biol Chem* 1999;**274**:5170–84.
42. Xia G, Chen J, Tiwari V, Ju W, Li JP, Malmstrom A, et al. Heparan sulfate 3-*O*-sulfotransferase isoform 5 generates both an antithrombin-binding site and an entry receptor for herpes simplex virus, type 1. *J Biol Chem* 2002;**277**:37912–9.
43. Xu D, Tiwari V, Xia G, Clement C, Shukla D, Liu J. Characterization of heparan sulphate 3-*O*-sulphotransferase isoform 6 and its role in assisting the entry of herpes simplex virus type 1. *Biochem J* 2005;**385**:451–9.
44. Shukla D, Liu J, Blaiklock P, Shworak NW, Bai X, Esko JD, et al. A novel role for 3-*O*-sulfated heparan sulfate in herpes simplex virus 1 entry. *Cell* 1999;**99**:13–22.
45. Vanpouille C, Deligny A, Delehedde M, Denys A, Melchior A, Lienard X, et al. The heparin/heparan sulfate sequence that interacts with cyclophilin B contains a 3-*O*-sulfated N-unsubstituted glucosamine residue. *J Biol Chem* 2007;**282**:24416–29.
46. Wages DS, Staprans I, Hambleton J, Bass NM, Corash L. Structural characterization and functional effects of a circulating heparan sulfate in a patient with hepatocellular carcinoma. *Am J Hematol* 1998;**58**:285–92.
47. Hosseini G, Liu J, de Agostini AI. Characterization and hormonal modulation of anticoagulant heparan sulfate proteoglycans synthesized by rat ovarian granulosa cells. *J Biol Chem* 1996;**271**:22090–9.
48. Hasan S, Hosseini G, Princivalle M, Dong JC, Birsan D, Cagide C, et al. Coordinate expression of anticoagulant heparan sulfate proteoglycans and serine protease inhibitors in the rat ovary: a potent system of proteolysis control. *Biol Reprod* 2002;**66**:144–58.
49. Princivalle M, Hasan S, Hosseini G, de Agostini AI. Anticoagulant heparan sulfate proteoglycans expression in the rat ovary peaks in preovulatory granulosa cells. *Glycobiology* 2001;**11**:183–94.
50. de Agostini AI, Dong JC, de Vantery Arrighi C, Ramus MA, Dentand-Quadri I, Thalmann S, et al. Human follicular fluid heparan sulfate contains abundant 3-*O*-sulfated chains with anticoagulant activity. *J Biol Chem* 2008;**283**:28115–24.
51. Xiao H, Miller SJ, Bang NU, Faulk WP. Protein-bound heparin/heparan sulfates in human adult and umbilical cord plasma. *Haemostasis* 1999;**29**:237–46.

52. de Agostini AI, Watkins SC, Slayter HS, Youssoufian H, Rosenberg RD. Localization of anticoagulantly active heparan sulfate proteoglycans in vascular endothelium: antithrombin binding on cultured endothelial cells and perfused rat aorta. *J Cell Biol* 1990;**111**:1293–304.
53. Shworak NW, HajMohammadi S, de Agostini AI, Rosenberg RD. Mice deficient in heparan sulfate 3-*O*-sulfotransferase-1: normal hemostasis with unexpected perinatal phenotypes. *Glycoconj J* 2002;**19**:355–61.
54. Marcum JA, McKenney JB, Rosenberg RD. Acceleration of thrombin–antithrombin complex formation in rat hindquarters via heparinlike molecules bound to the endothelium. *J Clin Invest* 1984;**74**:341–50.
55. Marcum JA, McKenney JB, Galli SJ, Jackman RW, Rosenberg RD. Anticoagulantly active heparin-like molecules from mast cell-deficient mice. *Am J Physiol* 1986;**250**:H879–88.
56. Dewerchin M, Herault JP, Wallays G, Petitou M, Schaeffer P, Millet L, et al. Life-threatening thrombosis in mice with targeted Arg48-to-Cys mutation of the heparin-binding domain of antithrombin. *Circ Res* 2003;**93**:1120–6.
57. Labarrere CA. Relationship of fibrin deposition in microvasculature to outcomes in cardiac transplantation. *Curr Opin Cardiol* 1999;**14**:133–9.
58. Christie PD, Edelberg JM, Picard MH, Foulkes AS, Mamuya W, Weiler-Guettler H, et al. A murine model of myocardial microvascular thrombosis. *J Clin Invest* 1999;**104**:533–9.
59. Weiler-Guettler H, Christie PD, Beeler DL, Healy AM, Hancock WW, Rayburn H, et al. A targeted point mutation in thrombomodulin generates viable mice with a prethrombotic state. *J Clin Invest* 1998;**101**:1983–91.
60. Lawson CA, Yan SD, Yan SF, Liao H, Zhou YS, Sobel J, et al. Monocytes and tissue factor promote thrombosis in a murine model of oxygen deprivation. *J Clin Invest* 1997;**99**:1729–38.
61. Scully MF, Ellis V, Kakkar VV. Heparan sulphate with no affinity for antithrombin III and the control of haemostasis. *FEBS Lett* 1988;**241**:11–4.
62. Bourin MC, Lundgren-Akerlund E, Lindahl U. Isolation and characterization of the glycosaminoglycan component of rabbit thrombomodulin proteoglycan. *J Biol Chem* 1990;**265**:15424–31.
63. Van't Veer C, Mann KG. Regulation of tissue factor initiated thrombin generation by the stoichiometric inhibitors tissue factor pathway inhibitor, antithrombin-III, and heparin cofactor-II. *J Biol Chem* 1997;**272**:4367–77.
64. Hockin MF, Jones KC, Everse SJ, Mann KG. A model for the stoichiometric regulation of blood coagulation. *J Biol Chem* 2002;**277**:18322–33.
65. Mann KG. Thrombin formation. *Chest* 2003;**124**:4S–10S.
66. Monroe DM, Roberts HR, Hoffman M. Platelet procoagulant complex assembly in a tissue factor-initiated system. *Br J Haematol* 1994;**88**:364–71.
67. Ishiguro K, Kojima T, Kadomatsu K, Nakayama Y, Takagi A, Suzuki M, et al. Complete antithrombin deficiency in mice results in embryonic lethality. *J Clin Invest* 2000;**106**:873–8.
68. Girardin EP, Hajmohammadi S, Birmele B, Helisch A, Shworak NW, de Agostini AI. Synthesis of anticoagulantly active heparan sulfate proteoglycans by glomerular epithelial cells involves multiple 3-*O*-sulfotransferase isoforms and a limiting precursor pool. *J Biol Chem* 2005;**280**:38059–70.
69. Li ZY, Hirayoshi K, Suzuki Y. Expression of *N*-deacetylase/sulfotransferase and 3-*O*-sulfotransferase in rat alveolar type II cells. *Am J Physiol Lung Cell Mol Physiol* 2000;**279**:L292–301.
70. Irving JA, Steenbakkers PJ, Lesk AM, Op den Camp HJ, Pike RN, Whisstock JC. Serpins in prokaryotes. *Mol Biol Evol* 2002;**19**:1881–90.
71. Irving JA, Pike RN, Lesk AM, Whisstock JC. Phylogeny of the serpin superfamily: implications of patterns of amino acid conservation for structure and function. *Genome Res* 2000;**10**:1845–64.
72. Jiang Y, Doolittle RF. The evolution of vertebrate blood coagulation as viewed from a comparison of puffer fish and sea squirt genomes. *Proc Natl Acad Sci USA* 2003;**100**:7527–32.

73. Shworak NW. Antithrombin. In: Aird WC, editor. *Endothelial medicine*. New York: Cambridge University Press; 2007. p. 960–72.
74. Medeiros GF, Mendes A, Castro RA, Bau EC, Nader HB, Dietrich CP. Distribution of sulfated glycosaminoglycans in the animal kingdom: widespread occurrence of heparin-like compounds in invertebrates. *Biochim Biophys Acta* 2000;**1475**:287–94.
75. Kamimura K, Rhodes JM, Ueda R, McNeely M, Shukla D, Kimata K, et al. Regulation of Notch signaling by *Drosophila* heparan sulfate 3-*O* sulfotransferase. *J Cell Biol* 2004;**166**:1069–79.
76. Isobe H, Okajima K, Uchiba M, Harada N, Okabe H. Antithrombin prevents endotoxin-induced hypotension by inhibiting the induction of nitric oxide synthase in rats. *Blood* 2002;**99**:1638–45.
77. Dickneite G, Kroez M. Treatment of porcine sepsis with high-dose antithrombin III reduces tissue edema and effusion but does not increase risk for bleeding. *Blood Coagul Fibrinolysis* 2001;**12**:459–67.
78. Minnema MC, Chang AC, Jansen PM, Lubbers YT, Pratt BM, Whittaker BG, et al. Recombinant human antithrombin III improves survival and attenuates inflammatory responses in baboons lethally challenged with *Escherichia coli*. *Blood* 2000;**95**:1117–23.
79. Okajima K. Antithrombin prevents endotoxin-induced pulmonary vascular injury by inhibiting leukocyte activation. *Blood Coagul Fibrinolysis* 1998;**9**(Suppl. 2):S25–37.
80. Uchiba M, Okajima K. Antithrombin III (AT III) prevents LPS-induced pulmonary vascular injury: novel biological activity of AT III. *Semin Thromb Hemost* 1997;**23**:583–90.
81. Wiedermann CJ, Hoffmann JN, Juers M, Ostermann H, Kienast J, Briegel J, et al. High-dose antithrombin III in the treatment of severe sepsis in patients with a high risk of death: efficacy and safety. *Crit Care Med* 2006;**34**:285–92.
82. Yamauchi T, Umeda F, Inoguchi T, Nawata H. Antithrombin III stimulates prostacyclin production by cultured aortic endothelial cells. *Biochem Biophys Res Commun* 1989;**163**:1404–11.
83. Uchiba M, Okajima K, Murakami K, Okabe H, Takatsuki K. Effects of antithrombin III (AT III) and Trp49-modified AT III on plasma level of 6-keto-PGF1 alpha in rats. *Thromb Res* 1995;**80**:201–8.
84. Oelschlager C, Romisch J, Staubitz A, Stauss H, Leithauser B, Tillmanns H, et al. Antithrombin III inhibits nuclear factor kappaB activation in human monocytes and vascular endothelial cells. *Blood* 2002;**99**:4015–20.
85. Esmon CT. Role of coagulation inhibitors in inflammation. *Thromb Haemost* 2001;**86**:51–6.
86. Rosenberg RD, Shworak NW, Liu J, Schwartz JJ, Zhang L. Heparan sulfate proteoglycans of the cardiovascular system. Specific structures emerge but how is synthesis regulated? *J Clin Invest* 1997;**100**:S67–75.
87. Hoffmann JN, Vollmar B, Laschke MW, Inthorn D, Kaneider NC, Dunzendorfer S, et al. Adverse effect of heparin on antithrombin action during endotoxemia: microhemodynamic and cellular mechanisms. *Thromb Haemost* 2002;**88**:242–52.
88. Warren BL, Eid A, Singer P, Pillay SS, Carl P, Novak I, et al. Caring for the critically ill patient. High-dose antithrombin III in severe sepsis: a randomized controlled trial. *JAMA* 2001;**286**:1869–78.
89. Harada N, Okajima K, Uchiba M, Kushimoto S, Isobe H. Antithrombin reduces ischemia/reperfusion-induced liver injury in rats by activation of cyclooxygenase-1. *Thromb Haemost* 2004;**92**:550–8.
90. Harada N, Okajima K, Kushimoto S, Isobe H, Tanaka K. Antithrombin reduces ischemia/reperfusion injury of rat liver by increasing the hepatic level of prostacyclin. *Blood* 1999;**93**:157–64.
91. Kaneider NC, Reinisch CM, Dunzendorfer S, Romisch J, Wiedermann CJ. Syndecan-4 mediates antithrombin-induced chemotaxis of human peripheral blood lymphocytes and monocytes. *J Cell Sci* 2002;**115**:227–36.

92. Uchiba M, Okajima K, Murakami K. Effects of various doses of antithrombin III on endotoxin-induced endothelial cell injury and coagulation abnormalities in rats. *Thromb Res* 1998;**89**:233–41.
93. Kaneider NC, Egger P, Dunzendorfer S, Wiedermann CJ. Syndecan-4 as antithrombin receptor of human neutrophils. *Biochem Biophys Res Commun* 2001;**287**:42–6.
94. Stevens RL, Colombo M, Gonzales JJ, Hollander W, Schmid K. The glycosaminoglycans of the human artery and their changes in atherosclerosis. *J Clin Invest* 1976;**58**:470–81.
95. Nishinaga M, Ozawa T, Shimada K. Homocysteine, a thrombogenic agent, suppresses anticoagulant heparan sulfate expression in cultured porcine aortic endothelial cells. *J Clin Invest* 1993;**92**:1381–6.
96. Paka L, Kako Y, Obunike JC, Pillarisetti S. Apolipoprotein E containing high density lipoprotein stimulates endothelial production of heparan sulfate rich in biologically active heparin-like domains. A potential mechanism for the anti-atherogenic actions of vascular apolipoprotein e. *J Biol Chem* 1999;**274**:4816–23.
97. Mochizuki H, Yoshida K, Gotoh M, Sugioka S, Kikuchi N, Kwon YD, et al. Characterization of a heparan sulfate 3-*O*-sulfotransferase-5, an enzyme synthesizing a tetrasulfated disaccharide. *J Biol Chem* 2003;**278**:26780–7.
98. Wiedermann Ch J, Romisch J. The anti-inflammatory actions of antithrombin—a review. *Acta Med Austriaca* 2002;**29**:89–92.

Endothelial Heparan Sulfate in Angiogenesis

Mark M. Fuster[*,†,1] and
Lianchun Wang[‡,1]

[*]*Department of Medicine, Division of Pulmonary and Critical Care, University of California San Diego, La Jolla, California, USA*

[†]*VA San Diego Healthcare System, La Jolla, California, USA*

[‡]*Department of Biochemistry and Molecular Biology, Complex Carbohydrate Research Center, University of Georgia, Athens, Georgia, USA*

Heparan sulfate (HS) is a linear polysaccharide composed of 50–200 glucosamine and uronic acid (glucuronic acid or iduronic acid) disaccharide repeats with epimerization and various sulfation modifications. HS is covalently attached to core proteins to form HS-proteoglycans. Most of the functions of HS-proteoglycans are mediated by their HS moieties. The biosynthesis of HS is initiated by chain polymerization and is followed by stepwise modification reactions,

[1]Both authors contributed equally

Progress in Molecular Biology
and Translational Science, Vol. 93
DOI: 10.1016/S1877-1173(10)93009-3

1877-1173/10 $35.00

including sulfation and epimerization. These modifications generate ligand-binding sites that modulate cell functions and activities of proteinases and/or proteinase inhibitors. HS is abundantly expressed in developing and mature vasculature, and understanding its roles in vascular biology and related human diseases is an area of intense investigation. In this chapter, we summarize the significant recent advances in our understanding of the roles of HS in developmental and pathological angiogenesis with a major focus on studies using transgenic as well as gene knockout/knockdown models in mice and zebrafish. These studies have revealed that HS critically regulates angiogenesis by playing a proangiogenic role, and this regulatory function critically depends on HS fine structure. The latter is responsible for facilitating cell-surface binding of various proangiogenic growth factors that in turn mediate endothelial growth signaling. In cancer, mouse studies have revealed important roles for endothelial cell-surface HS as well as matrix-associated HS, wherein signaling by multiple growth factors as well as matrix storage of growth factors may be regulated by HS. We also discuss important mediators that may fine-tune such regulation, such as heparanase and sulfatases; and models wherein targeting HS (or core protein) biosynthesis may affect tumor growth and vascularization. Finally, the importance of targeting HS in other human diseases wherein angiogenesis may play pathophysiologic (or even therapeutic) roles is considered.

I. Endothelial Heparan Sulfate in Developmental and Physiologic Angiogenesis

A. Angiogenesis in Health and Disease: An Introduction

Angiogenesis refers to the generation of new blood vessels from existing ones, a process that differs from vasculogenesis (the *de novo* formation of blood vessels from mesoderm or endothelial cells (EC) progenitors).[1–3] Angiogenesis occurs through an orchestrated sequence of events consisting of two general processes: formation of a nascent vascular network and subsequent maturation. To form a nascent vascular network, ECs first sprout from existing capillary plexus in response to angiogenic stimuli, degrade the extracellular matrix (ECM), and then migrate into avascular tissue, assembling into an endothelial network. The proliferation and migration of ECs are potently modulated by growth factors, such as vascular endothelial growth factor (VEGF). The maturation process includes stabilization of the newly formed vessel with the recruitment of mural cells (MC; including vascular smooth-muscle cells (VSMC) in arteries, arterioles, and veins; and pericytes in capillaries), remodeling (branching, pruning, and regression of vasculature), and vessel specialization that determines arterio-venous commitment as well as organ-specific networks.[4–8] The maturation process is critically modulated by growth factors, including platelet-derived growth factor-B (PDGF-B) and transforming growth factor-β (TGF-β).

Physiological angiogenesis occurs mainly during embryonic development and in postnatal growth of tissues such as the retina.[1,9] In adults, it takes place primarily in the female reproductive system and in the intestinal villa.[10,11] However, abnormal angiogenesis plays a critical role in the pathogenesis of many diseases such as cancer, ischemic vascular disorders, and diabetic retinopathy.[1,12] Tumor-associated neovascularization (tumor angiogenesis) is required not only for tumor growth, but also for metastasis. Therefore, blockage of tumor angiogenesis represents a promising approach for the development of new cancer therapeutics. Stimulation and promotion of a mature neovascular network, referred to as therapeutic angiogenesis, shows promise as a treatment for ischemic diseases such as ischemic coronary artery disease and stroke.[1,8,12,13]

B. Heparan Sulfate in Development

Heparan sulfate proteoglycans (HSPGs) are glycoconjugates composed of a core protein with one or more covalently attached heparan sulfate (HS) chains (Fig. 1).[14–16] HSPGs are present abundantly on the cell surface and in the ECM where they interact with numerous growth factors, growth factor binding proteins, extracellular proteases, protease inhibitors, chemokines, morphogens, and adhesive proteins.[14–17] These complex interactions regulate the activity, gradient formation, and stability of many ligand–receptor interactions. Most of the interactions of HSPGs with the ligands are mediated directly by the HS moieties.[14–16,18] HS is a linear polysaccharide composed of glucosamine and uronic acid (glucuronic acid, GlcA, or iduronic acid, IdoA) disaccharide repeats with various sulfation modifications (Fig. 1), and is typically 50–200 disaccharides in length. HS biosynthesis is initiated by copolymerases Ext-1 and Ext-2, which alternatively add GlcA and *N*-acetylglucosamine residues from their nucleotide sugar precursors (Fig. 1).[15,16] Following chain elongation, *N*-deacetylase/*N*-sulfotransferases (*N*dsts) act on discrete regions of the HS precursor, replacing *N*-acetyl groups with *N*-sulfates.[19] Regions of *N*-sulfation (NS-domain) then act as the substrate for additional modifications, including epimerization, and 2-*O*-, 6-*O*-, and 3-*O*-sulfation.[15,20] The modifications are generally incomplete and the enzymes involved have substrate specificities, which result in IdoA and *O*-sulfate residues occurring predominantly in the NS-domains. The sulfation pattern within the NS-domains, as well as their length and spacing, contribute to the structural heterogeneity of HS and create specific binding sites for protein ligands (Fig. 1).[21–23] Intriguingly, the distribution, length, and modification level of such domains appear to be tightly regulated in a tissue/cell-specific fashion,[24] suggesting that the regulatory role

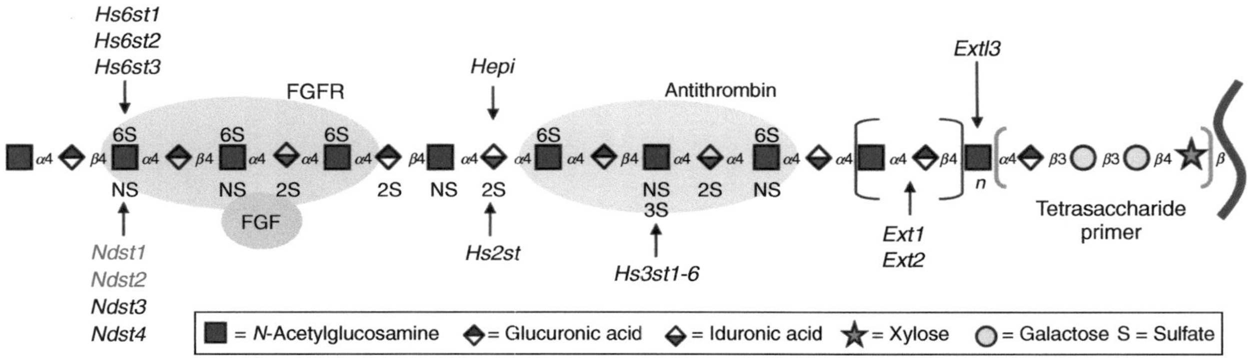

Fig. 1. HS structure and biosynthesis: Each sugar residue is depicted by a geometric symbol. Binding sites for ligands are defined by the arrangement of sulfate groups (NS, 2S, 3S, 6S) and uronic acid epimers (GlcA and IdoA). In mammals, as many as 26 enzymes (italicized) participate in the formation of HS chains. *N*-deacetylase/*N*-sulfotransferase (Ndst) initiates sulfation in clustered sites along HS chains, and isoenzymes 1 and 2 (of four expressed in mammals), highlighted in *red*, are expressed in ECs. Through the action of other sulfating enzymes, further sulfation modifications take place around sites of clustered *N*-sulfation. (For interpretation of the references to color in this figure legend, the reader is referred to the web version of this chapter.)

of HS occurs in a spatiotemporal manner by interacting with unique arrays of protein ligands in different tissues and at different developmental/pathological stages.[24,25] Loss-of-functional studies of HS biosynthetic genes in mice proved this concept by observing phenotypes ranging from early developmental defects (*Ext-1*, *Ext-2*), prenatal/neonatal lethality (*Ndst-1*, *2-O-sulfotransferase* (*Hs2st*), *C5-epimerase* (*Hepi*), *3-O-sulfotransferase 1* (*Hs3st-1*), *6-O-sulfotransferase 1* (*Hs6st-1*)), to a restricted heparin biosynthesis defect (*Ndst2*) (Table I). Further studies to elucidate the distinct molecular pathways modulated by HS, the HS structure–function relationship, and the underlying cellular and molecular mechanisms in the development of different organs remain as important tasks in the field.

C. Heparan Sulfate in Developmental Angiogenesis: Overview of Genetic Evidence

Blood vessels constitute the first organ in the embryo and form the largest network in the body. Therefore, the formation of new blood vessels accounts for one of the major events during embryonic development. Vascular formation is tightly controlled by the balance between pro- and antiangiogenic factors.[1] HS has been shown to facilitate both pro- and antiangiogenic factors *in vitro*.[31,49–52] These observations have spurred interest in the general role of HS in angiogenesis *in vivo*. This has recently been examined through genetic manipulation of HS biosynthetic genes in vertebrate model organisms of zebrafish and mice.

1. Zebrafish Studies[34]

HS has been shown to be an important modulator of patterning processes during organogenesis in development. RNA interference of the *Drosophila Hs6st* produces defective branching in the tracheal system.[53] Since *Drosophila* tracheogenesis and vertebrate vasculogenesis share many common molecules, such as FGFs, VEGF, and integrins, tracheogenesis in *Drosophila* is thought to be a close model for mammalian vascular development.[54] The tracheal branching defect observed in the *Hs6st* mutant *Drosophila* suggests that *HS6st* may modulate vascular branching in vertebrates.[53] This was tested in zebrafish by Chen et al.[34] In zebrafish, *Hs6st* exists in two isoforms (*Hs6st-1* and *-2*). *In situ* hybridization reveals dynamic and distinct expression patterns of these two genes during development. *Hs6st-2* is expressed in the cells surrounding the dorsal aorta and posterior cardinal vein, suggesting that *Hs6st-2* might have a role in vascular development or function. Indeed, *Hs6st-2* morphants exhibit abnormalities in the vascular branching morphogenesis of the caudal vein during embryonic development, showing vascular dysmorphogenesis ranging from reduced branching accompanying formation of large loops to a lack of

TABLE I

PHENOTYPES OF ZEBRAFISH OR MICE THAT ARE DEFICIENT IN HS BIOSYNTHETIC ENZYMES OR MICE DEFICIENT IN HS-BINDING ANGIOGENIC FACTOR OR THEIR RECEPTORS

Gene	Phenotype
MC-Ext-1−/− mice	Embryonic lethality associated with vascular patterning defects, edema, and hemorrhages during late gestation. MC recruitment in the skin with impaired vessel stability and variable diameters.[26]
Ndst-1−/− mice	Neonatal lethal, pulmonary atelectasis, cyanosis[27–29]; pericyte detachment and delayed migration due to impaired PDGF-B signaling and disruption of directed cell migration.[30]
EC-Ndst-1−/− mice	Impaired inflammatory response and tumor angiogenesis.[31,32]
Hepi−/− mice	Die shortly after birth with multiple developmental defects, such as skeletal malformations and kidney agenesis,[33] and transient delay of MC recruitment during embryonic vascular development.[30]
Hs6st-2 morphant	Abnormalities in the branching morphogenesis of the caudal vein during embryonic development. Hs6st-2 interacts synergistically with VEGF-A in angiogenesis *in vivo*.[34]
Hs6st-1−/− mice	Most die between E15.5 and perinatal stage. Approximately 50% reduction in microvessel density in the labyrinthine zone of placenta with a modest reduction in VEGF-A mRNA and protein.[35,36]
Hs3st-1−/− mice	No obvious procoagulant phenotype, background-specific lethality and intrauterine growth retardation, severely compromised fertility and disrupted maternal–fetal circulation in.[37,38]
VEGF120/120 mice	Impaired postnatal myocardial angiogenesis, cardiomyopathy and lethal cardiac failure, reduced vascular branching complexity and increased vascular caliber.[39–41]
VEGFR1 (flt-1)−/− mice	E8.5–9.5 Embryonic lethality exhibiting overgrowth of ECs and disorganized vasculature.[42]
VEGFR2 (flk-1)−/− mice	Homozygous: E8.5–9.5 Embryonic lethality, severely impaired hematopoietic and ECs; heterozygous appear normal.[43]
PDGF-Bret−/− mice	Vascular defect with pericyte detachment.[44]
PDGF-BB−/− mice	Embryonic lethality. MC loss which results in vascular defects including endothelial hyperplasia, capillary dilation, microaneurysms, vascular leakage, and hemorrhaging.[45,46]
PDGFR-β−/− mice	Similar to the PDGF-BB−/− mice.[47]
TGFRII-β−/− mice	Defective vascular development associated with reduced MC recruitment and impaired vessel stability.[48]

branching with disorganized and overlumenized vessels in place of the venous plexus. Although *Hs6st-1* has biochemical activity similar to *Hs6st-2*, vascular morphogenesis in *Hs6st-1* morpholino-injected embryos is normal, which is consistent with the lack of *Hs6st-1* expression in developing vasculature. To assess the nature of the vascular defects resulting from *Hs6st-2* deficiency, expression of the early vascular development marker *flk-1*, which encodes a VEGF receptor (VEGFR), was examined. Although primary formation of the axial vessels and initial sprouting of intersegmental vessels proceed normally in *Hs6st-2* morphants, the expression of the later vascular development markers tie-1 and tie-2 in the caudal vein are reduced during the time period of remodeling. This result correlates with the reduced branching complexity of the caudal vein plexus in the *Hs6st-2* morphants, and expression of *Hs6st-2* at a time point that overlaps angiogenic remodeling, further suggesting that *Hs6st-2* plays an essential role in vascular remodeling of the caudal vein plexus. Therefore, these observations demonstrate that HS is a critical controller of vascular morphogenesis in zebrafish. These observations also illustrate an important role for 6-*O*-sulfation of HS in vascular morphogenesis.

2. Mouse Studies

Within the last few years, several HS biosynthesis genes, including copolymerase and modification enzymes, have been genetically inactivated in mice, allowing us to determine the general roles as well as the fine structure of HS in the modulation of vascular development physiological conditions close to those in humans. So far, biological functions of HS in vascular development have been examined in mice lacking *Hs6st-1*, *Ndst-1*, *Hepi*, or percitype-specific *Ext-1* (Table I).

a. Hs6st-1−/− Mice[35,36]. Since 6-*O*-sulfation of HS has been shown to be critical for vascular development in zebrafish, Habuchi et al. further examined the biological functions of *Hs6st* in vascular development in mice, wherein *Hs6st* exists in three isoforms (*Hs6st-1-3*) and one alternatively spliced form (*Hs6st-2s*).[35] Biochemical and expression pattern analyses showed that although the substrate specificities of these isoforms overlap, their expression varies in a tissue and developmental stage-specific manner, suggesting the possibility that the three *Hs6sts* might not have overlapping functions *in vivo* as do those that occur with the *Ndsts*. Among the three *Hs6sts*, *Hs6st-1* carries out the 6-*O*-sulfation modification of HS in most tissues. The biological and physiological importance of *Hs6st-1* was determined by examining *Hs6st-1* null (*Hs6st-1−/−*) mice.[35,36] Most of the *Hs6st-1−/−* mice die between embryonic day 15.5 (E15.5) and the perinatal stage, and a low percentage of the mutant mice that survive are considerably smaller than their wild-type littermates (Table I). Some of the *Hs6st-1−/−* mice exhibited developmental abnormalities, including a ∼50 % reduction in the number of fetal microvessels in the labyrinthine zone of the developing placenta. In support of this, HS structure analysis showed that *Hs6st-1* plays a particularly important role in HS biosynthesis in the placenta. Interestingly, despite the apparent reduction of angiogenesis in the placentas of *Hs6st-1−/−* mice, vascularization in the embryo body, including in the yolk sac, appeared unaffected. These results indicate that the nutritional supply and gas exchange in the labyrinthine zone may be jeopardized in *Hs6st-1*-deficient placentas due to reduced angiogenesis. The reduction in the number of microvessels in the placentas of *Hs6st-1*-deficient embryos suggests the presence of ischemia, which at least partially contributes to the embryonic lethality phenotype of the *Hs6st-1−/−* mice. Taken together, these data demonstrate that 6-*O*-sulfation of HS is also critically required for vascular morphogenesis in mammals.

b. Ndst-1−/− and Hepi−/− Mice[29,32,33]. Ndst carries out the first step in the modification of HS biosynthesis by replacing the *N*-acetyl group of discrete *N*-acetylglucosamine resides with a sulfate (Fig. 1). This reaction creates the basis for all further modification of the chains in the NS-domains and determines the fine structure of the final HS. Therefore, genetic manipulation of *Ndst* expression represents an effective way to determine the HS structure–function relationship in the regulation of vascular morphogenesis. *Ndst* exists in four isoforms, *Ndst-1-4*. Both *Ndst-1* and *Ndst2* are ubiquitously expressed during development and in adult mice, while *Ndst3* and *Ndst4* are mostly expressed during development[55–57]. Surprisingly, inactivation of *Ndst2* only results in mast deficiency in the connective tissue in the mutant mice (Table I).[58] In contrast, mice lacking *Ndst-1* (*Ndst-1−/−*) have a severe phenotype and die between E14.5 and shortly after birth (Table I).[29] The dramatically reduced HS-sulfation modifications in *Ndst-1−/−* mice results in brain malformation and skeletal defects.[56,59] In contrast to wild-type hindbrains where vascular endothelium and pericytes are closely associated with tip cells at the outermost front of the growing vascular plexus, a detailed examination of developing vasculature in the E11.5 hindbrain reveals that pericytes in the *Ndst-1−/−* hindbrain frequently failed to reach the growing vascular front and pericytic processes frequently stretched away from the endothelium. This examination shows defective MC spreading and tight attachment during vascular maturation in *Ndst-1−/−* mice.[30] Quantification of the proportion of pericyte-endothelial apposition (pericyte coverage) in relation to the total area of the endothelium showed that when compared with wild-type littermates, the pericyte coverage is significantly reduced in the periphery, although not at the midline of the *Ndst-1−/−* hindbrain, thus indicating that the initial recruitment phase is more severely affected by *Ndst-1* deficiency. In summary, these observations established that *N*-sulfation of HS is critically required for MC recruitment in angiogenesis.

Staining of wild-type hindbrains with the anti-HS antibody HepSS-1 revealed that the most abundant HS deposition is on the abluminal endothelial surface, the interface between ECs and pericytes. In contrast, *Ndst-1−/−* embryos almost completely lacked endothelial HepSS-1 labeling. These results demonstrated that *Ndst-1* is required for endothelial HS-sulfation, and also suggested that this in turn is critical for pericyte recruitment and coverage in angiogenesis. In endothelium, *Ndst-1* is abundantly expressed. We recently generated EC-specific *Ndst-1* mutant (*EC-Ndst-1−/−*) mice. The targeted inactivation of *Ndst-1* results in endothelial HS with about 50% reduction in N-, 2-, and 6-*O*-sulfation, showing a globally reduced sulfate modification that is similar to the *Ndst-1−/−* mice.[31,32] In contrast to the *Ndst-1−/−* mice, the *EC-Ndst-1−/−* mice appear normal and fertile. Detailed examination revealed that vascular morphogenesis is slightly disrupted in some localized areas (Zhang and Wang, unpublished data), consistent with the hypothesis that sulfated domains on endothelial HS regulate vascular morphogenesis *in vivo*. However, it needs to be determined whether endothelial *Ndst-1* deficiency disrupts pericyte recruitment thus contributing to vascular dysmorphogenesis in the *EC-Ndst-1−/−* mice. Since disruption of the *Ndst-1* gene results in only a partial lesion in endothelial HS structure, additional experiments that disrupt endothelial *Ndst-1* and *Ndst2* (or the HS copolymerase *Ext-1* or *Ext-2*) may provide more clear data about the role of endothelial HS in vascular development.

Ndst-1 inactivation reduces multiple sulfation modifications (by ~50% overall).[31,32] Therefore, the findings upon examination of *Ndst-1−/−* embryos/mice implicate the highly sulfated NS-domains in vascular development, but do not specifically differentiate among requirements for defined sequence or overall charge density. Studies to determine the fine structure requirements for HS in angiogenesis have also merged. As described earlier, 6-*O*-sulfation of HS is critically required for vascular morphogenesis in mammals.[35,36] The role of *Hepi* in MC recruitment was examined too.[30] In HS biosynthesis, Hepi catalyzes the epimerization of GlcA to IdoA, which increases the flexibility of the sugar chain, and is also shown to be a prerequisite to some of the further steps in *O*-sulfation. Deletion of the *Hepi* gene leads to the formation of a mutant HS that is highly *N*-sulfated, but devoid of IdoA units, and is characterized by a severely perturbed fine structure.[33] Analysis of MC recruitment and coverage in *Hepi−/−* embryos revealed only a transient delay in MC recruitment in the hindbrains of E10.5 *Hepi*$^{-/-}$ embryos with no sustained defects.[30] Therefore, this study illustrated that unlike *N*- and 6-*O*-sulfation, C5-epimerization appears not to be critical for MC recruitment in vascular development.

c. Mural Cell-Specific Ext-1 Null (MC-Ext-1−/−) Mice[26]. The *Ext-1* gene encodes for the key glycosyltransferase that is essential for elongation of the HS disaccharide backbone chain (Fig. 1). Deletion of *Ext-1* leads to a complete loss of HS, and therefore can determine the general role of HS in physiological/

pathological processes.[60–62] To study the functional role of MC-derived HS in angiogenesis, *Ext-1* was selectively inactivated in mouse MCs (*MC-Ext-1−/−*).[26] Anti-HS antibody staining confirmed substantial loss of HS in the MC compartment of major arteries and veins, and in smaller vessels as well as MC progenitors in *MC-Ext-1−/−* embryos, allowing one to address the functional importance of MC-specific HS in vascular development. *MC-Ext-1−/−* embryos appear to develop normally by E11.5, but die between E13.5 and E19 with hemorrhages and edema in the skin due to increase vessel permeability.[26] Detailed whole-mount staining for ECs of the skin reveals striking abnormalities in superficial vessels, particularly around the spinal cord in the *MC-Ext-1−/−* embryos. In contrast to the wild-type controls that display an organized and hierarchical vascular pattern, vessels in *MC-Ext-1−/−* embryos appear random and chaotically branched with variable diameters, and frequently form glomeruloid structures resembling microaneurysms. Detailed examination further observed significant increases in vessel regression, diameter variability, and permeability in the *MC-Ext-1−/−* skin tissues.[26] Intriguingly, the central cardiovascular development and function appear normal in *MC-Ext-1−/−* embryos, indicating that the observed vascular defects in the skin reflect local defects associated with MC HS deficiency.

Examination of MCs in skin samples revealed that in the control embryos, MCs are closely attached to the endothelium and cover almost the entire vascular network. The MC-*Ext-1−/−* skin samples from identical regions showed significantly reduced MC coverage, with many vessels lacking properly attached MCs.[26] In parallel, examination of isolated NG2-positive cells, which represent MC progenitor cells not yet recruited to the vessels or MCs that have detached, showed that the isolated NG2 positive cells are more abundant in MC-*Ext-1−/−* skin than in the wild-type control.[26] Morphologically, the isolated NG2-positive MCs appear spindle-shaped in wild type, while they are conspicuously round in the mutant, revealing a MC–EC attachment defect. Taken together, these observations demonstrate that MC HS is critically required for MC cell recruitment as well as for maturation and stabilization of nascent vessels during development.

D. Heparan Sulfate in Developmental Angiogenesis: Modulation of Growth Factor Signaling

Although HS has been shown to facilitate the biological functions of both pro- and antiangiogenic factors, the study of HS mutants using both zebrafish and mouse models illustrates that HS plays a proangiogenic role during developmental angiogenesis. Therefore, the predominant effect of endothelial HS in this setting is to facilitate the actions of proangiogenic factors that promote developmental angiogenesis. So far, studies of *Hs6st-2−/−* zebrafish and mice lacking *Hs6st-1*, *Ndst-1*, *Hepi*, or MC-*Ext-1* reveal that HS critically modulates VEGF-A, PDGF-B, and TGF-β signaling to promote the various steps in developmental angiogenesis.

1. VEGF-A Signaling

Genetic studies established that VEGF-A is a master regulator of developmental angiogenesis as a result of its ability to modulate almost all aspects of the process, including EC differentiation, assembly, proliferation, and migration (Table I).[1,2,8,63] VEGF-A is expressed and secreted as multiple, homodimeric isoforms that are formed as a result of differential splicing of the VEGF pre-mRNA. The most abundantly expressed isoforms are VEGF120, VEGF164, VEGF188 in mice, and VEGF121, VEGF165, and VEGF189 in humans. Although all variants contain the same binding sites for the VEGFR, they differ in their affinity for HS because of the presence or absence of HS-binding domains that are encoded by exons 6 and 7.[40,41] *In vivo*, both domains independently mediate interactions with HS present on the cell surface and in the ECM. VEGF120, which lacks the two HS-binding domains, is freely diffusible in ECM, whereas VEGF164 and VEGF188 exhibit moderate and high binding affinity, respectively, resulting in partial or complete retention of these isoforms in HS-rich compartments.[39–41] Meanwhile, studies also observed that these isoforms differ in their mitogenicity, chemotactic properties, receptor-binding characteristics, and tissue-specific expression,[64–67] suggesting that HS may differentially modulate the biological functions of VEGF isoforms *in vivo*. This was tested by examination of mice expressing exclusively the VEGF120 isoform (VEGF120/120 mice).[39] The most dramatic phenotype of the VEGF120/120 mice is ischemic cardiomyopathy, resulting from the impaired postnatal myocardial angiogenesis. The VEGF120/120 mice ultimately die of cardiac failure.[39] Detailed examination of vascular development in VEGF120/120 embryos further revealed that, although heparin-binding VEGF-A is not essential for vasculogenesis or angiogenesis, the mutant shows reduced vascular branching complexity with increased microvessel caliber,[39–41] suggesting that the ability to bind HS is necessary for VEGF-A to regulate vascular branching. This observation also suggests that HS interacts with VEGF-A to modulate vascular development *in vivo*.

The regulatory role of HS in VEGF-A signaling *in vivo* was examined in zebrafish by injection or coinjection of morpholinos against *Hs6st-2* and VEGF-A.[34] Compared with a single injection of either *Hs6st-2* MO or *VEGF-A* MO alone, coinjection of *Hs6st-2* MO and *VEGF-A* MO resulted in a synergistic increase in the frequency of embryos with caudal vein branching defects. Meanwhile, coinjection of *Hs6st-2* and a four-base mismatch *VEGF-A*

MO abolished the synergy.[34] These results illustrate that *Hs6st-2* and *VEGF-A* interact *in vivo* during caudal vein formation, also establishing that HS facilitates VEGF-A signaling in angiogenesis *in vivo*.

Studies have also addressed the mechanism of how HS modulates VEGF-A signaling. The detailed analysis of vascular branching patterns, EC behavior, and VEGF-A protein distribution in wild-type mouse embryos and mouse mutants producing solely VEGF120 or VEGF188 revealed that the heparin-binding VEGF-A isoforms serve as spatially restricted stimulatory cues that elicit stereotypical branching behavior at the leading edge of the growing microvessel network,[39–41] an analogy to the role of HS localizing FGF and BMP in other branching organs. These observations strongly suggest that HS functions to maintain a VEGF-A concentration gradient that provides spatially restricted stimulatory cues to polarize and thereby guide sprouting ECs to initiate vascular branch formation. Meanwhile, *in vitro* EC culture studies established that HS functions *in cis* as a cell-surface coreceptor for VEGF-A signaling. Recent *in vitro* studies using chimeric cultures of embryonic stem cells defective in either HS production (combined deficiency of *Ndst-1* and *Ndst-2* resulting in cells deficient in *N*-sulfate, 2-*O*-sulfate, and 6-*O*-sulfate) or VEGFR2 synthesis demonstrated that VEGF signaling in ECs is fully supported by HS expressed *in trans* by adjacent MCs, suggesting a mechanism whereby HS *in trans* potentiates VEGF-A signaling during developmental angiogenesis.[68] However, VEGF-A signaling in ECs in MC-*Ext-1*−/− embryos is not altered,[26] indicating that MC HS is not required for VEGF-A signaling during vessel development *in vivo*. This observation also highlights the importance of determining the requirement of endothelial HS for VEGF-A signaling in vascular development *in vivo*.

2. PDGF-B Signaling

Deletion of PDGF-B and PDGF receptor (PDGFR-β) causes embryonic lethality resulting from vascular defects, including microaneurysms, vascular leakage, and hemorrhaging as a result of severe MC loss, establishing that PDGF signaling is pivotal for MC recruitment, vascular maturation, and stability.[44–46] Conditional endothelial inactivation of PDGF-B established that the endothelium provides the major source of PDGF-B required for the migration and proliferation of MCs.[46] PDGF-B protein is secreted as PDGF-BB homodimers. PDGF-BB binding to PDGFR-β on MCs leads to receptor dimerization and phosphorylation of PDGFR-β, activating multiple downstream signaling pathways, including PI3K and Erk, which then stimulate cell migration and proliferation. PDGF-B contains a conserved C-terminal sequence of basic amino acids (retention motif), which is structurally similar to the HS-interacting domain of VEGF-A and is thought to mediate PDGF-B binding to the cell surface or ECM.[69,70] Mice deficient in the PDGF-B retention motif (*PDGF-B$^{ret/ret}$*) display vascular defects associated with MC detachment (Table I),[44] which is reminiscent of the MC recruitment defects observed in *Ndst-1−/−* and *MC-Ext-1−/−* mice, suggesting that HS facilitates PDGF-B signaling to recruit MCs during developmental angiogenesis. Protein lysates from MC-*Ext-1−/−* skin samples showed a strong reduction in phosphorylated PDGFR-β, and loss of SHP2 and Erk1/2 activation.[26] These observations reveal that MC HS participates in MC recruitment by facilitating PDGFR-β signaling in skin. Interestingly, MCs in the CNS do not require HS production for effective PDGFR-β signaling and recruitment to the vessel wall. The differential requirement of HS may result from different modes of MC recruitment in different tissues. In CNS, the MC recruitment occurs via a longitudinal model, which relies on migration and proliferation of MCs along the abluminal endothelial surface, and may critically require endothelial HS to control PDGF-B retention and to function as a coreceptor for pericytic PDGFR-β activation *in trans*;[71] whereas the skin recruitment of MC to vessels involves induction of new progenitor cells from mesenchymal lineages. MC progenitor cells that reside at a distance from the vessel may cell-autonomously require HS on the cell surface for activation *in cis*.[71] Therefore, the spatial relationship between the signal-sending cell (the endothelium) and the signal-receiving cell (the MC) in skin is different from CNS during MC recruitment, explaining the difference in vascular phenotypes of skin and CNS in *MC-Ext-1−/−* mice.[26] Meanwhile, this prospective has been supported by studies on developmental and tumor angiogenesis showing that HS retains PDGF-B close to the EC surface to facilitate directed MC migration along the sprouting vessel and to mediate proper MC attachment to the vessel during angiogenesis.[44,46] Therefore, it would be interesting to see whether endothelial HS deficiency disrupts PDGF-B signaling and the PDGF-B-mediated MC recruitment, contributing to the vascular development defects in EC-*Ndst−/−* mice that we maintain in the laboratory.

The fine structure of HS required for PDGFR-β signaling *in vivo* was examined in *Ndst-1−/−* and *Hepi−/−* mice.[30] Reduction of *N*-sulfation due to deficiency in *Ndst-1* attenuated PDGF-B binding *in vitro* and led to pericyte detachment and delayed pericyte migration *in vivo*. Reduced *N*-sulfation also impaired PDGF-B signaling and directed cell migration, but not proliferation. In contrast, HS from *Hepi−/−* mutants, which is extensively *N*-sulfated, but lacks 2-*O* and 6-O sulfation, retained PDGF-B *in vitro* and showed that pericyte recruitment *in vivo* was only transiently delayed. On the basis of these *in vitro* as well as *in vivo* studies, it appears that pericyte recruitment requires HS with sufficiently extended and appropriately spaced *N*-sulfated domains to retain PDGF-B and to activate PDGFR-β signaling, whereas the specific sequence of monosaccharide and sulfate residues does not appear to be important for this interaction.[30] Currently, it is believed that HS functions to retain a PDGF-B gradient to direct MC progenitor cell migration and as a coreceptor for PDGFR-β signaling,[26] as seen for VEGF-A and FGF.

3. TGF-β Signaling

Ex vivo studies using ECs and 10T1/2 cell (MC progenitor cells) coculture illustrated that TGF-β is also critically involved in MC recruitment and differentiation.[72,73] This *ex vivo* observation was confirmed by genetic studies showing that deficiency in the TGF-β receptor II results in mouse embryonic lethality with defects in vascular development.[48] Meanwhile, mice bearing mutations in the TGF-β signaling pathway, such as TGFRII, Alk1, Alk5, endoglin, and SMAD1/5 mutants, exhibit defective vascular development associated with reduced MC recruitment and impaired vessel stability, demonstrating that TGF-β signaling is essentially required for vascular development. Detailed studies further revealed that TGF-β critically regulates EC and MC proliferation and differentiation, and mediates the *de novo* induction of MCs from the mesenchymal lineage state during embryonic development.[48] TGF-β also induces actin reorganization by activating Rho-GTPases through phosphorylation of SMAD 2 and 3,[74] which may modulate MC attachment to ECs as well as EC responses. The components of the TGF-β pathway, including TGF-β receptors, interact and cocluster directly with VE-cadherin at EC–EC junctions, suggesting that TGF-β signaling may promote vessel stabilization and quiescence.[75] TGF-β is known to bind HS,[76,77] and HS plays an important role in linking the latent TGF-β-binding protein LTBP1 to fibronectin,[78] providing a mechanism for TGF-β storage in the ECM. During the patterning process in *Drosophila*, HS has been proven to be essential for gradient formation and extracellular transport of the TGF-β family member Dpp, suggesting that HS may be important for the availability, potential gradient formation, and activity of TGF-β during the induction and recruitment of MCs. In MC-*Ext-1*−/− embryonic skin vasculature, TGF-β signaling is defective in ECs.[26] TGF-β is expressed by both ECs and MCs, and reciprocal signaling regulates MC induction from mesenchymal progenitors as well as various EC functions, including expression of stabilizing matrix components, cell proliferation, and differentiation. Therefore, the disruption of the TGF-β signaling in ECs may contribute to the vascular development defect in the MC-*Ext-1*−/− embryo.

4. Additional Signaling Pathways

In addition to VEGF, PDGF, and TGF-β signaling, other angiogenic pathways may also be modulated by HS during development. This may include FGF, ephrin, slit/roundabout, the netrin/UNC (uncoordinated) receptor, Wnt, hedgehog, and sprouty, as well as ECM components such as fibronectin and laminin. For example, while the *Hs6st-1*−/− mouse placenta had a normal level of *Wnt2* expression, HS from the same placenta showed attenuated binding affinity to Wnt2.[35] In agreement, *Hs6st-1* in zebrafish appears to control Wnt-dependent signaling pathways.[34] Wnt signaling has been known to induce VEGF-A expression, a strong inducer of angiogenesis.[79] Attenuated Wnt2–HS interaction in *Hs6st-1−/−* mice may cause decreased VEGF-A expression that is observed in the mutant placentas.[35] More work is needed to determine whether the interactions of HS with these growth factors, morphogens, and ECM components are biologically significant in angiogenesis *in vivo*.

E. Heparan Sulfate in Adult Physiological Angiogenesis

1. Reproductive Angiogenesis

The female reproductive organs (ovary, uterus, and placenta) are some of the few adult tissues that exhibit regular intervals of rapid growth. They are highly vascularized, and angiogenesis is an important component of the growth and function of these tissues. As with many other tissues, VEGFs and FGFs appear to be the major angiogenic factors in the female reproductive organs with apparently abundant coexpression of HS.[35,38] Therefore, it is highly possible that HS may critically modulate reproductive angiogenesis by facilitating VEGF and/or FGF signaling. Although so far no studies have specifically addressed this issue, the phenotypes displayed by *Hs3st-1−/−* and *Hs6st-1−/−* mice appear to strongly support this speculation.[35,37,38]

Hs3sts carry out the last step of the sulfate modifications that occur during HS biosynthesis (Fig. 1). This occurs through the addition of 3-*O*-sulfate groups to *N*-sulfated glucosamine resides. Six *Hs3st* isoforms have been identified with different tissue expression patterns and acceptor substrate specificities. The *Hs3st-1* is the predominant form generating antithrombin-binding sites in anticoagulant HS, particularly in ECs.[37] Unexpectedly, *Hs3st-1−/−* mice did not show a procoagulant phenotype, but instead suffered severely compromised fertility.[37,38] Phenotypic characterization revealed that the female *Hs3st-1−/−* mice exhibited defective reproductive performance in different degrees, including impaired ovarian function as well as intrauterine growth restriction that were linked to delayed placental development.[37,38] Histological analysis of implantation sites at mid-gestation (E9.5–11.5) further revealed that placental development in the *Hs3st-1−/−* maternal uterus was markedly delayed, with typically increased numbers of trophoblast giant cells, a rudimentary labyrinth, and no patent maternal–fetal circulation (Table I). The lack of blood circulation highlights that angiogenesis is disrupted at the implantation sites of the *Hs3st-1−/−* female uterus, which may contribute significantly to the compromised fertility of *Hs3st-1−/−* female mice.[38] However, further studies are required to determine whether the *Hs3st-1* deletion disrupts angiogenic signaling in the mutant mice. The study of *Hs6st-1−/−* mice has provided further supportive evidence as well.[35] The *Hs6st-1−/−* mice exhibit ~50% reduction in the number of fetal microvessels in the labyrinthine zone of the placenta relative to that in wild-type littermates. The reduced vascularization leads to hypoxia, signifying that altered angiogenesis may impair the function of the *Hs6st-1−/−* placentas, which, as a consequence, may contribute to growth retardation and lethality in the *Hs6st-1−/−* embryos. Moreover, the expression of VEGF-A mRNA and its protein is reduced in the *Hs6st-1−/−* placenta, highlighting that HS regulates VEGF-A signaling to modulate vascular development in the placenta.[35] However, our *EC-Ndst-1−/−* mutants develop normally with normal fertility,[31,32] indicating that a more severe HS structure alteration in the vasculature is required to rigorously test whether HS modulates reproductive angiogenesis.

2. Wound Angiogenesis

Wound healing is a complex tissue remodeling process, and may be conveniently divided into three phases: inflammatory, proliferative, and remodeling. Leukoctyes that infiltrate tissues during the inflammatory phase secrete proangiogenic growth factors and cytokines that recruit ECs to initiate angiogenesis, a major feature of the proliferative phase. Angiogenesis is essential for wound healing. Newly formed blood vessels participate in provisional granulation-tissue formation, and provide nutrition and oxygen to the growing tissues. Wound angiogenesis is tightly regulated by diverse HS-binding factors, including VEGF, FGF, TGF-β, PDGF-B, SDF-1, and MCP-1, which act in sequential, concerted, and synergistic manners, suggesting that HS may modulate the functions of these protein ligands to critically regulate wound angiogenesis. This hypothesis has been strongly supported by studies of *syndecan-4−/−* and heparinase-overexpressing transgenic mice.[80,81]

The syndecans are a family of transmembrane HS-proteoglycans. Syndecans have been shown to bind proangiogenic factors and cytokines to regulate the angiogenic process.[14] Syndecan-4 is detectable in the epidermis, but not in the dermis, of uninjured adult mouse skin. After skin injury, however, syndecan-4 is upregulated throughout the granulation tissue on fibroblasts and ECs, suggesting that syndecan-4 may regulate wound healing and related angiogenesis.[82] This was tested by the examination of *syndecan-4* mutant mice.[80] The *syndecan-4−/−* and *syndecan-4+/−* mice were viable, fertile, and macroscopically indistinguishable from wild-type littermates, but showed statistically significant delayed healing of skin wounds and impaired angiogenesis in the granulation tissue, thus demonstrating directly that syndecan-4 is an important cell-surface receptor in wound healing and wound angiogenesis. Most of the biological functions of syndecans are mediated by their HS moiety,[14] suggesting that HS critically regulates wound angiogenesis. This is alternatively supported by examining wound healing in heparinase-overexpressing mice.[81] Heparanase, an endoglycosidase that degrades HS in the ECM and at cell surface, releases HS-bound growth factors and converts them into bioactive molecules. Heparanase is expressed in wound granulation tissue, suggesting that heparinase may modulate the bioavailabilty and activation of a multitude of mediators capable of promoting EC migration, proliferation, and MC recruitment in the complex setting of wound healing. Indeed, heparanase-overexpressing transgenic mice showed a remarkably elevated wound angiogenesis response.[81] Consistent with this, topical application of recombinant heparanase significantly accelerated wound healing and markedly improved wound survival, which then proceeded with enhanced blood vessel maturation.[81] These observations clearly demonstrated that heparanase-accelerated tissue repair and skin survival occur through enhancement of the angiogenic response, further supporting the importance of HS in the modulation of wound angiogenesis. In a recent study, where tumor angiogenesis was altered in the EC- *Ndst-1−/−* state, we noted that wound angiogenesis was not obviously altered in comparison with that of wild-type littermates,[31] showing that endothelial *Ndst-1* may be dispensable for wound healing and wound angiogenesis. Given that endothelial *Ndst-1* ablation only partially alters HS structure, more dramatic lesions in HS structure may be required to determine the requirement and the role of HS in wound angiogenesis.

II. Endothelial Heparan Sulfate in Pathologic Angiogenesis

A. Tumor, Vasculature, Stroma, and Heparan Sulfate: An Introduction

A large body of work has now highlighted a number of important roles that HS plays in the formation, vascularization, and spread of solid tumors. It is important to recognize at the outset that an important biological principle is that solid tumors and their metastases must gain a vasculature if they are to grow into macroscopic tumors. The threshold for nonvascular growth is typically ~ 1 mm. Beyond that size, tumors must be remodeled through the process of angiogenesis in order to grow (and contribute to multiple downstream pathologic consequences). The focus herein is on the roles of vascular endothelial HS in the growth and remodeling of tumor neovasculature. A practical way to discuss HSPGs produced by both nascent as well as established tumor vasculature is by categorizing them into either EC-surface HSPGs that remain tethered as part of the endothelial glycocalyx or secreted HSPGs distributed in the perivascular ECM, including those that contribute to the content of vascular basement membranes. It should be noted that production of the latter might result from HSPGs secreted by not solely the vascular endothelium, but by tumor and/or stromal cells, including perivascular cells, in the same micro-environment. We shall consider these anatomical relationships as we discuss the forces that regulate tumor vascular HSPGs and their interactions with multiple tumor vascular effector molecules and consider the effects of targeting genetic alterations in HSPG biosynthesis to distinct cellular compartments in models of tumor angiogenesis.

B. Endothelial Heparan Sulfate and the Control of Tumor Proangiogenic Growth Factors

1. Tumor Hypoxia, Heparan Sulfate, and the Distribution of Tumor Proangiogenic Growth Factors

During states of hypoxia, expression of the hypoxia-inducible transcription factor (HIF-1) stimulates the expression of VEGF, which leads to responses that promote oxygen delivery to hypoxic tissues, such as increases in vascular permeability and capillary density as a result of angiogenesis.[83,84] In neoplasia, this process is stimulated by rapid clonal tumor growth, with associated hypoxia as the tumor mass rapidly outgrows its blood supply. In addition, data from mice bearing an endothelial-specific genetic alteration in HIF-1alpha shows that HIF-1 expression is also important in sustaining an autocrine signaling loop involving VEGF and the major endothelial receptor VEGFR-2.[85] Through its ability to bind distinct isoforms of VEGF, endothelial HS may play important roles in regulating matrix as well as EC-surface distribution of VEGF during this process. The biosynthesis of VEGF, also known as VEGF-A to distinguish it from other VEGF homologs involved in cardiac and lymphatic development, is posttranscriptionally regulated. A variety of splice variants, including VEGF145, VEGF165, and VEGF189, retain basic-amino acid rich C-terminal exons (6 and/or 7) that confer heparin-binding ability.[86] HS expressed on EC-surface proteoglycans of growing tumor microvasculature as well as proteoglycans secreted into the ECM has the ability to differentially bind, and thereby control gradients, of these potent proangiogenic VEGF isoforms.[87] This is in addition to its ability to control gradients of several other growth factors (to be discussed further), allowing the distribution and/or mobilization of several important proangiogenic factors to be modulated by the production and distribution of HS. In the case of VEGF-A, the distribution of heparin-binding species of VEGF (e.g., VEGF165 or VEGF189) is distinct from that of non-HS binding species such as VEGF121, wherein the latter may diffuse to a further distance from tumor cells, thereby affecting vasculature at a greater distance from VEGF-secreting tumor nests.[87] Moreover, the development of hypoxia within the center of rapidly growing tumors may affect both the biosynthesis of HS and the way in which HS interacts with these factors, thereby adding an additional layer growth factor modulation. For example, hypoxia increased FGF2-binding sites on HS as well as the action of FGF-2 over that observed in normoxic condx.[88] Moreover, with respect to VEGF, it is interesting that cancer microenvironment changes in pH appear to affect the expression of VEGF splice variants, with greater expression of VEGF121 by tumor cells exposed to a more acidemic environment.[89] This appears to be associated with p38 signal activation and SR protein (splicing factor) expression and phosphorylation. Since hypoxia results in anaerobic metabolism and tissue acidemia, it is possible that differential release of distinct VEGF isoforms varies with distance from the hypoxic tumor center. Thus, the hypoxic tumor center may produce a more "diffusible" form of VEGF-A (121 isoform), leaving the tumor center with a relative paucity of (matrix-bound) VEGF. Whether this is a major factor contributing to a progressive lack of blood supply and necrosis of the tumor center remains to be determined. Nevertheless, a common theme in the pathophysiology of tumor angiogenesis is the establishment of proangiogenic growth factor gradients that appear to be induced and controlled by both hypoxia as well as the expression of HS, with the latter possibly having profound effects on the distribution of major growth factors and the consequent patterns of tumor vascularization.

2. Multiple Heparin-Binding Growth Factors and the "Redundancy" of Proangiogenic Effectors

The targeting of VEGF-A in human tumors using a humanized anti-VEGF antibody has become standard practice for advanced-stage tumors of the lung and colon.[90,91] The effects of this on tumor vasculature include regression as well as narrowing and smooth-muscle investment of the tumor's tortuous vasculature network, with the latter resulting from the effects of high tumor VEGF-A levels on tumor-associated endothelium.[8] In lung cancer, adding this form of anti-VEGF therapy to combination chemotherapy results in a modest survival advantage[91]; however, a variety of mechanisms may limit efficacy and duration of the effect, including induction of other proangiogenic factors such as PDGF and FGF-2 (reviewed in Ref. 92). Targeting HS may alter receptor binding by several heparin-binding proangiogenic growth factors that may be induced in tumors (including HGF, PDGF, FGF-2, among others). Thus, it is possible that the escape from VEGF-blockade as a result of such redundant tumor growth mechanisms may be subverted by targeting endothelial HS biosynthesis. One method to achieve this may be through the introduction of polysulfated heparin mimetics or polysuflonated compounds (e.g., suramin) that may compete for binding to FGF-2[93] and possibly multiple growth factor–HS interactions during tumor angiogenesis. Yet another potential form of anti-VEGF resistance is the persistence of basement membrane "ghosts" following VEGF blockade.[94,95] It has been demonstrated that tumor endothelium may grow back to fill these channels once VEGF-therapy is stopped or held.[95] Basement membranes are endowed with abundant secreted HS-proteoglycans that may bind and store high levels of FGF-2, VEGF-A, and other proangiogenic factors. It is thus possible that targeting HS in this setting may prevent such regrowth following anti-VEGF therapy.

3. Regulation of Endothelial Heparan Sulfate in Cancer: Heparanase and Sulfatases

a. Heparanase. Storage of growth factors bound to HS in tumor ECM as well as vascular basement membranes provides a "bank" of proangiogenic factors that may be mobilized by the release of heparanase by tumor cells.[96] Mammalian heparanase is an endo-beta-D glucuronidase that degrades HS ubiquitously expressed on cell-surface as well as matrix-associated proteoglycans. Homozygous transgenic mice that overexpress heparanase demonstrate both a profound reduction in the size of HS chains as well as excess branching of mammary gland ducts and enhanced neovascularization of mammary ducts as well as hair follicles. The overexpression of heparanase by tumors may activate tumor angiogenesis through a variety of mechanisms in addition to promoting the release of growth factor-decorated HS fragments.[96,97] It now appears that transgenic (including tumor-induced) expression of heparanase leads to increased HS *N*- and *O*-sulfation, which in turn strongly promotes the formation of FGF-1 or FGF-2—FGFR-1 ternary complexes.[98] This may also promote HS coreceptor activity for other vascular endothelial growth factors. In addition, the action of heparanase in myeloma as well as other tumors promotes the shedding of syndecan-1, which in turn complexes with VEGF-A in the ECM that in turn facilitate growth and migration of ECs through the combined ability of VEGF-A/syndecan-1 complexes to activate endothelial VEGFRs and engage integrins though an integrin-binding region on the syndecan-1 core protein.[99,100] Accordingly, inhibition of heparinase has become appealing as a form of antiangiogenic therapy. Along those lines, a variety of nonanticoagulant forms of heparin have been developed as competitive inhibitors of mammalian heparinase.[101] Novel inhibitors of heparanase that may inhibit tumor angiogenesis include polysulfated penta and tetrasaccharide glycosides[102] as well as *N*-acetylated glycol-split heparin species.[103]

b. Sulfatases. A growing literature has revealed the importance of another family of enzymes that modifies the structure of mature HS chains in carcinoma: the extracellular sulfatases Sulf-1 and Sulf-2. As endosulfatases, these enzymes remove internal glucosamine 6-*O*-sulfate modifications within HS.[104] A variety of tumors, including ovarian, breast, and hepatocellular carcinomas, appear to downregulate Sulf-1 expression, which in turn appears to upregulate autocrine and/or paracrine tumor-cell signaling by a variety of

HS-binding tumor growth factors, including FGF-2, HGF, and HB-EGF.[105,106] The importance of these enzymes in tumor angiogenesis has been examined in a few preliminary studies. Curiously, while short hairpin RNA-mediated inhibition of Sulf-1 in human ECs was associated with increased FGF-2, HGF, and VEGF165-mediated endothelial proliferation,[106] another study demonstrated that the exogenous addition of purified recombinant Sulf-2 promoted angiogenesis in the chick chorioallantoic membrane assay.[107] One explanation for these apparently contradictory findings may center on the spatial context or site of action of the enzyme (i.e., removal of glucosamine 6-*O*-sulfate on ECM HS vs. cell-surface HS). Specifically, the effect may vary with respect to how the enzyme is manipulated during the experiment. In this manner, it is possible that inhibiting or silencing endothelial production of the enzyme may exert its dominant effect on cell-surface HS, thereby promoting HS glucosamine 6-*O*-sulfate-dependent receptor activation by FGF-2,[101,106] whereas exogenous addition of enzyme to the angiogenesis system may have preferential effects on reducing glucosamine 6-*O*-sulfation on ECM HS, thereby increasing the availability of matrix-bound growth factor for binding to endothelial growth receptors.[107] Further studies are needed to define the specific effects of the sulfatases on tumor endothelium and understand the interplay of those effects with the well-characterized roles of sulfatase expression on tumor-cell growth and behavior in the same microenvironment.

C. Targeting HSPG Mutations to Distinct Vascular Cells in Cancer Models: Lessons Learned

1. Models Targeting Endothelial Heparan Sulfate Biosynthesis

Data acquired from both developmental as well as tumor-based studies in mice employing endothelial-targeted genetic alterations in HS biosynthesis have improved our understanding of the importance of HS in tumor angiogenesis. One approach is to attempt the targeting of HS polymerization in the mouse, with the aim to block HS chain assembly on all HS-proteoglycans either systemically or targeted to the endothelium. The goal would then be to examine tumor growth and vascularization on the mutant background. This has been limited by either arrest of development during gastrulation (including lack of mesoderm development) in embryos harboring a homozygous-null mutation in the HS copolymerase *Ext-1*,[62] or embryonic lethality encountered upon attempting to drive the *Ext-1* mutation specifically in ECs by crossing conditional (floxed) *Ext-1* mutant mice onto the *Tie2Cre* transgenic background (unpublished data). Using chimeric cultures of embryonic stem cells bearing alterations in either the HS or VEGFR-2 biosynthetic pathways, Jakobsson and others[68] were able to demonstrate that endothelial signaling in proliferating vasculature may be fully supported by HS acting *in trans* by adjacent supporting smooth-muscle perivascular cells. The degree to which this mechanism might contribute to endothelial growth and remodeling during tumor angiogenesis is unknown; however, targeting biosynthesis of the major HS sulfating enzyme *N*-deacetylase/*N*-sulfotransferase-1 (Ndst-1) to ECs (i.e., *Ndst-1* f/f *Tie2Cre* mutants) is sufficient to alter tumor growth and angiogenesis in adult mutants.[31] Mutant tumor vasculature demonstrated abnormalities in branching as well as alterations in its ability to bind and signal in response to FGF-2 and VEGF165. Interestingly, our group has since discovered that marked inhibition of tumor growth angiogenesis is achieved in mice bearing the same endothelial-targeted mutation on a transgenic spontaneous mammary carcinoma background (M. M. Fuster, unpublished data). It thus appears that the cell-autonomous action of appropriately sulfated tumor endothelial HS is required to facilitate the action of major proangiogenic tumor growth factors in tumor angiogenesis. More work is needed to better understand the additional importance of HS produced by other cells, including tumor, mural, and other stromal cells, during tumor angiogenesis.

2. Models Targeting Pericyte Heparan Sulfate Biosynthesis

A number of studies now highlight important functions served by HS produced by vascular pericytes during endothelial sprouting and remodeling. However, most of these studies have focused on vascular growth and remodeling during development and not tumor angiogenesis *per se*. Nevertheless, understanding the genetic importance of "periendothelial" HS in this context may provide valuable clues for future studies examining the genetic importance of peircyte HS in tumor angiogenesis. A key heparin-binding growth factor that has been implicated in the recruitment of pericytes onto angiogenic vasculature is endothelial PDGF-BB. Deleting the C-terminal HS-binding motif of PDGF-BB results in impaired vascular retention of the growth factor and impaired pericyte recruitment. Studies examining pericyte recruitment in gene-targeted mice bearing alterations in HS biosynthesis, including *Ndst-1* mutants, showed that recruitment of pericytes requires glucosamine *N*-sulfated domains to activate the receptor PDGF-R-beta by retained PDGF-BB.[30] In carcinomas, the effect of altering PDGF binding and function through the targeting of HS may be complex. First, in tumor models, PDGF-BB appears to stimulate pericyte recruitment through induction of the endothelial chemokine stromal-derived factor-1 alpha (SDF1alpha), and CXCR4-mediated pericyte chemotaxis.[108] Although this may stimulate the release of proangiogenic mediators released from pericytes,[109] it also supports vessel maturation by investing newly formed endothelial tubes with VSMC pericytes. Interestingly, the latter phenomenon in tumors appears to occur during "normalization" or "maturation" of tumor vessels as a result of anti-VEGF therapy in cancer,[8] wherein pericyte investment occurs during a "straightening" and narrowing of the tortuous tumor vasculature during VEGF blockade. This in turn appears to result in reduced tumor interstitial pressure and more favorable hemodynamic conditions for the delivery chemotherapy. In addition, a recent study demonstrated that VEGF was found to negatively regulate the effects of PDGF on pericytes through the formation of novel receptor complexes made up of activated VEGFR-2 and PDGF-R-beta.[110] In a fibrosarcoma model, VEGF(−/−) sarcomas were associated with a reduction in VEGFR-2/PDGF-R-beta complexes when compared with that in VEGF(+/+) sarcomas, and this was in turn associated with tumor vessel maturation.[110] Thus, although endothelial HS alterations may attenuate angiogenic growth signaling in ECs, the effects of altering HS on pericytes during tumor angiogenesis may be more complex, wherein an inhibitory effect on the (VEGF-mediated) release of proangiogenic growth factors by perictyes may be balanced by an alteration in PDGF-mediated pericyte recruitment, and the associated maintenance of a tortuous, immature vessel that lacks smooth-muscle investment. Further genetic studies targeting pericyte HS during tumor angiogenesis may reveal the "balance" of these factors on tumor vascular growth and hemodynamics.

3. Models Targeting the Scaffolding of Heparan Sulfate: Core Proteins

Studying the effects of altering HS biosynthesis in endothelium has led to important insights with respect to the coreceptor- and growth factor-sequestering functions of HS in tumor angiogenesis, and several interesting patterns have emerged from genetic studies targeting HS proteoglycan core proteins. As protein scaffolds for HS chains on EC membranes, members of the syndecan and glypican core protein families (reviewed in Ref. [111]) may serve in a cell-autonomous manner to facilitate receptor binding and signaling by major proangiogenic growth factors to growing tumor endothelium. Athymic mice lacking glypican-1 demonstrate decreased tumor angiogenesis in response to intrapancreatic implantation of PANC-1 or T3M4 human pancreatic cancer cells. Syndecan-4 knockout mice demonstrate defects in wound angiogenesis,[112] although the specific importance of syndecan-4 in tumor angiogenesis has not been reported. The effects of syndecan-4 deletion on endothelial behavior may not solely be due to the associated reduction in EC-surface HS (i.e., as a coreceptor), but may also have to do with the finding that syndecan-4 plays an important role in focal adhesion kinase phosphorylation and cell migration.[113] In tumor models, genetic studies have also recently revealed a domain within the endothelial syndecan-1 core protein that directly associates with alpha(v)beta(3) and alpha(v)beta(5) integrins, and facilitates integrin-mediated tumor angiogenesis.[114] Finally, secreted HS-proteoglycans (such as perlecan, collagen-XVIII, agrin) residing on basement membranes and/or periendothelial matrix may serve dual functions: although they may sequester tumor proangiogenic growth factors (e.g., FGF-2, VEGF, or PDGF) onto tumor vascular basement membrane and/or matrix HS chains, thereby facilitating tumor angiogenesis, the core proteins may also be proteolytically cleaved, with the release of angiostatic C-terminal fragments. Genetic alteration of the perlecan NH_2-terminal HS-attachment sites resulted in altered tumor and wound angiogenesis.[115] Interestingly, perlecan-associated HS also may serve as a "rescue" mechanism during targeting of the tumor VEGF pathway. Specifically, an increase in perlecan and heparanase expression has been noted in regions of angiogenesis recovery in VEGFR-2 targeted hepatoblastoma xenografts.[116] But then, proteolytic cleavage of the C-terminal domains of perlecan as well as collagen-XVIII, for example, leads to the release of the angiostatic fragments endorepellin and endostatin, respectively. These fragments may counter HS proteoglycan-mediated angiostimulatory processes through their actions on integrins (alpha2beta1 in the case of endorepellin[117]) or through downregulation the VEGF signaling cascade and stimulation of thrombospondin-1 (in the case of endostatin[118]).

D. Other Forms of Pathologic Angiogenesis and Endothelial Heparan Sulfate

1. Lessons Learned from Proliferative Retinopathy and VEGF

Given the broad roles of HS in modulating the activity of multiple growth factors in tumor angiogenesis, in addition to the unique importance of HS proteoglycan core proteins in mediating distinct pathophysiologic steps in endothelial growth and migration, it is useful to consider how targeting HS might factor into the treatment of other angioproliferative diseases. In the last 10 years, anti-VEGF therapy, in the form of the humanized monoclonal antibody bevacizumab, has been FDA-approved for use in combination with chemotherapy in the treatment of colon cancer[90] and non-small cell lung cancer.[91] Prior to this, elevated levels of VEGF-A were found in the aqueous and vitreous humor of human eyes in the setting of diabetic retinopathy as well as adult macular degeneration (AMD), particularly exudative (or "wet") AMD.[119,120] These findings established the basis for the intravitreous injectable treatments pegaptinib (Macugen), a pegylated oligonucleotide aptamer that binds VEGF165, and ranibizumab (Lucentis), a recombinant humanized Fab that binds all VEGF-A isoforms.[121,122] Interestingly, these revolutionary AMD treatments not only slowed the loss of vision in such patients, but also led to an unexpected sustained gain of vision in a number of patients.[119] Although one may consider a role for targeting HS in the light of limitations in efficacy as well as mechanisms of resistance to anti-VEGF therapy in tumors,[92] targeting HS may also be a relevant topic in the consideration of current challenges in proliferative retinopathy, including combined therapies to improve duration of responses (e.g., beyond 24 months of anti-VEGF treatment in AMD) as well as minimize side effects.[123] Indeed, other proangiogenic pathways involving HS-binding growth factors appear to act in concert with VEGF-A during vascular progression in proliferative retinopathies, including AMD as well as diabetic retinopathy.[124–126] Thus, the application of novel agents that either compete with HS to simultaneously limit receptor activation by multiple growth factors (e.g., heparinoids or polysulfonated mimetics), or target HS biosynthesis (e.g., enzyme small-molecule inhibitors), may serve to limit levels of the anti-VEGF agent(s) while possibly prolonging action as well as limiting toxicity.

2. Novel Targeting of Angiogenesis in Other Chronic Inflammatory Conditions: Targeting Heparan Sulfate?

In addition to cancer and proliferative vascular retinopathies, pathologic angiogenesis also occurs during inflammatory disorders, with particularly well-characterized examples occurring in rheumatoid arthritis and psoriasis. The HS proteoglycan core proteins syndecan-2, syndecan-3, and glypican-4 appear to be upregulated in the endothelium of inflamed synovial tissues in both rheumatoid and psoriatic arthritis (when compared with synovial samples from normal or osteoarthritic joints), wherein intense angiogenesis contributes to progression of rheumatic joint pathology.[127] Interestingly, the upregulation of HS on multiple endothelial proteoglycan core proteins in such disorders may not only contribute to angiogenesis though its coreceptor action on proangiogenic growth factor receptors, but also through the ability of HS to concentrate the chemokine CXCL12 in the rheumatoid hyperplastic lining layer and endothelium. It appears that synoviocyte-derived CXCL12 is immobilized and accumulates on endothelial-cell HS, through which it promotes angiogenesis.[128] In this context, it may also be appealing to follow similar HS-targeting approaches as those proposed for proliferative vascular ocular disorders above, wherein HS itself may be targeted through competitive or small-molecule approaches. Given the simultaneous upregulation of a number of endothelial HS proteoglycan core proteins in rheumatoid arthropathy, targeting the (common) HS glycan chain may prove more promising than interference with specific core proteins.

E. Turning It Around—Therapeutic Angiogenesis and Endothelial Heparan Sulfate

1. The Actions of Proangiogenic Growth Factors in Cardiovascular Ischemic Remodeling

Insights on the genetic importance of HS in mediating angiogenesis as a pathologic process allow us to better investigate a new direction: Can one facilitate angiogenesis as a therapeutic process in ischemic disorders, and how might one manipulate HS in a manner that promotes therapeutic angiogenesis? First, the state of hypoxia, especially chronic hypoxia, upregulates the expression of hypoxia-inducible transcription factors (HIF-1 and HIF-2), which in turn upregulates VEGF-A expression[129,130] and vascular endothelial proliferation.[129] It is also interesting to note that cardiac ischemic models have shown that tissue upregulation of HS occurs as a result of the ischemic insult.[131,132] Although it is possible that the latter facilitates the action of proangiogenic growth factors, the cellular distribution as well as fine structure of upregulated HS in this setting are poorly understood. Nevertheless, direct delivery of proangiogenic growth factors, including FGFs and VEGF, to the myocardium has been demonstrated in animal models of coronary ischemia (employing intracoronary adenoviral-mediated gene transfer as well as novel ultrasound-mediated myocardial plasmid delivery), with resultant postischemic stimulation of angiogenesis, and improvements in regional myocardial blood flow and function.[133,134] It is possible that manipulating the production and structure of HS in the ischemic microenvironment (including that on the endothelium as well as ECM) may optimize the ability of HS to efficiently serve as a coreceptor such as growth factors, whether they are exogenously introduced or produced locally. This might also be achieved through targeted alterations in sulfatase and heparanase expression in ischemic tissues. Finally, upregulation or exogenous delivery of proangiogenic growth factors may have pathologic consequences, such as the Src-mediated vasogenic edema that results from VEGFR signaling pathway activation.[135] Thus, any strategy that augments specific growth factors, HS biosynthesis and/or fine structure, or both needs to be considered in light of not only its proangiogenic potential, but also its net effect on capillary leak and ischemic tissue edema.

2. Chronic Wound Healing and Modifying Wound Granulation

Wound healing is characterized by an active phase of granulation-tissue proliferation, typically within the first few days following acute injury. This form of inflammatory remodeling is characterized by intense angiogenesis, and a variety of factors may limit or halt this proliferative response, leading to chronic wound progression, tissue necrosis, infection, and other complications.[136] Angiopoietins, VEGF, FGFs, and TGF-β are involved in stimulating this process, and their actions are coordinated with those of upregulated laminins in the ECM as well as integrins (principally beta-1 and alpha-v subunits), mediating the interactions between migrating cells and the matrix.[137] Augmenting the expression or structural features of HS, which is present on wound endothelium, vascular basement membranes, and the ECM, may improve granulation-tissue angiogenesis and wound healing. One interesting approach is the exogenous addition or transgenic expression of heparanase, with associated enhanced wound epithelialization as well as wound angiogenesis and vessel maturation.[81] This may increase the availability of HS-bound growth factors, as occurs as a result of heparanase action during tumor angiogenesis, with subsequent stimulation of endothelial growth and angiogenesis. Genetic evidence points to matrix and basement membrane-associated HS expressed on perlecan as a particularly important form of HS presentation, contributing to granulation-associated angiogenesis.[115,138] It is possible that targeted upregulation of perlecan in chronic nonhealing wounds might augment angiogenesis and the regeneration of healthy granulation tissue, promoting vital wound healing and closure. It is also possible that manipulating the expression or delivery of heparanase (and possibly sulfatases) in this context might modify the structure of endothelial and/or matrix HS in the wound environment in a manner that maximizes granulation-associated angiogenesis through optimal interactions of wound endothelium with growth factors and integrin-mediated interactions with matrix proteins.

Acknowledgments

We thank Ms. Karen Howard for her careful reading and English version of this chapter. Our research has been supported by the American Heart Association (to L. W.) and NIH (R01HL093339 and P41RR005351 to L. W.); as well as the American Cancer Society (to M. M. F.) and the U.S. Department of Veterans Affairs (to M. M. F.).

References

1. Carmeliet P. Angiogenesis in life, disease and medicine. *Nature* 2005;**438**:932–6.
2. Coultas L, Chawengsaksophak K, Rossant J. Endothelial cells and VEGF in vascular development. *Nature* 2005;**438**:937–45.
3. Risau W. Mechanisms of angiogenesis. *Nature* 1997;**386**:671–4.
4. Suri C, Jones PF, Patan S, Bartunkova S, Maisonpierre PC, Davis S, et al. Requisite role of angiopoietin-1, a ligand for the TIE2 receptor, during embryonic angiogenesis. *Cell* 1996;**87**:1171–80.
5. Alva JA, Iruela-Arispe ML. Notch signaling in vascular morphogenesis. *Curr Opin Hematol* 2004;**11**:278–83.
6. Sato TN. Vascular development: molecular logic for defining arteries and veins. *Curr Opin Hematol* 2003;**10**:131–5.
7. Yancopoulos GD, Davis S, Gale NW, Rudge JS, Wiegand SJ, Holash J. Vascular-specific growth factors and blood vessel formation. *Nature* 2000;**407**:242–8.
8. Jain RK. Molecular regulation of vessel maturation. *Nat Med* 2003;**9**:685–93.
9. Drake CJ. Embryonic and adult vasculogenesis. *Birth Defects Res C Embryo Today* 2003;**69**:73–82.
10. Stappenbeck TS, Hooper LV, Gordon JI. Developmental regulation of intestinal angiogenesis by indigenous microbes via Paneth cells. *Proc Natl Acad Sci USA* 2002;**99**:15451–5.
11. Carmeliet P, Ferreira V, Breier G, Pollefeyt S, Kieckens L, Gertsenstein M, et al. Abnormal blood vessel development and lethality in embryos lacking a single VEGF allele. *Nature* 1996;**380**:435–9.

12. Folkman J. Angiogenesis in cancer, vascular, rheumatoid and other disease. *Nat Med* 1995;**1**:27–31.
13. Griffioen AW, Molema G. Angiogenesis: potentials for pharmacologic intervention in the treatment of cancer, cardiovascular diseases, and chronic inflammation. *Pharmacol Rev* 2000;**52**:237–68.
14. Bernfield M, Gotte M, Park PW, Reizes O, Fitzgerald ML, Lincecum J, et al. Functions of cell surface heparan sulfate proteoglycans. *Annu Rev Biochem* 1999;**68**:729–77.
15. Esko JD, Selleck SB. Order out of chaos: assembly of ligand binding sites in heparan sulfate. *Annu Rev Biochem* 2002;**71**:435–71.
16. Bishop JR, Schuksz M, Esko JD. Heparan sulphate proteoglycans fine-tune mammalian physiology. *Nature* 2007;**446**:1030–7.
17. Forsberg E, Kjellen L. Heparan sulfate: lessons from knockout mice. *J Clin Invest* 2001;**108**:175–80.
18. Rapraeger AC. Heparan sulfate-growth factor interactions. *Methods Cell Biol* 2002;**69**:83–109.
19. Kjellen L. Glucosaminyl *N*-deacetylase/*N*-sulphotransferases in heparan sulphate biosynthesis and biology. *Biochem Soc Trans* 2003;**31**:340–2.
20. Esko JD, Lindahl U. Molecular diversity of heparan sulfate. *J Clin Invest* 2001;**108**:169–73.
21. Stringer SE. The role of heparan sulphate proteoglycans in angiogenesis. *Biochem Soc Trans* 2006;**34**:451–3.
22. Presta M, Dell'Era P, Mitola S, Moroni E, Ronca R, Rusnati M. Fibroblast growth factor/fibroblast growth factor receptor system in angiogenesis. *Cytokine Growth Factor Rev* 2005;**16**:159–78.
23. Iozzo RV, San Antonio JD. Heparan sulfate proteoglycans: heavy hitters in the angiogenesis arena. *J Clin Invest* 2001;**108**:349–55.
24. Ledin J, Staatz W, Li JP, Gotte M, Selleck S, Kjellen L. Heparan sulfate structure in mice with genetically modified heparan sulfate production. *J Biol Chem* 2004;**279**:42732–41.
25. Allen BL, Filla MS, Rapraeger AC. Role of heparan sulfate as a tissue-specific regulator of FGF-4 and FGF receptor recognition. *J Cell Biol* 2001;**155**:845–58.
26. Stenzel D, Nye E, Nisancioglu M, Adams RH, Yamaguchi Y, Gerhardt H. Peripheral mural cell recruitment requires cell-autonomous heparan sulfate. *Blood* 2009;**114**:915–24.
27. Fan G, Xiao L, Cheng L, Wang X, Sun B, Hu G. Targeted disruption of NDST-1 gene leads to pulmonary hypoplasia and neonatal respiratory distress in mice. *FEBS Lett* 2000;**467**:7–11.
28. Pallerla SR, Pan Y, Zhang X, Esko JD, Grobe K. Heparan sulfate Ndst1 gene function variably regulates multiple signaling pathways during mouse development. *Dev Dyn* 2007;**236**:556–63.
29. Ringvall M, Ledin J, Holmborn K, van Kuppevelt T, Ellin F, Eriksson I. Defective heparan sulfate biosynthesis and neonatal lethality in mice lacking N-deacetylase/N-sulfotransferase-1. *J Biol Chem* 2000;**275**:25926–30.
30. Abramsson A, Kurup S, Busse M, Yamada S, Lindblom P, Schallmeiner E, et al. Defective *N*-sulfation of heparan sulfate proteoglycans limits PDGF-BB binding and pericyte recruitment in vascular development. *Genes Dev* 2007;**21**:316–31.
31. Fuster MM, Wang L, Castagnola J, Sikora L, Reddi K, Lee PH, et al. Genetic alteration of endothelial heparan sulfate selectively inhibits tumor angiogenesis. *J Cell Biol* 2007;**177**:539–49.
32. Wang L, Fuster M, Sriramarao P, Esko JD. Endothelial heparan sulfate deficiency impairs L-selectin- and chemokine-mediated neutrophil trafficking during inflammatory responses. *Nat Immunol* 2005;**6**:902–10.
33. Li JP, Gong F, Hagner-McWhirter A, Forsberg E, Abrink M, Kisilevsky R, et al. Targeted disruption of a murine glucuronyl C5-epimerase gene results in heparan sulfate lacking L-iduronic acid and in neonatal lethality. *J Biol Chem* 2003;**278**:28363–6.

34. Chen E, Stringer SE, Rusch MA, Selleck SB, Ekker SC. A unique role for 6-*O* sulfation modification in zebrafish vascular development. *Dev Biol* 2005;**284**:364–76.
35. Habuchi H, Nagai N, Sugaya N, Atsumi F, Stevens RL, Kimata K. Mice deficient in heparan sulfate 6-*O*-sulfotransferase-1 exhibit defective heparan sulfate biosynthesis, abnormal placentation, and late embryonic lethality. *J Biol Chem* 2007;**282**:15578–88.
36. Izvolsky KI, Lu J, Martin G, Albrecht KH, Cardoso WV. Systemic inactivation of Hs6st1 in mice is associated with late postnatal mortality without major defects in organogenesis. *Genesis* 2008;**46**:8–18.
37. HajMohammadi S, Enjyoji K, Princivalle M, Christi P, Lech M, Beeler D, et al. Normal levels of anticoagulant heparan sulfate are not essential for normal hemostasis. *J Clin Invest* 2003;**111**:989–99.
38. de Agostini A. An unexpected role for anticoagulant heparan sulfate proteoglycans in reproduction. *Swiss Med Wkly* 2006;**136**:583–90.
39. Carmeliet P, Ng YS, Nuyens D, Theilmeier G, Brusselmans K, Cornelissen I, et al. Impaired myocardial angiogenesis and ischemic cardiomyopathy in mice lacking the vascular endothelial growth factor isoforms VEGF164 and VEGF188. *Nat Med* 1999;**5**:495–502.
40. Ruhrberg C, Gerhardt H, Golding M, Watson R, Ioannidou S, Fujisawa H, et al. Spatially restricted patterning cues provided by heparin-binding VEGF-A control blood vessel branching morphogenesis. *Genes Dev* 2002;**16**:2684–98.
41. Gerhardt H, Golding M, Fruttiger M, Ruhrberg C, Lundkvist A, Abramsson A, et al. VEGF guides angiogenic sprouting utilizing endothelial tip cell filopodia. *J Cell Biol* 2003;**161**:1163–77.
42. Fong GH, Rossant J, Gertsenstein M, Breitman ML. Role of the Flt-1 receptor tyrosine kinase in regulating the assembly of vascular endothelium. *Nature* 1995;**376**:66–70.
43. Shalaby F, Rossant J, Yamaguchi TP, Gertsenstein M, Wu XF, Breitman ML, et al. Failure of blood-island formation and vasculogenesis in Flk-1-deficient mice. *Nature* 1995;**376**:62–6.
44. Lindblom P, Gerhardt H, Liebner S, Abramsson A, Enge M, Hellstrom M, et al. Endothelial PDGF-B retention is required for proper investment of pericytes in the microvessel wall. *Genes Dev* 2003;**17**:1835–40.
45. Lindahl P, Johansson BR, Leveen P, Betsholtz C. Pericyte loss and microaneurysm formation in PDGF-B-deficient mice. *Science* 1997;**277**:242–5.
46. Abramsson A, Lindblom P, Betsholtz C. Endothelial and nonendothelial sources of PDGF-B regulate pericyte recruitment and influence vascular pattern formation in tumors. *J Clin Invest* 2003;**112**:1142–51.
47. Hellstrom M, Gerhardt H, Kalen M, Li X, Eriksson U, Wolburg H, et al. Lack of pericytes leads to endothelial hyperplasia and abnormal vascular morphogenesis. *J Cell Biol* 2001;**153**:543–53.
48. Goumans MJ, Mummery C. Functional analysis of the TGFbeta receptor/Smad pathway through gene ablation in mice. *Int J Dev Biol* 2000;**44**:253–65.
49. Marneros AG, Olsen BR. Physiological role of collagen XVIII and endostatin. *FASEB J* 2005;**19**:716–28.
50. Clamp A, Blackhall FH, Henrioud A, Jayson GC, Javaherian K, Esko J, et al. The morphogenic properties of oligomeric endostatin are dependent on cell surface heparan sulfate. *J Biol Chem* 2006;**281**:14813–22.
51. Reis RC, Schuppan D, Barreto AC, Bauer M, Bork JP, Hassler G, et al. Endostatin competes with bFGF for binding to heparin-like glycosaminoglycans. *Biochem Biophys Res Commun* 2005;**333**:976–83.
52. Gaetzner S, Deckers MM, Stahl S, Lowik C, Olsen BR, Felbor U. Endostatin's heparan sulfate-binding site is essential for inhibition of angiogenesis and enhances in situ binding to capillary-like structures in bone explants. *Matrix Biol* 2005;**23**:557–61.

53. Kamimura K, Fujise M, Villa F, Izumi S, Habuchi H, Kimata K, et al. *Drosophila* heparan sulfate 6-*O*-sulfotransferase (dHS6ST) gene. Structure, expression, and function in the formation of the tracheal system. *J Biol Chem* 2001;**276**:17014–21.
54. Nakato H, Kimata K. Heparan sulfate fine structure and specificity of proteoglycan functions. *Biochim Biophys Acta* 2002;**1573**:312–8.
55. Ford-Perriss M, Guimond SE, Greferath U, Kita M, Grobe K, Habuchi H, et al. Variant heparan sulfates synthesized in developing mouse brain differentially regulate FGF signaling. *Glycobiology* 2002;**12**:721–7.
56. Grobe K, Ledin J, Ringvall M, Holmborn K, Forsberg E, Esko JD. Heparan sulfate and development: differential roles of the *N*-acetylglucosamine *N*-deacetylase/*N*-sulfotransferase isozymes. *Biochim Biophys Acta* 2002;**1573**:209–15.
57. Yabe T, Hata T, He J, Maeda N. Developmental and regional expression of heparan sulfate sulfotransferase genes in the mouse brain. *Glycobiology* 2005;**15**:982–93.
58. Forsberg E, Pejler G, Ringvall M, Lunderius C, Tomasini-Johansson B, Kusche-Gullberg M, et al. Abnormal mast cells in mice deficient in a heparin-synthesizing enzyme. *Nature* 1999;**400**:773–6.
59. Grobe K, Inatani M, Pallerla SR, Castagnola J, Yamaguchi Y, Esko JD. Cerebral hypoplasia and craniofacial defects in mice lacking heparan sulfate Ndst1 gene function. *Development* 2005;**132**:3777–86.
60. Inatani M, Irie F, Plump AS, Tessier-Lavigne M, Yamaguchi Y. Mammalian brain morphogenesis and midline axon guidance require heparan sulfate. *Science* 2003;**302**:1044–6.
61. Kraushaar DC, Yamaguchi Y, Wang L. Heparan sulfate is required for embryonic stem cells to exit form self-renewal. *J Biol Chem* 2009; **285**:5907–16.
62. Lin X, Wei G, Shi Z, Dryer L, Esko JD, Wells DE, et al. Disruption of gastrulation and heparan sulfate biosynthesis in EXT1-deficient mice. *Dev Biol* 2000;**224**:299–311.
63. Munoz-Chapuli R, Quesada AR, Angel Medina M. Angiogenesis and signal transduction in endothelial cells. *Cell Mol Life Sci* 2004;**61**:2224–43.
64. Soker S, Takashima S, Miao HQ, Neufeld G, Klagsbrun M. Neuropilin-1 is expressed by endothelial and tumor cells as an isoform-specific receptor for vascular endothelial growth factor. *Cell* 1998;**92**:735–45.
65. Neufeld G, Cohen T, Gengrinovitch S, Poltorak Z. Vascular endothelial growth factor (VEGF) and its receptors. *FASEB J* 1999;**13**:9–22.
66. Keyt BA, Berleau LT, Nguyen HV, Chen H, Heinsohn H, Vandlen R, et al. The carboxyl-terminal domain (111-165) of vascular endothelial growth factor is critical for its mitogenic potency. *J Biol Chem* 1996;**271**:7788–95.
67. Cheung N, Wong MP, Yuen ST, Leung SY, Chung LP. Tissue-specific expression pattern of vascular endothelial growth factor isoforms in the malignant transformation of lung and colon. *Hum Pathol* 1998;**29**:910–4.
68. Jakobsson L, Kreuger J, Holmborn K, Lundin L, Eriksson I, Kjellen L, et al. Heparan sulfate in trans potentiates VEGFR-mediated angiogenesis. *Dev Cell* 2006;**10**:625–34.
69. LaRochelle WJ, May-Siroff M, Robbins KC, Aaronson SA. A novel mechanism regulating growth factor association with the cell surface: identification of a PDGF retention domain. *Genes Dev* 1991;**5**:1191–9.
70. Ostman A, Andersson M, Betsholtz C, Westermark B, Heldin CH. Identification of a cell retention signal in the B-chain of platelet-derived growth factor and in the long splice version of the A-chain. *Cell Regul* 1991;**2**:503–12.
71. Armulik A, Abramsson A, Betsholtz C. Endothelial/pericyte interactions. *Circ Res* 2005;**97**:512–23.

72. Hirschi KK, Rohovsky SA, D'Amore PA. PDGF, TGF-beta, and heterotypic cell-cell interactions mediate endothelial cell-induced recruitment of 10T1/2 cells and their differentiation to a smooth muscle fate. *J Cell Biol* 1998;**141**:805–14.
73. Hirschi KK, Burt JM, Hirschi KD, Dai C. Gap junction communication mediates transforming growth factor-beta activation and endothelial-induced mural cell differentiation. *Circ Res* 2003;**93**:429–37.
74. Vardouli L, Vasilaki E, Papadimitriou E, Kardassis D, Stournaras C. A novel mechanism of TGFbeta-induced actin reorganization mediated by Smad proteins and Rho GTPases. *FEBS J* 2008;**275**:4074–87.
75. Rudini N, Felici A, Giampietro C, Lampugnani M, Corada M, Swirsding K, et al. VE-cadherin is a critical endothelial regulator of TGF-beta signalling. *EMBO J* 2008;**27**:993–1004.
76. Rider CC. Heparin/heparan sulphate binding in the TGF-beta cytokine superfamily. *Biochem Soc Trans* 2006;**34**:458–60.
77. Lyon M, Rushton G, Gallagher JT. The interaction of the transforming growth factor-betas with heparin/heparan sulfate is isoform-specific. *J Biol Chem* 1997;**272**:18000–6.
78. Chen Q, Sivakumar P, Barley C, Peters DM, Gomes RR, Farach-Carson MC, et al. Potential role for heparan sulfate proteoglycans in regulation of transforming growth factor-beta (TGF-beta) by modulating assembly of latent TGF-beta-binding protein-1. *J Biol Chem* 2007;**282**:26418–30.
79. Zhang X, Gaspard JP, Chung DC. Regulation of vascular endothelial growth factor by the Wnt and K-ras pathways in colonic neoplasia. *Cancer Res* 2001;**61**:6050–4.
80. Ishiguro K, Kadomatsu K, Kojima T, Muramatsu H, Tsuzuki S, Nakamura E, et al. Syndecan-4 deficiency impairs focal adhesion formation only under restricted conditions. *J Biol Chem* 2000;**275**:5249–52.
81. Zcharia E, Zilka R, Yaar A, Yacoby-Zeevi O, Zetser A, Metzger S, et al. Heparanase accelerates wound angiogenesis and wound healing in mouse and rat models. *FASEB J* 2005;**19**:211–21.
82. Gallo R, Kim C, Kokenyesi R, Adzick NS, Bernfield M. Syndecans-1 and -4 are induced during wound repair of neonatal but not fetal skin. *J Invest Dermatol* 1996;**107**:676–83.
83. Ryan HE, Lo J, Johnson RS. HIF-1 alpha is required for solid tumor formation and embryonic vascularization. *EMBO J* 1998;**17**:3005–15.
84. Shweiki D, Itin A, Soffer D, Keshet E. Vascular endothelial growth factor induced by hypoxia may mediate hypoxia-initiated angiogenesis. *Nature* 1992;**359**:843–5.
85. Tang N, Wang L, Esko J, Giordano FJ, Huang Y, Gerber HP. Loss of HIF-1alpha in endothelial cells disrupts a hypoxia-driven VEGF autocrine loop necessary for tumorigenesis. *Cancer Cell* 2004;**6**:485–95.
86. Robinson CJ, Stringer SE. The splice variants of vascular endothelial growth factor (VEGF) and their receptors. *J Cell Sci* 2001;**114**:853–65.
87. Grunstein J, Masbad JJ, Hickey R, Giordano F, Johnson RS. Isoforms of vascular endothelial growth factor act in a coordinate fashion to recruit and expand tumor vasculature. *Mol Cell Biol* 2000;**20**:7282–91.
88. Li J, Shworak NW, Simons M. Increased responsiveness of hypoxic endothelial cells to FGF2 is mediated by HIF-1alpha-dependent regulation of enzymes involved in synthesis of heparan sulfate FGF2-binding sites. *J Cell Sci* 2002;**115**:1951–9.
89. Elias AP, Dias S. Microenvironment changes (in pH) Affect VEGF alternative splicing. *Cancer Microenviron* 2008;**1**:131–9.
90. Hurwitz H, Fehrenbacher L, Novotny W, Cartwright T, Hainsworth J, Heim W, et al. Bevacizumab plus irinotecan, fluorouracil, and leucovorin for metastatic colorectal cancer. *N Engl J Med* 2004;**350**:2335–42.
91. Sandler A, Gray R, Perry MC, Brahmer J, Schiller JH, Dowlati A. Paclitaxel-carboplatin alone or with bevacizumab for non-small-cell lung cancer. *N Engl J Med* 2006;**355**:2542–50.

92. Jubb AM, Oates AJ, Holden S, Koeppen H. Predicting benefit from anti-angiogenic agents in malignancy. *Nat Rev Cancer* 2006;**6**:626–35.
93. Rusnati M, Urbinati C. Polysulfated/sulfonated compounds for the development of drugs at the crossroad of viral infection and oncogenesis. *Curr Pharm Des* 2009;**15**:2946–57.
94. Inai T, Mancuso M, Hashizume H, Baffert F, Haskell A, Baluk P. Inhibition of vascular endothelial growth factor (VEGF) signaling in cancer causes loss of endothelial fenestrations, regression of tumor vessels, and appearance of basement membrane ghosts. *Am J Pathol* 2004;**165**:35–52.
95. Mancuso MR, Davis R, Norberg SM, O'Brien S, Sennino B, Nakahara T, et al. Rapid vascular regrowth in tumors after reversal of VEGF inhibition. *J Clin Invest* 2006;**116**:2610–21.
96. Vlodavsky I, Friedmann Y. Molecular properties and involvement of heparanase in cancer metastasis and angiogenesis. *J Clin Invest* 2001;**108**:341–7.
97. Ilan N, Elkin M, Vlodavsky I. Regulation, function and clinical significance of heparanase in cancer metastasis and angiogenesis. *Int J Biochem Cell Biol* 2006;**38**:2018–39.
98. Escobar Galvis ML, Jia J, Zhang X, Jastrebova N, Spillmann D, Gottfridsson E, et al. Transgenic or tumor-induced expression of heparanase upregulates sulfation of heparan sulfate. *Nat Chem Biol* 2007;**3**:773–8.
99. Purushothaman A, Uyama T, Kobayashi F, Yamada S, Sugahara K, Rapraeger AC, et al. Heparanase-enhanced shedding of syndecan-1 by myeloma cells promotes endothelial invasion and angiogenesis. *Blood* 2010;**115**:2449–57.
100. Yang Y, Macleod V, Miao HQ, Theus A, Zhan F, Shaughnessy Jr. JD, et al. Heparanase enhances syndecan-1 shedding: a novel mechanism for stimulation of tumor growth and metastasis. *J Biol Chem* 2007;**282**:13326–33.
101. Lundin L, Larsson H, Kreuger J, Kanda S, Lindahl U, Salmivirta M, et al. Selectively desulfated heparin inhibits fibroblast growth factor-induced mitogenicity and angiogenesis. *J Biol Chem* 2000;**275**:24653–60.
102. Johnstone KD, Karoli T, Liu L, Dredge K, Copeman E, Li CP, et al. Synthesis and biological evaluation of polysulfated oligosaccharide glycosides as inhibitors of angiogenesis and tumor growth. *J Med Chem* 2010;**53**:1686–99.
103. Vlodavsky I, Ilan N, Naggi A, Casu B. Heparanase: structure, biological functions, and inhibition by heparin-derived mimetics of heparan sulfate. *Curr Pharm Des* 2007;**13**:2057–73.
104. Tang R, Rosen SD. Functional consequences of the subdomain organization of the sulfs. *J Biol Chem* 2009;**284**:21505–14.
105. Lai JP, Chien JR, Moser DR, Staub JK, Aderca I, Montoya DP, et al. hSulf1 sulfatase promotes apoptosis of hepatocellular cancer cells by decreasing heparin-binding growth factor signaling. *Gastroenterology* 2004;**126**:231–48.
106. Narita K, Staub J, Chien J, Meyer K, Bauer M, Friedl A, et al. HSulf-1 inhibits angiogenesis and tumorigenesis in vivo. *Cancer Res* 2006;**66**:6025–32.
107. Morimoto-Tomita M, Uchimura K, Bistrup A, Lum DH, Egeblad M, Boudreau N, et al. Sulf-2, a proangiogenic heparan sulfate endosulfatase, is upregulated in breast cancer. *Neoplasia* 2005;**7**:1001–10.
108. Song N, Huang Y, Shi H, Yuan S, Ding Y, Song X, et al. Overexpression of platelet-derived growth factor-BB increases tumor pericyte content via stromal-derived factor-1alpha/CXCR4 axis. *Cancer Res* 2009;**69**:6057–64.
109. Pietras K, Pahler J, Bergers G, Hanahan D. Functions of paracrine PDGF signaling in the proangiogenic tumor stroma revealed by pharmacological targeting. *PLoS Med* 2008;**5**:e19.
110. Greenberg JI, Shields DJ, Barillas SG, Acevedo LM, Murphy E, Huang J, et al. A role for VEGF as a negative regulator of pericyte function and vessel maturation. *Nature* 2008;**456**:809–13.
111. Bishop JR, Stanford KI, Esko JD. Heparan sulfate proteoglycans and triglyceride-rich lipoprotein metabolism. *Curr Opin Lipidol* 2008;**19**:307–13.

112. Echtermeyer F, Streit M, Wilcox-Adelman S, Saoncella S, Denhez F, Detmar M, et al. Delayed wound repair and impaired angiogenesis in mice lacking syndecan-4. *J Clin Invest* 2001;**107**:R9–R14.
113. Wilcox-Adelman SA, Denhez F, Goetinck PF. Syndecan-4 modulates focal adhesion kinase phosphorylation. *J Biol Chem* 2002;**277**:32970–7.
114. Beauvais DM, Ell BJ, McWhorter AR, Rapraeger AC. Syndecan-1 regulates alphavbeta is blocked by synstatin, a novel peptide inhibitor3 and alphavbeta5 integrin activation during angiogenesis. *J Exp Med* 2009;**206**:691–705.
115. Zhou Z, Wang J, Cao R, Morita H, Soininen R, Chan KM, et al. Impaired angiogenesis, delayed wound healing and retarded tumor growth in perlecan heparan sulfate-deficient mice. *Cancer Res* 2004;**64**:4699–702.
116. Kadenhe-Chiweshe A, Papa J, McCrudden KW, Frischer J, Bae JO, Huang J, et al. Sustained VEGF blockade results in microenvironmental sequestration of VEGF by tumors and persistent VEGF receptor-2 activation. *Mol Cancer Res* 2008;**6**:1–9.
117. Woodall BP, Nystrom A, Iozzo RA, Eble JA, Niland S, Krieg T, et al. Integrin alpha2beta1 is the required receptor for endorepellin angiostatic activity. *J Biol Chem* 2008;**283**:2335–43.
118. Abdollahi A, Hahnfeldt P, Maercker C, Grone HJ, Debus J, Ansorge W, et al. Endostatin's antiangiogenic signaling network. *Mol Cell* 2004;**13**:649–63.
119. Crawford Y, Ferrara N. VEGF inhibition: insights from preclinical and clinical studies. *Cell Tissue Res* 2009;**335**:261–9.
120. Aiello LP, Avery RL, Arrigg PG, Keyt BA, Jampel HD, Shah ST, et al. Vascular endothelial growth factor in ocular fluid of patients with diabetic retinopathy and other retinal disorders. *N Engl J Med* 1994;**331**:1480–7.
121. Gragoudas ES, Adamis AP, Cunningham Jr ET, Feinsod M, Guyer DR. Pegaptanib for neovascular age-related macular degeneration. *N Engl J Med* 2004;**351**:2805–16.
122. Ferrara N, Damico L, Shams N, Lowman H, Kim R. Development of ranibizumab, an anti-vascular endothelial growth factor antigen binding fragment, as therapy for neovascular age-related macular degeneration. *Retina* 2006;**26**:859–70.
123. Rodrigues EB, Farah ME, Maia M, Penha FM, Regatieri C, Melo GB, et al. Therapeutic monoclonal antibodies in ophthalmology. *Prog Retin Eye Res* 2009;**28**:117–44.
124. Stahl A, Paschek L, Martin G, Feltgen N, Hansen LL, Agostini HT. Combinatory inhibition of VEGF and FGF2 is superior to solitary VEGF inhibition in an in vitro model of RPE-induced angiogenesis. *Graefes Arch Clin Exp Ophthalmol* 2009;**247**:767–73.
125. Jonas JB, Neumaier M. Vascular endothelial growth factor and basic fibroblast growth factor in exudative age-related macular degeneration and diffuse diabetic macular edema. *Ophthalmic Res* 2007;**39**:139–42.
126. Simo R, Carrasco E, Garcia-Ramirez M, Hernandez C. Angiogenic and antiangiogenic factors in proliferative diabetic retinopathy. *Curr Diabetes Rev* 2006;**2**:71–98.
127. Patterson AM, Cartwright A, David G, Fitzgerald O, Bresnihan B, Ashton BA, et al. Differential expression of syndecans and glypicans in chronically inflamed synovium. *Ann Rheum Dis* 2008;**67**:592–601.
128. Pablos JL, Santiago B, Galindo M, Torres C, Brehmer MT, Blanco FJ, et al. Synoviocyte-derived CXCL12 is displayed on endothelium and induces angiogenesis in rheumatoid arthritis. *J Immunol* 2003;**170**:2147–52.
129. Marti HH. Angiogenesis—a self-adapting principle in hypoxia. *EXS* 2005;**Part 2**:163–80.
130. Lee SH, Wolf PL, Escudero R, Deutsch R, Jamieson SW, Thistlethwaite PA. Early expression of angiogenesis factors in acute myocardial ischemia and infarction. *N Engl J Med* 2000;**342**:626–33.

131. Mataveli FD, Han SW, Nader HB, Mendes A, Kanishiro R, Tucci P, et al. Long-term effects for acute phase myocardial infarct VEGF165 gene transfer cardiac extracellular matrix remodeling. *Growth Factors* 2009;**27**:22–31.
132. Rehm M, Bruegger D, Christ F, Conzen P, Thiel M, Jacob M, et al. Shedding of the endothelial glycocalyx in patients undergoing major vascular surgery with global and regional ischemia. *Circulation* 2007;**116**:1896–906.
133. Giordano FJ, Ping P, McKirnan MD, Nozaki S, DeMaria AN, Dillmann WH, et al. Intracoronary gene transfer of fibroblast growth factor-5 increases blood flow and contractile function in an ischemic region of the heart. *Nat Med* 1996;**2**:534–9.
134. Fujii H, Sun Z, Li SH, Wu J, Fazel S, Weisel RD, et al. Ultrasound-targeted gene delivery induces angiogenesis after a myocardial infarction in mice. *JACC Cardiovasc Imaging* 2009;**2**:869–79.
135. Weis S, Shintani S, Weber A, Kirchmair R, Wood M, Cravens A, et al. Src blockade stabilizes a Flk/cadherin complex, reducing edema and tissue injury following myocardial infarction. *J Clin Invest* 2004;**113**:885–94.
136. Bauer SM, Bauer RJ, Velazquez OC. Angiogenesis, vasculogenesis, and induction of healing in chronic wounds. *Vasc Endovascular Surg* 2005;**39**:293–306.
137. Li J, Zhang YP, Kirsner RS. Angiogenesis in wound repair: angiogenic growth factors and the extracellular matrix. *Microsc Res Tech* 2003;**60**:107–14.
138. Knox SM, Whitelock JM. Perlecan: how does one molecule do so many things? *Cell Mol Life Sci* 2006;**63**:2435–45.

Hepatic Heparan Sulfate Proteoglycans and Endocytic Clearance of Triglyceride-Rich Lipoproteins

Erin M. Foley
and Jeffrey D. Esko

Department of Cellular and Molecular Medicine, Glycobiology Research and Training Center, University of California, San Diego, La Jolla, California, USA

Hypertriglyceridemia, characterized by the accumulation of triglyceride-rich lipoproteins in the blood, affects 10–20% of the population in western countries and increases the risk of atherosclerosis, coronary artery disease, and pancreatitis. The etiology of hypertriglyceridemia is complex, and much interest exists in identifying and characterizing the biological and environmental factors that affect the synthesis and turnover of plasma triglycerides. Genetic studies in mice have recently identified that heparan sulfate proteoglycans are a class of receptors that mediate the clearance of triglyceride-rich lipoproteins in the liver. Heparan sulfate proteoglycans are expressed by endothelial cells that line the hepatic sinusoids and the underlying hepatocytes, and are present in the perisinusoidal space (space of Disse). This chapter discusses the dependence of lipoprotein binding on heparan sulfate structure and the identification of hepatocyte syndecan-1 as the primary proteoglycan that mediates triglyceride-rich lipoprotein clearance.

Progress in Molecular Biology
and Translational Science, Vol. 93
DOI: 10.1016/S1877-1173(10)93010-X

I. Lipoprotein Metabolism

Lipoproteins transport triglycerides, cholesterol, and phospholipids through the blood for distribution and use throughout the body. These particles consist of noncovalent associations of lipid and protein and contain a hydrophobic core of cholesterol ester and triglyceride enveloped by a monolayer of phospholipids, unesterified cholesterol, and apolipoproteins. Lipoprotein classes differ in density, size, function, and lipid and apolipoprotein content.

The triglyceride-rich lipoproteins are functionally defined as the fraction of plasma lipoproteins that have a density of less than 1.006 g/mL. Triglyceride-rich lipoproteins arise from both exogenous (dietary) and endogenous (liver) sources. Dietary triglyceride is packaged into large, buoyant chylomicrons, which contain apoB48, apoCII, and apoE, by the mucosal cells that line the intestine. In circulation, chylomicrons encounter lipoprotein lipase (LPL), which is anchored to the endothelial surface by its receptor GPIHBP1.[1] LPL hydrolyzes the triglycerides in the core of these particles into free fatty acids, which can then be used for energy production or storage. The inner core is partially depleted of triglycerides by lipolysis, resulting in the formation of chylomicron remnants. These smaller remnant particles pass through the fenestrated endothelium of the liver sinusoids and are sequestered in the space of Disse. Here, the remnant lipoproteins become enriched in apoE, undergo further processing by LPL and hepatic lipase (HL), and are cleared by one of several endocytic receptors.[2]

The liver packages endogenous fats into very low density lipoproteins (VLDLs) that contain apoB100 (and apoB48 in the mouse). VLDLs also encounter LPL in the circulation and, after hydrolysis, are converted into VLDL remnants and intermediate density lipoproteins (IDLs). Like chylomicron remnants, VLDL remnants enter into the hepatic space of Disse and may be cleared from the circulation. IDLs that remain in the circulation are further processed by HL and, in humans, by the cholesterol ester transfer protein (CETP), resulting in the formation of cholesterol-rich low density lipoproteins (LDL). The liver also synthesizes high density lipoproteins (HDL) that transport surplus cholesterol from the periphery back to the liver.[3] This process is called reverse cholesterol transport.

In the liver, several endocytic receptors mediate the clearance of triglyceride-rich lipoproteins from circulation. The first to be identified was the LDL receptor (Ldlr), initially described by Brown and Goldstein.[4] Mice, rabbits, and humans with inactivating mutations in the *Ldlr* gene have high levels of cholesterol and exhibit a modest accumulation of triglycerides,[5–7] suggesting that Ldlr is responsible for some, but not all, triglyceride-rich lipoprotein clearance. Herz and colleagues later identified another group of endocytic receptors bearing structural homology to Ldlr, the Ldlr-related proteins (Lrps).[8,9] Though Lrps bind and internalize remnants enriched in apoE,[10] they also have diverse functions in signaling, neurotransmission, and cellular entry of toxins and viruses.[11] The liver makes Lrp1 and Lrp5, but these proteins are also widely expressed in other tissues.[9,12] Systemic Lrp1 null mice are not viable because Lrp1 is essential for embryo implantation.[13] Liver-specific ablation of Lrp1 does not result in remnant particle accumulation unless compounded with Ldlr deficiency.[14] Systemic Lrp5 null mice are viable, but do not accumulate remnant particles unless they are subjected to a high fat diet.[15] Vldlr, another member of the Lrp family, mediates clearance of apoE-rich triglyceride-rich lipoproteins in the periphery (heart, muscle, adipose tissue, macrophages) but is not expressed in the liver.[16] *Vldlr*$^{-/-}$ mice have normal levels of plasma lipids, and the role of Vldlr in triglyceride-rich lipoprotein clearance becomes apparent only after severely stressing triglyceride metabolism.[17–19]

The lipolysis-stimulated lipoprotein receptor (Lsr) is also a candidate receptor, though its exact role in triglyceride-rich lipoprotein clearance is still under study. Upon activation by free fatty acids, Lsr undergoes a conformational change and binds to apoB- and apoE-containing lipoproteins.[20,21] Liver-specific knockdown of Lsr expression leads to hypertriglyceridemia in the fed state, but fasting triglyceride levels are normal.[22] Heterozygous *Lsr*$^{+/-}$ mice have delayed postprandial clearance of triglyceride-rich lipoproteins,[23] but systemic inactivation of *Lsr* leads to embryonic lethality,[24] alluding to perhaps other roles for this protein. Other receptors of note include the Scavenger Receptor BI (SR-B1). Though some evidence suggests that SR-B1 can mediate clearance of VLDLs and chylomicrons *in vivo*,[25–27] its major role is in HDL clearance.[28]

II. Syndecan-1 and Remnant Lipoprotein Clearance

A large body of work has suggested that heparan sulfate proteoglycans act in the clearance of triglyceride-rich lipoproteins. These studies, pioneered in the laboratories of K.J. Williams and R.W. Mahley, reported decreased lipoprotein uptake in cells treated with heparinases that degrade heparan sulfate chains, pharmacological agents that manipulate glycosaminoglycan biosynthesis, or in Chinese hamster ovary (CHO) cell mutants lacking specific heparan sulfate biosynthetic enzymes.[29–37] Upon infusion of mice with lactoferrin, heparinase, or heparan sulfate mimetics, which bind, degrade, or compete with heparan sulfate respectively, reduced plasma clearance rates and hepatic uptake of labeled VLDL were observed as well.[38–40] Furthermore, it was shown that heparan sulfate was localized on hepatocyte basal membranes facing the space of Disse in the liver by using immunoelectron microscopy.[41,42]

In tissue culture models, lipoproteins can bind to several proteoglycans. Fuki et al. have described increased cell association and uptake of LPL-enriched lipoproteins by CHO cells transfected with expression constructs for the transmembrane proteoglycans syndecan-1, -2, and -4, though the greatest effect was seen in the syndecan-1 expressing cells.[43] They also showed that WiDr cells, a variant colon carcinoma line expressing only the secreted proteoglycan perlecan, were able to bind, internalize, and degrade LPL -enriched lipoproteins.[44] In another study, transfection of an antisense oligonucleotide to syndecan-1 and treatment with an antibody against syndecan-1 decreased the binding of remnant particles to human HepG2 hepatoma cells by 50–70%.[45]

A recent study in syndecan-1 knockout mice ($Sdc1^{-/-}$) has provided genetic evidence that syndecan-1 is the primary heparan sulfate proteoglycan that mediates the clearance of triglyceride-rich lipoproteins *in vivo*.[46] In the fasted state, Sdc1-deficient mice accumulate triglycerides in approximately twofold excess compared to the wild type, but have normal cholesterol levels (Table I). The accumulated particles consist of chylomicron remnants, VLDLs, and IDLs containing apoB100, apoB48, apoE, apoAI, and apoCs. The clearance of intestinally derived chylomicrons is delayed in $Sdc1^{-/-}$ mice, and restoring syndecan-1 expression in the liver, using adenovirus, corrects the postprandial clearance defect *in vivo*. Though these results disagree with a previous study in which infection with an adenovirus containing syndecan-1 caused hypertriglyceridemia,[49] the disparity may have been caused by differences in viral load between the two studies, timing of lipid analysis relative to viral infection, or acute liver injury caused by the virus. Importantly, syndecan-1 facilitates uptake and degradation of labeled VLDL in cultured hepatocytes. Taken together, these findings suggest that syndecan-1 is the long sought-after receptor that works in parallel with Ldlr in the clearance of triglyceride-rich lipoproteins *in vivo*.

TABLE I
FASTING LIPID LEVELS IN HEPARAN SULFATE BIOSYNTHETIC AND PROTEOGLYCAN MUTANT MICE

Genotype	Triglycerides (mg/dL)	Cholesterol (mg/dL)	References
Wild type	38–58	70–98	46–48
$Ndst1^{f/f}AlbCre^{+}$	92 ± 35	81 ± 9	47
$Hs2st^{f/f}AlbCre^{+}$	100 ± 10	98 ± 4	48
$Hs6st1^{f/f}AlbCre^{+}$	80 ± 27		48
$Sdc1^{-/-}$	95 ± 11	70 ± 1	46
$Sdc3^{-/-}$	49 ± 8		46
$Sdc4^{-/-}$	59 ± 11		46

Abbreviations: Ndst1, *N*-acetyl-D-glucosamine *N*-deacetylase/*N*-sulfotransferase 1; Alb, albumin; Cre, Cre recombinase; Sdc, Syndecan; Hs2st, uronyl 2-*O*-sulfotransferase; Hs6st1, glucosaminyl6-*O*-sulfotransferase 1.

III. Syndecan-1 Structure and Regulation

Syndecan-1 is a type-I transmembrane proteoglycan, containing three extracellular attachment sites for heparan sulfate and two for chondroitin sulfate (Fig. 1; Ref. 51). Occupation levels of these sites varies in different cell types,[52] but its most active form contains heparan sulfate at all three N-terminal sites.[53,54] Syndecan-1 was first described by Bernfield and colleagues in mammary epithelial cells[55,56] and has diverse functions in cell adhesion and motility,[57] tumorigenesis,[58] and microbial infection.[59] Syndecan-1 contains a juxtamembrane cleavage site,[60] allowing protease-mediated shedding of the ectodomain and glycosaminoglycan chains.[61]

As syndecan-1 plays a major role in the clearance of triglycerides, the regulation of its expression and activity are of great interest. Syndecan-1 expression is partially controlled by nuclear hormone receptors. These receptors are activated by small, lipophilic ligands and subsequently translocate into the nucleus to alter gene expression.[62] The farnesoid-X-receptor (FXR) has been

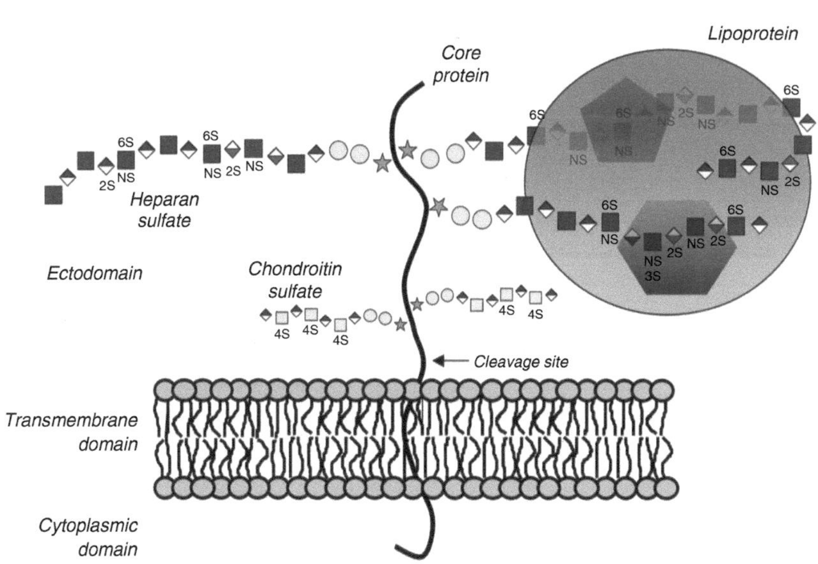

FIG. 1. Syndecan-1 is a type I transmembrane protein bearing up to three heparan sulfate and two chondroitin/dermatan sulfate chains. It undergoes proteolytic cleavage, resulting in shedding of the large extracellular ectodomain bearing the GAG chains. The short cytoplasmic tail interacts with a number of cytosolic proteins and may play a role in Sdc1 endocytosis. A lipoprotein is shown binding to the heparan sulfate chains via the interaction of short sulfated domains with apolipoproteins or lipases depicted by the large hexagon and pentagon. The GAG chains were rendered using the symbol nomenclature described in Ref. 50.

shown to bind a nonconsensus DR-1 element in the syndecan-1 promoter and to increase expression of the *SDC1* gene in HepG2 human hepatoma cells.[63] FXR agonists increased binding, uptake, and degradation of radiolabeled methylated LDL, a chemically modified form of LDL that no longer binds to Ldlr or Lrp. Though FXR modulates expression of other factors involved in lipoprotein metabolism, including apoCII[64] and apoE,[65] regulation of syndecan-1 by FXR might also contribute to lipoprotein uptake in humans. Syndecan-1 expression in prostate cells is also controlled by peroxisome proliferator-activated receptor gamma (PPARγ), another nuclear hormone receptor.[66] Syndecan-1 mRNA was increased in the prostate gland of animals fed an n-3 polyunsaturated fatty acid-enriched diet. Similar effects were noted in the human prostate cancer cell line, PC-3. This effect was mimicked by a PPARγ agonist, troglitazone, and inhibited in the presence of a PPARγ antagonist. PPARγ-mediated regulation of syndecan-1 was also seen in human breast cancer cells treated with n-3 fatty acid-enriched LDLs.[67] Whether PPARs influence syndecan-1 expression in hepatocytes is not known. These findings suggest that the ability of dietary n-3 fatty acids to ameliorate hypertriglyceridemia in patients may be partially due to the enhanced expression of syndecan-1.

Syndecan-1 function is also regulated by proteolytic shedding. Proteolysis of syndecan-1 at a specific juxtamembrane cleavage site between amino acids A243 and S244 releases the extracellular ectodomain containing the glycosaminoglycan chains.[60] Various metalloproteinases mediate syndecan-1 ectodomain shedding, including MMP-7 in lung epithelial cells,[68] MMP-9 in HeLa cells and primary macrophages,[69] and MMP-14 in HEK293T and HT1080 cells.[70] Even though syndecan-1 shedding can occur constitutively, shedding is induced following the activation of both G-protein coupled receptors and protein tyrosine kinases by specific agonists, including thrombin and epidermal growth factor.[71] Other shedding agonists include bacterial virulence factors[72] and phorbol esters such as phorbol myristic acid (PMA).[73] Although the biological significance of syndecan-1 shedding is unclear, it is often thought to attenuate syndecan-1 function. For example, syndecan-1 shedding protects against tissue damage during the inflammatory response by modulating the availability of chemokines.[74]

Although syndecan-1 shedding has not been characterized in the liver, syndecan-1 ectodomains accumulate in conditioned media from cultured primary hepatocytes (Yiping Deng and J. D. Esko, unpublished observations). Syndecan-1 shedding from hepatocytes may be a mechanism to downregulate its activity as a cell-surface endocytic receptor, but this process would increase the local concentration of free ectodomains in the space of Disse. Such free ectodomains might sequester triglyceride-rich lipoproteins in circulation and prevent their diffusion out of the space, or they might compete with cell-surface syndecan-1 or other receptors for lipoprotein binding. Thus, the functional significance of syndecan-1 shedding in the liver and the identification of the hepatic "sheddase" require further study. Small molecules or antibodies that inhibit syndecan shedding may prove to be useful research tools or may have pharmacological relevance in the treatment of dyslipidemias.

Syndecan-1 is also regulated in an "inside-to-outside" manner via its cytoplasmic domain. The 34-amino acid cytoplasmic domain consists of two constant regions (C1 and C2) that are conserved among the four syndecan family members.[51] The membrane-proximal region (C1) is thought to mediate syndecan dimerization and ligand binding.[75,76] The C2 region interacts with cytoskeletal proteins[77] and terminates in an EYFA domain that can bind the PDZ (postsynaptic density protein, Drosophila disc large tumor suppressor, and zonula occludens-1 protein) domain-containing proteins.[78,79] C1 and C2 flank a syndecan-specific variable region (V) that is believed to direct individual functions of the family members. The cytoplasmic domains of all syndecans contain four invariant tyrosine residues. Some studies have suggested that tyrosine phosphorylation promotes syndecan-1 shedding.[80,81] Another study reported that mutations in the tyrosine residues abrogates shedding, but that this inhibition occurred in a phosphorylation-independent manner.[82] Also, these investigators found that the small GTPase Rab5 directly interacts with the cytoplasmic domain and acts as a molecular switch that regulates syndecan-1 shedding. It remains to be determined if cytoplasmic domain signaling regulates syndecan-1 endocytosis as well.

IV. Structural Features of Heparan Sulfate Required for Lipoprotein Binding and Clearance

Liver heparan sulfate is unique in its structure. The chains contain approximately 1.34 sulfates per disaccharide,[83] which is roughly twice the amount observed in other mouse and human tissues but less than heparin (2–2.5 sulfates per disaccharide).[84] Liver heparan sulfate is rich in *N*-sulfated glucosamine and 2-*O*-sulfated iduronic acid, and contains a high proportion of trisulfated disaccharides (D2S6, disaccharide code described in Ref. 85). The average chain is approximately 40–60 disaccharides long (M_r 14,000–22,000),[83,86] which is shorter than heparan sulfate chains found in other tissues. These modified sugar residues are not equally dispersed throughout the chain and the highly sulfated disaccharides are clustered furthest (distal) from the protein core.[83]

Studies using mutant mice with defects in heparan sulfate biosynthesis have identified structural features of the chains that are required for lipoprotein binding. One study focused on the enzyme *N*-acetylglucosamine *N*-deacetylase/*N*-sulfotransferase-1 (Ndst1), which removes acetyl groups from subsets of *N*-acetylglucosamine residues and adds sulfate to the free amino groups.

Systemic inactivation of Ndst1 results in perinatal lethality due to brain, lung and other defects.[87–89] To study the effect of inactivating Ndst1 in triglyceride-rich lipoprotein clearance, a mouse bearing a conditional allele of Ndst1 was cross-bred to mice expressing the bacterial Cre recombinase under the control of the liver specific albumin promoter (AlbCre). $Ndst1^{f/f}AlbCre^{+}$ mice were viable and fertile[47] and have a phenotype that is essentially identical to $Sdc1^{-/-}$ mice[46]; they accumulate fasting triglycerides approximately twofold (Table I) and show delayed clearance of postprandial chylomicrons and injected human VLDL.

The deletion of Ndst1 in hepatocytes decreased overall sulfation of heparan sulfate by approximately 50%.[47] Significant changes in disaccharide composition were also observed, as initial *N*-deacetylation and *N*-sulfation of *N*-acetylglucosamine residues are coupled to subsequent uronyl 2-*O*-sulfation and glucosaminyl 6-*O*-sulfation.[90] *N*- and 2-*O*-sulfation of heparan sulfate, isolated from highly purified $Ndst1^{f/f}AlbCre^{+}$ hepatocytes, was decreased significantly compared to wild-type cells (decreased from 38 sulfate groups/100 disaccharides to 18 for *N*-sulfates and 17 sulfate groups/100 disaccharides to 4 for 2-*O*-sulfate groups, respectively). However, the abundance of 6-*O*-sulfated disaccharides was not significantly affected (31 sulfate groups/100 disaccharides in mutants vs. 26 in the wild type).

Accumulation of triglyceride-rich lipoproteins in $Ndst1^{f/f}Cre^{+}$ mice might have been caused by specific changes in heparan sulfate fine structure or simply by an overall reduction in charge. To address this question, plasma triglyceride levels were measured in mice with hepatocyte specific inactivation of either uronyl 2-*O*-sulfotransferase ($Hs2st^{f/f}AlbCre^{+}$) or glucosaminyl 6-*O*-sulfotransferase-1 ($Hs6st1^{f/f}AlbCre^{+}$).[48] As expected, $Hs2st^{f/f}AlbCre^{+}$ mice have dramatically reduced levels of 2-*O*-sulfated disaccharides (4 sulfate groups/100 disaccharides in mutants vs. 21 in wild-type littermates). These mice accumulate plasma triglycerides (Table I) and have delayed clearance of intestinally derived chylomicrons and injected human VLDLs to approximately the same extent as observed in $Ndst1^{f/f}AlbCre^{+}$ and $Sdc1^{-/-}$ mice. In contrast, the $Hs6st1^{f/f}AlbCre^{+}$ mice do not exhibit any changes in plasma triglycerides. The lack of phenotype might reflect a partial reduction in glucosamine 6-*O*-sulfation in $Hs6st1^{f/f}AlbCre^{+}$ hepatocytes due to the expression of other Hs6st isoforms or the enrichment of trisulfated disaccharides.[91] However, 6-*O*-desulfated heparin competed for lipoprotein binding, at the surface of primary hepatocytes, to the same extent as unmodified heparin; whereas heparinoids lacking *N*-sulfate and 2-*O*-sulfate groups did not block binding. Taken together, these data demonstrate that plasma triglyceride-rich lipoprotein binding and uptake depends less on overall charge of the chains, and more on uronic acid 2-*O*-sulfation and glucosamine *N*-sulfation, with overall glucosamine 6-*O*-sulfation making little contribution. However, it has been suggested that specific 6-*O*-sulfate groups might be important.[91a]

V. Apolipoprotein and Lipase Interactions with Heparan Sulfate

The binding of heparan sulfate to protein ligands occurs through electrostatic interactions between the negatively charged sulfates and carboxyl groups on the polysaccharide with their complementary positively charged domains in the protein. These interactions are often highly specific; binding sites usually consist of 8–12 sugar residues with patterned arrangements of sulfated sugars and uronic acid epimers.[90] This raises the question as to the identity of the apolipoprotein ligand on remnant lipoproteins that mediates binding to heparan sulfate. The binding sequences for LPL and apoE have been identified (Table II).[92,93,102,107] However, care must be taken in interpreting these findings, since the physiological relevance of these interactions to remnant clearance has not been established in vivo through the study of mutants.

Interestingly, the diameter of a remnant triglyceride-rich lipoprotein particle is 40–80 nm,[83,86] and is approximately equal to the extended length of a heparan sulfate chain in the liver. Given their relatively large size, it is possible that lipoproteins require multiple interactions with heparan sulfate to bind with sufficient affinity to stabilize the lipoprotein–proteoglycan complex before endocytosis. This might be achieved by simultaneous engagement with multiple heparan chains—a "proteoglycan hug" (Fig. 1). The unique ability of syndecan-1 to bind and internalize remnant lipoproteins might therefore stem from the fact that its ectodomain contains three sites for heparan sulfate attachment. Syndecan-1 can also multimerize,[76] and thus increase the local concentration of heparan sulfate chains around a ligand. Notably, several studies have reported that syndecan-1 clustering precedes endocytosis,[43,108] suggesting that lipoproteins may bind to and cluster multiple syndecan-1 molecules before internalization.

TABLE II
APOLIPOPROTEINS AND LIPASES THAT BIND HEPARIN/HEPARAN SULFATE (FROM REF. 48)

Ligand	Binding specificity	References
ApoE	Sequences rich in GlcNS6S-IdoA2s; heparin binding domains mapped	92–95
ApoB	Unknown, but interacts with heparan sulfate and heparin; heparin binding domains mapped	96–98
Hepatic lipase	Heparin oligosaccharides; heparin binding domains mapped	99–101
Lipoprotein lipase	Affinity for sequences rich in GlcNS6S-IdoA2S, modestly sulfated sequences	102,103
Endothelial lipase	Interacts with heparan sulfate	104,105
ApoA-V	Interacts with heparin	106

Many triglyceride-rich lipoprotein-associated proteins, including apoE, apoB100, apoB48, LPL, and HL bind heparin. Whether the binding of lipoproteins to heparan sulfate is mediated by a specific protein ligand has not been conclusively demonstrated. One theory, known as the "secretion-capture" hypothesis, suggests that apoE produced by hepatocytes enriches lipoproteins in the space of Disse and acts as a ligand for clearance through proteoglycans.[109] Indeed, heparan sulfate chains isolated from proteoglycans produced by $Ndst1^{f/f}AlbCre^{+}$ hepatocytes show decreased binding to apoE. However, some genetic studies do not support the conclusion that apoE is the primary ligand for heparan sulfate-mediated clearance. ApoE deficient mice have elevated levels of triglyceride-rich lipoproteins, but interbreeding these mice with $Ldlr^{-/-}$ mice showed that nonfasting plasma triglyceride levels were similar in $Ldlr^{-/-}apoE^{-/-}$ compound mutants and single mutants deficient in Ldlr or apoE.[110] The lack of an additive effect in these compound mutants suggests that apoE-dependent clearance of remnant particles was mediated mostly via Ldlr.

ApoB, a key structural component of both the intestine- and liver-derived lipoproteins, might also serve as a ligand for syndecan-1-mediated uptake. In humans, full length apoB100 is made by the liver while truncated apoB48 is made in enterocytes of the small intestine as a result of mRNA editing. In mice, both these forms are made in the liver. Veniant and colleagues made mutant mice that expressed only one of the apoB isoforms.[111] Both "apoB48-only" and "apoB100-only" mice produced and cleared VLDLs and chylomicrons. However, when each of these mice was bred with Ldlr-deficient mice, fasting plasma triglyceride accumulated to a greater extent when apoB48 was lacking (i.e., in the apoB100-only mice). Thus, syndecan-1 may mediate clearance via apoB48, an idea that can be tested by breeding syndecan-1 deficient mice with apoB48-only mice.

LPL is thought to have a "bridging function," independent of its enzymatic activity. LPL is made by muscle and adipose tissue, transported to the capillary lumen, and anchored there by GPIHBPI.[1] LPL can become associated with lipoproteins in the blood whence it travels with them back to the liver. Heeren et al. showed that triglyceride-rich lipoproteins isolated from patients with catalytically inactive LPL were cleared after injection into mice and enrichment with inactive LPL further enhanced hepatic clearance by ~35%. Furthermore, catalytically inactive LPL expressed in muscle in transgenic mice also caused greater hepatic uptake by a comparable amount.[112] In other studies, LPL-enriched chylomicrons were shown to bind to HepG2 cells 30-fold better than nonenriched particles, and heparan lyase treatment abolished this enhanced binding.[113,114] Similar observations have been made in syndecan-1 transfected CHO cells using methylated LDL enriched with LPL.[43]

Further genetic studies using mice deficient in various combinations of the lipoprotein receptors and ligands will be necessary to characterize interactions between heparan sulfate and lipoproteins.

VI. Syndecan-1 is the Primary Proteoglycan Receptor in the Liver

Mouse hepatocytes contain mRNA for eight heparan sulfate proteoglycans including the membrane proteoglycans syndecan-1, -2, and -4, glypican-1 and -4, and the secreted proteoglycans perlecan, agrin, and collagen XVIII.[46] Of the available mutant mice, only syndecan-1[46] and collagen XVIII deficient mice accumulate fasting triglycerides (J. R. Bishop, K. Moulton, and J. D. Esko, unpublished data). Crossbreeding mutants defective in syndecan-1 and Ndst1 (*Sdc1*$^{-/-}$*Ndst1*$^{f/f}$*AlbCre*$^{+}$) does not accentuate triglyceride accumulation beyond levels observed in each single mutant, indicating that syndecan-1 is the primary proteoglycan receptor mediating binding, uptake, and degradation of triglyceride-rich lipoproteins in the liver.[46] Characterization of the systemic collagen XVIII null mouse supports this conclusion, as hyperchylomicronemia in this mouse is not caused by clearance defects in the liver, but instead by the altered secretion of LPL and altered lipolysis in peripheral tissues (J. R. Bishop, K. Moulton, and J. D. Esko, unpublished data).

Though selective ablation of liver heparan sulfate causes a twofold accumulation of triglyceride-rich lipoproteins, mice with an endothelial deficiency in Ndst1 expression (*Ndst1*$^{f/f}$*Tie2Cre*$^{+}$) do not accumulate triglycerides.[47] This finding supports the idea that only hepatocyte heparan sulfate binds to lipoprotein particles and mediates their clearance. The liver contains other cell types, including stellate cells and Kupffer cells. Both cell types produce heparan sulfate proteoglycans,[115–117] but their contribution to lipoprotein sequestration, processing, and eventual clearance is unknown.

As syndecan-1 is found in other cell types, we wonder why hepatic syndecan-1 mediates clearance, while syndecan-1 in other tissues does not. On the hepatocyte surface, syndecan-1 is specifically localized to the space of Disse.[41,42,46] Thus, it is poised in a unique environment where the relative concentration of both lipoproteins and ligands is high. The local concentration of syndecan-1 may be substantially higher here, than in other tissues. Another possibility is that the relatively high levels of sulfation observed in liver heparan sulfate create more binding sites for the triglyceride-rich lipoproteins. Liver HS is especially rich in *N*- and 2-*O*-sulfation, modifications that are required for lipoprotein binding and uptake.[47,48] The degree of substitution of the core protein could also be a variable.

VII. Relative Contribution of Syndecan-1 to Clearance

As noted above, there are at least three major receptors involved in the hepatic clearance of triglyceride-rich lipoproteins: Ldlr, Lrp, and syndecan-1. The relative contribution of each receptor to lipoprotein clearance has not been determined *in vivo*. Before the identification of syndecan-1 as a receptor, pharmacological inhibition of Ldlr family members was used to address this question. Mice injected with an Ldlr-blocking antibody exhibited a 45% decrease in rapid removal of remnant particles from the plasma and also a 45% decrease in liver uptake.[118] Treatment with exogenous receptor associated protein (RAP), which blocks binding to Lrp, decreased remnant clearance by about 55% and reduced liver uptake by 70%. However, RAP also binds to heparan sulfate proteoglycans and with low affinity to the Ldlr, making these results hard to interpret. Indeed, co-treatment with both the Ldlr-blocking antibody and RAP did not result in an additive accumulation, suggesting that RAP may have decreased uptake by blocking multiple receptors.

Additional insight has been gained by the analyses of mice bearing paired mutations in triglyceride-rich lipoprotein receptors. Crossbreeding $Ndst1^{f/f}AlbCre^{+}$ mice with $Ldlr^{-/-}$ mice resulted in a larger accumulation of triglycerides than observed in either of the individual mutant mice,[47] suggesting that these two receptors work independently and may partially compensate for one another. Furthermore, cholesterol levels were accentuated in the compound mutant compared to mice lacking only the Ldlr. Thus, HSPGs may participate in the clearance of cholesterol-rich lipoproteins, independent of Ldlr.

The number of proteoglycan binding sites for VLDL on the hepatocyte surface is approximately 100-times greater than the number of Ldlr molecules.[46] Based on this difference in receptor number, it seems as though syndecan-1 may have a greater contribution to clearance. However, endocytosis of Ldlr occurs by clathrin-mediated pathways and is rapid, with receptors having a half-life of approximately 5–10 min on the cell surface.[119] Syndecan-1 lacks the hallmarks of a classical clathrin-mediated receptor, and has been reported to enter the cell by a lipid raft pathway with relatively slow kinetics ($t_{1/2} \sim 20$–60 min).[108,120–122] Thus, compared to Ldlr, syndecan-1 constitutes a low-affinity, high-capacity uptake system. The fact that both $Sdc1^{-/-}$ and $Ldlr^{-/-}$ mice accumulate triglycerides suggests that they cannot fully compensate for one another, possibly because they recognize different subsets of particles which differ in their apolipoprotein composition or size. Nevertheless, the synergistic effect of compounding mutations in Ldlr and Sdc1 would suggest that they also clear a common pool of triglyceride-rich lipoproteins. The differences in receptor preference and the nature of the subsets of particles remain unknown.

Several groups have proposed that proteoglycans mediate a multi-step internalization process involving triglyceride-rich lipoprotein binding and subsequent transfer to the Lrp before endocytosis.[123,124] It is still unclear whether such "lipoprotein handoff" occurs, but ample evidence suggests that syndecan-1 and other heparan sulfate proteoglycans can act independently as endocytic receptors for a variety of natural and synthetic ligands.[125–127] Syndecan-1 has been shown to mediate uptake of apoE-VLDL, independent of the Lrp, in a Ldlr-deficient cell line,[121] and Lrp mediates endocytosis of nonlipoprotein ligands independent of heparan sulfate proteoglycans.[128] Additionally, mice with a liver specific ablation of Lrp1 do not accumulate triglycerides unless stressed with a high fat diet,[14] but syndecan-1 deficient mice fed normal chow accumulate triglycerides.[46] It is likely, therefore, that the two receptors work independently to clear triglycerides under normal conditions.

VIII. Conclusion

Hypertriglyceridemia increases the risk of atherosclerosis, coronary artery disease, and pancreatitis and thus much interest exists in understanding the geneitc and envrionmental factors that contribute to dyslipidemia.[129–134] Genetic data showing that heparan sulfate proteoglycans act as independent receptors for the clearance of triglyceride-rich lipoproteins in the liver provides new insights into the role of hepatic receptors and opens up many new avenues for research. For example, changes in heparan sulfate abundance and structure may be an underlying cause of human dyslipidemias caused by hormone therapy, drug treatment, obesity, and metabolic syndrome. Some familial human dyslipidemias may be explained by polymorphisms in genes encoding proteoglycan core proteins and enzymes involved in heparan sulfate biosynthesis. Based on the findings that have emerged from studies of syndecan-1 and heparan sulfate-deficient mice, we predict that studies of other genes in the system will yield new insights into the role of heparan sulfate proteoglycans in lipid clearance and other aspects of lipid metabolism in extrahepatic tissues.

References

1. Beigneux AP, Davies BS, Gin P, Weinstein MM, Farber E, Qiao X, et al. Glycosylphosphatidylinositol-anchored high-density lipoprotein-binding protein 1 plays a critical role in the lipolytic processing of chylomicrons. *Cell Metab* 2007;**5**:279–91.
2. Havel RJ, Hamilton RL. Hepatic catabolism of remnant lipoproteins: where the action is. *Arterioscler Thromb Vasc Biol* 2004;**24**:213–5.
3. Trigatti BL, Krieger M, Rigotti A. Influence of the HDL receptor SR-BI on lipoprotein metabolism and atherosclerosis. *Arterioscler Thromb Vasc Biol* 2003;**23**:1732–8.

4. Brown MS, Goldstein JL. A receptor-mediated pathway for cholesterol homeostasis. *Science* 1986;**232**:34–47.
5. Rubinsztein DC, Cohen JC, Berger GM, van der Westhuyzen DR, Coetzee GA, Gevers W. Chylomicron remnant clearance from the plasma is normal in familial hypercholesterolemic homozygotes with defined receptor defects. *J Clin Invest* 1990;**86**:1306–12.
6. Kita T, Goldstein JL, Brown MS, Watanabe Y, Hornick CA, Havel RJ. Hepatic uptake of chylomicron remnants in WHHL rabbits: a mechanism genetically distinct from the low density lipoprotein receptor. *Proc Natl Acad Sci USA* 1982;**79**:3623–7.
7. Ishibashi S, Perrey S, Chen Z, Osuga J, Shimada M, Ohashi K, et al. Role of the low density lipoprotein (LDL) receptor pathway in the metabolism of chylomicron remnants. A quantitative study in knockout mice lacking the LDL receptor, apolipoprotein E, or both. *J Biol Chem* 1996;**271**:22422–7.
8. Kowal RC, Herz J, Goldstein JL, Esser V, Brown MS. Low density lipoprotein receptor-related protein mediates uptake of cholesteryl esters derived from apoprotein E-enriched lipoproteins. *Proc Natl Acad Sci USA* 1989;**86**:5810–4.
9. Herz J, Hamann U, Rogne S, Myklebost O, Gausepohl H, Stanley KK. Surface location and high affinity for calcium of a 500-kd liver membrane protein closely related to the LDL-receptor suggest a physiological role as lipoprotein receptor. *EMBO J* 1988;**7**:4119–27.
10. Beisiegel U, Weber W, Ihrke G, Herz J, Stanley KK. The LDL-receptor-related protein, LRP, is an apolipoprotein E-binding protein. *Nature* 1989;**341**:162–4.
11. May P, Woldt E, Matz RL, Boucher P. The LDL receptor-related protein (LRP) family: an old family of proteins with new physiological functions. *Ann Med* 2007;**39**:219–28.
12. Kim DH, Inagaki Y, Suzuki T, Ioka RX, Yoshioka SZ, Magoori K, et al. A new low density lipoprotein receptor related protein, LRP5, is expressed in hepatocytes and adrenal cortex, and recognizes apolipoprotein E. *J Biochem* 1998;**124**:1072–6.
13. Herz J, Clouthier DE, Hammer RE. LDL receptor-related protein internalizes and degrades uPA-PAI-1 complexes and is essential for embryo implantation. *Cell* 1992;**71**:411–21.
14. Rohlmann A, Gotthardt M, Hammer RE, Herz J. Inducible inactivation of hepatic LRP gene by cre-mediated recombination confirms role of LRP in clearance of chylomicron remnants. *J Clin Invest* 1998;**101**:689–95.
15. Fujino T, Asaba H, Kang MJ, Ikeda Y, Sone H, Takada S, et al. Low-density lipoprotein receptor-related protein 5 (LRP5) is essential for normal cholesterol metabolism and glucose-induced insulin secretion. *Proc Natl Acad Sci USA* 2003;**100**:229–34.
16. Sakai J, Hoshino A, Takahashi S, Miura Y, Ishii H, Suzuki H, et al. Structure, chromosome location, and expression of the human very low density lipoprotein receptor gene. *J Biol Chem* 1994;**269**:2173–82.
17. Frykman PK, Brown MS, Yamamoto T, Goldstein JL, Herz J. Normal plasma lipoproteins and fertility in gene-targeted mice homozygous for a disruption in the gene encoding very low density lipoprotein receptor. *Proc Natl Acad Sci USA* 1995;**92**:8453–7.
18. Goudriaan JR, Espirito Santo SM, Voshol PJ, Teusink B, van Dijk KW, van Vlijmen BJ, et al. The VLDL receptor plays a major role in chylomicron metabolism by enhancing LPL-mediated triglyceride hydrolysis. *J Lipid Res* 2004;**45**:1475–81.
19. Espirito Santo SM, Rensen PC, Goudriaan JR, Bensadoun A, Bovenschen N, Voshol PJ, et al. Triglyceride-rich lipoprotein metabolism in unique VLDL receptor, LDL receptor, and LRP triple-deficient mice. *J Lipid Res* 2005;**46**:1097–102.
20. Mann CJ, Khallou J, Chevreuil O, Troussard AA, Guermani LM, Launay K, et al. Mechanism of activation and functional significance of the lipolysis-stimulated receptor. Evidence for a role as chylomicron remnant receptor. *Biochemistry* 1995;**34**:10421–31.

21. Bihain BE, Yen FT. Free fatty acids activate a high-affinity saturable pathway for degradation of low-density lipoproteins in fibroblasts from a subject homozygous for familial hypercholesterolemia. *Biochemistry* 1992;**31**:4628–36.
22. Narvekar P, Berriel Diaz M, Krones-Herzig A, Hardeland U, Strzoda D, Stohr S, et al. Liver-specific loss of lipolysis-stimulated lipoprotein receptor triggers systemic hyperlipidemia in mice. *Diabetes* 2009;**58**:1040–9.
23. Yen FT, Roitel O, Bonnard L, Notet V, Pratte D, Stenger C, et al. Lipolysis stimulated lipoprotein receptor: a novel molecular link between hyperlipidemia, weight gain, and atherosclerosis in mice. *J Biol Chem* 2008;**283**:25650–9.
24. Mesli S, Javorschi S, Berard AM, Landry M, Priddle H, Kivlichan D, et al. Distribution of the lipolysis stimulated receptor in adult and embryonic murine tissues and lethality of $LSR^{-/-}$ embryos at 12.5 to 14.5 days of gestation. *Eur J Biochem* 2004;**271**:3103–14.
25. Out R, Hoekstra M, de Jager SC, de Vos P, van der Westhuyzen DR, Webb NR, et al. Adenovirus-mediated hepatic overexpression of scavenger receptor class B type I accelerates chylomicron metabolism in C57BL/6J mice. *J Lipid Res* 2005;**46**:1172–81.
26. Out R, Kruijt JK, Rensen PC, Hildebrand RB, de Vos P, Van Eck M, et al. Scavenger receptor BI plays a role in facilitating chylomicron metabolism. *J Biol Chem* 2004;**279**:18401–6.
27. Van Eck M, Hoekstra M, Out R, Bos IS, Kruijt JK, Hildebrand RB, et al. Scavenger receptor BI facilitates the metabolism of VLDL lipoproteins *in vivo*. *J Lipid Res* 2008;**49**:136–46.
28. Krieger M. Scavenger receptor class B type I is a multiligand HDL receptor that influences diverse physiologic systems. *J Clin Invest* 2001;**108**:793–7.
29. Krapp A, Ahle S, Kersting S, Hua Y, Kneser K, Nielsen M, et al. Hepatic lipase mediates the uptake of chylomicrons and β-VLDL into cells via the LDL receptor-related protein (LRP). *J Lipid Res* 1996;**37**:926–36.
30. Williams KJ, Fless GM, Petrie KA, Snyder ML, Brocia RW, Swenson TL. Mechanisms by which lipoprotein lipase alters cellular metabolism of lipoprotein(a), low density lipoprotein, and nascent lipoproteins. Roles for low density lipoprotein receptors and heparan sulfate proteoglycans. *J Biol Chem* 1992;**267**:13284–92.
31. Feussner G, Scharnagl H, Scherbaum C, Acar J, Dobmeyer J, Lohrmann J, et al. Apolipoprotein E5 (Glu212–>Lys): increased binding to cell surface proteoglycans but decreased uptake and lysosomal degradation in cultured fibroblasts. *J Lipid Res* 1996;**37**:1632–45.
32. Ji ZS, Brecht WJ, Miranda RD, Hussain MM, Innerarity TL, Mahley RW. Role of heparan sulfate proteoglycans in the binding and uptake of apolipoprotein E-enriched remnant lipoproteins by cultured cells. *J Biol Chem* 1993;**268**:10160–7.
33. Al-Haideri M, Goldberg IJ, Galeano NF, Gleeson A, Vogel T, Gorecki M, et al. Heparan sulfate proteoglycan-mediated uptake of apolipoprotein E-triglyceride-rich lipoprotein particles: a major pathway at physiological particle concentrations. *Biochemistry* 1997;**36**:12766–72.
34. Huff MW, Miller DB, Wolfe BM, Connelly PW, Sawyez CG. Uptake of hypertriglyceridemic very low density lipoproteins and their remnants by HepG2 cells: the role of lipoprotein lipase, hepatic triglyceride lipase, and cell surface proteoglycans. *J Lipid Res* 1997;**38**:1318–33.
35. Ji ZS, Pitas RE, Mahley RW. Differential cellular accumulation/retention of apolipoprotein E mediated by cell surface heparan sulfate proteoglycans. Apolipoproteins E3 and E2 greater than e4. *J Biol Chem* 1998;**273**:13452–60.
36. Seo T, Al-Haideri M, Treskova E, Worgall TS, Kako Y, Goldberg IJ, et al. Lipoprotein lipase-mediated selective uptake from low density lipoprotein requires cell surface proteoglycans and is independent of scavenger receptor class B type 1. *J Biol Chem* 2000;**275**:30355–62.
37. Medh JD, Fry GL, Bowen SL, Ruben S, Wong H, Chappell DA. Lipoprotein lipase- and hepatic triglyceride lipase-promoted very low density lipoprotein degradation proceeds via an apolipoprotein E-dependent mechanism. *J Lipid Res* 2000;**41**:1858–71.

38. Ji ZS, Sanan DA, Mahley RW. Intravenous heparinase inhibits remnant lipoprotein clearance from the plasma and uptake by the liver: *in vivo* role of heparan sulfate proteoglycans. *J Lipid Res* 1995;**36**:583–92.
39. Mortimer BC, Beveridge DJ, Martins IJ, Redgrave TG. Intracellular localization and metabolism of chylomicron remnants in the livers of low density lipoprotein receptor-deficient mice and apoE-deficient mice. Evidence for slow metabolism via an alternative apoE-dependent pathway. *J Biol Chem* 1995;**270**:28767–76.
40. Windler E, Greeve J, Robenek H, Rinninger F, Greten H, Jackle S. Differences in the mechanisms of uptake and endocytosis of small and large chylomicron remnants by rat liver. *Hepatology* 1996;**24**:344–51.
41. Hayashi K, Hayashi M, Jalkanen M, Firestone JH, Trelstad RL, Bernfield M. Immunocytochemistry of cell surface heparan sulfate proteoglycan in mouse tissues. A light and electron microscopic study. *J Histochem Cytochem* 1987;**35**:1079–88.
42. Roskams T, Moshage H, De Vos R, Guido D, Yap P, Desmet V. Heparan sulfate proteoglycan expression in normal human liver. *Hepatology* 1995;**21**:950–8.
43. Fuki IV, Kuhn KM, Lomazov IR, Rothman VL, Tuszynski GP, Iozzo RV, et al. The syndecan family of proteoglycans. Novel receptors mediating internalization of atherogenic lipoproteins *in vitro*. *J Clin Invest* 1997;**100**:1611–22.
44. Fuki IV, Iozzo RV, Williams KJ. Perlecan heparan sulfate proteoglycan–a novel receptor that mediates a distinct pathway for ligand catabolism. *J Biol Chem* 2000;**275**:25742–50.
45. Zeng BJ, Mortimer BC, Martins IJ, Seydel U, Redgrave TG. Chylomicron remnant uptake is regulated by the expression and function of heparan sulfate proteoglycan in hepatocytes. *J Lipid Res* 1998;**39**:845–60.
46. Stanford KI, Bishop JR, Foley EM, Gonzales JC, Niesman IR, Witztum JL, et al. Syndecan-1 is the primary heparan sulfate proteoglycan mediating hepatic clearance of triglyceride-rich lipoproteins in mice. *J Clin Invest* 2009;**119**:3236–45.
47. MacArthur JM, Bishop JR, Wang L, Stanford KI, Bensadoun A, Witztum JL, et al. Liver heparan sulfate proteoglycans mediate clearance of triglyceride-rich lipoproteins independently of LDL receptor family members. *J Clin Invest* 2007;**117**:153–64.
48. Stanford KI, Wang L, Castagnola J, Song D, Bishop JR, Brown JR, et al. Heparan sulfate 2-*O*-sulfotransferase is required for triglyceride-rich lipoprotein clearance. *J Biol Chem* 2010;**285**:286–94.
49. Cortes V, Amigo L, Donoso K, Valencia I, Quinones V, Zanlungo S, et al. Adenovirus-mediated hepatic syndecan-1 overexpression induces hepatocyte proliferation and hyperlipidaemia in mice. *Liver Int* 2007;**27**:569–81.
50. Varki A, Cummings R, Esko JD, Freeze H, Stanley P, Bertozzi CR, Hart GW, Etzler ME. *Essentials of glycobiology*. 2nd ed. New York: Cold Spring Harbor Laboratories Press; 2009.
51. Bernfield M, Kokenyesi R, Kato M, Hinkes MT, Spring J, Gallo RL, et al. Biology of the syndecans: a family of transmembrane heparan sulfate proteoglycans. *Annu Rev Cell Biol* 1992;**8**:365–93.
52. Sanderson RD, Bernfield M. Molecular polymorphism of a cell surface proteoglycan: distinct structures on simple and stratified epithelia. *Proc Natl Acad Sci USA* 1988;**85**:9562–6.
53. Kokenyesi R, Bernfield M. Core protein structure and sequence determine the site and presence of heparan sulfate and chondroitin sulfate on syndecan-1. *J Biol Chem* 1994;**269**:12304–9.
54. Langford JK, Stanley MJ, Cao DJ, Sanderson RD. Multiple heparan sulfate chains are required for optimal syndecan-1 function. *J Biol Chem* 1998;**273**:29965–71.
55. Rapraeger A, Jalkanen M, Endo E, Koda J, Bernfield M. The cell surface proteoglycan from mouse mammary epithelial cells bears chondroitin sulfate and heparan sulfate glycosaminoglycans. *J Biol Chem* 1985;**260**:11046–52.
56. Saunders S, Jalkanen M, O'Farrell S, Bernfield M. Molecular cloning of syndecan, an integral membrane proteoglycan. *J Cell Biol* 1989;**108**:1547–56.

57. Xian X, Gopal S, Couchman JR. Syndecans as receptors and organizers of the extracellular matrix. *Cell Tissue Res* 2010;**339**:31–46.
58. Alexander CM, Reichsman F, Hinkes MT, Lincecum J, Becker KA, Cumberledge S, et al. Syndecan-1 is required for Wnt-1-induced mammary tumorigenesis in mice. *Nat Genet* 2000;**25**:329–32.
59. Park PW, Pier GB, Hinkes MT, Bernfield M. Exploitation of syndecan-1 shedding by *Pseudomonas aeruginosa* enhances virulence. *Nature* 2001;**411**:98–102.
60. Wang Z, Gotte M, Bernfield M, Reizes O. Constitutive and accelerated shedding of murine syndecan-1 is mediated by cleavage of its core protein at a specific juxtamembrane site. *Biochemistry* 2005;**44**:12355–61.
61. Jalkanen M, Rapraeger A, Saunders S, Benfield M. Cell surface proteoglycan of mouse mammary epithelial cells is shed by cleavage of its matrix-binding ectodomain from its membrane-associated domain. *J Cell Biol* 1987;**105**:3087–96.
62. Mangelsdorf DJ, Thummel C, Beato M, Herrlich P, Schutz G, Umesono K, et al. The nuclear receptor superfamily: the second decade. *Cell* 1995;**83**:835–9.
63. Anisfeld AM, Kast-Woelbern HR, Meyer ME, Jones SA, Zhang Y, Williams KJ, et al. Syndecan-1 expression is regulated in an isoform-specific manner by the farnesoid-X receptor. *J Biol Chem* 2003;**278**:20420–8.
64. Kast HR, Nguyen CM, Sinal CJ, Jones SA, Laffitte BA, Reue K, et al. Farnesoid X-activated receptor induces apolipoprotein C-II transcription: a molecular mechanism linking plasma triglyceride levels to bile acids. *Mol Endocrinol* 2001;**15**:1720–8.
65. Mak PA, Kast-Woelbern HR, Anisfeld AM, Edwards PA. Identification of PLTP as an LXR target gene and apoE as an FXR target gene reveals overlapping targets for the two nuclear receptors. *J Lipid Res* 2002;**43**:2037–41.
66. Edwards IJ, Sun H, Hu Y, Berquin IM, O'Flaherty JT, Cline JM, et al. *In vivo* and *in vitro* regulation of syndecan 1 in prostate cells by n-3 polyunsaturated fatty acids. *J Biol Chem* 2008;**283**:18441–9.
67. Sun H, Berquin IM, Owens RT, O'Flaherty JT, Edwards IJ. Peroxisome proliferator-activated receptor gamma-mediated up-regulation of syndecan-1 by n-3 fatty acids promotes apoptosis of human breast cancer cells. *Cancer Res* 2008;**68**:2912–9.
68. Chen P, Abacherli LE, Nadler ST, Wang Y, Li Q, Parks WC. MMP7 shedding of syndecan-1 facilitates re-epithelialization by affecting alpha(2)beta(1) integrin activation. *PLoS ONE* 2009;**4**:e6565.
69. Brule S, Charnaux N, Sutton A, Ledoux D, Chaigneau T, Saffar L, et al. The shedding of syndecan-4 and syndecan-1 from HeLa cells and human primary macrophages is accelerated by SDF-1/CXCL12 and mediated by the matrix metalloproteinase-9. *Glycobiology* 2006;**16**:488–501.
70. Endo K, Takino T, Miyamori H, Kinsen H, Yoshizaki T, Furukawa M, et al. Cleavage of syndecan-1 by membrane type matrix metalloproteinase-1 stimulates cell migration. *J Biol Chem* 2003;**278**:40764–70.
71. Subramanian SV, Fitzgerald ML, Bernfield M. Regulated shedding of syndecan-1 and -4 ectodomains by thrombin and growth factor receptor activation. *J Biol Chem* 1997;**272**:14713–20.
72. Popova TG, Millis B, Bradburne C, Nazarenko S, Bailey C, Chandhoke V, et al. Acceleration of epithelial cell syndecan-1 shedding by anthrax hemolytic virulence factors. *BMC Microbiol* 2006;**6**:8.
73. Fitzgerald ML, Wang ZH, Park PW, Murphy G, Bernfield M. Shedding of syndecan-1 and -4 ectodomains is regulated by multiple signaling pathways and mediated by a TIMP-3-sensitive metalloproteinase. *J Cell Biol* 2000;**148**:811–24.
74. Hayashida K, Parks WC, Park PW. Syndecan-1 shedding facilitates the resolution of neutrophilic inflammation by removing sequestered CXC chemokines. *Blood* 2009;**114**:3033–43.

75. Tkachenko E, Rhodes JM, Simons M. Syndecans: new kids on the signaling block. *Circ Res* 2005;**96**:488–500.
76. Dews IC, Mackenzie KR. Transmembrane domains of the syndecan family of growth factor coreceptors display a hierarchy of homotypic and heterotypic interactions. *Proc Natl Acad Sci USA* 2007;**104**:20782–7.
77. Carey DJ, Bendt KM, Stahl RC. The cytoplasmic domain of syndecan-1 is required for cytoskeleton association but not detergent insolubility. Identification of essential cytoplasmic domain residues. *J Biol Chem* 1996;**271**:15253–60.
78. Grootjans JJ, Zimmermann P, Reekmans G, Smets A, Degeest G, Durr J, et al. Syntenin, a PDZ protein that binds syndecan cytoplasmic domains. *Proc Natl Acad Sci USA* 1997;**94**:13683–8.
79. Gao Y, Li M, Chen W, Simons M. Synectin, syndecan-4 cytoplasmic domain binding PDZ protein, inhibits cell migration. *J Cell Physiol* 2000;**184**:373–9.
80. Ott VL, Rapraeger AC. Tyrosine phosphorylation of syndecan-1 and -4 cytoplasmic domains in adherent B82 fibroblasts. *J Biol Chem* 1998;**273**:35291–8.
81. Reiland J, Ott VL, Lebakken CS, Yeaman C, McCarthy J, Rapraeger AC. Pervanadate activation of intracellular kinases leads to tyrosine phosphorylation and shedding of syndecan-1. *Biochem J* 1996;**319**:39–47.
82. Hayashida K, Stahl PD, Park PW. Syndecan-1 ectodomain shedding is regulated by the small GTPase Rab5. *J Biol Chem* 2008;**283**:35435–44.
83. Lyon M, Deakin JA, Gallagher JT. Liver heparan sulfate structure. A novel molecular design. *J Biol Chem* 1994;**269**:11208–15.
84. Gallagher JT, Walker A. Molecular distinctions between heparan sulphate and heparin. Analysis of sulphation patterns indicates that heparan sulphate and heparin are separate families of *N*-sulphated polysaccharides. *Biochem J* 1985;**230**:665–74.
85. Lawrence R, Lu H, Rosenberg RD, Esko JD, Zhang L. Disaccharide structure code for the easy representation of constituent oligosaccharides from glycosaminoglycans. *Nat Methods* 2008;**5**:291–2.
86. Oldberg Å, Kjellén L, Höök M. Cell-surface heparan sulfate. Isolation and characterization of a proteoglycan from rat liver membranes. *J Biol Chem* 1979;**254**:8505–10.
87. Ringvall M, Ledin J, Holmborn K, Van Kuppevelt T, Ellin F, Eriksson I, et al. Defective heparan sulfate biosynthesis and neonatal lethality in mice lacking *N*-deacetylase/*N*-sulfotransferase-1. *J Biol Chem* 2000;**275**:25926–30.
88. Fan G, Xiao L, Cheng L, Wang X, Sun B, Hu G. Targeted disruption of NDST-1 gene leads to pulmonary hypoplasia and neonatal respiratory distress in mice. *FEBS Lett* 2000;**467**:7–11.
89. Grobe K, Inatani M, Pallerla SR, Castagnola J, Yamaguchi Y, Esko JD. Cerebral hypoplasia and craniofacial defects in mice lacking heparan sulfate Ndst1 gene function. *Development* 2005;**132**:3777–86.
90. Esko JD, Lindahl U. Molecular diversity of heparan sulfate. *J Clin Invest* 2001;**108**:169–73.
91. Habuchi H, Nagai N, Sugaya N, Atsumi F, Stevens RL, Kimata K. Mice deficient in heparan sulfate 6-*O*-sulfotransferase-1 exhibit defective heparan sulfate biosynthesis, abnormal placentation, and late embryonic lethality. *J Biol Chem* 2007;**282**:15578–88.
91a. Williams KJ, Chen K. Recent insights into factors affecting remnant lipoprotein uptake. *Curr Opin Lipidol.* 2010;**21**:218–28.
92. Libeu CP, Lund-Katz S, Phillips MC, Wehrli S, Hernaiz MJ, Capila I, et al. New insights into the heparan sulfate proteoglycan-binding activity of apolipoprotein E. *J Biol Chem* 2001;**276**:39138–44.
93. Dong J, Peters-Libeu CA, Weisgraber KH, Segelke BW, Rupp B, Capila I, et al. Interaction of the N-terminal domain of apolipoprotein E4 with heparin. *Biochemistry* 2001;**40**:2826–34.
94. Cardin AD, Weintraub HJ. Molecular modeling of protein-glycosaminoglycan interactions. *Arteriosclerosis* 1989;**9**:21–32.

95. Bazin HG, Marques MA, Owens 3rd AP, Linhardt RJ, Crutcher KA. Inhibition of apolipoprotein E-related neurotoxicity by glycosaminoglycans and their oligosaccharides. *Biochemistry* 2002;**41**:8203–11.
96. Flood C, Gustafsson M, Richardson PE, Harvey SC, Segrest JP, Boren J. Identification of the proteoglycan binding site in apolipoprotein B48. *J Biol Chem* 2002;**277**:32228–33.
97. Goldberg IJ, Wagner WD, Pang L, Paka L, Curtiss LK, DeLozier JA, et al. The NH2-terminal region of apolipoprotein B is sufficient for lipoprotein association with glycosaminoglycans. *J Biol Chem* 1998;**273**:35355–61.
98. Borén J, Olin K, Lee I, Chait A, Wight TN, Innerarity TL. Identification of the principal proteoglycan-binding site in LDL - A single-point mutation in apo-B100 severely affects proteoglycan interaction without affecting LDL receptor binding. *J Clin Invest* 1998;**101**:2658–64.
99. Merchant ZM, Erbe EE, Eddy WP, Patel D, Linhardt RJ. Effect of very low molecular weight heparin-derived oligosaccharides on lipoprotein lipase release in rabbits. *Atherosclerosis* 1986;**62**:151–8.
100. Yu W, Hill JS. Mapping the heparin-binding domain of human hepatic lipase. *Biochem Biophys Res Commun* 2006;**343**:659–65.
101. Sendak RA, Berryman DE, Gellman G, Melford K, Bensadoun A. Binding of hepatic lipase to heparin: identification of specific heparin-binding residues in two distinct positive charge clusters. *J Lipid Res* 2000;**41**:260–8.
102. Parthasarathy N, Goldberg IJ, Sivaram P, Mulloy B, Flory DM, Wagner WD. Oligosaccharide sequences of endothelial cell surface heparan sulfate proteoglycan with affinity for lipoprotein lipase. *J Biol Chem* 1994;**269**:22391–6.
103. Larnkjaer A, Nykjaer A, Olivecrona G, Thogersen H, Ostergaard PB. Structure of heparin fragments with high affinity for lipoprotein lipase and inhibition of lipoprotein lipase binding to alpha 2-macroglobulin-receptor/low-density-lipoprotein-receptor-related protein by heparin fragments. *Biochem J* 1995;**307**(Pt 1):205–14.
104. Jaye M, Lynch KJ, Krawiec J, Marchadier D, Maugeais C, Doan K, et al. A novel endothelial-derived lipase that modulates HDL metabolism. *Nat Genet* 1999;**21**:424–8.
105. Fuki IV, Blanchard N, Jin W, Marchadier DH, Millar JS, Glick JM, et al. Endogenously produced endothelial lipase enhances binding and cellular processing of plasma lipoproteins via HSPG-mediated pathway. *J Biol Chem* 2003;**278**:34331–8.
106. Lookene A, Beckstead JA, Nilsson S, Olivecrona G, Ryan RO. Apolipoprotein A-V-heparin interactions: implications for plasma lipoprotein metabolism. *J Biol Chem* 2005;**280**:25383–7.
107. Spillmann D, Lookene A, Olivecrona G. Isolation and characterization of low sulfated heparan sulfate sequences with affinity for lipoprotein lipase. *J Biol Chem* 2006;**281**:23405–13.
108. Fuki IV, Meyer ME, Williams KJ. Transmembrane and cytoplasmic domains of syndecan mediate a multi-step endocytic pathway involving detergent-insoluble membrane rafts. *Biochem J* 2000;**351**:607–12.
109. Mahley RW, Ji ZS, Brecht WJ, Miranda RD, He D. Role of heparan sulfate proteoglycans and the LDL receptor-related protein in remnant lipoprotein metabolism. *Ann NY Acad Sci* 1994;**737**:39–52.
110. Ishibashi S, Herz J, Maeda N, Goldstein JL, Brown MS. The two-receptor model of lipoprotein clearance: tests of the hypothesis in "knockout" mice lacking the low density lipoprotein receptor, apolipoprotein E, or both proteins. *Proc Natl Acad Sci USA* 1994;**91**:4431–5.
111. Farese Jr RV, Veniant MM, Cham CM, Flynn LM, Pierotti V, Loring JF, et al. Phenotypic analysis of mice expressing exclusively apolipoprotein B48 or apolipoprotein B100. *Proc Natl Acad Sci USA* 1996;**93**:6393–8.
112. Heeren J, Niemeier A, Merkel M, Beisiegel U. Endothelial-derived lipoprotein lipase is bound to postprandial triglyceride-rich lipoproteins and mediates their hepatic clearance *in vivo*. *J Mol Med* 2002;**80**:576–84.

113. Beisiegel U, Weber W, Bengtsson-Olivecrona G. Lipoprotein lipase enhances the binding of chylomicrons to low density lipoprotein receptor-related protein. *Proc Natl Acad Sci USA* 1991;**88**:8342–6.
114. Eisenberg S, Sehayek E, Olivecrona T, Vlodavsky I. Lipoprotein lipase enhances binding of lipoproteins to heparan sulfate on cell surfaces and extracellular matrix. *J Clin Invest* 1992;**90**:2013–21.
115. Roskams T, Rosenbaum J, De Vos R, David G, Desmet V. Heparan sulfate proteoglycan expression in chronic cholestatic human liver diseases. *Hepatology* 1996;**24**:524–32.
116. Kovalszky H, Gallai M, Armbrust T, Ramadori G. Syndecan-1 gene expression in isolated rat liver cells (hepatocytes, Kupffer cells, endothelial and Ito cells). *Biochem Biophys Res Commun* 1994;**204**:944–9.
117. Weiner OH, Zoremba M, Gressner AM. Gene expression of syndecans and betaglycan in isolated rat liver cells. *Cell Tissue Res* 1996;**285**:11–6.
118. de Faria E, Fong LG, Komaromy M, Cooper AD. Relative roles of the LDL receptor, the LDL receptor-like protein, and hepatic lipase in chylomicron remnant removal by the liver. *J Lipid Res* 1996;**37**:197–209.
119. Brown MS, Anderson RG, Goldstein JL. Recycling receptors: the round-trip itinerary of migrant membrane proteins. *Cell* 1983;**32**:663–7.
120. Tkachenko E, Lutgens E, Stan RV, Simons M. Fibroblast growth factor 2 endocytosis in endothelial cells proceed via syndecan-4-dependent activation of Rac1 and a Cdc42-dependent macropinocytic pathway. *J Cell Sci* 2004;**117**:3189–99.
121. Wilsie LC, Gonzales AM, Orlando RA. Syndecan-1 mediates internalization of apoE-VLDL through a low density lipoprotein receptor-related protein (LRP)-independent, non-clathrin-mediated pathway. *Lipids Health Dis* 2006;**5**:23.
122. Thankamony SP, Knudson W. Acylation of CD44 and its association with lipid rafts are required for receptor and hyaluronan endocytosis. *J Biol Chem* 2006;**281**:34601–9.
123. Mahley RW, Ji ZS. Remnant lipoprotein metabolism: key pathways involving cell-surface heparan sulfate proteoglycans and apolipoprotein E. *J Lipid Res* 1999;**40**:1–16.
124. Wilsie LC, Orlando RA. The low density lipoprotein receptor-related protein complexes with cell surface heparan sulfate proteoglycans to regulate proteoglycan-mediated lipoprotein catabolism. *J Biol Chem* 2003;**278**:15758–64.
125. Williams KJ, Fuki IV. Cell-surface heparan sulfate proteoglycans: dynamic molecules mediating ligand catabolism. *Curr Opin Lipidol* 1997;**8**:253–62.
126. Belting M. Heparan sulfate proteoglycan as a plasma membrane carrier. *Trends Biochem Sci* 2003;**28**:145–51.
127. Elson-Schwab L, Garner OB, Schuksz M, Crawford BE, Esko JD, Tor Y. Guanidinylated neomycin delivers large, bioactive cargo into cells through a heparan sulfate-dependent pathway. *J Biol Chem* 2007;**282**:13585–91.
128. Warshawsky I, Herz J, Broze Jr GJ, Schwartz AL. The low density lipoprotein receptor-related protein can function independently from heparan sulfate proteoglycans in tissue factor pathway inhibitor endocytosis. *J Biol Chem* 1996;**271**:25873–9.
129. Stalenhoef AF, de Graaf J. Association of fasting and nonfasting serum triglycerides with cardiovascular disease and the role of remnant-like lipoproteins and small dense LDL. *Curr Opin Lipidol* 2008;**19**:355–61.
130. Yuan G, Al-Shali KZ, Hegele RA. Hypertriglyceridemia: its etiology, effects and treatment. *CMAJ* 2007;**176**:1113–20.
131. Nordestgaard BG, Benn M, Schnohr P, Tybjaerg-Hansen A. Nonfasting triglycerides and risk of myocardial infarction, ischemic heart disease, and death in men and women. *JAMA* 2007;**298**:299–308.

132. Bansal S, Buring JE, Rifai N, Mora S, Sacks FM, Ridker PM. Fasting compared with nonfasting triglycerides and risk of cardiovascular events in women. *JAMA* 2007;**298**:309–16.
133. Langsted A, Freiberg JJ, Nordestgaard BG. Fasting and nonfasting lipid levels: influence of normal food intake on lipids, lipoproteins, apolipoproteins, and cardiovascular risk prediction. *Circulation* 2008;**118**:2047–56.
134. Fujioka Y, Ishikawa Y. Remnant lipoproteins as strong key particles to atherogenesis. *J Atheroscler Thromb* 2009; **16**:145–54.

Serglycin Proteoglycan Deletion in Mouse Platelets: Physiological Effects and Their Implications for Platelet Contributions to Thrombosis, Inflammation, Atherosclerosis, and Metastasis

Barbara P. Schick

Department of Medicine, Cardeza Foundation for Hematologic Research, Thomas Jefferson University, Philadelphia, Pennsylvania, USA

Progress in Molecular Biology and Translational Science, Vol. 93
DOI: 10.1016/S1877-1173(10)93011-1

1877-1173/10 $35.00

Serglycin is found in all nucleated hematopoietic cells and platelets, blood vessels, various reproductive and developmental tissues, and in chondrocytes. The serglycin knockout mouse has demonstrated that this proteoglycan is required for proper generation and function of secretory granules in several hematopoietic cells. The effects on platelets are profound, and include diminishing platelet aggregation responses and formation of platelet thrombi. This chapter will review cell-specific aspects of serglycin structure, its gene regulation, cell and tissue localization, and the effects of serglycin deletion on hematopoietic cell granule structure and function. The effects of serglycin knockout on platelets are described and discussed in detail. Rationales for further investigations into the contribution of serglycin to the known roles of platelets in thrombosis, inflammation, atherosclerosis, and tumor metastasis are presented.

I. Introduction

It has long been known that serglycin is the proteoglycan that is packaged into secretory granules and vesicles in platelets, megakaryocytes and all myeloid cells. Numerous *in vitro* studies have suggested that the functions of serglycin include packaging of proteins into the granules, protection of molecules, for example proteases or growth factors with which they are secreted as complexes, modulation of the activity of these proteins by binding to them until the proteins are delivered to their targets, and facilitation of the delivery of these proteins to their targets after secretion. Heparan sulfate proteoglycans have been implicated in the establishment of chemokine gradients. Such gradients are thought to be involved in immune function, coagulation, embryonic implantation, and developmental processes, as well as in the development of atherosclerosis or metastasis. A cell critical to all of these processes is the blood platelet. Whether and how chondroitin sulfate serglycin/protein complexes, whether bound to cells or matrix or free in solution, might be responsible for establishing these gradients are as yet unexplored questions. As these complexes are presumably the initiation point of delivery of these proteins, it is logical to presume that they are important in establishing such gradients.

This chapter will review current knowledge on the very interesting, cell-specific, regulation of serglycin expression and structure in hematopoietic and other cells. Platelet serglycin will be the central focus, and will be examined in the light of current knowledge of serglycins in other cells. The deletion of serglycin leads to remarkable changes in platelet function, *in vitro* and *in vivo*. These functional changes will be discussed in the context of the complex mechanisms involved in platelet activation. The implications of these findings for physiological events such as thrombosis and normal blood coagulation, as well as inflammation, atherosclerosis, and tumor metastasis, in which platelets play a significant role, will be explored.

Platelets are the first line of defense for sealing a wound following an injury. They interact with the vessel wall and with other blood cells during this process. They also interact with the proteins of the coagulation system during the process of normal blood clotting in response to injury, and in the harmful corollary, thrombosis, which can lead to serious disease or even death. Platelets are derived by fragmentation of the cytoplasm of the parent cell, the bone marrow megakaryocyte, by a process which yields several thousand platelets per megakaryocyte, but this process is not completely defined as yet. As platelets are derived from the cytoplasm, they are anucleate. All structures of the mature platelet are presumably completely organized before the release of platelets from the megakaryocyte. The platelet structures most important to the considerations in this chapter are the α-granules, which are the presumed sites of storage of serglycin. The α-granules contain several hundred proteins, some of which bind to serglycin and many have significant biological activity. α-Granule constituents interact in a highly complex manner with components of the dense granules, which contain serotonin, ADP, ATP, and various other molecules, the surface membrane, which is a highly intercalated structure may also contain serglycin proteoglycan, the cytoskeletal components, and various signaling pathways. The loss of serglycin affects these processes in unexpected ways.

II. Historical Perspective: Cloning, and Cell and Tissue Localization of Serglycin

Serglycin was the first proteoglycan gene to be cloned. Serglycin mRNA was cloned initially from the rat L2 yolk sac tumor,[1,2] on the basis of the amino acid sequence of the chondroitin sulfate proteoglycan that was copiously secreted from these cells. The L2 cells are derived from the rat parietal endoderm.[3] The translated protein had a molecular weight of 18 kDa, and is the smallest known proteoglycan core protein. The name serglycin was chosen to reflect the 42 amino acid serine/glycine repeat region, which was deduced from the cDNA. Each serine of a ser/gly repeat is a potential glycosaminoglycan (GAG) attachment site. This structure, which allows clustering of GAGs near the center of the protein, is unique to serglycin. The core protein of this proteoglycan is unique; there is no known similarity with other proteoglycans or proteins.

For a number of years, it had been known that hematopoietic cells contained chondroitin sulfate proteoglycans. The presence of mucopolysaccharides in platelets, in a proteoglycan form, was recognized at least forty years ago.[4] The presence of clusters of GAG chains had been inferred from a number of biochemical studies of the hematopoietic proteoglycans. Thus, serglycin, with its ser/gly repeats being a likely site of clustered GAG attachment, was the logical candidate for identification as the hematopoietic proteoglycan. Indeed, cDNAs for serglycin core proteins were subsequently cloned from mouse[5] and human hematopoietic cells, beginning with HL60 promyelocytic[6,7] and HEL erythroleukemic cells[8]; and the sequence of the purified platelet proteoglycan was found to be identical to that deduced from the HEL cDNA.[9] The mouse and human cDNAs reveal a ser/gly region with only eight repeats, and the interspecies sequence homology is greatest in the N-terminal half of the molecule. Transcripts of the serglycin gene have been detected in a variety of human, nonerythroid hematopoietic cells and in multipotential hematopoietic tumor cell lines by Northern blotting or by *in situ* hybridization.[10–14] Serglycin is the only proteoglycan identified thus far in the granules of the hematopoietic cells and is best known as "the hematopoietic proteoglycan." Serglycin is found in the protease-containing granules in mast cells (heparin is assembled exclusively on the serglycin core protein in specific subsets of mast cells) and eosinophils,[15] and in platelet α-granules; in these cells, serglycin and the granule contents are stored until the cell is activated. In contrast to such storage, serglycin is constitutively secreted from lymphocytes[10] and a number of cultured hematopoietic cell lines.[11,12]

In recent years it has been shown that serglycin is also expressed in nonhematopoietic cells. We have shown serglycin mRNA expression and the presence of the serglycin core protein in hitherto unrecognized sites: endothelial cells,[16] murine uterine mesometrial decidua during mid-pregnancy,[17] murine parietal endoderm,[17] murine fetal liver but not yolk sac hematopoietic cells,[17] and in murine embryonic stem (ES) cells.[18] Serglycin is also synthesized by chondrocytes.[19] The significance of serglycin in these cells has not yet been established.

III. Structure of Serglycin

A. Cell-Specific Variation in Serglycin Proteoglycan Size

The overall size of the serglycin proteoglycans can vary from 135 to >600 kDa on SDS-PAGE, with a Kav of 0.08–0.25 on Sepharose CL-6B columns. This variation is due to differences in the size of GAG chains synthesized by different particular cell types and in the number of GAG chains.[10] We isolated proteoglycans from guinea pig and human megakaryocytes and platelets,[20] rat platelets,[21] megakaryocytic cell lines,[11,12] and HL60 cells (unpublished) and found a range of GAG chain lengths from 10 to 60 kDa. U937-1 cell serglycin has 5 kDa GAG chains.[22]

B. Cell-Specific Variation in Glycosaminoglycan Type, Size, and Number

Mast cells synthesize serglycin from oversulfated chondroitin sulfates or heparin or heparan sulfates depending on the tissue of residence of the cell. It should be noted that mast cell serglycin is the only source of true heparin. The type of GAG chain appears to be coordinately regulated with the type of chymase that the particular mast cell produces.[23] The mouse mastocytoma cell line, MST, serglycin has both chondroitin sulfate and heparin GAGs, including a CS/heparin hybrid molecule.[24] The GAG chains in human, rat, and mouse platelets are exclusively chondroitin-4-sulfate, but the megakaryocytic tumor cells HEL and CHRF-288-11 also synthesize a hybrid chondroitin sulfate/heparan sulfate serglycin.[11,12,25] The uterine decidual and HUVEC serglycins are chondroitin sulfate[16,17] and these intact proteoglycans and their GAG chains are smaller than their platelet counterparts.

A recent paper has shed some light on the mechanisms that enable specificity in the sulfation patterns of mast cell serglycin.[26] Duelli *et al.* showed that expression of not only the serglycin core protein, but also of two sulfotransferases involved in the sulfation of chondroitin sulfate, chondroitin 4-sulfotransferase (C4ST)-1, and GalNAc(4S)-6-*O*-sulfotransferase (GalNAc4S6ST), are closely linked to mast cell maturation. In contrast, the expression of chondroitin

6-sulfotransferase correlated negatively with mast cell maturation. The expression of *N*-deacetylase/*N*-sulfotransferase (NDST)-2, a key enzyme in heparin synthesis, also correlated strongly with mast cell maturation, whereas the expression of the NDST-1 isoform was approximately equal at all stages of maturation. The effect of mast cell activation by either calcium ionophore or IgE ligation on these enzymes was investigated by biosynthetic radiolabeling of GAGs. Mast cell activation caused an increase in the expression of the serglycin core protein, C4ST-1, and GalNAc4S6ST, and was accompanied by an increase in the secretion of chondroitin sulfate. In contrast, NDST-2 was downregulated after mast cell activation, suggesting that mast cell activation modulates the nature of the GAG chains attached to the serglycin core protein. These studies show that the synthesis and sulfation of serglycin and its GAG chains are regulated, at the gene level, as a function of both mast cell maturation and activation.

The GAG chains in the mast cells are very short compared to those of other hematopoietic cells. In other cell culture studies, the length of the GAG chains has been shown to vary depending on the type of stimulus provided to the cells. No studies have been done on the regulation of the chain elongation enzymes.

The number of potential and the actual GAG chains is species-specific: mouse and human serglycin have a 16-amino acid ser/gly repeat, while that of the rat is 42 amino acids. From GAG chain length (MW) and overall proteoglycan size determined by gel filtration or on SDS-PAGE, it has been estimated that the human platelet serglycin core protein bears about four GAG chains, and that the rat yolk sac tumor serglycin has about 14 GAG chains.[27] Thus, not all the potential GAG attachment sites are utilized.

C. Cell-Specific Variation in Apparent Core Protein Size

The core protein is, most commonly, analyzed after chondroitinase digestion, which is known to leave a hexasaccharide stub (about 1.5 kDa) at each GAG attachment site. These stubs will, of course, influence the apparent size of the core protein on SDS-PAGE gels. The deduced sequence of the HL60 cDNA gives a MW of 14,100 Da after cleavage of the signal peptide.[7] However, the apparent size of the chondroitinase-digested core protein on SDS-PAGE is at least twice of what is expected from the cDNA coding sequence.[12,28] This may be due to an unusual secondary structure of the core protein, a possible dimerization in the absence of the GAG chains (we have found that an N-terminal peptide derived from serglycin dimerizes via the cysteine residue),[29] or due to the presence of the hexasaccharide stubs, which remain at the sites of attachment of GAG chains after chondroitinase or heparitinase digestion. The size of the *nonglycosylated* core protein, immunoprecipitated from HL-60 cells after a brief pulse with ^{35}S-methionine, was 20 kDa[30]; as the molecule was processed further over a 1 h period, the apparent immunoprecipitability was lost, either due to proteolysis and loss of the ^{35}S-methionines or to masking of the epitope by the growing GAG chains.[30] The recombinant, human serglycin protein that we have prepared runs at about 22 kDa on SDS-PAGE,[16] which is in close agreement with the size of the HL60 nonglycosylated core protein.[30] The apparent size of the chondroitinase-derived core protein is cell-specific, and several forms may be present in a single cell type. We have found core proteins of MW 39 and 36 kDa after chondroitinase digestion of guinea pig megakaryocyte and platelet proteoglycans; the smaller molecule is synthesized in the most mature megakaryocytes.[20,31] Core proteins of MW 27, 31, and 38 kDa have been found in HEL cells, depending on growth conditions.[11] We have found that CHRF-288-11 cells synthesize only the 31 kDa serglycin core.[12] We have confirmed the identity of the core proteins as serglycin in several of these cells by Western blotting using an antibody generated in our laboratory.[16] It is intriguing that Perin *et al.*[28] found only a 27 kDa core protein after chondroitinase digestion of proteoglycans from normal human platelets and we have confirmed this in our studies.

There is also some evidence that proteolysis of the core protein can occur within the cell.[11,30,32] We have speculated that proteolysis releases biologically active peptides; a possible candidate would be the high-homology N-terminal end of the ser/gly repeat region which is able to bind heparin.[33]

D. Cell-Specific Regulation of Serglycin GAG Mass

We have found that intact serglycin from rat and human platelets have about the same overall size as determined by SDS-PAGE and Sepharose CL-6B elution, but the rat platelet serglycin GAG chains are only about half as long as human serglycin GAGs.[21] This implies that the rat molecule has a larger number of GAG chains. Functionally, this would suggest that the α-granules need a serglycin molecule with a defined mass of GAG chains, but can achieve this in different ways, that is by establishing an inverse relationship between GAG chain number and chain length. Remarkably, the megakaryocytic tumor cells and U937 cells respond to stimuli such as phorbol myristate acetate (PMA) by changing their GAG chain length in exactly this way.[11,12]

IV. Binding of Biologically Active Proteins to Serglycin: Cell-Specific Binding and Functional Implications

The serglycin free core protein has not been shown to bind to other proteins.[12,21,34–36] Several studies on the binding of intact chondroitin sulfate serglycins to other proteins have shown that the overall structure of the intact proteoglycan is essential for high-affinity binding. Several binding studies have indicated that the intact chondroitin sulfate serglycin proteoglycan binds to other proteins as strongly as heparin, but chondroitin sulfate GAG chains isolated from serglycins do not have any significant binding capability.[12,33,36–38] Proteins which bind to the chondroitin sulfate form of serglycin include extracellular matrix components (*fibronectin and collagen*),[12,21,37] growth factors/cytokines/chemokines (*platelet factor 4,* MIP 1α BMP-like protein),[39] the membrane protein/proteoglycan *CD44,*[38] and lysozyme.[39,40] Other candidate proteins include heparin-binding proteins such as IL-8,[41] tissue plasminogen activator (tPA), thrombospondin, Alzheimer's basic protein precursor, and lactoferrin. All these proteins, except CD44, are secreted by either constitutive or regulated pathways. The heparin serglycin proteoglycan has been postulated to bind with specific chymases, which are different from those bound to mast cell chondroitin sulfate serglycin, and to regulate the function of these enzymes after secretion.[23] Disruption of the serglycin heparin GAG structure, *in vivo*, by deletion of the NDST-2 enzyme (heparin NDST 2) has borne out these predictions, in that, both granule protein content and function in the heparin-serglycin mast cells were abnormal.[34,35] Such a cell-specific experiment was feasible because heparin synthesis is restricted to a subpopulation of mast cells. It is, unfortunately, not possible to assess the role of the much more ubiquitous chondroitin sulfate form of serglycin by chondroitin sulfate GAG disruption *in vivo* because chondroitin sulfate is essentially present in the diverse proteoglycans of many cells.

Among the known serglycin-binding proteins, the chemokine Platelet Factor 4 (PF4) (CXCL4) is of special interest. PF4 is synthesized only by the megakaryocytes. Serglycin is well known as the "carrier for PF4." This important and unique chemokine is released from activated platelets, binds to and activates neutrophils,[42–51] and is associated with the development of atherosclerotic plaques.[52,53] Platelets interact with monocytes and other leukocytes during coagulation and inflammation. These interactions are modulated by chemokines and cytokines; platelet cytokines, for example RANTES and MIP 1α may affect other blood cells[49,50,54,55]; and PF4 from platelets activates neutrophils. A better understanding of these interactions may provide insight into blood cell interactions during inflammatory disease, asthma and AIDS,[55] in which these chemokines are major players, or to the role of platelets in chemokine-related events in tumor metastasis.

We, along with others, hypothesize that the cell-specific granule protein/serglycin interactions are critical in granule development. Proper packaging of correctly synthesized serglycin and its binding protein ligands, and controlled release of these complexes from the granules are likely to be critical for normal tissue function.

V. Regulation of Serglycin Expression

A. Background

A number of studies have shown that changes in the level of serglycin mRNA expression are induced by various agents, which are, presumably, reflected in the amount of core protein synthesized. The level of synthesis of the core protein appears to affect the structure and, most likely, the function of the proteoglycan through the ability of the cell to alter the mean GAG chain length or GAG type, and therefore, the overall hydrodynamic size of the molecule. There is often an inverse relationship between mRNA expression (which is presumably reflected in the amount of core protein synthesized) and GAG chain length: when serglycin mRNA expression is increased the GAG size is decreased, and vice versa. This change in structure could affect the packaging of proteins into the secretory granulesand their ultimate secretion and function. In some cases, the cells may have a mechanism by which the total mass of GAGs placed onto serglycin is regulated, which is distributed among the available amount of core protein,[21] so that a constant mass of GAG is present within the granule. A recent study has shown that the regulation of sulfation is dependent on the state of maturation or activation of mast cells.[26] A clinically relevant study has shown that the level of serglycin expression is associated with the degree of chemotherapy drug resistance in malignant hematopoietic cell lines.[56] If a change in mRNA expression is associated with a change in GAG structure, then secretory granule proteins that are involved in multidrug resistance might also be affected. This tight regulation of the synthesis and structure of serglycin suggests that these molecules may have important functions.

B. Constitutive and Induced Cell-Specific Regulation of Serglycin Expression

Serglycin synthesis is constitutively upregulated in normal megakaryocytes[20,31] and leukocytes[14] during granule formation and cell maturation; and changes in GAG chain length occur during megakaryocyte maturation.[31] This latter finding may represent the synthesis of different types of granules during early -to- late megakaryocyte development. We have found that the normal upregulation of serglycin synthesis that occurs in megakaryocytes during cytoplasmic maturation does not occur in unstimulated HEL cells, even though at least three other α-granule proteins are constitutively upregulated as the control cultures mature.[57] PMA induces the upregulation of serglycin in HEL cells[11,12] along with the megakaryocytic proteins, PF4,[57–59] P-selectin,[57] GPIIb/IIIa,[60] PDGF-A,[61] and PDGF-B.[14] Serglycin expression is induced in U937-1 cells upon stimulation with retinoic acid and PMA.[22] PMA stimulates serglycin synthesis in HEL and CHRF cells but inhibits serglycin synthesis in HL-60 and other myeloid cell lines.[62–64] The observed effects of PMA are transcriptional in nature.[13] We have found that the effect of cAMP differs among the cell types. Serglycin is upregulated by dibutyryl (db) cAMP in HEL, CHRF, and HL60 cells.[65] Serglycin expression is induced in F9 teratocarcinoma cells by all-*trans* retinoic acid with or without dbcAMP but not by dbcAMP alone.[66] In contrast, cAMP is needed along with retinoic acid to upregulate serglycin expression in ES cells.[18] We have found that neither retinoic acid nor dexamethasone affect serglycin mRNA expression in HEL, CHRF, or HL-60 cells, and that retinoic acid does not affect its expression in murine mastocytoma (MST)[67] cells (our data, unpublished). Eklund *et al.*[68] showed that glucocortocoids inhibit the cytokine-induced proliferation of mast cells and expression of TNF-α and chymases, but increased the expression of serglycin and carboxypeptidase A (CPA).

C. Structure and Potential Regulatory Elements of the Murine and Human Serglycin Genes

The murine and human serglycin genes have been cloned.[6,30,69] About 1900 bp of the 5′-flanking regions of the human[6,30] and 1200 bp of the mouse[30,69,70] serglycin genes have been sequenced. The nucleotides from − 1 to − 119 are 96% conserved, and the − 120 to − 331 sequence is 65% conserved. An A-rich region at − 600 in the mouse gene sequence is analogous to that at − 560 in the human sequence. Murine − 417/−358 has a 53% homology with human − 506/−448. The mouse gene contains several known transcription factor binding sequences: yeast GCN4 at − 265/−260; a heat shock gene element GGGAAA at − 247/−242; an *ets* site, CAGGAA, at − 85/−80, the CRE at − 70/−65; and the glucocorticoid response element (GRE) at − 63/−58; these are similarly placed in the human gene. Both genes have a number of CANNTG (E-box) sequences, including the mouse CAGATG at − 241/−236, which is found in inverse orientation at − 522 in the human gene.

D. Cell-Specific Regulation of Serglycin Promoter Elements

Avraham *et al.*[71] have studied a 500 bp of the murine promoter that was transfected into rat RBL-2 cells. The region − 250 to − 190 contained a negative regulatory element, and the region − 118 to − 81 contained a positive regulatory element. A truncation of the *ets* site at − 81 resulted in the nearly complete loss of activity, suggesting that the *ets* site might be the only critical regulatory element under the conditions used in those experiments. The binding of nuclear factors to the potential *ets* site was not evaluated. We have studied the role of the human serglycin promoter in human hematopoietic tumor cell lines.[65] The neighboring *ets* and CRE sites in the promoter of the human gene appear to be responsible for regulating the expression of serglycin in these cells. We have also studied the role of the murine promoter in murine ES cells, F9 teratocarcinoma cells, and in mastocytoma cells. In contrast to the human hematopoietic cells and rat RBL2 cells, the *ets* site appears to be nonfunctional in ES, F9, and MST cells. We found no evidence of the negative or positive regulatory regions of the murine promoter as described by Avraham *et al.*[71] in the human or murine cells that we have used. The same modifications of the TATA-like region which caused 70% loss of activity of the murine promoter[71] did not affect the activity of the human promoter.[65] These differences may be due to species difference (human promoter in human cells, murine promoter in rat cells vs. murine cells), or the types of cells that we had used. We also used the same reporter expression and transfection systems as Avraham *et al.*

E. DNase I Hypersensitive Sites

DNase hypersensitivity assays have provided information on the regulation of many viral and cellular genes.[72–74] These sites are generally found at positions that are proximal to, or within the promoters, of actively transcribed genes. It is thought that a gene that is expressed in a particular cell type will have tissue-specific hypersensitive sites. A majority of the DHSS that are associated with genes play a functional role in determining the position and frequency of transcription initiation or termination, and are related to hormonal status.[73,75] We have found cell-specific serglycin DHSS in the promoter and intronic regions of hematopoietic cells[76,77] and endothelial cells and, strikingly, a site unique to neutrophils in exon 2, which may result in a shortened core protein[76] (Fig. 1). These sites may account for the observed differences in the levels of endogenous expression, regulation of expression by differentiation-inducing agents, and in cell-specific expression. This information may also provide insights into chromatin behavior under different growth conditions.

Several of the DHSS appear to be involved in regulating the level of endogenous and PMA-induced changes in serglycin expression in the hematopoietic cell lines.[76] The presence of DHSS III, which is unique to HL-60 resting cells, may explain the high degree of serglycin expression in resting HL-60 cells relative to HEL and CHRF. The presence of DHSS VI may be related to the endogenous high expression of serglycin in HL-60 cells and the previously reported increase in serglycin expression in PMA-treated CHRF cells.[12,77] The existance of DHSS VII may be related to the increase in serglycin expression in HEL cells. DHSS IV, V, and X–XV are found only in PMA-treated HL-60 cells, and may have a combined effect on the suppression of serglycin expression by PMA in these cells. It is interesting to compare the DHSS of normal cells and the tumor cells used in these studies. The DHSS VI at +5325 bp in HUVEC DNA was seen in resting and PMA-treated HL-60

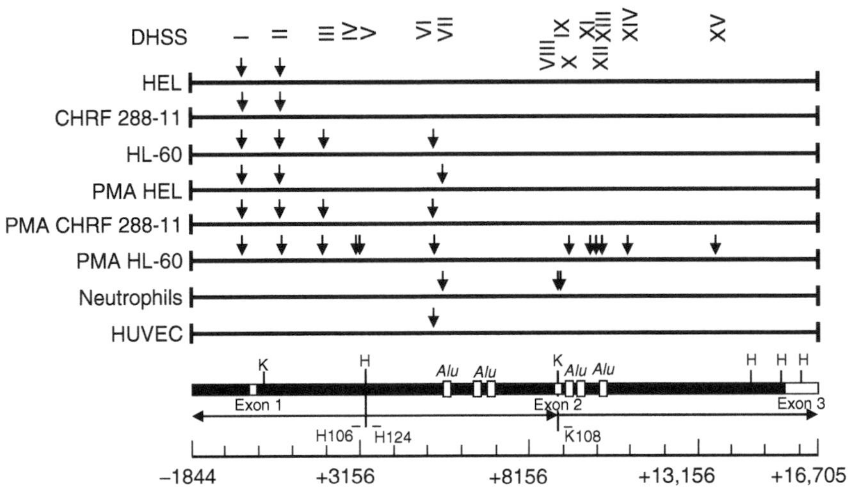

FIG. 1. DHSS map of human serglycin proteoglycan gene in HEL, CHRF 288-11, HL-60, neutrophils, and HUVEC. *Top*: the DHSS are identified in *roman numerals*. *Vertical arrows* denote the locations of the 15 DHSS in resting and PMA-treated cells. *Bottom*: restriction enzyme sites (*H*, *Hin*dIII; *K*, *Kpn*I) for probes H106, H124, and K108 are indicated. *Horizontal arrows* indicate the direction of DNase I mapping. Coding exons (*small open boxes*) and regions where DHSS are associated with *Alu* repeats (*large open boxes*) are shown. Numbering is from Ref. 76.

cells and PMA-treated CHRF cells, suggesting that a common transcriptional mechanism may be involved; but this DHSS was not seen in neutrophil DNA. The absence of this band from neutrophil DNA may be due to the greater maturity of the blood cells compared to HL-60-like bone marrow progenitors.

Some of the DHSS that lie in regions of the human serglycin gene are of interest because of their homology with the murine serglycin gene locus and also because they are present in regions which are rich in transcription factor binding sites that may be involved in regulation of serglycin expression. For example, DHSS I at − 276 bp is within the region that is 65% conserved between mouse and human serglycin and is only 101 bases upstream from the proximal promoter region that is 95% conserved.[6,30,69] It is also about 200 bp upstream from the critical *ets* and CRE sites, which were found to be the two major regulatory elements for the constitutive expression of serglycin gene in HEL and CHRF cells.[77] Similarly, DHSS II is found at only 151 bp 5′ to a region in Intron 1 which has high homology to a similarly placed region of Intron 1 in the murine gene locus, and is the only intronic region with > 75% homology. This homology was revealed by a search of the Celera human and murine data bases. This region is rich in potential binding sites for the PEA-1 and AP-1 transcription factors. DHSS III, which is unique to HL-60 cells and PMA-treated CHRF cells and is located in Intron 1 at + 2075 bp,[77] contains multiple potential E-box sites, motifs of the sequence CANNTG that bind to the basic helix-loop-helix proteins, and may be related to the high basal expression of serglycin in resting HL-60 cells relative to resting HEL or CHRF cells.

Several of the DHSS in Intron 1 of the serglycin gene lie within or very close to *Alu* repeats. DHSS VII at + 5675 bp lies in an *Alu* region between nucleotides + 5479 and + 5756, and DHSS X lies in the *Alu* region 9333–9622 bp. DHSS XI at 9871 bp and DHSS XII at + 10,001 bp are between two closely apposed *Alu* regions 9627–9759 bp and 10,392–10,519 bp. Our study is the first to implicate both cell-specific and differentiation-specific *Alu*-related DHSS in the regulation of the expression of a gene.

The DHSS found in the DNA of normal neutrophils were different from those of the HL-60 cell line, the latter would be representative of immature myeloid cells. Also striking was the mapping of DHSS VIII to Exon 2 in the neutrophils and DHSS IX to a proximal region of Intron 2.

VI. Serglycin in Hematopoietic Cells

A. Platelets

Platelet α-granules contain serglycin and many proteins that are involved in platelet activation during thrombus formation, inflammation, and atherosclerosis. Only one natural model of a defect in serglycin structure has been described[21]: we have described abnormal platelet chondroitin sulfate proteoglycans that involve shortening of GAG chain length and a decrease in total GAG mass per cell by about 60% in Wistar Furth macrothrombocytopenic (platelets are abnormally large and are severely reduced in number) rats. This syndrome is associated with the loss of α-granule proteins. The abnormal serglycin may account for the low content of α-granule proteins in these cells. This rat platelet disorder is similar to a human disorder known as gray platelet syndrome. α-Granules appear to be unstable in the human gray platelet syndrome. Premature release of α-granule growth factors or chemokines from megakaryocytes into the marrow in this disorder, as well as in myeloproliferative disorders and myelofibrosis is thought to cause the lethal marrow fibrosis which is commonly found in these disorders. We have shown that proteoglycan synthesis in normal megakaryocytes appears to be directed towards storage of these molecules in α-granules during production of normal platelets, and that 85% of the platelet proteoglycans are released by thrombin treatment.[31] In contrast, we, and others, have found that the human hematopoietic tumor cell lines constitutively secrete the bulk of their proteoglycans.[11,12] Similarly, Wickenhauser *et al.*[78] have shown that lysozyme is constitutively released by megakaryocytes from patients with polycythemia vera but not from normal megakaryocytes. This is of interest because lysozyme interacts with chondroitin sulfate serglycin,[39,40] suggesting that lysozyme is released as a complex with serglycin. The basis for the differences in secretion patterns between normal and malignant cells is not understood. It is possible that the serglycin/protein complexes are not formed properly, are directed to constitutive secretory vesicles instead of storage granules, or that the storage granules have defective storage and/or release mechanisms.

Serglycin abnormalities may be associated with platelet disorders. We have studied patients with essential thrombocythemia (ET) and polycythemia vera (PV). These patients have 2–4× normal platelet counts. We also found a unique low-MW component of serglycin in the platelets of both the patients with ET but not in any of the 10 patients with PV. All ET and PV platelets had 50% reduction in proteoglycan mass per platelet (B. P. Schick, unpublished).

Platelets have long been implicated in the development of atherosclerosis. The basic platelet α-granule protein PF4[79] has been found in human atherosclerotic plaques,[80] and could serve as a growth factor for arterial smooth muscle cells. If serglycin is a vehicle for the delivery of these proteins to the blood vessels, then the absence of this proteoglycan might prevent deposition of these proteins within the blood vessels. The presence of PF4 has not been assessed in the murine models of atherosclerosis.

B. Mast Cells

The importance of correctly synthesized serglycin proteoglycan for proper secretory granule assembly has been dramatically made evident through work with mast cells. Much work has focused on the role of the heparin form of serglycin that is made only in mast cells, and is the only biological source of heparin.[81–83] Two studies have emphatically shown the importance of the overall structure of *heparin* serglycin in the organization and function of mast cell granules.[34,35] These studies used a model that had a targeted gene knockout of NDST-2, a critical sulfotransferase needed for heparin biosynthesis, and were thus able to specifically target mast cell serglycin. Both studies showed severe derangement of the granules and their function, and are consistent with *in vitro* data which had shown the need for a specific heparin serglycin conformation for binding to chymases.[23,84] Serglycin from mast cells may be involved in blood coagulation. Pejler and colleagues have provided data on the role of the heparin serglycin proteoglycan in regulating the activity of the chymases that may affect the coagulation system, including cleavage of thrombin and on the effects of protamine and protection of chymase from inhibition by serpins Also, heparin, in the form of a GAG released from mast cell serglycin, has been found to be responsible for the generation of kinins in allergic reactions subsequent to the activation of the contact system.[85,86]

Mast cell chondroitin sulfate serglycins are oversulfated and bind to proteases that are different from those in the heparin-bearing mast cells.[82,87–89] Hunt *et al.*[90] proposed that uterine mast cell MCP-9 uniquely complexes with the serglycin chondroitin sulfate proteoglycan. Some of these proteases (MCPs) form tetramers in their active state,[83] and this is reminiscent of PF4, which forms a tetrameric structure and is associated with the platelet chondroitin sulfate-form of serglycin.[91]

C. Neutrophils

Gullberg *et al.*[92] have proposed that serglycin proteoglycans (chondroitin sulfate) cause aggregation of specific proteins required for packaging into neutrophil granules. Cowland[93] determined mRNA levels of serglycin at different stages of differentiation and for a number of different proteins that are known to be sequestered into the three types of granules which are found in the mature cells. Serglycin was synthesized at all stages and was presumed to be localized to all the three types of granules, but the granules were not visualized for serglycin content. However, by direct proteoglycan analysis of isolated granules, we found evidence of serglycin core protein residence only in the secondary and tertiary granules.[76] It is possible that early serglycin mRNA synthesis is associated with constitutive secretion from the immature cells via processes that were not detectable by the methods of Cowland *et al.*,[93] similar to what we, and others, have seen in immature hematopoietic cell lines. It has been shown that the secondary granules of leukocytes, but not the primary granules, are deficient in patients with the gray platelet syndrome.[94] These investigators speculated that serglycin might be associated with this defect.

D. Lymphocytes

EBV infection, a model for infection or inflammatory stimuli, increases chondroitin sulfate serglycin mRNA expression in T-lymphocytes.[95] This is consistent with early studies that showed secretion of serglycin from lymphocytes in culture.[96]

E. NK Cells

Serglycin may be involved in apoptosis of NK cells. Granzyme B has been shown to be associated with serglycin in natural killer cells. The evidence for a significant role for natural killer cell chondroitin sulfate serglycin in Granzyme B-mediated apoptosis has been presented.[97,98] Studies with cytotoxic T-lymphocytes (CTLs) from serglycin knockout mice suggest that serglycin is not needed for Granzyme B-related apoptosis.[99]

F. Hematopoietic Tumors

Serglycin may affect drug metabolism in malignant hematopoietic cells. Chondroitin sulfate serglycin is more highly expressed in chemotherapy drug-resistant hematopoietic tumor cell lines than in the drug-sensitive parental cell lines.[56] Mast cell heparin serglycin may serve as a carrier for doxorubicin and gentamicin since it enables an increase in the uptake of these drugs by both macrophages and smooth muscle cells.[100]

Niemann *et al.*[101] characterized serglycin expression in several hematological malignancies using immunohistochemistry and ELISA. They found that serglycin could be used as a selective marker of immature myeloid cells, labeling acute myelogenous leukemia (AML) but not Philadelphia chromosome-negative chronic myeloproliferative disorders. Serglycin was localized to immature blasts in the marrow from patients with AML but not from acute lymphocytic leukemia (ALL); plasma serglycin was 15-fold higher in AML compared to ALL patients. Patients with three types of marrow myeloproliferative disorders contained serglycin in amounts that were equivalent to controls. However, there still may be serglycin abnormalities in at least one myeloproliferative disorder. We studied patients with ET and PV. These patients have 2–4× normal platelet counts. We observed a unique low-MW component of serglycin in platelets of both of two patients with ET but none in 10 patients with PV. All ET and PV platelets had 50% reduced proteoglycan mass per platelet (B. P. Schick, unpublished). Another hematologic malignancy, multiple myeloma, has been associated with overexpression of serglycin. Serglycin mRNA was detected in six MM cell lines. Serglycin was present both intracellularly and on the cell surface. Immunoblotting of bone marrow aspirates from 30% of the study patients with newly diagnosed MM showed an increase in serglycin. Serglycin isolated from these cells was found to compromise bone mineralization by inhibiting the crystal growth rate of hydroxyapatite.[102]

G. Serglycin in Blood Coagulation

Proteoglycans with amino acid compositions, protease resistance, and sizes consistent with serglycin have been found to interfere with fibrin clot formation[103] and to bind to C1q.[104] Thus, delivery of serglycin to sites of injury might affect fibrin polymerization.

VII. Serglycin in Nonhematopoietic Cells

A. Endothelial and Vascular Smooth Muscle Cells

Recent studies from our lab have shown the presence of serglycin mRNA and protein in human umbilical vein endothelial cells.[16] It was reported that some endothelial cell proteoglycans bind to tPA[105] but the specific proteoglycans were not identified. We have shown, by confocal microscopy, that tPA is distributed in the same pattern as serglycin throughout HUVEC. It was shown that heparin serglycin, but not free heparin, causes the formation of dynamic expanding and regressing lacunae in confluent endothelial monolayers.[106] Heparin serglycin released from rat serosal mast cells inhibits proliferation of rat aortic smooth muscle cells in culture.[107] Kulseth *et al.*[108] showed that serglycin mRNA increases in HUVEC after treatment with TNF-α and IL1-β. We[16] have shown that TNF-α-treated HUVEC secrete virtually all of their serglycin. Thus, TNF-α may cause increased secretion of serglycin, and presumably also its interaction partners, from the endothelium into the surrounding blood and tissues *in vivo*. Vascular smooth muscle cells also synthesize and secrete serglycin.[109]

B. Reproduction and Development: Parietal Endoderm, Embryonic Stem Cells, Uterine Decidua

We have identified the presence of serglycin in the mouse uterine decidual cells and embryonic parietal endoderm during mid-stage pregnancy,[17] and in ES cells.[18] The predominant fate of serglycin in these tissues and cells appears to be constitutive secretion. Serglycin is localized to cytoplasmic vesicles in these cells. Serglycin is not seen in the blood islands during yolk sac hematopoiesis in mice, but appears during fetal liver hematopoiesis.[17] As ES cells synthesize and secrete serglycin, and are essentially the inner cell mass of the blastocyst, we hypothesize that serglycin may be involved in modulating communication between the inner cell mass and the trophectodermal cells during implantation. Decidual serglycin may be involved later, in communication between the decidua and the placenta. A recent paper has reported that serglycin is present in mouse oocytes and that it is upregulated in fertilized zygotes,[110] suggesting that serglycin may play an important role in fertility. tPA is a significant secretory product of the parietal endoderm outgrowth from F9 teratocarcinoma cells during retinoic acid/dbcAMP-induced differentiation[111]; this is the same time that serglycin is induced and tPA is also secreted by murine uterine decidua,[112] which also secrete serglycin.[17]

C. Chondrocytes

Matrix metalloproteinase (MMP)-13 activity is necessary for normal skeletal development and it plays a central role in cartilage degeneration associated with osteoarthritis (OA). Several chondrocyte proteoglycans, that is decorin, syndecan 4, and serglycin, were identified as binding partners of MMP-13. Serglycin mRNA expression in chondrocytes and the localization of serglycin with MMP-13 in the cytoplasmic granules have been demonstrated. The authors have suggested that this interaction plays a role in the regulation of cartilage degradation.[113]

VIII. The Serglycin Knockout Mouse: Mast Cells, Lymphocytes, and Leukocytes

A. Background

Studies of mast cells from mice in which an enzyme specific to heparin synthesis was deleted[34,35] convincingly showed the need for heparin serglycin for the formation of normal mast cell granule structure. However, these studies could not address the relevance of serglycin in cells which generate chondroitin sulfate serglycin proteoglycans. An analogous knockout of a gene required for chondroitin sulfate synthesis would likely be lethal, as this GAG is present in proteoglycans of virtually all cells. To address the problem of understanding the function of serglycin, Abrink *et al.*[114] created a most valuable tool, a serglycin knockout mouse, so that effect of the absence of serglycin in all cells that normally synthesize it could be evaluated. Interestingly, despite the previous reports of serglycin expression in ES cells[18] and the parietal endoderm,[115] these mice did not show any early developmental abnormalities and reproduce copiously. The serglycin knockout has been the subject of interesting studies from several laboratories. Each study has addressed questions related to mRNA expression, protein synthesis, storage, degradation, and secretion of cell-specific proteins that were previously found to be associated with serglycin *in vivo*. Cells from the serglycin knockout mice that have been studied include peritoneal, connective tissue and bone marrow mast cells, splenic and peritoneal macrophages, CTLs, and neutrophils. The questions have been addressed at the level of *in vitro* culture of cells obtained from these animals. Some physiological *in vivo* challenges have also been reported. The data obtained from these studies are in agreement with the predictions from *in vitro* studies of normal cells and the NDST-2 knockout described above.

We studied platelets and megakaryocytes from these mice, and performed *in vivo* studies to explore the effect of loss of serglycin on blood coagulation.

In this section, data from nonplatelet cells from the serglycin knockout will be described. Section IX, below, will describe our work with platelets and hemostasis in this knockout model.

B. Mast Cells

In their initial report on the serglycin knockout, Abrink *et al.*[114] reported an absence of strongly staining peritoneal mast cells. Mast cells were only about 0.2% of the peritoneal cell population, whereas in WT animals they represent about 4%. Surprisingly, the c-kit receptor, which is normally abundant on the surface of mast cells, was absent in the SG−/− cells. Mast cell proteases (MCPs) and tryptases were absent from ear and peritoneal mast cells, but mRNA analysis showed that all these genes were transcribed. The investigators cultured bone marrow with IL3 to stimulate bone marrow mast cell (mBMMC) production. At early time points, empty vesicles appeared in SG+/+, +/−, and −/− cultures. Later, vesicles of SG+/+ and +/− cells filled with protein, but the vesicles of the SG−/− cells remained empty. Chloroesterase (CE) positivity followed granule density. All cell types expressed mRNA for the proteases. The absence of CE activity in SG−/− mBMMC was dramatically different from the NDST-2−/− cells in which early-stage mBMMC granules, which contained intact chondroitin sulfate serglycin, matured normally. The authors concluded that the early stages of granule formation are not dependent on serglycin.

A study from the same laboratory[116] addressed the role of serglycin in the synthesis, accumulation, storage and secretion of proteases in mBMMC cultured over a period of about a month. Electron microscopy of the SG−/− cells showed the presence of vacuoles that lacked an electron-dense core but contained proteins. Transcripts for mMCP-6 and CPA were detected from day 5 but transcripts for mMCP-5 began to appear at day 12 in both genotypes. In SG+/+ cells, mMCP-5 and mMCP-6 were detected from day 12, and were maximally expressed at 26 days. CPA was detected at 5 days and plateaued at 12 days. In the −/− cells, no mMCP-5 was present and mMCP6 was much lower than normal. mMCP5 accumulated intracellularly when cells were treated with NH_4Cl to inactivate lysosomal proteases and granule cathepsins. Therefore, mMCP5 is degraded in SG−/− mBMMC by proteases with low pH optima. mMCP5 was not degraded by proteasomes. CPA content was about the same as normal. However, only the pro-form was found, indicating that serglycin is necessary for generating the mature protease, probably via cathepsin E, which was also absent.

The secretion of proteases was also evaluated in this model. mMCP5 was present at low levels in the culture medium of SG+/+ cells after prolonged culture, but absent from medium from SG−/− cells despite the presence of its mRNA transcript. mMCP6 was clearly detected from day 14 in the medium of both cell genotypes. CPA was secreted from +/+ cells but only the pro-CPA from was secreted by −/− cells. Pro-CPA was secreted by early WT cells before serglycin accumulation began. A23187 caused secretion of mMCP6 and CPA into culture medium. Thus mMCP6 was sorted into releasable secretory vesicles despite the absence of SG.

SG, in normal cells, was produced well before proteases accumulated, so levels of stored proteases were not directly related to the amount of SG core protein mRNA being expressed. Protease accumulation coincided with the appearance of more highly charged GAG chains at the later time points. Thus, protease accumulation is associated with an increase in the synthesis of sulfated PGs and an increased charge density of the GAGs. MCPs can enter the granules without depending on SG, but they must interact with SG in order to be retained.

C. Macrophages

Zernichow *et al.*[117] studied macrophages cultured from adherent cells derived from peritoneal and spleen cells. They observed no notable differences in morphology by either May Grunwald Giemsa staining or electron microscopy. They focused on lysozyme, MIP 1α, IL-1α, and TNF-α. TNF-α is not known to bind to serglycin. Lysozyme and MIP 1α have been previously shown to bind to serglycin.[10,87,118] Adherent macrophages contained nonserglycin proteoglycans but only serglycin was secreted to the medium. They found that heparan sulfate GAGs were attached to the macrophage serglycin core. GAGs attached to SG and other PGs had similar charge densities. The major portion of GAGs released by SG−/− cells were free GAG chains. The major secretory products from SG+/+ cells eluted from Sepharose CL-6B at Kav 0.3, and from SG−/− cells at Kav 0.75.

The cells could be activated with LPS. Lysozyme secretion was not affected by the lack of serglycin, nor were there consistent differences with MMP-9, MIP-1α or IL-1α. This is surprising since both lysozyme and MIP-1α bind to serglycin.[118] However, this finding may be due to the use of serglycin with much longer GAG chains in the binding studies[118] than the very short chains associated with macrophage serglycin. Notably, the levels of TNF-α were markedly higher in the conditioned medium from SG−/− peritoneal and splenic macrophages compared to those from SG+/+, and the heterozygotes had intermediate levels of secretion. Low-level constitutive secretion was seen in all cell types. The increase in TNF-α in the medium after LPS was due to increased secretion, rather than increased synthesis. Any contribution of mast cells to TNF-α secretion was ruled out. It is possible that the increased secretion of TNF-α is secondary to the loss of serglycin and is related to the absence of another protein or proteins that are normally found in these granules.

D. Cytotoxic T-Lymphocytes

Grujic *et al.*[99] investigated the effect of serglycin loss on the packaging of granzymes and perforin into secretory granules of the CTLs. Previous studies of human CTLs have shown that Granzyme B is secreted as a complex with serglycin and that it is important for the rapid induction of apoptosis through both caspase-dependent and -independent pathways.[119–122] Electron dense granules were absent from these CTLs. Serglycin was found to be necessary for the storage of Granzyme B but not for the storage of Granzyme A or perforin. Computer modeling of the structures of the two granzymes suggested that both granzymes have clearly defined positively charged regions but they are located in different regions in the two proteins; this may explain the selective interaction of Granzyme B, but not Granzyme A, with serglycin.

Serglycin did not appear to be necessary for cytotoxic activities of CTL cells in an *in vitro* assay, but the authors have proposed means by which serglycin might still be critical *in vivo*.

In a later *in vivo* study,[123] the role of serglycin in the CD8+ T-cell response to viral infection was studied. Wild-type mice and serglycin-deficient mice were infected with lymphocytic choriomeningitis virus (LCMV). Wt and SG(−/−) mice cleared 10(3) PFU of highly invasive LCMV with the same kinetics, and the CD8(+) T lymphocytes from WT and SG(−/−) animals did not differ in GrB, perforin, IFN-γ, or TNF-α content. However, when a less invasive LCMV strain was used, SG(−/−) GrB(+) CD8(+) T cells contained approximately 30% less GrB than WT GrB(+) CD8(+) T cells. Interestingly, the contraction of the antiviral CD8(+) T cell response to the highly invasive LCMV was markedly delayed in SG(−/−) mice; and a similar response was seen after infection with the vesicular stomatitis virus. This delayed contraction was associated with the sustained proliferation of Ag-specific CD8(+) T cells in SG(−/−) mice. Moreover, WT LCMV-specific CD8(+) T cells from TCR318 transgenic mice expanded much more extensively in virus-infected SG(−/−) mice than in matched WT mice, indicating that the delayed contraction represents a T cell extrinsic phenomenon. Therefore, serglycin appears to be important for regulating the kinetics of antiviral CD8(+) T cell responses.

E. Neutrophils and Lymphocytes

Granule proteins are essential for the ability of neutrophils to kill bacteria. It has been reported that neutrophils from the serglycin −/− mice lack elastase but not the other neutrophil granule proteins such as myeloperoxidase,[124] even though elastase mRNA was present. Both, cell counts and neutrophil ultrastructure appeared normal. An *in vivo* study was performed, in which it was observed that the virulence of intraperitoneally injected Gram-negative bacteria (Klebsiella pneumonia) was enhanced in the serglycin knockout mice compared to controls. Similar results have been previously reported in the elastase knockout mice.

Wernersson *et al.*[125] showed that mice lacking SG spontaneously develop an enlargement of multiple lymphoid organs, including the spleen, Peyer's patches (PP), and bronchus-associated lymphoid tissue. In the spleen, the proportion of $CD4^+$ cells was decreased, and the $CD45RC^+$ leukocyte population was increased, indicating an expansion of the naïve lymphocytes. In the PP, there was a general increase in cellularity, but the relative proportions of leukocyte populations remained similar. Serum levels of inflammatory cytokines were not increased. The number of peritoneal mast cells was not affected, but the intensity of c-kit staining was reduced significantly. However, the number of peritoneal macrophages was decreased markedly in the older serglycin-null mice. Older serglycin-deficient mice had a more pronounced inflammatory response than did wild-type mice during induction of airway inflammation by bacterial lipopolysaccharide. These data suggest that serglycin deficiency has multiple, and age-related effects on the lymphoid system.

In summary, the mast cells in the SG−/− mice have been found to have defects in development of electron-dense granules. One or more granule proteins specific to three types of mast cells, macrophages, CTLs, or neutrophils have been found to be absent or greatly reduced. The mice reproduce and develop, and age normally, but show abnormalities in response to *in vivo* challenges to infectious or inflammatory stimuli.

IX. The Serglycin Knockout Mouse: Platelets, Megakaryocytes, and Thrombus Formation

A. Background and Rationale for the Studies

Our studies below describe the striking effects of serglycin deletion on platelet function. It will be helpful here to describe the aspects of platelet function that our study addressed. Platelets initiate hemostasis, the cessation of bleeding, by forming a plug at the site of vascular injury. Platelets also are involved in pathological thrombosis by similar or other mechanisms. The collagen in vessel walls is exposed when a vessel is injured, and this is a potent inducer of platelet aggregation. This process involves formation of a single layer of platelets at the site of injury, and additional platelets are recruited to form the growing plug. There are mechanisms in place for stabilization of the plug; if such mechanisms are deficient, then plugs become unstable and disaggregate, and rebleeding can occur. On the other hand, a balance must exist so that platelet plugs that would go on to form occlusive thrombi can disaggregate in a timely fashion so as to not cause ischemic injury or death.

Platelets contain two distinct types of secretory granules. The α-granules, which are the putative site of serglycin storage, contain numerous proteins including growth factors, chemokines, and thrombogenic proteins, such as thrombospondin, fibronectin, and the von Willebrand Factor (vWF); these are also found in plasma and promote platelet activation and aggregation. The dense granules contain small molecules such as ADP and ATP, histamine and serotonin. When platelets are activated with strong agonists, these granules are first centralized and then brought to the membrane surface by cytoskeletal rearrangements, thence their contents are secreted. The α-granule membrane fuses with the platelet plasma membrane, exposing its inner surface to the outside, and thus aiding in the recruitment of more platelets to the plug.

In the initial phase of platelet aggregation, activation of platelets by substances such as ADP, TXA2 (both of which come from platelets) or thrombin (the terminal enzyme of the blood coagulation cascade) causes a change in conformation of the membrane receptor complex αIIbβ3, known also as GPIIbIIIa, by a process called "inside-out" signaling. This conformational change produces an activated complex that can bind to ligands that further enhance platelet activation by an "outside-in" signaling mechanism, which includes phosphorylation of the tyrosine residues in the cytoplasmic domain of the β3 integrin. The activation of αIIbβ3 is thought to be the result of the final common pathway of all agonists, and the binding of fibrinogen or vWF is thought to be the major event in platelet–platelet interactions. Several signal transduction pathways have been implicated, including Eph kinases and ephrins, Akt proteins, Rab 1b, PI3 Kinase, and a variety of G proteins that are coupled to receptors on the platelets. Small Rho family GTPases (RhoA, Cdc42, and Rac1) are involved in cytoskeletal rearrangements through formation of stress fibers, filopodia, and lamellipodia, respectively. These cytoskeletal rearrangements are necessary for the centralization of the granules and their movement toward the membrane to secrete the physiologically active substances.

A critical event in the activation process and its amplification is the release of the proteins from the α-granules following fusion of the granule membrane with the surface membrane. This opens the granule, resulting in the exposure of the inner surface of the α-granule to the outside of the platelet. Proteins released from the α-granule are known to bind back to the platelet surface; plasma proteins also bind to the activated surface. Chemokines such as PF4 and RANTES, which are stored in α-granules, may also bind to the platelets, and if released, involve in the activation of leukocytes during *in vivo* aggregation. Chemokine mediated activation of platelets has been associated with GAGs on the platelet surface, but the core protein to which these GAGs are bound have not been defined. Ca++ mobilization is an important factor in the activation process and it has also been associated with platelet surface proteoglycans.[126] It is not known whether other chemokine receptors are involved.

Another important contributor to platelet aggregation is the dense granule, which also releases its contents upon activation. The primary molecules involved in dense granule release are ADP, ATP, serotonin, and calcium. Serglycin is not known to be contained within dense granules; however, as will be shown below, the absence of serglycin results in abnormal dense granule function.

The formation of a clot via the interaction of platelets and the coagulation system is very complex. The deletion of a number of α-granule or platelet signaling proteins in knockout mice has led to models in which the platelet response to various aggregating agents is reduced *in vitro* and often these mice show defects in *in vivo* clot formation or stability in several of the commonly used protocols for induction of arterial thrombosis. Knockouts of α-granule proteins, and the signal transduction proteins and pathways, which lead to decreased aggregation and unstable clot formation include PF4,[127] Akt 1 and 2,[128,129] $G\alpha q$ and $G\alpha_{13}$,[130] Rap1b,[131] vitronectin,[132] Gas6,[133–135] Mer receptor tyrosine kinase (Gas6 receptor),[134] and P13Kγ,[136] and P-selectin.[137] As will be discussed below, the deletion of serglycin leads to a phenotype which is deficient coagulation and resembles many of these knockouts.

All studies on platelets from the serglycin knockout were reported in Woulfe *et al.*[138] All work with animals was approved by the Institutional Animal Care and Use Committee of the Thomas Jefferson University.

B. Experimental Protocols and Results

1. Evaluation of Platelet and Megakaryocyte Morphology[138]

Platelets were isolated from blood obtained by cardiac puncture from anesthetized mice. Whole-blood platelet counts were similar (~ 4–7×10^8 ml^{-1}) in SG−/− and WT animals. We first investigated whether the absence of serglycin would result in any obvious abnormalities in α-granules. α-Granules of normal platelets have a characteristic appearance, including an eccentric electron-dense nucleoid. Transmission electron microscopic (TEM) analysis (Fig. 2) did not reveal any gross defects in the SG−/− platelet α-granule structure. We found a similar range of α and dense granules per platelet from WT and SG−/− platelets. The presence of electron-dense nucleoids in most α-granules of both phenotypes suggested that the absence of serglycin did not affect aggregation of the nucleoid proteins.

However, we found two striking and unexpected abnormalities in platelets and megakaryocytes of the SG−/− mice. The first was the appearance of unusual scroll-like membranous inclusions which were decorated with glycogen particles (Fig. 2A, panels a–c, f). These inclusions were present in the SG−/− platelets of varying size and degrees of granulation. These structures have been previously described as "cigars" in platelets of patients suffering from the Medich syndrome, a gray platelet-like disorder,[139] and in the Wistar Furth heritable macrothrombocytopenic rat platelets[139]; the latter disorder is characterized by abnormal platelet proteoglycan GAG structure and a bleeding diathesis.[21] The second phenomenon was the emperipolesis of neutrophils (taking up of the entire neutrophil) within the cytoplasm of megakaryocytes of all stages of maturation from the SG−/− (Fig. 2A, panels d–g) but

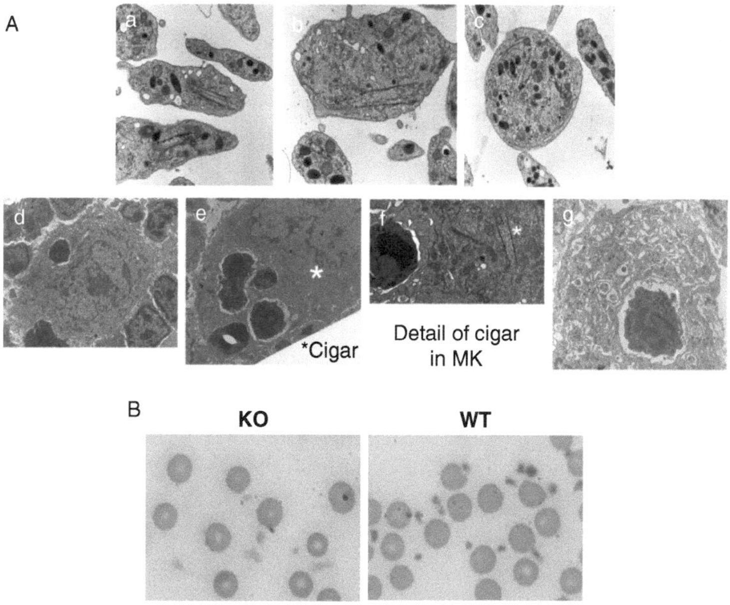

FIG. 2. MKs and platelets from SG−/− mice. This figure and Figures 3–8 are reprinted from Ref. 138. (A). Electron micrographs of SG−/− platelets and megakaryocytes. *Panel a*: 2 average size platelets with scroll-like membranes called "cigars"; elongated intracellular membranous structures decorated with glycogen particles. *Panel b*: Large hypogranulated platelet and small platelet with cigars. *Panel c*: Large heavily granulated platelet with cigar. No cigars were seen in normal platelets. *Panel d*: Immature MK with neutrophil. *Panel e*: Mature MK with several neutrophils; "Cigar" is just to left of *. *Panel f*: Detail showing "cigar" from e; neutrophil is the one directly left of * in e. *Panel g*: Neutrophil surrounded by demarcation membrane system in mature MK. α-Granules had nucleoids, but many granules were elongated (Fig. 6A, center; Fig. 6B, platelet on right). *Magnification (at microscope objective)*: Platelets (a) 4200×; (b) 5500×; (c) 4200×. *MK*: (d, e) 1000×; (g) 1600×. (B). Wright-Giemsa stained blood smear. Photographs were taken electronically and are presented in their original form. Magnification was 1000×. Note identical appearance of red blood cells in the WT and KO smears.

not the WT mice. As many as eight neutrophils were observed within a single large megakaryocyte. Emperipolesis also occurs in megakaryocytes from the gray platelet syndrome[140–142] and in myelofibrosis.[141,142] It is not understood how the loss of serglycin might contribute to either the "cigar"-like formations or to the degree of emperipolesis and what the consequences of these phenomena might be with regard to platelet function.

In contrast to the normal appearance of platelet granules under electron microscopy, the staining of platelet granules differed greatly between the SG−/− and WT platelets. Wright-Giemsa staining showed intense staining and granulation in WT platelets (Fig. 2B), whereas SG−/− platelets were more spread out and much less bright without a granular appearance, reminiscent of the human gray platelets.[143–145] Giant platelets were observed in smears from both WT and SG−/− animals. Giant platelets from SG−/− mice were occasionally larger than the RBC.

2. Proteoglycans and GAGs in WT and SG−/− Plateletsy[138]

Proteoglycans were extracted from whole platelets with a solution containing 8 M urea/0.2% Triton X-100. The proteoglycans were purified by DEAE-Sephacel chromatography. The size of the intact proteoglycans was determined by SDS-PAGE using 4–20% gradient gels. The intact PGs from the WT platelets migrated at the top of the gel with apparent MW > 250 kDa, as determined by SDS-PAGE analysis and Alcian blue staining (Fig. 3A, B, and E), and were similar in size to serglycin from human platelets but had two major components rather than one. No intact PGs were detected in platelets from the SG−/− mice (Fig. 3A and B). GAG chains that were released from the protein core by treatment with 0.2 M NaOH are shown in Fig. 3B. The WT platelet GAGs ranged in size 15–45 kDa. A faint smear of free GAGs was seen at the same size range in some WT and KO samples (Fig. 3B) before NaOH digestion. This smear likely represents residual plasma GAGs trapped in the platelet pellet, since its intensity was greatly increased in the NaOH digest of WT but remained unchanged after NaOH digestion of the SG−/− proteoglycans.

The core protein of the WT platelet chondroitinase ABC-digested PGs is shown in Fig. 3C. The protein was identified by immunoblotting. The core protein appeared, as expected, at about 28 kDa, along with an additional band at ~ 20–25 kDa and two minor bands of < 10 kDa. The latter are likely to be SG degradation products. Figure 3D shows intact proteoglycans from two SG+/− mice. A uniformly stained smear extended from the top of the gel to just above 250 kDa, in contrast to the biphasic character of the smear in the WT PGs.

Figure 3E depicts an experiment that was designed to isolate membrane proteoglycans by a wash of the DEAE column with 1% Triton X-100. This experiment was done because others have shown that platelet surface PGs are involved in platelet activation.[55,126,146] Proteoglycans were observed in the TX-100 wash in WT and SG+/− but not the SG−/− eluates. The size of the TX-eluted PG from the SG+/− platelets appears to be larger than that of the WT platelets.

3. Platelet Function Defects[138]

a. Serglycin Deficiency Causes Defective Platelet Aggregation[138]. To determine whether serglycin deficiency adversely affects platelet function, platelet aggregation was compared in platelet rich plasma prepared from WT

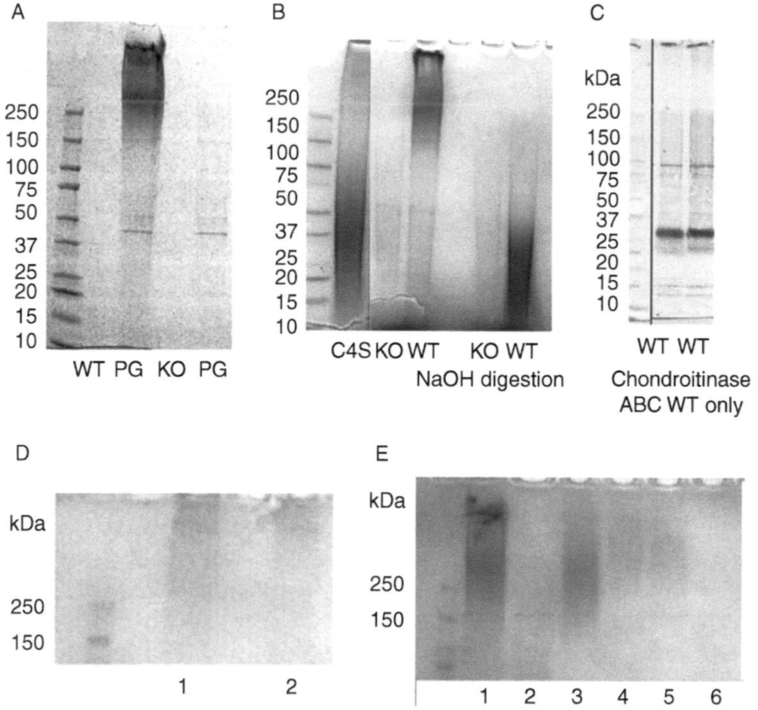

FIG. 3. Proteoglycans, GAGs, and serglycin core proteins of platelets. Samples were analyzed on 4–20% gradient SDS-PAGE gels. Eluates of 4×10^8 platelets were applied to each lane. Proteoglycans and GAGs are stained with Alcian blue and appear as smears on the gels. No samples were applied to lanes that are not marked. (A) The 0.8 M NaCl DEAE-Sephacel eluates from WT and KO PGs are shown. Gel also stained with Coomassie blue revealed several sharp protein bands. A high Mr proteoglycan with two components was present in the WT platelets. No proteoglycans were visible on the gel of KO platelets. (B) Intact and 0.2 M NaOH-digested proteoglycans. C4S is shark chondroitin 4-sulfate. *Lanes left to right*: KO PGs, WT PGs, blank lane, KO GAGs, WT GAGs after NaOH digestion. (C) Immunoblot of SG core protein from 2.3×10^8 WT platelets per lane, one mouse per lane. (D) *Lanes 1 and 2*: PGs from SG+/− mice. (E) *Lane 1*: Intact PGs from 0.8 M NaCl eluate from WT platelets. *Lane 2*: 0.8 M NaCl eluate from KO platelets. *Lane 3*: 0.8 M/0.2% TX-100 eluate, WT platelets. *Lanes 4–5*: 0.8 M NaCl/0.2% TX-100 eluate from SG+/− platelets. *Lane 6*: 0.8 M NaCl/0.2% TX-100 eluate from KO platelets.

and SG−/− animals. Platelet aggregation studies are performed in aggregometers which are instruments that measure light transmission through a solution. Platelet-rich plasma is opaque, and as the platelets form aggregates the solution allows greater and greater light transmission. Tracings are recorded over several minutes after addition of the aggregating agents. Aggregation defects at low concentrations of the agonists for thrombin or collagen receptors are frequently indicative of a defect in secretion or secretory granule content.[129] Aggregations of WT and SG−/− platelets were compared simultaneously using identical concentrations of agonists in a two-channel aggregometer. SG−/− platelets showed a markedly defective aggregation response to low concentrations of collagen Type 1 and the PAR4 agonist peptide AYPGKF compared to WT platelets (Fig. 4). It is notable that the SG−/− platelets showed a greater tendency to disaggregate at concentrations of collagen < 10μg/ml compared to WT platelets. In contrast, WT and SG−/− platelets aggregated identically in response to all concentrations of ADP tested, including at very low concentrations that induced only 10–20% aggregation (Fig. 4A).

b. Serglycin Deficiency Causes Defective Platelet Dense Granule Secretion[138]. The standard means of studying platelet activation is to measure the degree of serotonin release from agonist-treated platelets. Platelets are incubated with 3H-serotonin, which is taken up rapidly into the dense granules and equilibrates with endogenous serotonin. An aliquot of this plasma is taken to measure the amount of serotonin uptake. The remainder of the platelets are treated with agonist, whence serotonin released in to the medium is quantitated. Serotonin uptake was normal but serotonin release was markedly reduced in platelets from SG−/− mice in response to AYPGKF (Fig. 3B). In two experiments, ATP release was reduced by half in SG−/− platelets (not shown), which is consistent with the serotonin secretion results.

c. Surface Membrane Effects: Flow Cytometric Evaluations of Platelet Activation[138]. Platelet agonists induce a signaling cascade that results in an activation-dependent conformational change in the major platelet membrane receptor, αIIbβ3. Fibrinogen binding to the activated αIIbβ3 is a central, critical step in platelet aggregation. We thus assessed if the observed defects in the aggregation of SG−/− platelets were associated with reduced fibrinogen binding and therefore activation of αIIbβ3. The incubation of platelets with increasing concentrations of AYPGKF caused a proportional increase in fibrinogen binding to WT platelets. Fibrinogen binding was reduced in SG−/− platelets, relative to WT platelets, both at low and intermediate concentrations of agonist (0.4 and 0.8 mM AYPGKF). Even at a concentration of 1.2 mM AYPGKF, the WT platelets were completely activated, whereas fewer than 50% of the SG−/− platelets were positive (Fig. 5A). Thus, the absence of SG results in reduced αIIbβ3 activation at low peptide concentrations. At 1.5 mM AYPGKF, however, SG−/− platelets bound to amounts of fibrinogen that were equivalent to that of WT platelets (not shown).

To further evaluate platelet activation and availability of the αIIbβ3 complex, surface exposure of αIIb was determined by flow cytometry using an anti-CD41 (CD41 is αIIb) antibody. Exposure was increased after platelets were

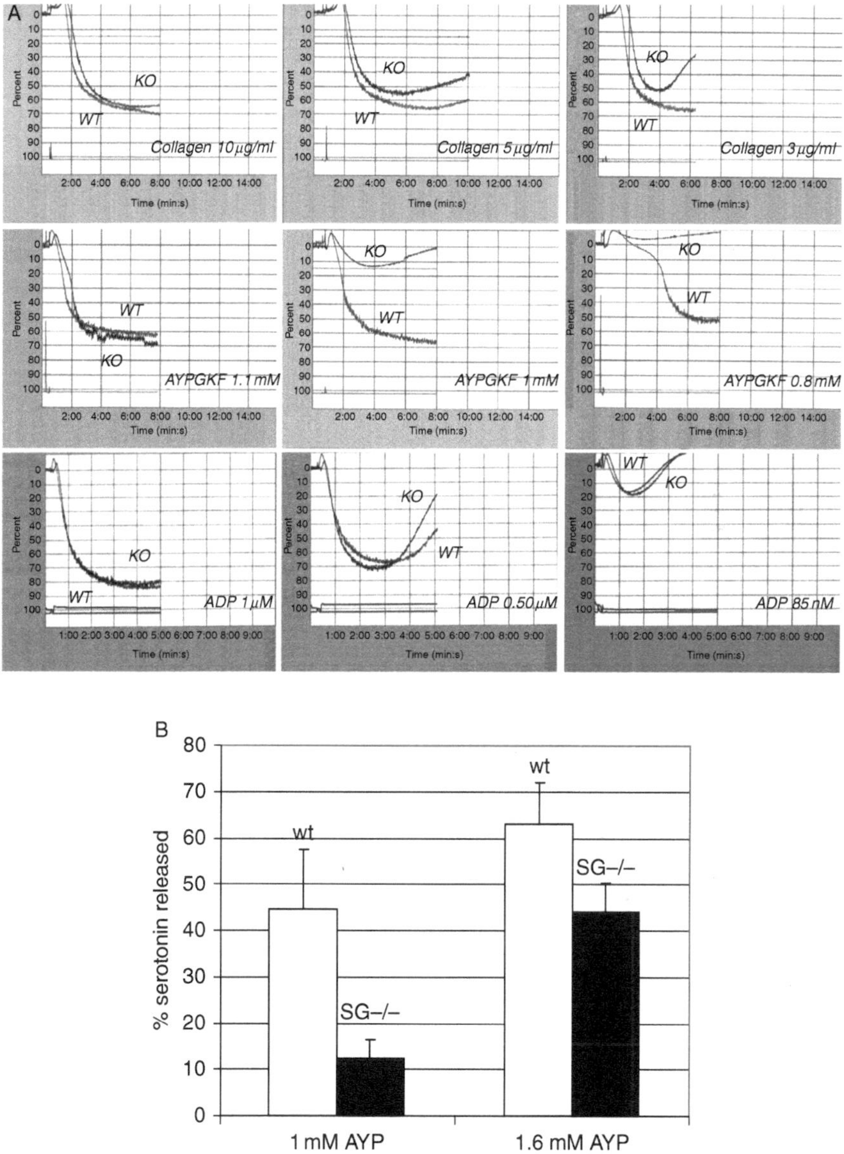

FIG. 4. Aggregation and secretion. (A) *Aggregation of WT and SG KO platelets*: WT and KO platelets were run simultaneously in a dual channel aggregometer with the concentrations of agents indicated on each panel. Results shown are representative of 2–3 independent experiments with very similar results. (B). *Serotonin release*: PAR4 agonist-induced release of serotonin from WT and SG–/– platelets was measured after loading platelets with ^{3}H-serotonin. The mean percentage release (± S.D.) from three separate experiments is shown.

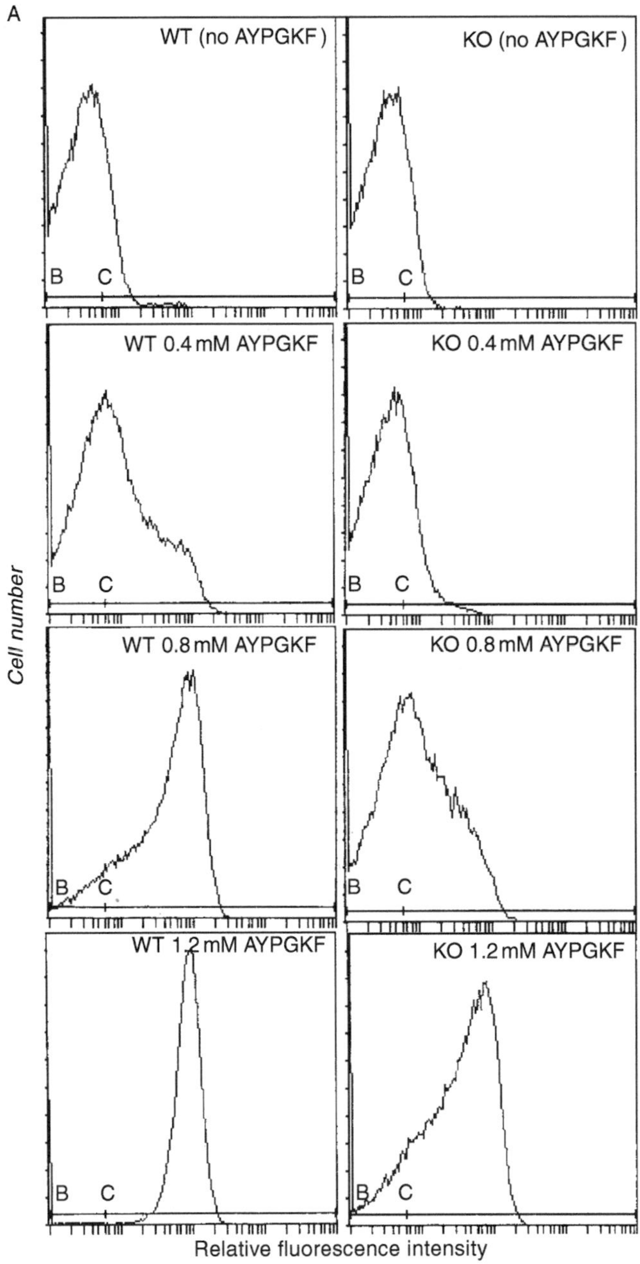
A
WT (no AYPGKF)
KO (no AYPGKF)
WT 0.4 mM AYPGKF
KO 0.4 mM AYPGKF
WT 0.8 mM AYPGKF
KO 0.8 mM AYPGKF
WT 1.2 mM AYPGKF
KO 1.2 mM AYPGKF
B
C
Cell number
Relative fluorescence intensity

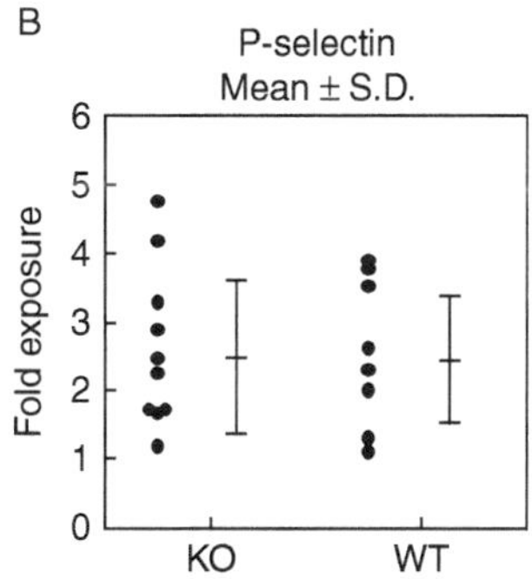

FIG. 5. Surface labeling of αIIbβ3 and P-selectin. (A). *Fibrinogen binding to platelets activated with AYPGKF*: Platelets from WT and SG−/− mice were treated with the PAR4 agonist peptide AYPGKF at the concentrations indicated, followed by incubation with FITC-labeled fibrinogen as described in the text. Binding was quantitated by flow cytometry. Each assay represents 50,000 platelets. Results shown are representative of two independent experiments with very similar results. (B). Exposure of P-selectin. Platelets were treated with 1 U/ml thrombin and stained with anti-P-selectin antibody as described in the text. Each point represents platelets from a single mouse.

treated with either thrombin or AYPGKF. Exposure of αIIb was similar in both WT and SG−/− platelets. Therefore, the defect in fibrinogen binding appears to be due to a defect in the activation of integrin αIIbβ3, rather than a defect in agonist-induced surface exposure of the integrin.

Another measurement of platelet activation is the display of P-selectin on the platelet surface. P-selectin is a protein integral to the inner surface of the α-granule membrane. When platelets are activated, the α-granule membrane fuses to the plasma membrane and the granule opens to the external environment, thus exposing the inner surface of the granule membrane. P-selectin display was similar in WT and SG−/− platelets treated with 0.2–1 U/ml thrombin (shown for 1 U/ml, Fig. 5B), and 0.2–0.5 U/ml thrombin (not shown). As all of the platelet P-selectin is localized to the α-granule membrane, serglycin deficiency does not appear to alter activation-dependent α-granule inner membrane exposure. It should be noted that about half the platelet αIIb is localized to the inner surface of the α-granule membrane. Therefore, the results of the P-selectin and aIIb are consistent and show that the absence of serglycin does not alter the exposure of the inner surface of the α-granule membrane in activated platelets.

4. DEFECTIVE CAROTID ARTERY THROMBUS FORMATION[138]

Many investigators have studied carotid artery thrombosis, *in vivo*, in mice as a measure of platelet function and the role of endothelial injury in thrombosis. In order to assess if the absence of serglycin affects thrombus formation *in vivo*, mice were subjected to $FeCl_3$-induced carotid artery thrombosis.[129,147] Mice were anesthetized, the carotid artery exposed and treated briefly with $FeCl_3$-impregnated filter paper. The vessel was then monitored with a Doppler probe to measure blood flow. A decrease in blood flow indicates thrombus formation and can be measured. One might observe either no change in blood flow, cessation of blood flow indicating complete occlusion of the artery by the thrombus, or intermediate reductions in flow. If the thrombus is stable, the flow will not return to normal. If the thrombus is unstable, as it breaks down, the blood flow increases and may even return to normal. Figure 6 summarizes the results of studies with seven WT and nine SG−/− male mice after treatment of

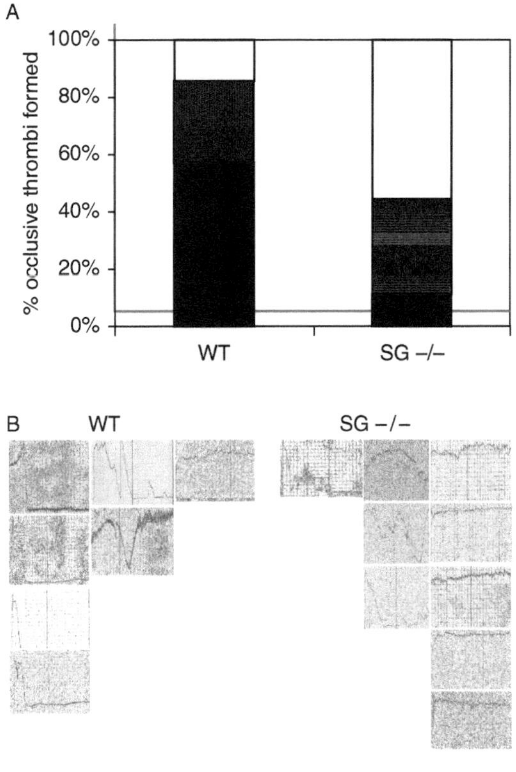

Fig. 6. Thrombus formation in the carotid artery after $FeCl_3$ treatment. Experiments were carried out as described in the text. (A) The histogram shows the distribution of responses to $FeCl_3$ treatment. *Black*: Stable thrombi. *Hatched*: Unstable thrombi. *White*: No thrombi. (B) Tracings for all mice tested using 2 min 15 s exposure to $FeCl_3$ are shown. Similar results were found for mice exposed to $FeCl_3$ for 2 min 30 s. Normal flow rate is ∼1 ml/min. A flow rate of zero indicates complete occlusion of the artery by the thrombus.

the carotid artery with $FeCl_3$ for 2 min 15 s. Most of the WT mice (4/7) developed stable occlusive thrombi by 2–4 min after $FeCl_3$ treatment. In contrast, only one of the nine SG−/− mice formed stable occlusive thrombi under the same conditions. Only one of seven WT mice failed to form thrombi after injury, whereas five of nine SG−/− mice failed to form thrombi. The tracings of the individual flow rates are shown on the right.

5. Reduction in α-Granule Proteins[138]

PF4, a major component of the platelet α-granule, has long been known to be associated with the platelet secretory granule proteoglycan.[4] However, the *in vivo* functional significance of this interaction has not been determined. In order to address the role of serglycin in PF4 storage, we analyzed platelets from WT and SG−/− platelets for the presence of PF4 by immunoblot analysis. As shown in Fig. 7A, the levels of PF4 were dramatically reduced in platelets from the SG−/− animals. The protein control was the actin band of a Coomassie-stained gel run with identical aliquots of the samples on the blot. The content of two other α-granule proteins, β-thromboglobulin and PDGF, was similarly reduced in the SG−/− platelets (Fig. 7B and C). To determine whether the low SG−/− platelet content of PF4 might be due to its secretion in blood, we measured PF4 in plasma and serum. However, PF4 was profoundly reduced in both plasma and serum from the SG−/− animals (Fig. 7D). During the preparation of blood serum, platelets are activated and their α-granule contents released. The approximately 15-fold higher level of PF4 in WT serum is consistent with the observed decrease in the content of PF4 of SG −/− versus the WT platelet lysates.

To determine if the reduction in PF4 in platelets and plasma could be explained by the reduced transcription of PF4, we evaluated mRNA expression in the spleen and bone marrow cells of both WT and SG−/− mice. Aliquots of the same RNA extract were subjected to RT-PCR for PF4 and serglycin expression, using actin as a control. PF4 is synthesized only by megakaryocytes and not by mature platelets, whereas serglycin may be expressed by a multitude of cell types found within both the tissues. As shown in Fig. 8, similar levels of PF4 mRNA were found in samples from WT and SG−/− mice, both in bone marrow and in the spleen. Serglycin expression was also detected both in the bone marrow and the spleen from WT animals but, as expected, not in samples from SG−/− animals. These results indicate that PF4 mRNA is expressed to the same extent in WT and SG−/− animals. Presumably, even though the protein is synthesized it is not incorporated into the platelets. It is also not accumulated in plasma in these mice (Fig. 7B), unlike the gray platelet and other α-granule defects in humans in which several platelet chemokines are elevated in plasma.[144] If the protein was secreted into the plasma by the SG−/− megakaryocytes, we would have detected it. A likely explanation for

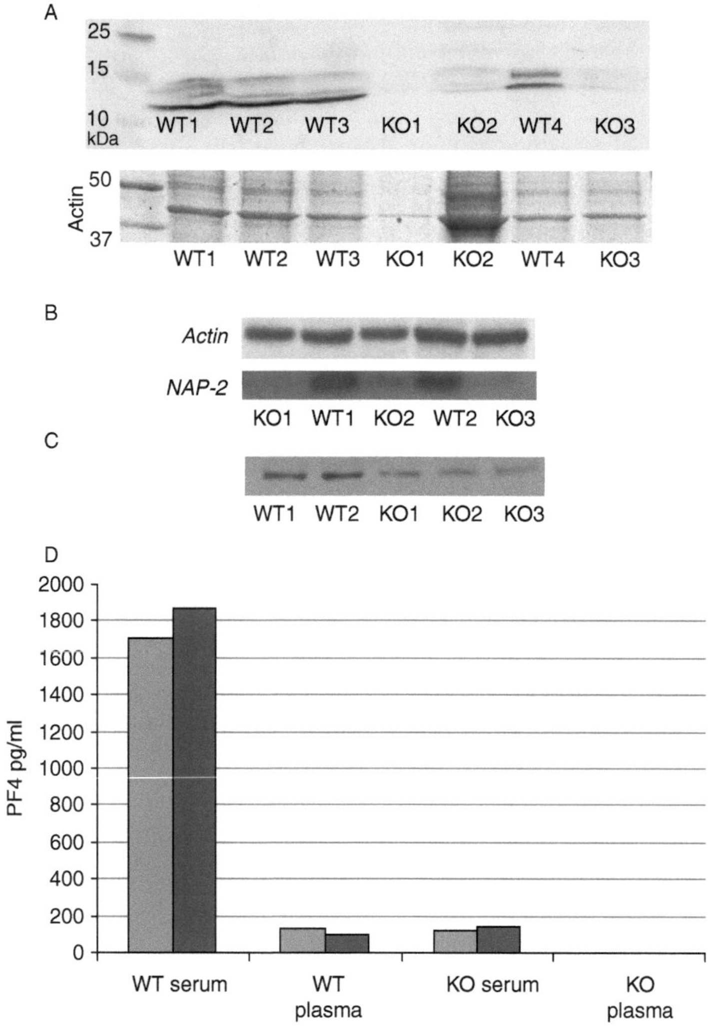

Fig. 7. α-Granule proteins of platelets, plasma and serum in WT and SG−/− mice. (A). Immunoblot of platelet PF4. Proteins from 4×10^7 platelets per sample (each sample represents a single mouse) were subjected to SDS-PAGE followed by immunoblot analysis for PF4. The upper panel in (A) shows Coomassie staining of the same samples as in the lower panel; note the actin band. (B) Immunoblot of NAP-2. Proteins from equal numbers of WT and KO platelets were applied to the gel. NAP-2 was identified by immunoblotting (*lower panel*). The same blot was probed with anti-actin (*upper panel*). Detection was by ECL. (C) Immunoblot of PDGF. Aliquots of same samples shown in Fig. 6B were applied to the gels. Labeling of PDGF was detected by ECL. (D) ELISA for quantitation of PF4 in plasma and serum. Half of the blood sample from each mouse was anticoagulated (plasma) and half was allowed to clot (serum). Two WT and two SG−/− mice were tested.

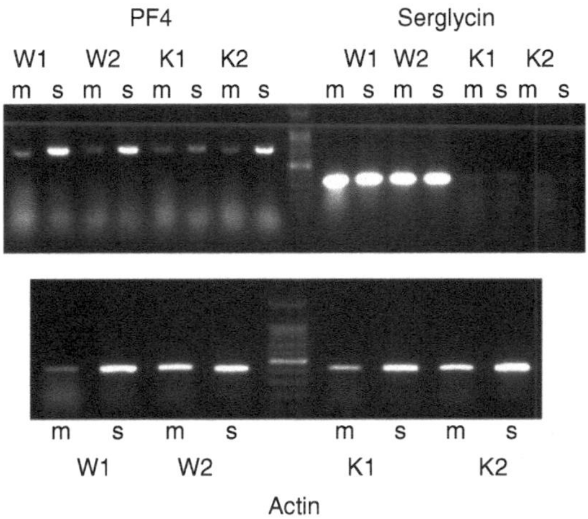

FIG. 8. mRNA expression of PF4, serglycin and actin in bone marrow and spleen of WT and SG−/− cells. mRNA was extracted from bone marrow (m) and spleen (s) of two WT (W1 and W2) and two KO (K1 and K2) mice. The gel shows the RT-PCR products. *Top panel*: mRNA for PF4 and SG; marker is the 100 bp ladder. *Bottom panel*: Actin mRNA.

this observation is that megakaryocytes synthesize PF4 but either degrade it within the cell, or secrete it into the marrow matrix where it is degraded. The reduced level of PF4 protein in SG−/− platelets is most likely due to a defect in the storage capacity of the α-granules rather than a lower rate of PF4 synthesis.

X. Significance

A. What are the Possible Mechanisms for the Dramatic Effects of the Absence of Serglycin on Platelets and Megakaryocytes?

1. Proteoglycan Content of SG−/− Platelets

In the above studies we have demonstrated that the serglycin proteoglycan plays a crucial role in normal platelet function, since its absence results in defects in platelet function. No proteoglycans were detected in SG−/− platelets and thus serglycin appears to be the only proteoglycan in platelets. The identity of the putative surface proteoglycan had not been determined previously. A proteoglycan which elutes in the hydrophobic fraction from DEAE-Sephacel columns is present in the WT platelets but absent in the SG−/−

platelets. These data suggest that serglycin is the putative membrane proteoglycan in normal platelets. This is unexpected, but it has been reported that serglycin is present on the surface of rat L2 yolk sac tumor cells.[37] We speculate that serglycin interacts with hydrophobic proteins on the platelet surface and forms a complex that is not broken up by treatment of the platelets with 8 M urea. The absence of both the surface proteoglycan as well as the granule serglycin could be involved in the defects in platelet function that we have observed.

2. Ultrastructural Abnormalities

We studied platelet and megakaryocyte morphology both by electron microscopy and the conventional Wright-Giemsa staining. We were interested primarily in comparing the effect of serglycin deletion on α-granule structure with its effects on secretory granule structure of other cells from this knockout mouse. To our surprise, EM ultrastructural examination revealed an essentially normal structure of the SG−/− α-granules, including the presence of the normally observed eccentric electron-dense nucleoids within the granules. Such structures are thought, in other cells, to contain serglycin/protein complexes, and therefore we expected that these nucleoids would be absent. Our results demonstrate that serglycin is not necessary for the aggregation of proteins within the α-granule nucleoid. However, two very striking and unexpected phenomena were observed in the photomicrographs. SG−/− platelets and megakaryocytes displayed unusual scroll-like membranous inclusions, reminiscent of those found in Medich platelet syndrome in humans[139] which is a disease characterized by excessive bleeding, thrombocytopenia and giant platelets with a gray platelet-like appearance. Such similar membrane inclusions are also found in the Wistar Furth hereditary macrothrombocytopenic rat platelets.[139] These rats also have a bleeding defect. We have shown their platelets to have abnormally small serglycin GAG chains.[21] We therefore suggest that these scroll-like membranous inclusions are related to defects in serglycin synthesis. We do not understand the mechanism by which these structures are formed, or their functional significance, but it is apparent that the defect arises in the megakaryocyte (Fig. 2A, panel f).

Another unusual feature of the SG−/− megakaryocytes was the large degree of neutrophil emperipolesis. Megakaryocytes are known to have the ability to incorporate neutrophils into their large cytoplasmic structures; however, in general this phenomenon is rare. In our study, extensive emperipolesis was observed, with as many as eight neutrophils within a single megakaryocyte. Interestingly, this phenomenon has also been observed in the gray platelet syndrome,[140] a disease in which the α-granules are severely defective and the patients suffer from bone marrow myelofibrosis, as well as in patients with marrow myelofibrosis from other causes.[141] It is not known whether serglycin is absent in or has an abnormal structure in gray platelets. It has been suggested that emperipolesis of neutrophils by megakaryocytes is related to the premature secretion of chemokines and growth factors from α-granules that are being formed with in the megakaryocytes, thus contributing to the development of marrow myelofibrosis.[141] We propose that this inappropriate secretion of chemokines may be due to defects in serglycin synthesis. It will be of interest in the future to determine whether myelofibrotic disorders can be related to defects in serglycin expression or its GAG structure.

3. Functional Defects

The most critical information from our study is that the lack of serglycin resulted in notable defects in platelet function. These defects were shown in two situations: *in vitro* aggregation response to several agonists, and *in vivo* disruption of thrombus formation. We demonstrated a greatly reduced platelet aggregation, *in vitro*, in response to both collagen and the PAR4 thrombin receptor agonist peptide, thus demonstrating that platelets are functionally compromised when they are lacking in serglycin. In contrast, platelet aggregation was normal in response to ADP. Aggregation abnormalities in response to collagen and the PAR4 agonist peptide have been shown in platelets from a number of mice in which genes involved in platelet activation have been deleted. However, generally in these mice, the aggregation response to ADP is also reduced. Interestingly, the pattern observed in the serglycin knockout, that is reduced response to collagen and peptide but normal response to ADP, has not been shown in any other mouse platelet protein knockout models. Normally, aggregation *in vitro* is irreversible. However, the SG−/− platelets showed a great tendency towards disaggregation within a short time after the aggregation response was maximal. Stabilization of the platelet plug is critical for hemostasis, and reduced stability of platelet aggregates may therefore lead to impaired hemostasis in response to an injury. Our results indicate that serglycin has an important role in both the platelet aggregation phase as well as in the subsequent stabilization process.

Unexpectedly, we also observed reduced serotonin secretion and ATP release in SG−/−platelets in response to the agonists, suggesting that SG affects platelet aggregation also through an effect on dense granule function.

A number of mice in which the genes for platelet proteins have been deleted also show defects in thrombus formation in response to $FeCl_3$-induced carotid injury or in other models of thrombosis (reviewed Ref. 148). Knockouts of α-granule proteins, signal transduction proteins and the pathways which lead to decreased aggregation and unstable clot formation include PF4,[127] Akt 1 and 2,[128,129] Gαq and $G\alpha_{13}$,[130] Rap1b,[131] vitronectin,[132] Mer receptor tyrosine kinase Gas6 receptor,[134] PLCβ and PI3Kγ,[136] and P-selectin.[137] The *in vivo* thrombosis response towards $FeCl_3$ is defective in SG−/− mice, as indicated by a greatly reduced tendency to form occlusive thrombi. In most animals, either no thrombus was formed, or the thrombus was unstable, as evidenced by the resumption of blood flow following initial partial vessel occlusion. These results are consistent with the effects on platelet aggregation and the increased tendency of the aggregated SG−/− platelets to disaggregate at low agonist concentrations. The observed effect on thrombus formation may also result from altered interactions of platelets with the endothelium or with other blood cells. Both endothelial and vascular smooth muscle cells synthesize and secrete serglycin, and the loss of vessel serglycin might also play a part in the reduced ability of the SG−/− platelets to adhere to the vessel wall.

We hypothesize that the effects of serglycin deletion on platelet aggregation, and thrombus formation and stability might be due to the loss of physiologically relevant proteins from the α-granules. SG−/− platelets are deficient in PF4, the basic protein β-thromboglobulin, and PDGF. Mice lacking PF4 display similar defects in *in vivo* thrombosis following $FeCl_3$-induced injury.[127] However, the PF4 knockout mice contain normal β-thromboglobulin. It appears plausible that the defects seen in SG−/− mice can be attributed, at least in part, to the lack of stored PF4, and that the serglycin–PF4 complex may be a functional entity.

How might the absence of serglycin result in platelet aggregation and thrombus stabilization defects? A major component of the aggregation process is the induction of a conformational change and thereby activation of αIIbβ3, a major platelet membrane receptor,[149] by inside-out signaling. The activated αIIbβ3 is then able to bind fibrinogen and initiate outside-in signaling to amplify the platelet aggregation response.[120] The absence of serglycin is associated with a reduction in the activation of the integrin, as evidenced by a large right-shift in the concentration curve of fibrinogen binding in response to AYPGKF. These results suggest that defects exist in the series of events that lead to activation of αIIbβ3 on the platelet surface. We propose that these defects are related to the absence of α-granule proteins such as PF4 which are thought to bind back to the platelet surface after release, thereby reducing the amplification of aggregation which normally occurs through surface receptors. Clemetson has proposed that surface GAGs promote chemokine activation of platelets.[55] PF4, most likely, binds to membrane proteoglycans. Additional proteins likely to be missing from SG−/− platelets are α-granule chemokines such as RANTES or MIP-1α which have been shown to bind to serglycin.[10,118,150] Other proteins thought to be released from α-granules to support aggregation, for example fibrinogen, fibronectin (known to bind to serglycin), and PF V, might also be reduced in SG−/− platelets. The absence of any of these proteins, secondary to the lack of serglycin, might compromise regulatory interactions in the interplay between outside-in and inside-out signaling which require the binding of any of a number of proteins to the platelet surface, including the newly exposed inner surface of the α-granule membrane. The apparent reduction in dense granule secretion, as evidenced by reduced secretion of serotonin and ATP, is also likely to interfere with the normal amplification process, and may be secondary to the absence of specific α-granule proteins.

4. Granulogenesis and the Content of α-Granule Proteins in SG−/− Platelets

A major interest in serglycin research has been to outline its role in the granulogenesis of all hematopoietic cell types. The previous studies described above have shown that serglycin is essential for storage of various secretory granule enzymes in mast cells, CTLs, and neutrophils. In some cells, granule ultrastructure is affected by the lack of serglycin (mast cells, CTLs) while in others the granule morphology is retained (macrophages, neutrophils). We have identified three secretory granule proteins which appear to depend on the presence of serglycin for storage in the granule: PF4, β-thromboglobulin, and PDGF. The interaction of these proteins with serglycin is probably based on electrostatic interactions between the basically-charged surface regions of the proteins and the strongly anionic serglycin GAG chainsand it is reasonable to assume that this interaction is essential for bringing and/or retaining these proteins into the α-granule. It has been shown that the degree of serglycin dependence for the storage of chymases is closely correlated with the net basic charge of the respective chymase.[151] PF4 binds to serglycin with $K_d \sim 300$ nM (B. P. Schick, unpublished), compared to 840 nM for Granzyme B.[120] PDGF has also been found to bind to PG core proteins[152,153] and GAGs.[154]

5. Conclusion

In summary, our results thus point to an important role for serglycin in the regulation of platelet aggregation and thrombus formation *in vivo*. Our data are clearly compatible with a model where activation of αIIbβ3 at low agonist concentrations is defective in the absence of serglycin, most likely due to an impairment in the amplifying effects of the release of α-granule and dense granule contents. These substances are likely to interact with the surface membrane so as to enhance signaling activities and cytoskeletal rearrangements necessary for proper platelet function. The inability of SG−/− platelets to store PF4 and the absence of PF4 from plasma likely contribute to the impaired platelet function *in vivo* and *in vitro*.

XI. Future Directions in Platelet Serglycin Research

A. Further Definition of α-Granule Protein Content of SG−/− Platelets

We identified three α-granule proteins, which are almost completely absent from SG−/− platelets. These proteins as are most of the α-granule proteins are synthesized by the megakaryocytes. There may be other proteins of equal importance. A complete analysis of platelet α-granule proteins and their ability to bind to serglycin should be undertaken. Various proteomics techniques have been used on subsets of platelet proteins.[155–161] Coppinger *et al.*[162] have identified 300 proteins in the thrombin releasate from human platelets The releasates of activated platelets should represent the proteins within the granules.[162] Proteins from the α-granule that are of particular interest because they are known to be involved in hemostasis are PF4, fibrinogen, fibronectin, thrombospondin, vitronectin, and vWF. Gas6 is a recently characterized α-granule protein. It is found in mouse platelets, but may[133] or may not[163] be found in human platelets, although it is found in human plasma.[163] The knockout of Gas6 or its receptor or the blocking of Gas6 with antisera leads to defects in platelet aggregation and *in vivo* thrombus formation, effects which are very similar to those we have observed in the serglycin −/− mice.[133–135] RANTES and MIP-1α are both found in platelets. These chemokines have been associated with asthma and inflammation.[164–177] Knockouts of α-granule proteins, which lead to decreased aggregation and unstable clot formation include PF4,[127] vitronectin,[132] and P-selectin.[137]

Platelets are thought to accumulate fibrinogen, Factor V, and albumin into their α-granules primarily through uptake from plasma.[178–183] This mechanism of incorporation of plasma proteins into secretory granules is unique to platelets. These proteins are secreted when platelets are activated and likely support the aggregation process. It would be of interest to determine whether the loss of serglycin results in differential loss of serglycin-binding proteins that are synthesized by the megakaryocytes compared to those known to be taken up from plasma, at least in part, by platelets. Such studies would provide important information on proteins which may not yet be recognized to contribute to platelet function. They may identify important proteins whose loss is secondary to the loss of a serglycin-binding protein.

Wagner *et al.*[184] have summarized their studies on the interactions of the pathways leading to thrombosis and inflammation, and have concluded that platelets play a central role in inflammation via many of the components of pathways that result in thrombosis. The endothelium is intimately involved in both these processes. Serglycin expression is increased in activated endothelial and smooth muscle cells.[109,185] The role of serglycin in the interaction between the vessel wall and platelets needs to be explored and can be easily addressed by a cross-bone marrow transplant between SG−/− and WT mice. Thus, thrombosis could be studied in the context of SG−/− platelets and an SG+/+ endothelium, and vice versa.

B. Need for Conditional Serglycin Knockout

An important contribution to the cell-specific study of serglycin function would be generation of a conditional knockout of the serglycin gene. Of potential interest would be conditional knockout of serglycin in platelets, endothelial cells, neutrophils, or lymphocytes. Such knockouts would sort out the contribution of serglycin and its associated proteins in each of these cells to thrombosis and its corollaries, inflammation, atherosclerosis, and metastasis; this is especially so for proteins that are common to several of the cells in question, for example RANTES and other chemokines.

C. Inflammation and Atherosclerosis

Recent reviews have described in detail the mechanisms by which platelets are thought to be involved in inflammation[184] and atherosclerosis.[186] In brief, inflammation involves complex reactions between the platelets, the endothelium, and the leukocytes. Whether through infection or other mechanisms, inflammation involves endothelial activation, which in turn results in platelet activation, release of proinflammatory chemokines, and leukocyte activation and extravasation. These activation processes involve interactions among these cells mediated by selectins and their ligands, adhesion proteins and adhesion receptors on membranes, as well as a variety of growth factors, cytokines, chemokines, and coagulation factors released from these cells. It would be informative to conduct a series of experiments with different types of inflammatory challenges and induced infection to assess the effects of serglycin loss, and the loss of its conjugate proteins. This information would highlight the specific involvement of different blood cells and blood vessel cells during the different stages in the processes of inflammation and atherosclerosis.

Atherogenesis is considered to be a product of inflammation. Inflammation involves a complex interplay of chemokines released from the platelets and the vessel wall, and attraction of leukocytes and monocytes to the inflamed vessel. Monocytes mature into macrophages, and are responsible for the lipid accumulation in the atherosclerotic plaque. Chemical mediators known to be important include RANTES, MIP 1α, platelet activating factor, TGF-β, PDGF, serotonin, matrix-degrading proteases, and the CD-40 ligand. A major indicator of the role of platelets in atherosclerosis is the deposition of PF4, a platelet-specific protein, in human atherosclerotic lesions. PF4 is a chemoattractant for monocytes, and promotes their differentiation into macrophages. PF4 may also promote the retention of lipoproteins within the plaque.

A platelet proteoglycan with serglycin-like characteristics inhibits LDL uptake by macrophages, and could play a role in the inhibition of atherosclerosis.[187]The close association of PF4 with serglycin, and its near absence from the platelets and plasma of SG−/− mice, suggests that the SG−/− mice would be excellent candidates for studies of experimental atherosclerosis.

D. Tumor Growth and Metastasis

Studies mentioned above have shown a potential role for serglycin in the adverse sequelae of multiple myeloma[102] and its involvement in leukemia.[101] We should also consider the possibility that platelet serglycin may be involved in solid tumor metastasis. Many studies have suggested that platelets are involved in tumor metastasis,[188–192] but the mechanism by which this may occur is not clear. The possibilities include binding of tumor cells to the surface of activated platelets via GP IIbIIIa, thrombospondin secreted from platelets and bound back to the surface, P-selectin, or by attachment of tumor cells to platelet microparticles. These processes, in any case, are most likely dependent on platelet activation, which results in fusion of the α-granule membrane with the platelet surface membrane, the exposure of activation-induced surface proteins, and secretion of chemokines and other proteins. The presence or absence of serglycin in platelets may also influence the balance of tumor growth-promoting chemokines, growth factors and cytokines that are delivered to the environment of tumor cells which are related to tumor growth. Also relevant is the release of the CXC chemokines, PF4 and Platelet Basic Protein, which may affect angiogenesis and therefore blood supply to the tumor, and the CC chemokine MIP-1α. If a serglycin deficit produces changes in the platelet α-granule protein content or secretion resulting in decreased platelet activation *in vivo*, adhesion to and transport of tumor cells is also likely be decreased. The effects of other proteoglycans such as decorin and heparan sulfate proteoglycans[193–205] on metastasis have been studied, and the effects can be either increased or decreased metastasis or angiogenesis. Thus, it is reasonable to suggest that the presence or absence of serglycin, which interacts with chemokines related to metastasis and presumably delivers them to their sites of action, will play a role in tumor growth and/or metastasis. The SG−/− mouse would be an excellent animal model for assessing the role of platelet serglycin and its associated proteins in tumor metastasis. The serglycin/chemokine interactions could conceivably be a target for anti-metastasis drug development.

Acknowledgments

The work from the author's laboratory was supported by NIH grants RO1-HL-29282 and R21-GM-070630, and by RO1-HL081241 to D. S. Woulfe.

References

1. Bourdon MA, Oldberg A, Pierschbacher M, Ruoslahti E. Molecular cloning and sequence analysis of a chondroitin sulfate proteoglycan cDNA. *Proc Natl Acad Sci USA* 1985;**82**:1321–5.
2. Bourdon MA, Shiga M, Ruoslahti E. Identification from cDNA of the precursor form of a chondroitin sulfate proteoglycan core protein. *J Biol Chem* 1986;**261**:12534–7.
3. Wewer U. Characterization of a rat yolk sac carcinoma cell line. *Dev Biol* 1982;**93**:416–21.
4. Barber AJ, Kaser-Glanzmann R, Jakabova M, Luscher EF. Characterization of a chondroitin 4-sulfate proteoglycan carrier for heparin neutralizing activity (platelet factor 4) released from human blood platelets. *Biochim Biophys Acta* 1972;**286**:312–29.
5. Avraham S, Stevens RL, Nicodemus CF, Gartner MC, Austen KF, Weis JH. Molecular cloning of a cDNA that encodes the peptide core of a mouse mast cell secretory granule proteoglycan and comparison with the analogous rat and human cDNA. *Proc Natl Acad Sci USA* 1989;**86**:3763–7.
6. Humphries D, Nicodemus C, Schiller V, Stevens R. The human serglycin gene-nucleotide sequence and methylation pattern in human promyelocytic leukemia HL-60 cells and T-lymphoblast molt-4 cells. *J Biol Chem* 1992;**267**:13558–63.
7. Stevens R, Avraham S, Gartner M, Bruns G, Austen KF, Weis J. Isolation and characterization of a cDNA that encodes the peptide core of the secretory granule proteolgycan of human promyelocytic leukemia HL-60 cells. *J Biol Chem* 1987;7287–91.
8. Alliel PM, Perin JP, Maillet P, Bonnet F, Rosa JP, Jolles P. Complete amino acid sequence of a human platelet proteoglycan. *FEBS Lett* 1988;**236**:123–6.
9. Perin JP, Bonnet F, Maillet P, Jolles P. Characterization and N-terminal sequence of human platelet proteoglycan. *Biochem J* 1988;**255**:1007–13.
10. Kolset SO, Gallagher JT. Proteoglycans in hematopoietic cells. *Biochim Biophys Acta* 1990;**1032**:191–211.
11. Schick BP, Senkowski-Richardson S. Proteoglycan synthesis in human erythroleukaemia (HEL) cells. *Biochem J* 1992;**282**:651–8.
12. Schick BP, Jacoby JA. Serglycin and betaglycan proteoglycans are expressed in the megakaryocytic cell line CHRF 288-11 and normal human megakaryocytes. *J Cell Physiol* 1995;**165**:96–106.
13. Stellrecht CM, Frazier G, Selvanayagams C, Chao LY, Lee A, Saunders GF. Transcriptional regulation of a hematopoietic proteoglycan core protein gene during hematopoiesis. *J Biol Chem* 1993;**268**:4078–84.
14. Stellrecht CM, Mars WM, Miwa H, Beran M, Saunders GF. Expression pattern of a hematopoietic proteoglycan core protein gene during human hematopoiesis. *Differentiation* 1991;**48**:127–35.
15. Metcalfe DD, Litvin J, Wasserman SI. The isolation, identification and characterization of sulfated glycosaminoglycans synthesized *in vitro* by human eosinophils. *Biochim Biophys Acta* 1982;**715**:196–204.
16. Schick BP, Gradowski JF, San Antonio JD. Synthesis, secretion and subcellular localization of serglycin proteoglycan in human endothelial cells. *Blood* 2001;**97**:449–58.
17. Ho H-CK, McGrath KE, Palis J, Schick BP, Brodbeck KC. Serglycin proteoglycan synthesis in the murine uterine decidua and early embryo. *Biol Reprod* 2001;**64**:1667–76.
18. Schick BP, Ho HC, Brodbeck KC, Wrigley CW, Klimas J. Serglycin proteoglycan expression and synthesis in embryonic stem cells. *Biochim Biophys Acta* 2003;**1593**:259–67.
19. Zhang X, Kidder GM, Zhang C, Khamsi F, Armstrong DT. Expression of plasminogen activator genes and enzymatic activities in rat preimplantation embryos. *J Reprod Fertil* 1994;**101**:235–40.

20. Schick PK, Schick BP, Williams-Gartner K. Characterization of guinea pig megakaryocyte subpopulations at different phases of maturation prepared with a Celsep separation system. *Blood* 1989;**73**:1801–8.
21. Schick BP, Pestina TI, San Antonio JD, Stenberg PE, Jackson CW. Decreased serglycin proteoglycan size is associated with the plate*let al*pha granule storage defect in Wistar Furth hereditary macrothrombocytopenic rats. Serglycin binding affinity to type I collagen is unaltered. *J Cell Physiol* 1997;**172**:87–93.
22. Kulseth MA, Mustorp SL, Uhlin-Hansen L, Oberg F, Kolset SO. Serglycin expression during monocytic differentiation of U937-1 cells. *Glycobiology* 1998;**8**:747–53.
23. Sali A, Matsumoto R, McNeil HP, Karplus M, Stevens RL. Three-dimensional models of four mouse mast cell chymases. *J Biol Chem* 1993;**268**:9023–34.
24. Lidholt K, Eriksson I, Kjellen L. Heparin proteoglycans synthesized by mouse mastocytoma contain chondroitin sulphate. *Biochem J* 1995;**311**:233–8.
25. Kolset SO, Ivh I, Ivervatn A, Nilsson K. Differentiation-associated changes in the expression of chondroitin sulfate proteoglycan in induced U-397 cells. *Cancer Res* 1988;**48**:6103–8.
26. Duelli A, Ronnberg E, Waern I, Ringvall M, Kolset SO, Pejler G. Mast cell differentiation and activation is closely linked to expression of genes coding for the serglycin proteoglycan core protein and a distinct set of chondroitin sulfate and heparin sulfotransferases. *J Immunol* 2009;**183**:7073–83.
27. Oldberg A, Hayman EG, Ruoslahti E. Isolation of a chondroitin sulfate proteoglycan from a rat yolk sac tumor and immunochemical demonstration of its cell surface localization. *J Biol Chem* 1981;**256**:10847–52.
28. Perin J-P, Bonnet F, Maillet P, Jolles P. Characterization and N-terminal sequence of human plalelet proteoglycan. *Biochem J* 1988;**255**:1007–13.
29. Verrecchio A, Germann MW, Schick BP, Kung B, Twardowski T, San Antonio JD. Design of peptides with high affinities for heparin and endothelial cell proteoglycans. *J Biol Chem* 2000;**275**:7701–7.
30. Nicodemus CF, Avraham S, Austen KF, Purdy S, Jablonski J, Stevens RL. Characterization of the human gene that encodes the peptide core of secretory granule proteoglycans in promyelocytic leukemia HL-60 cells and analysis of the translated product. *J Biol Chem* 1990;**265**:5889–96.
31. Schick BP, Walsh CW, Jenkins-West T. Sulfated proteoglycans and sulfated proteins in guinea pig megakaryocytes and platelets *in vivo*: relevance to megakaryocyte maturation and platelet activation. *J Biol Chem* 1988;**263**:1052–62.
32. Schick BP, Eras JL, Mintz PS. Phosphorothioate oligonucleotides cause degradation of secretory but not intracellular serglycin proteoglycan core protein in a sequence-independent manner in human megakaryocytic tumor cells. *Antisense Res Dev* 1995;**5**:59–65.
33. Verrecchio A, Germann MW, Schick BP, Kung B, Twardowski T, SanAntonio JD. Design of peptides with high affinities for heparin and endothelial cell proteoglycans. *J Biol Chem* 2000;**271**:7701–7.
34. Forsberg E, Pejler G, Ringvall M, Lunderius C, Tomasini-Johansson B, Kusche-Gullberg M, et al. Abnormal mast cells in mice deficient in a heparin-synthesizing enzyme. *Nature* 1999;**400**:773–6.
35. Humphries DE, Wong GW, Friend DS, Gurish MF, Qiu W-T, Huang C, et al. Heparin is essential for the storage of specific granule proteases in mast cells. *Nature* 1999;**400**:769–72.
36. Toyama-Sorimachi N, Sorimachi H, Tobita Y, Kitamura F, Yagita H, Suzuki K, et al. A novel ligand for CD44 is serglycin, a hematopoietic cell lineage-specific proteoglycan. Possible involvement in lymphoid cell adherence and activation. *J Biol Chem* 1995;**270**:7437–44.
37. Brennan MJ, Oldberg A, Hayman EG, Ruoslahti E. Effect of a proteoglycan produced by rat tumor cells on their adhesion to fibronectin-collagen substrata. *Cancer Res* 1983;**43**:4302–7.

38. Toyama-Sorimachi N, Kitamura F, Habuchi H, Tobita Y, Kimata K, Miyasaka M. Widespread expression of chondroitin sulfate-type serglycins with CD44 binding ability in hematopoietic cells. *J Biol Chem* 1997;**272**:26714–9.
39. Kolset SO, Mann DM, Uhlin-Hansen L, Winberg JO, Ruoslahti E. Sergylcin-binding proteins in activated macrophages and platelets. *J Leukoc Biol* 1996;**59**:545–54.
40. Lemansky P, Hasilik A. Chondroitin sulfate is involved in lysosomal transport of lysozyme in U937 cells. *J Cell Sci* 2001;**114**:345–52.
41. Witt DP, Lander AD. Differential binding of chemokines to glycosaminoglycan subpopulations. *Curr Biol* 1994;**4**:394–400.
42. Clark-Lewis I, Dewald B, Geiser T, Moser B, Baggiolini M. Platelet factor 4 binds to interleukin 8 receptors and activates neutrophils when its N terminus is modified with Glu-Leu-Arg. *Proc Natl Acad Sci USA* 1993;**90**:3574–7.
43. Bebawy ST, Gorka J, Hyers TM, Webster RO. *In vitro* effects of platelet factor 4 on normal human neutrophil functions. *J Leukoc Biol* 1986;**39**:423–34.
44. Strieter RM, Polverini PJ, Arenberg DA, Kunkel SL. The role of CXC chemokines as regulators of angiogenesis. *Shock* 1995;**4**:155–60.
45. Aziz KA, Cawley JC, Zuzel M. Platelets prime PMN via released PF4: mechanism of priming and synergy with GM-CSF. *Br J Haematol* 1995;**91**:846–53.
46. Kasper B, Brandt E, Ernst M, Petersen F. Neutrophil adhesion to endothelial cells induced by platelet factor 4 requires sequential activation of Ras, Syk, and JNK MAP kinases. *Blood* 2006;**107**:1768–75.
47. McColl SR, Clark-Lewis I. Inhibition of murine neutrophil recruitment *in vivo* by CXC chemokine receptor antagonists. *J Immunol* 1999;**163**:2829–35.
48. Aziz KA, Cawley JC, Treweeke AT, Zuzel M. Sequential potentiation and inhibition of PMN reactivity by maximally stimulated platelets. *J Leukoc Biol* 1997;**61**:322–8.
49. Baltus T, von Hundelshausen P, Mause SF, Buhre W, Rossaint R, Weber C. Differential and additive effects of platelet-derived chemokines on monocyte arrest on inflamed endothelium under flow conditions. *J Leukoc Biol* 2005;**78**:435–41.
50. Mack M, Pfirstinger J, Weber C, Weber KS, Nelson PJ, Rupp T, et al. Chondroitin sulfate A released from platelets blocks RANTES presentation on cell surfaces and RANTES-dependent firm adhesion of leukocytes. *Eur J Immunol* 2002;**32**:1012–20.
51. Tanaka Y, Adams DH, Hubscher S, Hirano H, Siebenlist U, Shaw S. T-cell adhesion induced by proteoglycan-immobilized cytokine MIP-1 beta. *Nature* 1993;**361**:79–82.
52. Nassar T, Sachais BS, Akkawi S, Kowalska MA, Bdeir K, Leitersdorf E, et al. Platelet factor 4 enhances the binding of oxidized low-density lipoprotein to vascular wall cells. *J Biol Chem* 2003;**278**:6187–93.
53. Sachais BS, Higazi AA, Cines DB, Poncz M, Kowalska MA. Interactions of platelet factor 4 with the vessel wall. *Semin Thromb Hemost* 2004;**30**:351–8.
54. Weber C. Platelets and chemokines in atherosclerosis: partners in crime. *Circ Res* 2005;**96**:612–6.
55. Clemetson KJ, Clemetson JM, Proudfoot AE, Power CA, Baggiolini M, Wells TN. Functional expression of CCR1, CCR3, CCR4, and CXCR4 chemokine receptors on human platelets. *Blood* 2000;**96**:4046–54.
56. Beyer-Sehlmeyer G, Hiddemann W, Wormann B, Bertram J. Suppressive subtractive hybridisation reveals differential expression of serglycin, sorcin, bone marrow proteoglycan and prostate-tumour-inducing gene I (PTI-1) in drug resistant and sensitive tumour cell lines of hematopoietic origin. *Eur J Cancer* 1999;**35**:1735–42.
57. Schick BP, Thornton RD. Expression of mRNA for serglycin core protein and other platel*et al*pha granule proteins is increased in human erythroleukemia cells by phorbol myristate acetate. *Leukemia* 1993;**7**:1955–9.

58. Doi T, Greenberg S, Rosenberg R. Structure of the rat platelet factor 4 gene: a marker for megakaryocyte differentiation. *Mol Cell Biol* 1987;**7**:898–904.
59. Eisman R, Surrey S, Ramachandran B, Schwartz E, Poncz M. Structural and functional comparison of the genes for human platelet factor 4 and PF4-alt. *Blood* 1990;336–44.
60. Tabilio A, Rosa JP, Testa U, Kieffer N, Nurden AT, Del Canizo MC, et al. Expression of platelet membrane glycoproteins and alpha-granule proteins by a human erythroleukemia cell line (HEL). *EMBO J* 1984;**3**:453–9.
61. Takimoto Y, Wang ZY, Kobler K, Deuel TF. Promoter region of the human platelet-derived growth factor A-chain gene. *Proc Natl Acad Sci USA* 1991;**898**:1686–90.
62. Bentley S, Kirby S. Biosynthesis of proteochondroitin sulfate by HL-60 human promyelocytic cells. *J Leukoc Biol* 1989;**45**:46–54.
63. Whyzmuzis CA, Wu JM, Danishefsky I. Changes in proteoglycans core protein (PCP) mRNA levels during HL-60 cell differentiation. *Biochem Biophys Res Commun* 1993;**194**:118–25.
64. Maillet P, Alliel PM, Mitjavila MT. Expression of the serglycin gene in human leukemic cell lines. *Leukemia* 1992;**6**:1143–7.
65. Schick BP, Petrushina I, Brodbeck KC, Castronuevo P. Promoter regulatory elements and DNase-I hypersensitive sites in human erythroleukemia (HEL), CHRF 288-11 and HL-60 cells. *J Biol Chem* 2001;**276**:24726–35 (Originally published as JBC Papers in press online May 24721, 22001).
66. Grover A, Edwards SA, Bourdon M, Adamson ED. Proteoglycan-19, laminin and collagen type IV production is correlated with the levels of mRNA in F9 cell aggregates differentiating in the presence or absence of cyclic AMP. *Differentation* 1987;**36**:138–44.
67. Montgomery RI, Lidholt K, Flay NW, Liang J, Vertel B, Lindahl U, et al. Stable heparin-producing cell lines derived from the Furth murine mastocytoma. *Proc Natl Acad Sci USA* 1992;**89**:11327–31.
68. Eklund KK, Humphries DE, Zia Z, Ghildyal N, Friend DS, Gross V, et al. Glucocorticoids inhibit the cytokine-induced proliferation of mast cells, the high affinity IgE receptor-mediated expression of TNF-alpha, and the IL-10 induced expression of chymases. *J Immunol* 1997;**158**:4373–80.
69. Avraham S, Austen KF, Nicodemus CF, Gartner MC, Stevens RL. Cloning and characterization of the mouse gene that encodes the peptide core of secretory granule proteoglycans and expression of this gene in transfected rat-1 fibroblasts. *J Biol Chem* 1989;**264**:16719–26.
70. Angerth T, Huang RY, Aveskogh M, Pettersson I, Kjellen L, Hellman L. Cloning and structural analysis of a gene encoding a mouse mastocytoma proteoglycan core protein; analysis of its evolutionary relation to three cross hybridizing regions in the mouse genome. *Gene* 1990;**93**:235–40.
71. Avraham S, Avraham H, Austen KF, Stevens RL. Negative and positive cis-acting elements in the promoter of the mouse gene that encodes the serine/glycine-rich peptide core of secretory granule proteoglycans. *J Biol Chem* 1992;**267**:610–7.
72. Wu C. Analysis of hypersensitive sites in chromatin. *Methods Enzymol* 1989;**170**:269–89.
73. Gross DS, Gerrard WT. Nuclease hypersensitive sites in chromatin. *Annu Rev Biochem* 1989;**57**:159–95.
74. Ginder G. Analysis of DNA methylation and DNase I sensitivity in gene-specific regions of chromatin. In: Benz EJ, editor. *Methods in hematology: molecular genetics***20**:New York: Churchill Livingston Inc.; 1989. p. 111–23.
75. Leroy-Viard K, Vinit MA, Lecointe N, Mathieu-Mahul D, Romeo PH. Distinct DNase-I hypersensitive sites are associated with TAL-1 transcription in erythroid and T-cell lines. *Blood* 1995;**84**:3819–27.
76. Castronuevo P, Thornton MA, McCarthy LE, Klimas J, Schick BP. DNase I hypersensitivity patterns of the serglycin proteoglycan gene in resting and phorbol 12-myristate 13-acetate-

stimulated human erythroleukemia (HEL), CHRF 288-11, and HL-60 cells compared with neutrophils and human umbilical vein endothelial cells. *J Biol Chem* 2003;**278**:48704–12.

77. Schick BP, Petrushina I, Brodbeck KC, Castronuevo P. Promoter regulatory elements and DNase I-hypersensitive sites involved in serglycin proteoglycan gene expression in human erythroleukemia, CHRF 288-11, and HL-60 cells. *J Biol Chem* 2001;**276**:24726–35.
78. Wickenhauser C, Thiele J, Schmitz B, Frimpong S, Neumann I, Schramm K, et al. Polycythemia vera megakaryocytes store and release lysozyme to a higher extent than megakaryocytes in secondary polycythemia (polyglobuly). *Leuk Res* 1999;**23**:299–306.
79. Sachais BS, Kuo A, Nassar T, Morgan J, Kariko K, Williams KJ, et al. Platelet factor 4 binds to low-density lipoprotein receptors and disrupts the endocytic machinery, resulting in retention of low-density lipoprotein on the cell surface. *Blood* 2002;**99**:3613–22.
80. De Meyer GR, De Cleen DM, Cooper S, Knaapen MW, Jans DM, Martinet W, et al. Platelet phagocytosis and processing of beta-amyloid precursor protein as a mechanism of macrophage activation in atherosclerosis. *Circ Res* 2002;**90**:1197–204.
81. Pejler G, Sadler JE. Mechanism by which heparin proteoglycan modulates mast cell chymase activity. *Biochemistry* 1999;**38**:12187–95.
82. Pejler G. Mast cell chymase in complex with heparin proteoglycan is regulated by protamine. *FEBS Lett* 1996;**383**:170–4.
83. Pejler G, Berg L. Regulation of rat mast cell protease 1 activity. Protease inhibition is prevented by heparin proteoglycan. *Eur J Biochem* 1995;**233**:192–9.
84. Matsumoto R, Sali A, Ghildyal N, Karplus M, Stevens RL. Packaging of proteases and proteoglycans in the granules of mast cells and other hematopoietic cells. A cluster of histidines on mouse mast cell protease 7 regulates its binding to heparin serglycin proteoglycans. *J Biol Chem* 1995;**270**:19524–31.
85. Noga O, Brunnee T, Schaper C, Kunkel G. Heparin, derived from the mast cells of human lungs is responsible for the generation of kinins in allergic reactions due to the activation of the contact system. *Int Arch Allergy Immunol* 1999;**120**:310–6.
86. Brunnee T, Reddigari SR, Shibayama Y, Kaplan AP, Silverberg M. Mast cell derived heparin activates the contact system: a link to kinin generation in allergic reactions. *Clin Exp Allergy* 1997;**27**:653–63.
87. Kolset SO, Prydz K, Pejler G. Intracellular proteoglycans. *Biochem J* 2004;**379**:217–27.
88. Pejler G, Abrink M, Ringvall M, Wernersson S. Mast cell proteases. *Adv Immunol* 2007;**95**:167–255.
89. Pejler G, Maccarana M. Interaction of heparin with rat mast cell protease 1. *J Biol Chem* 1994;**269**:14451–6.
90. Hunt JE, Friend DS, Gurish MF, Feyfant E, Sali A, Huang C, et al. Mouse mast cell protease 9, a novel member of the chromosome 14 family of serine proteases that is selectively expressed in uterine mast cells. *J Biol Chem* 1997;**272**:29158–66.
91. Barber A, Kaser-Glanzmann R, Jakabova M, Luscher E. Characterization of a chondroitin 4-sulfate proteoglycan carrier for heparin neutralizing activity (platelet factor 4) released from human blood platelets. *Biochem Biophys Acta* 1972;**286**:312–29.
92. Gullberg U, Andersson E, Garwicz D, Lindmark A, Olsson I. Biosynthesis, processing and sorting of neutrophil proteins: insight into neutrophil granule development. *Eur J Haemtol* 1997;**58**:137–53 [Review].
93. Cowland JB, Borregaard N. The individual regulation of granule protein mRNA levels during neutrophil maturation explains the heterogeneity of neutrophil granules. *J Leukoc Biol* 1999;**66**:989–95.
94. Drouin A, Favier R, Masse JM, Debili N, Schmitt A, Elbim C, et al. Newly recognized cellular abnormalities in the gray platelet syndrome. *Blood* 2001;**98**:1382–91.

95. Birkenbach M, Josefsen K, Yalamanchili R, Lenoir G, Kieff E. Epstein-Barr virus-induced genes: first lymphocyte-specific G protein- coupled peptide receptors. *J Virol* 1993;**67**:2209–20.
96. Levitt D, Ho PL. Induction of chondroitin sulfate proteoglycan synthesis and secretion in lymphocytes and monocytes. *J Cell Biol* 1983;**97**:351–8.
97. Galvin JP, Spaeny-Dekking LH, Wang B, Seth P, Hack CE, Froelich CJ. Apoptosis induced by granzyme B-glycosaminoglycan complexes: implications for granule-mediated apoptosis *in vivo*. *J Immunol* 1999;**162**:5345–50.
98. Raja SM, Wang B, Dantuluri M, Desai UR, Demeler B, Spiegel K, et al. Cytotoxic cell granule-mediated apoptosis: Characterization ofthe macromolecular complex of granzyme B with serglycin. *J Biol Chem* 2002;**277**(51):49523–30.
99. Grujic M, Braga T, Lukinius A, Eloranta ML, Knight SD, Pejler G, et al. Serglycin-deficient cytotoxic T lymphocytes display defective secretory granule maturation and granzyme B storage. *J Biol Chem* 2005;**280**(39):33411–8.
100. Decorti G, Klugmann FB, Crivellato E, Malusa N, Furlan G, Candussio L, et al. Biochemical and microscopic evidence for the internalization of drug-containing mast cell granules by macrophages and smooth muscle cells. *Toxicol Appl Pharmacol* 2000;**169**:269–75.
101. Niemann CU, Kjeldsen L, Ralfkiaer E, Jensen MK, Borregaard N. Serglycin proteoglycan in hematologic malignancies: a marker of acute myeloid leukemia. *Leukemia* 2007;**21**:2406–10.
102. Theocharis AD, Seidel C, Borset M, Dobra K, Baykov V, Labropoulou V, et al. Serglycin constitutively secreted by myeloma plasma cells is a potent inhibitor of bone mineralization *in vitro*. *J Biol Chem* 2006;**281**(46):35116–28.
103. Galanakis DK, Gebrehiwet B. A unique property of a plasma proteoglycan, the C1q inhibitor. An anticoagulant state resulting from its binding to fibrinogen. *J Clin Invest* 1994;**93**:303–10.
104. Silvestri L, Baker JR, Roden L, Stroud RM. The C1q inhibitor in serum is a chondroitin-4-sulfate proteoglycan. *J Biol Chem* 1981;**256**:7383–7.
105. Bohm T, Geiger M, Binder BR. Plasminogen activator-binding proteoglycans from human umbilical vein endothelial cells. *Thromb Vascular Biol* 1996;**16**:665–72.
106. Lagunoff D, Rickard A. Mast cell granule heparin proteoglycan induces lacunae in confluent endothelial cell monolayers. *Am J Pathol* 1999;**154**:1591–600.
107. Wang Y, Kovanen PT. Heparin proteoglycans released from rat serosal mast cells inhibit proliferation of rat aortic smooth muscle cells in culture. *Circ Res* 1999;**84**:74–83.
108. Kulseth MA, Kolset SO, Ranheim T. Stimulation of serglycin and CD44 mRNA expression in endothelial cells exposed to TNF-alpha and IL-1 alpha. *Biochim Biophys Acta* 1999;**1428**:225–32.
109. Lemire JM, Chan CK, Bressler S, Miller J, LeBaron RG, Wight TN. Interleukin-1beta selectively decreases the synthesis of versican by arterial smooth muscle cells. *J Cell Biochem* 2007;**101**:753–66.
110. Yao YQ, Xu JS, Lee WM, Yeung WS, Lee KF. Identification of mRNAs that are up-regulated after fertilization in the murine zygote by suppression subtractive hybridization. *Biochem Biophys Res Commun* 2003;**304**:60–6.
111. Rickles RJ, Darrow AL, Strickland S. Molecular cloning of complementary DNA to mouse tissue plasminogen activator mRNA and its expression during F9 teratocarcinoma cell differentiation. *J Biol Chem* 1988;**263**(3):1563–9.
112. Kennedy TG, Ross HE, Barbe GJ, Shu MA, Zhang X. Secretion of plasminogen activator by cultured rat endometrial stromal cells from uteri differentially sensitized for the decidual cell reaction. *Mol Reprod Dev* 1998;**49**:268–76.
113. Zhang L, Yang M, Yang D, Cavey G, Davidson P, Gibson G. Molecular interactions of MMP-13 C-terminal domain with chondrocyte proteins. *Connect Tissue Res* 2010;**51**(3):230–9.

114. Abrink M, Grujic M, Pejler G. Serglycin is essential for maturation of mast cell secretory granule. *J Biol Chem* 2004;**279**:40897–905.
115. Ho H-CK, McGrath KE, Brodbeck KC, Palis J, Schick BP. Serglycin proteoglycan synthesis in the murine uterine decidua and early embryo. *Biol Reprod* 2001;**64**(6):1667–76.
116. Henningsson F, Hergeth S, Cortelius R, Abrink M, Pejler G. A role for serglycin proteoglycan in granular retention and processing of mast cell secretory granule components. *FEBS J* 2006;**273**(21):4901–12.
117. Zernichow L, Abrink M, Hallgren J, Grujic M, Pejler G, Kolset SO. Serglycin is the major secreted proteoglycan in macrophages and has a role in the regulation of macrophage tumor necrosis factor-alpha secretion in response to lipopolysaccharide. *J Biol Chem* 2006;**281**:26792–801.
118. Kolset SO, Mann DM, Uhlin-Hansen L, Winberg JO, Ruoslahti E. Serglycin-binding proteins in activated macrophages and platelets. *J Leukoc Biol* 1996;**59**:545–54.
119. Metkar SS, Wang B, Aguilar-Santelises M, Raja SM, Uhlin-Hansen L, Podack E, et al. Cytotoxic cell granule-mediated apoptosis: perforin delivers granzyme B-serglycin complexes into target cells without plasma membrane pore formation. *Immunity* 2002;**16**:417–28.
120. Raja SM, Wang B, Dantuluri M, Desai UR, Demeler B, Spiegel K, et al. Cytotoxic cell granule-mediated apoptosis. Characterization of the macromolecular complex of granzyme B with serglycin. *J Biol Chem* 2002;**277**:49523–30.
121. Raja SM, Metkar SS, Honing S, Wang B, Russin WA, Pipalia NH, et al. A novel mechanism for protein delivery: granzyme B undergoes electrostatic exchange from serglycin to target cells. *J Biol Chem* 2005;**280**:20752–61.
122. Veugelers K, Motyka B, Frantz C, Shostak I, Sawchuk T, Bleackley RC. The granzyme B-serglycin complex from cytotoxic granules requires dynamin for endocytosis. *Blood* 2004;**103**:3845–53.
123. Grujic M, Christensen JP, Sorensen MR, Abrink M, Pejler G, Thomsen AR. Delayed contraction of the CD8+ T cell response toward lymphocytic choriomeningitis virus infection in mice lacking serglycin. *J Immunol* 2008;**181**:1043–51.
124. Niemann CU, Abrink M, Pejler G, Fischer RL, Christensen EI, Knight SD, et al. Neutrophil elastase depends on serglycin proteoglycan for localization in granules. *Blood* 2007;**109**(10):4478–86.
125. Wernersson S, Braga T, Sawesi O, Waern I, Nilsson KE, Pejler G, et al. Age-related enlargement of lymphoid tissue and altered leukocyte composition in serglycin-deficient mice. *J Leukoc Biol* 2009;**85**:401–8.
126. Steiner M. Role of glycosaminoglycans in calcium metabolism of human platelets. *Biochim Biophys Acta* 1986;**886**:406–10.
127. Eslin DE, Zhang C, Samuels KJ, Rauova L, Zhai L, Niewiarowski S, et al. Transgenic mice studies demonstrate a role for platelet factor 4 in thrombosis: dissociation between anticoagulant and antithrombotic effect of heparin. *Blood* 2004;**104**:3173–80.
128. Chen J, Somanath PR, Razorenova O, Chen WS, Hay N, Bornstein P, et al. Akt1 regulates pathological angiogenesis, vascular maturation and permeability *in vivo*. *Nat Med* 2005;**11**:1188–96.
129. Woulfe D, Jiang H, Morgans A, Monks R, Birnbaum M, Brass LF. Defects in secretion, aggregation, and thrombus formation in platelets from mice lacking Akt2. *J Clin Invest* 2004;**113**:441–50.
130. Moers A, Wettschureck N, Gruner S, Nieswandt B, Offermanns S. Unresponsiveness of platelets lacking both Galpha(q) and Galpha(13). Implications for collagen-induced platelet activation. *J Biol Chem* 2004;**279**:45354–9.

131. Chrzanowska-Wodnicka M, Smyth SS, Schoenwaelder SM, Fischer TH, White 2nd GC. Rap1b is required for normal platelet function and hemostasis in mice. *J Clin Invest* 2005;**115**:680–7.
132. Reheman A, Gross P, Yang H, Chen P, Allen D, Leytin V, et al. Vitronectin stabilizes thrombi and vessel occlusion but plays a dual role in platelet aggregation. *J Thromb Haemost* 2005;**3**:875–83.
133. Angelillo-Scherrer A, de Frutos P, Aparicio C, Melis E, Savi P, Lupu F, et al. Deficiency or inhibition of Gas6 causes platelet dysfunction and protects mice against thrombosis. *Nat Med* 2001;**7**:215–21.
134. Chen C, Li Q, Darrow AL, Wang Y, Derian CK, Yang J, et al. Mer receptor tyrosine kinase signaling participates in platelet function. *Arterioscler Thromb Vasc Biol* 2004;**24**:1118–23.
135. Ishimoto Y, Nakano T. Release of a product of growth arrest-specific gene 6 from rat platelets. *FEBS Lett* 2000;**466**:197–9.
136. Lian L, Wang Y, Draznin J, Eslin D, Bennett JS, Poncz M, et al. The relative role of PLCbeta and PI3Kgamma in platelet activation. *Blood* 2005;**106**:110–7.
137. Yokoyama S, Ikeda H, Haramaki N, Yasukawa H, Murohara T, Imaizumi T. Platelet P-selectin plays an important role in arterial thrombogenesis by forming large stable platelet-leukocyte aggregates. *J Am Coll Cardiol* 2005;**45**:1280–6.
138. Woulfe DS, Lilliendahl JK, August S, Rauova L, Kowalska MA, Abrink M, et al. Serglycin proteoglycan deletion induces defects in platelet aggregation and thrombus formation in mice. *Blood* 2008;**111**:3458–67.
139. White JG. Medich giant platelet disorder: a unique alpha granule deficiency I. Structural abnormalities. *Platelets* 2004;**15**:345–53.
140. Falik-Zaccai TC, Anikster Y, Rivera CE, Horne 3rd MK, Schliamser L, Phornphutkul C, et al. A new genetic isolate of gray platelet syndrome (GPS): clinical, cellular, and hematologic characteristics. *Mol Genet Metab* 2001;**74**:303–13.
141. Schmitt A, Jouault H, Guichard J, Wendling F, Drouin A, Cramer EM. Pathologic interaction between megakaryocytes and polymorphonuclear leukocytes in myelofibrosis. *Blood* 2000;**96**:1342–7.
142. Cashell AW, Buss DH. The frequency and significance of megakaryocytic emperipolesis in myeloproliferative and reactive states. *Ann Hematol* 1992;**64**:273–6.
143. Breton-Gorius J, Vainchenker W, Nurden A, Levy-Toledano S, Caen J. Defective alpha-granule production in megakaryocytes from gray platelet syndrome: ultrastructural studies of bone marrow cells and megakaryocytes growing in culture from blood precursors. *Am J Pathol* 1981;**102**:10–9.
144. Nurden AT, Nurden P. The gray platelet syndrome: clinical spectrum of the disease. *Blood Rev* 2007;**21**:21–36.
145. Smith MP, Cramer EM, Savidge GF. Megakaryocytes and platelets in alpha-granule disorders. *Baillieres Clin Haematol* 1997;**10**:125–48.
146. Steiner M. Platelet surface glycosaminoglycans are an effective shield for distinct platelet receptors. *Biochim Biophys Acta* 1987;**931**:286–93.
147. Fay WP, Parker AC, Ansari MN, Zheng X, Ginsburg D. Vitronectin inhibits the thrombotic response to arterial injury in mice. *Blood* 1999;**93**:1825–30.
148. Sachs UJ, Nieswandt B. *In vivo* thrombus formation in murine models. *Circ Res* 2007;**100**:979–91.
149. Brass LF, Zhu L, Stalker TJ. Minding the gaps to promote thrombus growth and stability. *J Clin Invest* 2005;**115**:3385–92.
150. von Hundelshausen P, Petersen F, Brandt E. Platelet-derived chemokines in vascular biology. *Thromb Haemost* 2007;**97**:704–13.

151. Braga T, Grujic M, Lukinius A, Hellman L, Abrink M, Pejler G. Serglycin proteoglycan is required for secretory granule integrity in mucosal mast cells. *Biochem J* 2007;**403**:49–57.
152. Gohring W, Sasaki T, Heldin CH, Timpl R. Mapping of the binding of platelet-derived growth factor to distinct domains of the basement membrane proteins BM-40 and perlecan and distinction from the BM-40 collagen-binding epitope. *Eur J Biochem* 1998;**255**:60–6.
153. Goretzki L, Burg MA, Grako KA, Stallcup WB. High-affinity binding of basic fibroblast growth factor and platelet-derived growth factor-AA to the core protein of the NG2 proteoglycan. *J Biol Chem* 1999;**274**:16831–7.
154. Fager G, Camejo G, Olsson U, Ostergren-Lunden G, Lustig F, Bondjers G. Binding of platelet-derived growth factor and low density lipoproteins to glycosaminoglycan species produced by human arterial smooth muscle cells. *J Cell Physiol* 1995;**163**:380–92.
155. Maguire PB, Foy M, Fitzgerald DJ. Using proteomics to identify potential therapeutic targets in platelets. *Biochem Soc Trans* 2005;**33**:409–12.
156. Maguire PB, Moran N, Cagney G, Fitzgerald DJ. Application of proteomics to the study of platelet regulatory mechanisms. *Trends Cardiovasc Med* 2004;**14**:207–20.
157. Gnatenko DV, Perrotta PL, Bahou WF. Proteomic approaches to dissect platelet function: one-half of the story. *Blood* 2006;**108**(13):3983–91.
158. Macaulay IC, Carr P, Gusnanto A, Ouwehand WH, Fitzgerald D, Watkins NA. Platelet genomics and proteomics in human health and disease. *J Clin Invest* 2005;**115**:3370–7.
159. Maguire PB, Fitzgerald DJ. Platelet proteomics. *J Thromb Haemost* 2003;**1**:1593–601.
160. McRedmond JP, Park SD, Reilly DF, Coppinger JA, Maguire PB, Shields DC, et al. Integration of proteomics and genomics in platelets: a profile of platelet proteins and platelet-specific genes. *Mol Cell Proteomics* 2004;**3**:133–44.
161. Thadikkaran L, Siegenthaler MA, Crettaz D, Queloz PA, Schneider P, Tissot JD. Recent advances in blood-related proteomics. *Proteomics* 2005;**5**:3019–34.
162. Coppinger JA, Cagney G, Toomey S, Kislinger T, Belton O, McRedmond JP, et al. Characterization of the proteins released from activated platelets leads to localization of novel platelet proteins in human atherosclerotic lesions. *Blood* 2004;**103**:2096–104.
163. Balogh I, Hafizi S, Stenhoff J, Hansson K, Dahlback B. Analysis of Gas6 in human platelets and plasma. *Arterioscler Thromb Vasc Biol* 2005;**25**:1280–6.
164. Chvatchko Y, Proudfoot AE, Buser R, Juillard P, Alouani S, Kosco-Vilbois M, et al. Inhibition of airway inflammation by amino-terminally modified RANTES/CC chemokine ligand 5 analogues is not mediated through CCR3. *J Immunol* 2003;**171**:5498–506.
165. Schuh JM, Blease K, Bruhl H, Mack M, Hogaboam CM. Intrapulmonary targeting of RANTES/CCL5-responsive cells prevents chronic fungal asthma. *Eur J Immunol* 2003;**33**:3080–90.
166. John AE, Berlin AA, Lukacs NW. Respiratory syncytial virus-induced CCL5/RANTES contributes to exacerbation of allergic airway inflammation. *Eur J Immunol* 2003;**33**:1677–85.
167. Folkard SG, Westwick J, Millar AB. Production of interleukin-8, RANTES and MCP-1 in intrinsic and extrinsic asthmatics. *Eur Respir J* 1997;**10**:2097–104.
168. Pype JL, Dupont LJ, Menten P, Van Coillie E, Opdenakker G, Van Damme J, et al. Expression of monocyte chemotactic protein (MCP)-1, MCP-2, and MCP-3 by human airway smooth-muscle cells. Modulation by corticosteroids and T-helper 2 cytokines. *Am J Respir Cell Mol Biol* 1999;**21**:528–36.
169. Schuh JM, Blease K, Hogaboam CM. The role of CC chemokine receptor 5 (CCR5) and RANTES/CCL5 during chronic fungal asthma in mice. *FASEB J* 2002;**16**:228–30.
170. Blease K, Mehrad B, Standiford TJ, Lukacs NW, Gosling J, Boring L, et al. Enhanced pulmonary allergic responses to Aspergillus in CCR2−/− mice. *J Immunol* 2000;**165**:2603–11.

171. Gonzalo JA, Lloyd CM, Wen D, Albar JP, Wells TN, Proudfoot A, et al. The coordinated action of CC chemokines in the lung orchestrates allergic inflammation and airway hyperresponsiveness. *J Exp Med* 1998;**188**:157–67.
172. Wagner L, Yang OO, Garcia-Zepeda EA, Ge Y, Kalams SA, Walker BD, et al. Beta-chemokines are released from HIV-1-specific cytolytic T-cell granules complexed to proteoglycans. *Nature* 1998;**391**:908–11.
173. Speyer CL, Gao H, Rancilio NJ, Neff TA, Huffnagle GB, Sarma JV, et al. Novel chemokine responsiveness and mobilization of neutrophils during sepsis. *Am J Pathol* 2004;**165**:2187–96.
174. Rot A, Krieger M, Brunner T, Bischoff SC, Schall TJ, Dahinden CA. RANTES and macrophage inflammatory protein 1 alpha induce the migration and activation of normal human eosinophil granulocytes. *J Exp Med* 1992;**176**:1489–95.
175. Kuschert GS, Coulin F, Power CA, Proudfoot AE, Hubbard RE, Hoogewerf AJ, et al. Glycosaminoglycans interact selectively with chemokines and modulate receptor binding and cellular responses. *Biochemistry* 1999;**38**:12959–68.
176. Hoogewerf AJ, Kuschert GS, Proudfoot AE, Borlat F, Clark-Lewis I, Power CA, et al. Glycosaminoglycans mediate cell surface oligomerization of chemokines. *Biochemistry* 1997;**36**:13570–8.
177. Gonzalo JA, Lloyd CM, Kremer L, Finger E, Martinez AC, Siegelman MH, et al. Eosinophil recruitment to the lung in a murine model of allergic inflammation. The role of T cells, chemokines, and adhesion receptors. *J Clin Invest* 1996;**98**:2332–45.
178. Camire RM, Pollak ES, Kaushansky K, Tracy PB. Secretable human platelet-derived factor V originates from the plasma pool. *Blood* 1998;**92**:3035–41.
179. Bouchard BA, Williams JL, Meisler NT, Long MW, Tracy PB. Endocytosis of plasma-derived factor V by megakaryocytes occurs via a clathrin-dependent, specific membrane binding event. *J Thromb Haemost* 2005;**3**:541–51.
180. Gould WR, Simioni P, Silveira JR, Tormene D, Kalafatis M, Tracy PB. Megakaryocytes endocytose and subsequently modify human factor V *in vivo* to form the entire pool of a unique platelet-derived cofactor. *J Thromb Haemost* 2005;**3**:450–6.
181. Handagama PJ, Shuman MA, Bainton DF. Incorporation of intravenously injected albumin, immunoglobulin G, and fibrinogen in guinea pig megakaryocyte granules. *J Clin Invest* 1989;**84**:73–82.
182. Handagama PJ, Bainton DF. Incorporation of a circulating protein into alpha granules of megakaryocytes. *Blood Cells* 1989;**15**:59–72.
183. Handagama PJ, George JN, Shuman MA, McEver RP, Bainton DF. Incorporation of a circulating protein into megakaryocyte and platelet granules. *Proc Natl Acad Sci USA* 1987;**84**:861–5.
184. Wagner DD, Burger PC. Platelets in inflammation and thrombosis. *Arterioscler Thromb Vasc Biol* 2003;**23**:2131–7.
185. Kulseth MA, Kolset SO, Ranheim T. Stimulation of serglycin and CD44 mRNA expression in endothelial cells exposed to TNF-alpha and IL-1alpha. *Biochim Biophys Acta* 1999;**1428**:225–32.
186. Gawaz M, Langer H, May AE. Platelets in inflammation and atherogenesis. *J Clin Invest* 2005;**115**:3378–84.
187. Mas-Oliva J, Arnold KS, Wagner WD, Phillips DR, Pitas RE, Innerarity T. Isolation and characterization of a platelet-derived macrophage-binding proteoglycan. *J Biol Chem* 1994;**269**:10177–83.
188. Erpenbeck L, Schon MP. Deadly allies: the fatal interplay between platelets and metastasizing cancer cells. *Blood* 2010;**115**(17):3427–36.
189. Borsig L. The role of platelet activation in tumor metastasis. *Expert Rev Anticancer Ther* 2008;**8**:1247–55.

190. Chen M, Geng JG. P-selectin mediates adhesion of leukocytes, platelets, and cancer cells in inflammation, thrombosis, and cancer growth and metastasis. *Arch Immunol Ther Exp (Warsz)* 2006;**54**:75–84.
191. Mousa SA. Antithrombotics in thrombosis and cancer. *Hamostaseologie* 2005;**25**:380–6.
192. Sierko E, Wojtukiewicz MZ. Platelets and angiogenesis in malignancy. *Semin Thromb Hemost* 2004;**30**:95–108.
193. Kelly T, Miao HQ, Yang Y, Navarro E, Kussie P, Huang Y, et al. High heparanase activity in multiple myeloma is associated with elevated microvessel density. *Cancer Res* 2003;**63**:8749–56.
194. Yang Y, Yaccoby S, Liu W, Langford JK, Pumphrey CY, Theus A, et al. Soluble syndecan-1 promotes growth of myeloma tumors *in vivo*. *Blood* 2002;**100**:610–7.
195. Sanderson RD, Borset M. Syndecan-1 in B lymphoid malignancies. *Ann Hematol* 2002;**81**:125–35.
196. Yang Y, Macleod V, Bendre M, Huang Y, Theus AM, Miao HQ, et al. Heparanase promotes the spontaneous metastasis of myeloma cells to bone. *Blood* 2005;**105**:1303–9.
197. Sanderson RD, Yang Y, Suva LJ, Kelly T. Heparan sulfate proteoglycans and heparanase–partners in osteolytic tumor growth and metastasis. *Matrix Biol* 2004;**23**:341–52.
198. Seidel C, Borset M, Hjertner O, Cao D, Abildgaard N, Hjorth-Hansen H, et al. High levels of soluble syndecan-1 in myeloma-derived bone marrow: modulation of hepatocyte growth factor activity. *Blood* 2000;**96**:3139–46.
199. Zhu JX, Goldoni S, Bix G, Owens RT, McQuillan DJ, Reed CC, et al. Decorin evokes protracted internalization and degradation of the epidermal growth factor receptor via caveolar endocytosis. *J Biol Chem* 2005;**280**:32468–79.
200. Reed CC, Waterhouse A, Kirby S, Kay P, Owens RT, McQuillan DJ, et al. Decorin prevents metastatic spreading of breast cancer. *Oncogene* 2005;**24**:1104–10.
201. Tralhao JG, Schaefer L, Micegova M, Evaristo C, Schonherr E, Kayal S, et al. *In vivo* selective and distant killing of cancer cells using adenovirus-mediated decorin gene transfer. *FASEB J* 2003;**17**:464–6.
202. Grant DS, Yenisey C, Rose RW, Tootell M, Santra M, Iozzo RV. Decorin suppresses tumor cell-mediated angiogenesis. *Oncogene* 2002;**21**:4765–77.
203. Santra M, Eichstetter I, Iozzo RV. An anti-oncogenic role for decorin. Down-regulation of ErbB2 leads to growth suppression and cytodifferentiation of mammary carcinoma cells. *J Biol Chem* 2000;**275**:35153–61.
204. Fjeldstad K, Kolset SO. Decreasing the metastatic potential in cancers - targeting the heparan sulfate proteoglycans. *Curr Drug Targets* 2005;**6**:665–82.
205. Samoszuk MK, Kanakubo E, Chan JK. Degranulating mast cells in fibrotic regions of human tumors and evidence that mast cell heparin interferes with the growth of tumor cells through a mechanism involving fibroblasts. *BMC Cancer* 2005;**5**:121.

Congenital Disorders of Glycosylation with Emphasis on loss of Dermatan-4-Sulfotransferase

Lijuan Zhang,*
Thomas Müller,†
Jacques U. Baenziger,‡ and
Andreas R. Janecke†

*Department of Pathology and Immunology, Washington University School of Medicine, St. Louis, Missouri, USA
†Department of Pediatrics II, Innsbruck Medical University, Anichstrasse 35, Innsbruck, Austria
‡Department of Pathology, Washington University School of Medicine, St. Louis, Missouri, USA

The autosomal, recessively inherited, adducted thumb-clubfoot syndrome (ATCS) represents a generalized connective tissue disorder with congenital malformations, contractures of thumbs and feet, and a typical facial appearance. Cognitive development is normal in ATCS patients during childhood. ATCS is caused by homozygous nonsense and missense mutations in *CHST14* which encodes an *N*-acetylgalactosamine 4-*O*-sulfotransferase 1 (D4ST1) that catalyzes the 4-*O*-sulfation of *N*-acetylgalactosamine in the repeating iduronic acid-α-1,3-*N*-acetylgalactosamine disaccharide sequence to form dermatan sulfate (DS). ATCS mutations lead to either a decrease or a loss of D4ST1 activity, as revealed by absence of DS and an excess of chondroitin sulfate (CS) in patient's fibroblasts. Either of these effects or their combination might cause the observed clinical symptoms by altering the physiological pattern of dermatan and CS chains on their corresponding proteoglycans (PGs). ATCS is the

Progress in Molecular Biology
and Translational Science, Vol. 93
DOI: 10.1016/S1877-1173(10)93012-3

only recognized disorder resulting from a defect that is specific to DS biosynthesis, and thus represents another class of the congenital glycosylation disorders. Congenital disorders of glycosylation (CDG) include all genetic diseases that result from defects in the synthesis of glycans. These disorders cause a wide range of human diseases, with examples emanating from all medical subspecialties. ATCS is the first human disorder that emphasizes a role for DS in human development and extracellular matrix maintenance.

I. Introduction

The glycome, defined as all the sugar chains (glycans) synthesized by an organism, is estimated to be 10–10^4 times larger than the proteome in different species.[1,2] Glycosylation, the addition of glycans to proteins and lipids, can result in a complex array of glycoforms on a single carrier protein or lipid. In eukaryotes, 11 biosynthetic pathways are known to link glycans to proteins and lipids.[3] To-date, genetic disorders affecting six of these eleven pathways have been identified. Many of the causal mutations affect biosynthetic enzymes, but a number of them also affect chaperones and Golgi-trafficking complex subunits. The resulting impairment of one or several glycosylation pathways leads to altered glycan synthesis and structure, and generally to the impairment of several organ systems in affected patients.[3,4]

The developmental differentiation of cell types and tissue-specific assembly require extensive cell migrations and interactions with structures present in the extracellular environment, which is a complex array of biopolymers, commonly termed extracellular matrices. An example of an extracellular matrix is the basement membrane which separates distinct cell types. Important components of extracellular matrices, such as basement membranes, include laminins and collagens, and a number of proteoglycans (PGs), these are proteins with covalently attached glycosaminoglycan (GAG) polysaccharide side chains. These protein-bound GAGs are classified, according to their sugar composition, as keratan sulfate(s) (KS), chondroitin sulfate(s) (CS), dermatan sulfate(s) (DS), and heparan sulfate(s) (HS) (Glycosaminoglycan (GAG) Biosynthesis and GAG-Binding Proteins). The GAGs exhibit complex modification patterns, consisting predominantly of sulfations of hydroxyl groups and epimerizations of specific carbon atoms of individual sugar molecules. Hyaluronan represents another GAG, which shows no sulfation or epimerization, and is not attached to a protein to form a PG during biosynthesis.[5] There are a considerable number of PGs that bear more than one kind of GAG side chains.

While the roles of HS and CS in development and particularly in growth factor signaling have been extensively studied,[6–11] the biological function of DS is far from well understood. Moreover, mutations affecting the biosynthesis of HS and CS PGs have been described as the cause of several human hereditary diseases[2,12]; while the adducted thumb-clubfoot syndrome (ATCS, see below) represents the sole DS biosynthesis defect.[13] However, four GAG biosynthetic enzymes are necessary for synthesizing the linkage region (attaching GAGs to the PG core protein) of HS, CS, and DS. One of these, galactosyltransferase I encoded by *B4GALT7*, is mutated in patients with the very rare, progeroid form of the Ehlers–Danlos syndrome (EDS).[14,15] All three known human *B4GALT7* mutations have been shown to result in hypomorphs, and null mutations are most likely not compatible with human embryonic development, and is analogous to mice lacking another linkage-synthesizing enzyme, glucuronyltransferase I (Ref. 16 and Mice deficient in Glucuronyl C5-Epimerase: An Enzyme Converting Glucuronic Acid to Iduronic Acid in Heparan Sulfate/ Heparin Biosynthesis). A global deficiency and/or abnormal ratio of different GAGs on carrier proteins are supposed to cause the symptoms of this generalized connective tissue disorder. Indeed, symptoms of the progeroid form of EDS overlap with those of ATCS, an autosomal recessive, generalized connective tissue disorder with congenital, distal contractures, a variable degree of congenital malformations, but with normal cognitive development.[17–20] ATCS is caused due to a deficiency of D4ST1 that is essential for synthesis of the sulfated disaccharide, iduronic acid-*N*-acetylgalactosamine-4-*O*-sulfate. D4ST1 deficiency compromises the generation of DS from chondroitin.[13] To-date, ATCS represents the only human disorder related to a defect specific to DS biosynthesis. ATCS patients have a deficiency of DS in tissues that normally produce DS, providing an opportunity to gain an insight into the physiological function of this GAG and its PGs in humans.

II. Human Hereditary Disorders of Glycosylation

Glycosylation is an enzyme-directed site-specific process that links saccharides to produce glycans, which in turn are attached to proteins, lipids, or other organic molecules. Glycosylation represents both co- and posttranslational modifications. Such modifications can greatly increase the functions of a single gene (product). Glycans have a broad variety of structural and functional roles in membrane and secreted proteins. The majority of proteins synthesized in the rough ER undergo glycosylation. Glycosylation also occurs in the cytoplasm and nucleus as an *O*-GlcNAc modification.

Human hereditary disorders of glycosylation, generally referred to as congenital disorders of glycosylation (CDG), are a large family of genetic diseases resulting from defects in the synthesis or in the attachment of glycans, and have been related to a number of classes of glycans: N-linked glycans attached to a nitrogen of asparagine, O-linked glycans attached to the hydroxy oxygen of serine, threonine, tyrosine, hydroxylysine, or hydroxyproline, or to oxygens on lipids such as ceramide; phosphoglycans linked through the phosphate of a phosphoserine; C-linked glycans, a rare form of glycosylation where a sugar is added to a carbon on a tryptophan side chain, and glypiation which is the addition of a GPI anchor which links proteins to lipids through glycan linkages.[3]

More than 40 human CDGs are currently known, most of which are *N*-glycosylation defects, followed by *O*-Ser/Thr glycosylation defects.[1,4] The variability in clinical manifestations has been reported within one CDG subtype and even among affected siblings, but factors which contribute to variability of the clinical phenotype are largely unknown.[21]

A. *N*-Glycosylation Defects

Distinct patterns of glycoprotein synthesis take place in different cell compartments: in the cytoplasm, in the membrane of the endoplasmic reticulum (ER), in the Golgi complex, and in the nucleus. *N*-Glycosylation, characteristic of eukaryotic cells, starts on the cytoplasmic surface of the ER membrane by the addition of two *N*-acetylglucosamines and five mannoses, one by one, to a dolichylphosphate. Triggered by a still unknown mechanism, the whole complex is subsequently flipped to the lumenal side of the membrane, where further four mannoses and three glucoses are added. As the dolichylphosphate is released, the whole oligosaccharide is transferred to an asparagine residue of the polypeptide chain that is in the process of assembling into protein complexes. While proper protein folding is taking place, the three glucoses and one mannose are trimmed away by different enzymes. The glycoprotein is then transferred to the Golgi complex. Further mannose trimming occurs and subsequently *N*-acetylglucosamine, fucose, galactose, sialic acid and other moieties are sequentially added to the glycoprotein by different enzymes. Biosynthesis defects concerning steps in the cytoplasm and in the ER are referred to as CDG-I disorders, and defects in subsequent reactions as CDG-II disorders.[1,22]

Isoelectric focusing of serum transferrin is currently used as a relatively inexpensive screening test to detect abnormalities in *N*-glycosylation.[23] However, false-negative results have been obtained in the fetal and neonatal period, and also in some proven CDG individuals; and false-positive results have been obtained in secondary glycosylation defects (galactosemia, fructosemia, alcohol abuse, hemolytic-uremic syndrome), and in transferrin protein polymorphisms.[23] Some defects appear to require tissue-specific tests for detection.

Methods used for further CDG subtyping include lipid-linked oligosaccharide or protein-linked glycan analysis using metabolic labeling and various mass-spectrometric techniques. Confirmation of the diagnosis made is by enzymatic analysis in fibroblasts or leukocytes and/or mutation analysis. CDG classification is essential for genetic counseling and mutation detection is required for prenatal testing.[1,23,24] Based on overlapping features with known CDG forms and abnormal screening test results, a number of patients have been tentatively diagnosed with as yet unidentified glycosylation defects.[25,26]

CDG generally presents in early infancy, can involve any organ system, and can extensively vary in severity both between patients of the same and of different subgroups; and CDG most often progresses over time. Accordingly, CDG should be considered in every patient with an unexplained syndrome or with a suspected but unclassified metabolic disease. The presence of psychomotor delay, an abnormal fat distribution, and coagulation abnormalities in one patient might be considered a characteristic pattern of CDG. Additional common symptoms of CDG patients include eye movement abnormalities, retinitis pigmentosa, hormonal dysregulation, dysmorphic features and diarrhea.[25,26]

CDG Ib represents a notable exception within all CDG, presenting with treatable protein-losing enteropathy in the absence of mental retardation, and oral mannose supplementation representing an efficient treatment (Table I).[27]

B. *O*-Glycosylation Defects

There are CDG that affect only one or a few organ systems,[26] for example, the congenital muscle dystrophies (CMD) in association with migration disorders of the brain.[28,29] The latter are caused by mutations in genes that are involved in the synthesis of only a specific protein and/or a specific class of glycans: the six genes involved in the CMDs appear to act primarily on α-dystroglycan. Three of these genes have been shown to have glycosyltransferase activity, viz, the genes encoding protein *O*-mannosyltransferases 1 and 2 (POMT1, POMT2) and protein *O*-mannosyl β1,2-*N*-acetylglucosaminyltransferase I (POMGNTI). The function of the other three genes remains to be defined. Unlike the *N*-glycosylation defects, these O-linked disorders possess distinctive muscle, eye, and brain phenotypes.[30]

Another example of an *O*-glycosylation disorder is the Peters'-plus syndrome. This rare autosomal recessive disorder involves multiple organ system defects caused by mutations in the β1,3-glucosyltransferase enzyme that adds glucose to Fucα1-*O*-Ser/Thr to form a disaccharide specific to thrombospondin type 1 repeats (TSRs). This disaccharide is only found on thrombospondin 1 and a few other TSR-containing proteins. Isoelectric focusing of the purely *O*-glycosylated serum apolipoprotein CIII is the currently most widely used screening test for the detection of some *O*-mucin-type glycosylation disorders (Tables II–IV).[31,32]

TABLE I
DEFECTS IN PROTEIN *N*-GLYCOSYLATION

Defective gene	Defective protein	OMIM
PMM2 (*CDG-Ia*)	Phosphomannomutase 2	601785
MPI (*CDG-Ib*)	Phosphomannose isomerase	602579
ALG6 (*CDG-Ic*)	Dol-P-Glc:Man_9-$GlcNAc_2$-P-P-Dol glucosyltransferase (glucosyltransferase 1)	603147
NOT56L (*CDG-Id*)	Dol-P-Man:Man_5-$GlcNAc_2$-P-P-Dol mannosyltransferase (mannosyltransferase 6)	601110
ALG 12 (*CDG-Ig*)	Dol-P-Man:Man_7-$GlcNAc_2$-P-P-Dol mannosyltransferase (mannosyltransferase 8)	607143
ALG 8 (*CDG-Ih*)	Dol-P-Glc:Glc_1-Man_9-$GlcNAc_2$-P-P-Dol glucosyltransferase (glucosyltransferase 2)	608104
ALG2 (*CDG-Ii*)	GDP-Man:Man_1-$GlcNAc_2$-P-P-Dol mannosyltransferase (mannosyltransferase 2)	607906
DPAGT1 (*CDG-Ij*)	UDP-GlcNAc:Dol-P-GlcNAc-P transferase	608093
HMT1 (*CDG-Ik*)	GDP-Man:$GlcNAc_2$-P-P-Dol mannosyltransferase (mannosyltransferase 1)	608540
DIBD1 (*CDG-Il*)	Dol-P-Man:Man_6-and Man_8-$GlcNA_2$-P-P-Dol mannosyltransferase (mannosyltransferase 7–9)	608776
RFT1 (*CDG-In*)	Flippase of $Man_5GlcNAc_2$-PP-Dol	611633
MGAT2 (*CDG-IIa*)	*N*-acetylglucosaminyltransferase 2	602616
GLS1 (*CDG-IIb*)	Glucosidase 1	606056
TUSC3	Oligosaccharyltransferase subunit	601385

C. *N*-Acetylgalactosamine 4-*O*-Sulfotransferase 1 (D4ST1) Deficiency

DS or CS B is a GAG that is found in skin as well as in blood vessels, heart valves, tendons, and lungs. Roles for DS have been suggested in coagulation, cardiovascular disease, carcinogenesis, infection, wound repair, and fibrosis (Vascular Dermatan Sulfate and Heparin Cofactor II). DS accumulates abnormally in mucopolysaccharidoses types I, II, VI, VII, caused by the deficiency of distinct, DS-degrading, lysosomal enzyme activities.

GAGs are long chains of repeating disaccharide subunits that are covalently bound to serine residues on a variety of core proteins. An O-linked tetrasaccharide linkage sequence, *O*-xylose–galactose–galactose–glucuronic acid, is common to heparin, HS, CS, and DS. The addition of a β-*N*-acetylgalactosamine (GalNAc) to this linkage sequence by GalNAc transferase I [MIM 602273] distinguishes CS and DS from HS. DS is formed from CS by two further modification reactions, that is, enzymatic steps that are specific to DS

TABLE II
DEFECTS IN PROTEIN *O*-GLYCOSYLATION

Defective gene	Defective protein	OMIM
O-Xylosylglycan synthesis		
EXT1/EXT2 (multiple cartilaginous exotoses)	Glucuronyltransferase/ *N*-acetylglucosaminyltransferase	608177/ 608210
B4GALT7	β-1,4-Galactosyltransferase 7	604327
O-*N*-Acetylgalactosaminylglycan synthesis		
GALNT3 (familial tumoral calcinosis)	Polypeptide *N*-acetylgalactosaminyltransferase 3	601756
O-Xylosyl/*N*-acetylgalactosaminylglycan synthesis		
SLC35D1 (Schneckenbecken dysplasia)	Solute carrier family 35 (UDP-GlcA/UDP-*N*-acetylgalactosamine dual transporter), member D1	610804
O-Mannosylglycan synthesis		
POMT1/POMT2 (cong. muscular dystrophy spectrum)	Protein-*O*-mannosyltransferase 1/protein-*O*-mannosyltransferase 2	607423
POMGNT1 (cong. muscular dystrophy spectrum)	Protein-*O*-mannose β-1,2-*N*-acetylglucosaminyltransferase	606822
FKTN (cong. muscular dystrophy spectrum)	Fukutin	607440
FKRP (cong. muscular dystrophy spectrum)	Fukutin-related protein	606596
LARGE (cong. muscular dystrophy spectrum)	*N*-Acetylglucosaminyltransferase-like protein	603590
O-Fucosylglycan synthesis		
SCDO3 (spondylocostal dysostosis type 3)	*O*-Fucose-specific β-1,3-*N*-acetylglucosaminyltransferase	602576
B3GALTL (Peters plus syndrome)	*O*-Fucose-specific β-1,3-glucosyltransferase	610308

TABLE III
DEFECTS IN GLYCOSPHINGOLIPID AND GLYCOSYLPHOSPHATIDYLINOSITOL ANCHOR GLYCOSYLATION

Defective gene	Defective protein	OMIM
SIAT9 (Amish infantile epilepsy)	Lactosylceramide α-2,3 sialyltransferase (GM_3 synthase)	609056
PIGM (glycosylphosphatidylinositol deficiency)	Phosphatidylinositolglycan, class M	610273

TABLE IV
DEFECTS IN MULTIPLE GLYCOSYLATION AND OTHER PATHWAYS

Defective gene	Defective protein	OMIM
DPM1 (*CDG-Ie*)	GDP-Man:Dol-P-mannosyltransferase (Dol-P-Man synthase 1)	603503
MPDU1 (*CDG-If*)	Lec35 (Man-P-Dol utilization 1)	608799
B4GALT1 (*CDG-IId*)	β-1,4-Galactosyltransferase 1	607091
GNE (hereditary inclusion body myopathy)	UDP-GlcNAc epimerase/kinase	600737
SLC35A1 (*CDG-IIf*) (CMP-sialic acid transporter deficiency)	CMP-sialic acid transporter	605634
SLC35C1 (*CDG-IIc*) (GDP-fucose transporter deficiency)	GDP-fucose transporter	605881
Dolichol pathway		
DK1 (*CDG-Im*)	Dolichol kinase	610768
COG[a] complex		
COG_7 defect (*CDG-IIe*)	Component of conserved oligomeric Golgi complex 7	606978
COG_1 defect (*CDG-IIg*)	Component of conserved oligomeric Golgi complex 1	606973
COG_8 defect	Component of conserved oligomeric Golgi complex 8	606979
V-ATPase		
ATP6VOA2 defect (cutis laxa type II)	V0 subunit A2 of vesicular H(+)-ATPase	611716

[a]Conserved oligomeric Golgi.

biosynthesis: (1) Epimerization of the CS C-5 hydroxyl of glucuronic acid (GlcA) to iduronic acid (IdoA) by glucuronyl C5-epimerases.[33] Two enzymes, epimerase 1 (DS-epi1) and DS-epi2, have been cloned that convert GlcA to IdoA during chondroitin/DS biosynthesis. (2) The addition of a sulfate to the C-4 hydroxyl of GalNAc prevents further epimerization of the adjacent GlcA or IdoA. Four GalNAc-4-sulfotransferases have been described that differentiate between CS and DS; chondroitin 4-sulfotransferases 1–3 (CHST11 [MIM 610128], CHST12 [MIM 610129], CHST13 [MIM 610124]) and dermatan-4-*O*-sulfotransferase 1 (CHST14 (D4ST1) [MIM 608429]).[34,35] *In vitro* studies, as well as the determination of the molecular basis for ATCS (see below), revealed that D4ST1 is responsible for the 4-*O*-sulfation of GalNAc adjacent to IdoA. D4ST1 is a key enzyme this is indispensable during the formation of important functional domains in DS, and cannot be compensated for by other 4-*O*-sulfotransferases. CHST12 has also been reported to transfer a sulfate on to the C-4 hydroxyl of GalNAc of both chondroitin and dermatan[36]; however, it does not demonstrate a marked preference for chondroitin or dermatan and requires high concentrations of the acceptor to obtain significant levels of sulfate transfer. In each case, 3′-phosphoadenosine 5′-phosphosulfate acts as the sulfate donor and transfer occurs within the *trans* network of the Golgi. The proportion of IdoA and GlcA and the degree of sulfation in DS vary considerably between tissues and, in the case of cultured cells, are influenced by the level of free sulfate in the culture medium.[37]

DS can be further modified by the addition of a sulfate at the C-2 hydroxyl of the IdoA and to a lesser extent, GlcA, and to the C-4 or C-6 hydroxyls of terminal GalNAc[38] by other sulfotransferases.

Both CS and DS are highly complex linear polysaccharides that are ubiquitously found in the extracellular matrix and on the cell surface. Given the vast expression repertoire of the GAG assembly enzymes, GAGs may have a sulfation pattern, chain length, and fine structure, which are unique to each cell.[39] Cell surface GAGs turn over within 1/8 and 1/3 of a cell cycle[40], meaning that their quantity and fine structures are capable of rapidly changing in response to a variety of environmental factors. Several DS functions, such as binding to growth factors, appear to be mediated by domains composed of alternating IdoA and 4-*O*-sulfated *N*-acetylgalactosamine residues, named 4-*O*-sulfated IdoA blocks.[41,42] Interactions between DS-PGs and particular target molecules have been described for their core proteins as well as their DS chains. Known DS-PGs include decorin, biglycan, versican, and endocan.[43] The classification of core proteins that carry KS, CS, DS, or HS is not always strict because there are a considerable number of proteins that bear more than one kind of a GAG side chain. Mature PGs have been implicated in a wide range of biological processes, including cell migration, proliferation and survival, and modulation of growth factor signaling. DS-PGs are widely distributed in mammalian tissue and are found in blood vessel walls, skin, tendon, sclera, cartilage, and undifferentiated mesenchymal tissue.[43–45]

In contrast to the extensively studied roles of HS and CS in development and growth factor signaling,[2,6,12,46] the biological function of DS is less well understood.[41,47]

The human congenital malformation disorder, ATCS [MIM 601776], results from deficiency of D4ST1, encoded by *CHST14* on chromosome 15q15, and gives an insight into the physiological functions of DS. Using the positional candidate approach, four distinct homozygous mutations in CHST14 in patients with ATCS from four families were identified. ATCS was initially considered to represent a “new” autosomal, recessively inherited, clinical entity following the publication of four separate clinical case reports of patients with a similar pattern of dysmorphic features, congenital malformations, distal contractures, but with normal cognitive development.[17–19] Eleven patients with ATCS from four consanguineous families of Austrian (family 1), Turkish (families 2 and 3), and Japanese (family 4) descent were identified in the original study.[17–20] While families 1–3 were initially given a descriptive diagnosis of ATCS, the patients from family 4 were originally reported as representing "a new type of distal arthrogryposis."[20]

ATCS represents a generalized connective tissue disorder with normal cognitive development. The pattern of symptoms comprises a severely wasted build with dry and translucent skin, brachycephaly, and facial characteristics such as a broad and flat forehead, hypertelorism, downward slanting palpebral fissures, malar flatness, retrognathia, and prominent ears. The anterior fontanel is large at birth and closure is delayed until after 2 years of age. Congenital contractures of the thumbs improve spontaneously within weeks, whereas congenital clubfeet require surgical treatment. Marked arachnodactyly and tapering of the fingers, as well as hypermobility of the small joints of the hands, feet, and the shoulders are present. Delayed wound healing, ecchymoses and hematoma-formation were also recorded. Detailed analyses of blood coagulation in an 8-year-old patient from family 1 revealed a prolonged bleeding time (9 min, reference range: <7 min), indicating the presence of a coagulopathy in addition to a form of generalized connective tissue weakness. Mild osteopenia is apparent in childhood; however, bone densitometry of the lumbar spine was normal in the 8-year-old proband of the Austrian family, but revealed osteopenia and osteoporosis in two Turkish siblings at age 15 and 6 years, respectively. Minor degrees of cranial ventricular enlargement were present in all five patients examined. Bluish sclerae and intermittent exotropia were observed in four patients. An atrial septal heart defect was noted in two patients, with mild coarctation of the aorta in one patient. One patient was found to have a horseshoe kidney and two patients had bilateral hydronephrosis. All examined male patients were found to have undescended testes at birth. A cleft palate and cleft lip/palate were present in sibs from a Japanese family, respectively. The absence of the gastrocolic omentum was observed in association with a spontaneous volvulus of the small intestine in one Austrian ATCS patient, and a common mesentery was present in his deceased brother. This patient was born at 31 weeks of gestation with heart and kidney defects and died at birth from respiratory failure. Three affected children from two Turkish families died before 4 months of age and one child died at age 6 years, but the causes of death were not recorded.

ATCS thus represents a recognizable, generalized connective tissue disorder with normal cognitive development.[13,17–20] It resembles both the progeroid and vascular types of EDS [MIM 130070 and 130050] and the Loeys–Dietz syndrome (LDS [MIM 609192]) because of the wasted build and the translucent skin. The dynamic nature of distal contractures in the presence of striking hypermobility of distal joints is reminiscent of collagen VI-related muscle disorders (Bethlem myopathy [MIM 158810]) during childhood.[48] Characteristic facial features are shared in part with LDS patients[49] and autosomal recessive cutis laxa type II, a type II *N*-glycosylation defect (ARCL2 [MIM 219200]).[50,51] Congenital malformations of a wide spectrum occur variably in ATCS and include the unusual coincidence of a common mesentery and absence of the gastrocolic omentum in a sib-pair.

A genome-wide linkage scan, using medium-density SNP array-based genotyping in four affected and 11 unaffected individuals from ATCS families 1–3, which used an autosomal recessive, fully penetrant model was carried out, and identified a single candidate region of 3.76 Mb on chromosome 15q15 (multipoint LOD score of $Z = 5.91$ [$\theta = 0.0$]). Of the 73 genes identified in the critical region in the human reference sequence, 2 were sequenced and a nonsense, and distinct missense mutations in *CHST14* (c.145delG (p.V49X), c.638G>C (p.R213P), c.[404C>G; 410T>A] (p.[R135G; L137Q]), c.878A>G; (p.Y293C)) were identified in each of four ATCS families, demonstrating that ATCS is genetically homogeneous. The mutation, p.V49X, causes early protein truncation. Arg-213, which is known to mediate hydrogen bonding to 3′-phosphoadenosine 5′-phosphosulfate, and Tyr-293 are highly conserved among members of the HNK1-family as well as a number of other carbohydrate sulfotransferase families. Residues Arg-135 and Leu-137 are present in a 4-amino acid sequence that is specific to D4ST1 among the HNK-1 sulfotransferases and is conserved among the D4ST1 orthologues. In each case, the mutation segregated with disease status in the family and was not present in control samples. The single CHST14 exon encodes a type II membrane protein of 376 amino acids, belonging to the family of HNK-1 sulfotransferases, with a 43-amino acid cytoplasmic domain and a 316-amino acid luminal domain. A single 2.4-kb transcript is ubiquitously present, but is most highly expressed in connective tissues, heart, liver, pancreas, and placenta.

Light and electron microscopy have revealed normal structure and ultrastructure, resepcctively, of a skin biopsy in a patient homozygous for the D4ST1 mutation p.R213P.

The findings of a homozygous truncating mutation (p.V49X) and of an altered intracellular processing of D4ST1 missense mutations by immunoblotting using recombinant D4ST1 suggest that deficiency (loss-of-function) is the underlying mechanism of ATCS. The absence of missense mutants in the culture medium suggests that the intracellular processing of each of these forms differs from that of wild type and that they may not reach the Golgi but be degraded instead.

Reduced levels of D4ST1 activity are predicted to result in decreased DS biosynthesis and increased CS content in tissues, which normally express CHST14. This was shown to be the case by comparing DS biosynthesis by cultured fibroblasts from a patient homozygous for p.R213P with normal controls using capillary HPLC coupled MS analysis of the delta-UA-containing disaccharides from CS, DS, and both CS and DS. As predicted, IdoA–GalNAc-4-SO_4 was not detected whereas levels of GlcA-4-SO_4 as well as GlcA–GalNAc and IdoA–GalNAc were increased in the absence of IdoA–GalNAc sulfation by D4ST1 *in vivo*. As the conversion of GlcA to IdoA by C5-epimerase is a reversible reaction that favors the GlcA form,[38] the synthetic products seen in ATCS fibroblasts support the hypothesis that the addition of sulfate to the C4 hydroxyl of GalNAc by D4ST1 prevents "back" epimerization of IdoA to GlcA. As a result, in the absence of sulfate addition to IdoA–GalNAc by D4ST1, DS is replaced by CS.

It appears that CS cannot generally substitute for functions specially associated with DS as D4ST1 deficiency causes ATCS. However, it remains a possibility that some or all known D4ST1 mutants retain some residual enzyme activity or that D4ST1 activity is compensated for by C4ST activity, at least in part, in some tissues.

From the findings in ATCS, it emerges that DS has not only plays a role in connective tissue maintenance but also in human embryonic development, as evidenced by the wide and variable spectrum of congenital malformations observed in ATCS patients.

III. Dermatan Sulfate Functions in Development, Growth Factor Signaling, and Wound Repair

CS, DS, and CS/DS hybrid chains are biologically active like HS, and structurally are the most complex species of the GAG family, as a considerable number of unique oligosaccharide sequences have been revealed. They exist at the cell surface and in the extracellular matrix, attached to particular core proteins to form CS/DS PGs.[52,53] They are thought to function as regulators of growth factors, cytokines, chemokines, adhesion molecules, and lipoproteins through interactions with the ligands of these proteins via specific saccharide domains. Structural GAG alterations have been implicated in cancer and atherosclerosis.

The biosynthesis of CS/DS is a process requiring at least 22 enzymes that are directly involved in the assembly and modification of the chain, starting with the activated monosaccharides and the sulfate donor, 3′-phosphoadenosine 5′-phosphosulfate. Multienzyme complexes are thought to accomplish these coordinated and fast reactions. The final structure of CS/DS is subject to different levels of regulation *in vivo*; however, how this regulation is accomplished is still poorly understood despite all of the enzymes having been cloned and studied *in vitro*.[54]

The IdoA content of CS/DS and further modification with sulfate depend both on the level of GlcA epimerase activity and on the level of GalNAc C-4 sulfotransferase activity in specific tissues. The IdoA content of the same CS/DS PG can differ depending on the cell type and tissue of origin.[37,55] The content of IdoA in CS/DS is also dependent on the developmental stage.[7,8]

The name CS/DS denotes the hybrid GlcA–GalNAc–IdoA nature of the chain. The distribution of IdoAs within the chain is not random, but they are either clustered together, forming long IdoA blocks, or they are interspersed among surrounding GlcAs.[56] DS-epi1 and DS-epi2, which are encoded by the genes Dse and Dsel (Dse-like) in the mouse, respectively, are present in organisms ranging from *Xenopus tropicalis* to humans, but are not found in worms and flies.[57]

GlcA to IdoA epimerization is closely followed by C-4 sulfation of the adjacent GalNAc, mainly or exclusively by D4ST1 in human skin and supposedly in other connective tissues.[58] DS biosynthesis can be followed by the action of up to seven further C-specific *O*-sulfotranferases, which transfer sulfate groups to C-2 of both IdoA and GlcA and to C-6 (or C-4/C-6) of GalNAc.[38] These modification reactions can produce structural variability in the CS/DS chain by affecting only specific regions of the available substrates. Specific sequences in the CS/DS chains that are responsible for binding to proteins and mediating biological effects have been identified.[59–61] For instance, (IdoA-2OS-GalNAc-4OS)3- and GalNAc-4/6-diOS-containing structures bind and activate heparin cofactor II, which is the major antithrombotic system in the subendothelial layer.[62] DS also shows a localized profibrinolytic activity.[63] D4ST1 deficiency would thus be predicted to cause both thrombophilia and delayed fibrinolysis resulting in the observed tendency to ecchymoses and hematoma-formation in ATCS patients. The IdoA/GlcA-2OS-GalNAc-6OS-containing structures bind to pleiotrophin, mediating their neuritogenic activity.[7,64]

IdoAGalNAc-4OS-containing structures bind to the basic fibroblast growth factor, and the complex has been shown to be active in wound healing.[41] Very recently, it has been shown that the development of the primary axis, mesoderm induction, and neuronal differentiation, thought to be related to FGF signaling in the *Xenopus* embryo, is dependent on endogenous DS.[65] The wide and variable spectrum of congenital malformations in ATCS patients may result from disturbed FGF signaling due to the lack of IdoAs within the CS chains. ATCS thus highlights a role for DS not only in connective tissue maintenance, but also in embryonic development.

CS/DS PGs are mainly found in the extracellular matrix, and classified into four families: lecticans (versican, aggrecan, brevican, and neurocan), collagens (e.g., collagen IX), basement membrane PGs (e.g., SMC3, collagen XV, and perlecan, containing both HS and CS/DS), and small leucine-rich repeat PGs.

The archetypical small leucine-rich repeat PG family members, decorin, biglycan, fibromodulin, and lumican, bind fibrillar collagens and affect collagen fibril and scaffold formation in connective tissues.[66,67] Decorin and biglycan are substituted with one and two CS/DS chains, respectively.

Decorin is involved in collagen type I fibril formation and matrix assembly in a wide range of connective tissues and binds near the C terminus of collagen monomers, thus delaying their accretion to the growing fibrils.[68] The role of the decorin CS/DS chain *in vivo* has not been explored, although *in vitro* studies suggest that IdoA promotes the binding of CS/DS to collagen[69] and that it is required for self-association of the CS/DS chains.[70,71]

An animal model for D4ST1 deficiency has not yet been reported. However, knock-out mice for the major epimerase in skin and connective tissue, DS-epi1-deficient mice, were smaller than their wild-type littermates, and their skin collagen architecture was altered; they have a larger diameter than the wild-type fibrils. A closer examination of DS-epi1 deficient animals may reveal that many other organs are also affected. Altered CS/DS chains were found on DS-PG decorin in skin, and were considered to affect collagen fibril formation and reduce the tensile strength of the DS-epi1-null skin. In Dse−/− mutant mouse skin, only 10% of the IdoA blocks remained in decorin/biglycan. On the other hand, isolated IdoA residues, that is, surrounded by GlcA, were not affected in skin decorin/biglycan. In skin versican and whole-body CS/DS, both isolated IdoA and IdoA blocks were downregulated.[54] The length of the CS/DS chains did not appear to be affected, in contrast to the effect of CS4ST1 ablation in mice.[72] In adult animals, DS-epi1 is the enzyme responsible for the majority of the epimerase activity in the spleen, kidneys, lungs, and skin. The remaining activity in these tissues can most likely be attributed to DS-epi2. In contrast to the abnormal fibril distribution in the skin of DS-epi1 deficient mice, ATCS patients display a normal fibril distribution on electron microscopy in their thin-appearing skin, even though a large amount of DS is substituted by CS.

Both the human D4ST1 and the mouse DS-epi1 deficiency contrast with the lack of IdoA in HS and the resulting perinatal mortality of HS epimerase-deficient mice, which lack kidneys and have altered lungs.[73] This points to different roles for IdoA in HS and CS/DS during embryo development; alternatively, the difference might be due to the fact that only one epimerase is present in HS biosynthesis whereas D4ST1 and DS-epi1 deficiencies might be partly compensated for by isoforms. The skin in DS-epi1 deficiency resembles decorin deficiency in mice; both have overlapping but nonidentical alterations in skin collagen fibril structure.[74] Decorin is an abundant CS/DS PG in skin, binds to collagen, and affects collagen fibril formation *in vitro*.[10] D4ST1 and DS-epi1 deficiencies have effects that are different from those seen with decreased galactosyltransferase I levels responsible for a variant of the EDS; this leads to an altered amount of decorin, both as a PG and as a nonglycanated core protein.[14] It has been suggested that the IdoA blocks carried by the decorin chain have a role in determining the structure and mechanical properties of collagen.[54,75] Although understanding the role of the CD/DS chains is complicated by their multiple potential interactions, the study of ATCS patients and of mouse models for DS-epi1 and Decorin deficiencies all highlight a role for decorin/biglycan IdoA blocks in the biophysical characteristics and mechanical properties of skin.

DS-PGs are involved in the generation of morphogen gradients in epithelia,[47] in particular, the DS-PG decorin has been shown to neutralize the activity of TGFB1.[76] A deficiency in or the substitution of the DS chains by CS on decorin implicates altered TGFB1 signaling in the pathogenesis of ATCS, as it was shown in other human connective tissue diseases such as the Loeys–Dietz and Marfan syndromes.[49] Altered TGFB1 signaling might lead to features shared by LDS and ATCS.[77] Intriguingly, a homozygous disruption of the C4st1 locus in mice led to a severe chondrodysplasia, and affected the balance of TGFB family signaling in the cartilage growth plate.[72]

IV. Conclusion Remarks

CDG cause a wide range of human diseases, in them, ATCS represents the first human disorder that emphasizes the roles D4ST1 play in human development and extracellular matrix maintenance. The impact of the loss of DS, due to D4ST1deficiencies in ACTS, will lead to a thorough understanding of the roles of DS and its interaction partners during development, growth factor signaling, and wound repair at the molecular level.

References

1. Freeze HH. Genetic defects in the human glycome. *Nat Rev Genet* 2006;**7**:537–51.
2. Bulow HE, Hobert O. The molecular diversity of glycosaminoglycans shapes animal development. *Annu Rev Cell Dev Biol* 2006;**22**:375–407.
3. Schachter H, Freeze HH. Glycosylation diseases: quo vadis? *Biochim Biophys Acta* 2009;**1792**:925–30.
4. Jaeken J, Hennet T, Freeze HH, Matthijs G. On the nomenclature of congenital disorders of glycosylation (CDG). *J Inherit Metab Dis* 2008;**31**:669–72.
5. Toole BP. Hyaluronan: from extracellular glue to pericellular cue. *Nat Rev Cancer* 2004;**4**:528–39.
6. Bishop JR, Schuksz M, Esko JD. Heparan sulphate proteoglycans fine-tune mammalian physiology. *Nature* 2007;**446**:1030–7.
7. Sugahara K, Mikami T. Chondroitin/dermatan sulfate in the central nervous system. *Curr Opin Struct Biol* 2007;**17**:536–45.
8. Rauch U, Kappler J. Chondroitin/dermatan sulfates in the central nervous system: their structures and functions in health and disease. *Adv Pharmacol* 2006;**53**:337–56.
9. Handley CJ, Samiric T, Ilic MZ. Structure, metabolism, and tissue roles of chondroitin sulfate proteoglycans. *Adv Pharmacol* 2006;**53**:219–32.
10. Prabhakar V, Sasisekharan R. The biosynthesis and catabolism of galactosaminoglycans. *Adv Pharmacol* 2006;**53**:69–115.
11. Young MF, Bi Y, Ameye L, Chen XD. Biglycan knockout mice: new models for musculoskeletal diseases. *Glycoconj J* 2002;**19**:257–62.
12. Hacker U, Nybakken K, Perrimon N. Heparan sulphate proteoglycans: the sweet side of development. *Nat Rev Mol Cell Biol* 2005;**6**:530–41.

13. Dundar M, Muller T, Zhang Q, Pan J, Steinmann B, Vodopiutz J, et al. Loss of dermatan-4-sulfotransferase 1 function results in adducted thumb-clubfoot syndrome. *Am J Hum Genet* 2009;**85**:873–82.
14. Seidler DG, Faiyaz-Ul-Haque M, Hansen U, Yip GW, Zaidi SH, Teebi AS, et al. Defective glycosylation of decorin and biglycan, altered collagen structure, and abnormal phenotype of the skin fibroblasts of an Ehlers–Danlos syndrome patient carrying the novel Arg270Cys substitution in galactosyltransferase I (beta4GalT-7). *J Mol Med* 2006;**84**:583–94.
15. Quentin E, Gladen A, Roden L, Kresse H. A genetic defect in the biosynthesis of dermatan sulfate proteoglycan: galactosyltransferase I deficiency in fibroblasts from a patient with a progeroid syndrome. *Proc Natl Acad Sci USA* 1990;**87**:1342–6.
16. Izumikawa T, Kanagawa N, Watamoto Y, Okada M, Saeki M, Sakano M, et al. Impairment of embryonic cell division and glycosaminoglycan biosynthesis in glucuronyltransferase-I-deficient mice. *J Biol Chem* 2010;**285**:12190–6.
17. Dundar M, Demiryilmaz F, Demiryilmaz I, Kumandas S, Erkilic K, Kendirci M, et al. An autosomal recessive adducted thumb-club foot syndrome observed in Turkish cousins. *Clin Genet* 1997;**51**:61–4.
18. Dundar M, Kurtoglu S, Elmas B, Demiryilmaz F, Candemir Z, Ozkul Y, et al. A case with adducted thumb and club foot syndrome. *Clin Dysmorphol* 2001;**10**:291–3.
19. Janecke AR, Unsinn K, Kreczy A, Baldissera I, Gassner I, Neu N, et al. Adducted thumb-club foot syndrome in sibs of a consanguineous Austrian family. *J Med Genet* 2001;**38**:265–9.
20. Sonoda T, Kouno K. Two brothers with distal arthrogryposis, peculiar facial appearance, cleft palate, short stature, hydronephrosis, retentio testis, and normal intelligence: a new type of distal arthrogryposis? *Am J Med Genet* 2000;**91**:280–5.
21. Freeze HH, Aebi M. Altered glycan structures: the molecular basis of congenital disorders of glycosylation. *Curr Opin Struct Biol* 2005;**15**:490–8.
22. Freeze HH. Congenital disorders of glycosylation: CDG-I, CDG-II, and beyond. *Curr Mol Med* 2007;**7**:389–96.
23. Marklova E, Albahri Z. Screening and diagnosis of congenital disorders of glycosylation. *Clin Chim Acta* 2007;**385**:6–20.
24. Lowe JB, Marth JD. A genetic approach to mammalian glycan function. *Annu Rev Biochem* 2003;**72**:643–91.
25. Grunewald S. Congenital disorders of glycosylation: rapidly enlarging group of (neuro)metabolic disorders. *Early Hum Dev* 2007;**83**:825–30.
26. Jaeken J, Matthijs G. Congenital disorders of glycosylation: a rapidly expanding disease family. *Annu Rev Genomics Hum Genet* 2007;**8**:261–78.
27. Harms HK, Zimmer KP, Kurnik K, Bertele-Harms RM, Weidinger S, Reiter K. Oral mannose therapy persistently corrects the severe clinical symptoms and biochemical abnormalities of phosphomannose isomerase deficiency. *Acta Paediatr* 2002;**91**:1065–72.
28. Godfrey C, Clement E, Mein R, Brockington M, Smith J, Talim B, et al. Refining genotype phenotype correlations in muscular dystrophies with defective glycosylation of dystroglycan. *Brain* 2007;**130**:2725–35.
29. Schachter H, Vajsar J, Zhang W. The role of defective glycosylation in congenital muscular dystrophy. *Glycoconj J* 2004;**20**:291–300.
30. Hewitt JE. Abnormal glycosylation of dystroglycan in human genetic disease. *Biochim Biophys Acta* 2009;**1792**:853–61.
31. Wopereis S, Grunewald S, Morava E, Penzien JM, Briones P, Garcia-Silva MT, et al. Apolipoprotein C-III isofocusing in the diagnosis of genetic defects in *O*-glycan biosynthesis. *Clin Chem* 2003;**49**:1839–45.
32. Wopereis S, Grunewald S, Huijben KM, Morava E, Mollicone R, van Engelen BG, et al. Transferrin and apolipoprotein C-III isofocusing are complementary in the diagnosis of *N*- and *O*-glycan biosynthesis defects. *Clin Chem* 2007;**53**:180–7.

33. Pacheco B, Malmstrom A, Maccarana M. Two dermatan sulfate epimerases form iduronic acid domains in dermatan sulfate. *J Biol Chem* 2009;**284**:9788–95.
34. Evers MR, Xia G, Kang HG, Schachner M, Baenziger JU. Molecular cloning and characterization of a dermatan-specific *N*-acetylgalactosamine 4-*O*-sulfotransferase. *J Biol Chem* 2001;**276**:36344–53.
35. Mikami T, Mizumoto S, Kago N, Kitagawa H, Sugahara K. Specificities of three distinct human chondroitin/dermatan *N*-acetylgalactosamine 4-*O*-sulfotransferases demonstrated using partially desulfated dermatan sulfate as an acceptor: implication of differential roles in dermatan sulfate biosynthesis. *J Biol Chem* 2003;**278**:36115–27.
36. Hiraoka N, Nakagawa H, Ong E, Akama TO, Fukuda MN, Fukuda M. Molecular cloning and expression of two distinct human chondroitin 4-*O*-sulfotransferases that belong to the HNK-1 sulfotransferase gene family. *J Biol Chem* 2000;**275**:20188–96.
37. Cheng F, Heinegard D, Malmstrom A, Schmidtchen A, Yoshida K, Fransson LA. Patterns of uronosyl epimerization and 4-/6-*O*-sulphation in chondroitin/dermatan sulphate from decorin and biglycan of various bovine tissues. *Glycobiology* 1994;**4**:685–96.
38. Kusche-Gullberg M, Kjellen L. Sulfotransferases in glycosaminoglycan biosynthesis. *Curr Opin Struct Biol* 2003;**13**:605–11.
39. Esko JD, Lindahl U. Molecular diversity of heparan sulfate. *J Clin Invest* 2001;**108**:169–73.
40. Yanagishita M, Hascall VC. Cell surface heparan sulfate proteoglycans. *J Biol Chem* 1992;**267**:9451–4.
41. Taylor KR, Rudisill JA, Gallo RL. Structural and sequence motifs in dermatan sulfate for promoting fibroblast growth factor-2 (FGF-2) and FGF-7 activity. *J Biol Chem* 2005;**280**:5300–6.
42. Mitsunaga C, Mikami T, Mizumoto S, Fukuda J, Sugahara K. Chondroitin sulfate/dermatan sulfate hybrid chains in the development of cerebellum. Spatiotemporal regulation of the expression of critical disulfated disaccharides by specific sulfotransferases. *J Biol Chem* 2006;**281**:18942–52.
43. Trowbridge JM, Gallo RL. Dermatan sulfate: new functions from an old glycosaminoglycan. *Glycobiology* 2002;**12**:117R–25R.
44. Rosenberg LC, Choi HU, Tang LH, Johnson TL, Pal S, Webber C, et al. Isolation of dermatan sulfate proteoglycans from mature bovine articular cartilages. *J Biol Chem* 1985;**260**:6304–13.
45. Sugahara K, Mikami T, Uyama T, Mizuguchi S, Nomura K, Kitagawa H. Recent advances in the structural biology of chondroitin sulfate and dermatan sulfate. *Curr Opin Struct Biol* 2003;**13**:612–20.
46. Garcia-Garcia MJ, Anderson KV. Essential role of glycosaminoglycans in Fgf signaling during mouse gastrulation. *Cell* 2003;**114**:727–37.
47. Li F, Shetty AK, Sugahara K. Neuritogenic activity of chondroitin/dermatan sulfate hybrid chains of embryonic pig brain and their mimicry from shark liver. Involvement of the pleiotrophin and hepatocyte growth factor signaling pathways. *J Biol Chem* 2007;**282**:2956–66.
48. Lampe AK, Bushby KM. Collagen VI related muscle disorders. *J Med Genet* 2005;**42**:673–85.
49. Loeys BL, Chen J, Neptune ER, Judge DP, Podowski M, Holm T, et al. A syndrome of altered cardiovascular, craniofacial, neurocognitive and skeletal development caused by mutations in TGFBR1 or TGFBR2. *Nat Genet* 2005;**37**:275–81.
50. Morava E, Wopereis S, Coucke P, Gillessen-Kaesbach G, Voit T, Smeitink J, et al. Defective protein glycosylation in patients with cutis laxa syndrome. *Eur J Hum Genet* 2005;**13**:414–21.
51. Kornak U, Reynders E, Dimopoulou A, van Reeuwijk J, Fischer B, Rajab A, et al. Impaired glycosylation and cutis laxa caused by mutations in the vesicular H+-ATPase subunit ATP6V0A2. *Nat Genet* 2008;**40**:32–4.
52. Seidler DG, Breuer E, Grande-Allen KJ, Hascall VC, Kresse H. Core protein dependence of epimerization of glucuronosyl residues in galactosaminoglycans. *J Biol Chem* 2002;**277**:42409–16.

53. Tiedemann K, Larsson T, Heinegard D, Malmstrom A. The glucuronyl C5-epimerase activity is the limiting factor in the dermatan sulfate biosynthesis. *Arch Biochem Biophys* 2001;**391**:65–71.
54. Maccarana M, Kalamajski S, Kongsgaard M, Magnusson SP, Oldberg A, Malmstrom A. Dermatan sulfate epimerase 1-deficient mice have reduced content and changed distribution of iduronic acids in dermatan sulfate and an altered collagen structure in skin. *Mol Cell Biol* 2009;**29**:5517–28.
55. Choi HU, Johnson TL, Pal S, Tang LH, Rosenberg L, Neame PJ. Characterization of the dermatan sulfate proteoglycans, DS-PGI and DS-PGII, from bovine articular cartilage and skin isolated by octyl-sepharose chromatography. *J Biol Chem* 1989;**264**:2876–84.
56. Fransson LA, Roden L. Structure of dermatan sulfate. II. Characterization of products obtained by hyaluronidase digestion of dermatan sulfate. *J Biol Chem* 1967;**242**:4170–5.
57. Maccarana M, Olander B, Malmstrom J, Tiedemann K, Aebersold R, Lindahl U, et al. Biosynthesis of dermatan sulfate: chondroitin-glucuronate C5-epimerase is identical to SART2. *J Biol Chem* 2006;**281**:11560–8.
58. Pacheco B, Maccarana M, Malmstrom A. Dermatan 4-*O*-sulfotransferase 1 is pivotal in the formation of iduronic acid blocks in dermatan sulfate. *Glycobiology* 2009;**19**:1197–203.
59. Nandini CD, Sugahara K. Role of the sulfation pattern of chondroitin sulfate in its biological activities and in the binding of growth factors. *Adv Pharmacol* 2006;**53**:253–79.
60. Scott JE, Thomlinson AM. The structure of interfibrillar proteoglycan bridges (shape modules') in extracellular matrix of fibrous connective tissues and their stability in various chemical environments. *J Anat* 1998;**192**(Pt 3):391–405.
61. Nomura Y. Structural change in decorin with skin aging. *Connect Tissue Res* 2006;**47**:249–55.
62. Tollefsen DM. Heparin cofactor II modulates the response to vascular injury. *Arterioscler Thromb Vasc Biol* 2007;**27**:454–60.
63. Castanon MM, Gamba C, Kordich LC. Insight into the profibrinolytic activity of dermatan sulfate: effects on the activation of plasminogen mediated by tissue and urinary plasminogen activators. *Thromb Res* 2007;**120**:745–52.
64. Bao X, Muramatsu T, Sugahara K. Demonstration of the pleiotrophin-binding oligosaccharide sequences isolated from chondroitin sulfate/dermatan sulfate hybrid chains of embryonic pig brains. *J Biol Chem* 2005;**280**:35318–28.
65. Hou S, Maccarana M, Min TH, Strate I, Pera EM. The secreted serine protease xHtrA1 stimulates long-range FGF signaling in the early Xenopus embryo. *Dev Cell* 2007;**13**:226–41.
66. Iozzo RV. The biology of the small leucine-rich proteoglycans. Functional network of interactive proteins. *J Biol Chem* 1999;**274**:18843–6.
67. Nareyeck G, Seidler DG, Troyer D, Rauterberg J, Kresse H, Schonherr E. Differential interactions of decorin and decorin mutants with type I and type VI collagens. *Eur J Biochem* 2004;**271**:3389–98.
68. Kalamajski S, Aspberg A, Oldberg A. The decorin sequence SYIRIADTNIT binds collagen type I. *J Biol Chem* 2007;**282**:16062–7.
69. Obrink B. A study of the interactions between monomeric tropocollagen and glycosaminoglycans. *Eur J Biochem* 1973;**33**:387–400.
70. Coster L, Fransson LA, Sheehan J, Nieduszynski IA, Phelps CF. Self-association of dermatan sulphate proteoglycans from bovine sclera. *Biochem J* 1981;**197**:483–90.
71. Fransson LA, Coster L, Malmstrom A, Sheehan JK. Self-association of scleral proteodermatan sulfate. Evidence for interaction via the dermatan sulfate side chains. *J Biol Chem* 1982;**257**:6333–8.
72. Kluppel M, Wight TN, Chan C, Hinek A, Wrana JL. Maintenance of chondroitin sulfation balance by chondroitin-4-sulfotransferase 1 is required for chondrocyte development and growth factor signaling during cartilage morphogenesis. *Development* 2005;**132**:3989–4003.

73. Li JP, Gong F, Hagner-McWhirter A, Forsberg E, Abrink M, Kisilevsky R, et al. Targeted disruption of a murine glucuronyl C5-epimerase gene results in heparan sulfate lacking L-iduronic acid and in neonatal lethality. *J Biol Chem* 2003;**278**:28363–6.
74. Danielson KG, Baribault H, Holmes DF, Graham H, Kadler KE, Iozzo RV. Targeted disruption of decorin leads to abnormal collagen fibril morphology and skin fragility. *J Cell Biol* 1997;**136**:729–43.
75. Ruhland C, Schonherr E, Robenek H, Hansen U, Iozzo RV, Bruckner P, et al. The glycosaminoglycan chain of decorin plays an important role in collagen fibril formation at the early stages of fibrillogenesis. *FEBS J* 2007;**274**:4246–55.
76. Hocking AM, Shinomura T, McQuillan DJ. Leucine-rich repeat glycoproteins of the extracellular matrix. *Matrix Biol* 1998;**17**:1–19.
77. Tiedemann K, Olander B, Eklund E, Todorova L, Bengtsson M, Maccarana M, et al. Regulation of the chondroitin/dermatan fine structure by transforming growth factor-beta1 through effects on polymer-modifying enzymes. *Glycobiology* 2005;**15**:1277–85.

Heparan Sulfate Proteoglycans in Amyloidosis

Xiao Zhang[*] and Jin-Ping Li[†]

[*]*Department of Public Health and Caring Sciences, Molecular Geriatrics, University of Uppsala, Uppsala, Sweden*

[†]*Department of Medical Biochemistry and Microbiology, University of Uppsala, Uppsala, Sweden*

Amyloidosis is a generic term for a group of diseases characterized by deposits in different organ systems of insoluble materials composed mainly of distinct fibrillar proteins named amyloid. Besides amyloid, heparan sulfate proteoglycan (HSPG), is commonly found in most amyloid deposits, suggesting that HS/HSPG may be functionally involved in the pathogenesis of amyloidosis. HS or HSPG is found to interact with a number of amyloid proteins, displaying a promoting effect on amyloid fibrilization *in vitro*. In addition, HS is reported to be involved in processing amyloid precursor proteins and mediate amyloid toxicity. Although little is known about the *in vivo* mechanisms regarding the codeposition of HS with amyloid proteins in different amyloid diseases, experiments carried out in animal models, especially in transgenic mouse model where HS molecular structure is modified, support an active role for HS in amyloidogenesis. Further experimental evidence is required to

Progress in Molecular Biology
and Translational Science, Vol. 93
DOI: 10.1016/S1877-1173(10)93013-5

strengthen these *in vivo* findings at a molecular level. Animal models that express mutant forms of HS due to knockout of the enzymes involved in glycosaminoglycan (GAG) biosynthesis are expected to provide valuable tools for studying the implications of HS, as well as other GAGs, in amyloid disorders.

I. Introduction

Amyloidosis (Pl. amyloidoses) is a group of diseases characterized by deposits of abnormal proteins in different organ systems. The diseases are clinically categorized into systemic and localized amyloidosis; the amyloid deposits in the former disease forms are derived from plasma proteins and in the latter from proteins produced by cells near the deposition site.[1,2]

The "abnormal proteins" implicated in different amyloidoses are typically named with the term "amyloid" or "amyloid protein" or "amyloid peptide," and so was the term "amyloidosis" derived. These proteins share some common features. Molecularly, most of the amyloid proteins are derived from a specific precursor protein following certain posttranslational modifications. Biochemically, these proteins misfold and aggregate to form insoluble fibrils. Pathologically, these misfolded proteins deposit in organs preferably targeted by different amyloid diseases. Histologically, the amyloid deposits are stained by Congo red, an aniline dye, showing apple-green birefringence under polarized light. In addition to the main amyloid protein component, amyloid fibrils *in vivo* are codeposited with many other substances; heparan sulfate proteoglycan (HSPG) is one such component commonly found in all types of amyloid deposit.[3]

HSPGs are ubiquitous macromolecules associated with the cell surface and extracellular matrix (ECM) of mammalian tissues.[4,5] HSPG consists of a protein core to which several linear heparan sulfate (HS) carbohydrate chains are covalently attached. HS interacts with a multitude of proteins and plays important roles not only in development and homeostasis, but also in pathological processes.[5,6] HS directly interacts with amyloid proteins *in vitro* and accelerates fibrillization of the proteins. Alteration of the HS structure or inhibition of the HS synthesis in animal and cell models results in attenuation of amyloidosis or amyloid toxicity. Evidence suggests an active role for HS in amyloidosis; however, the exact functions of HS/HSPG in the pathogenesis of different amyloidosis remain obscure. This chapter attempts to discuss different aspects of HS/HSPG in relation to the pathophysiology of major amyloidoses.

II. Amyloidosis

A. Amyloid

The term "amyloid" typically denotes a protein in its insoluble fibrillar form present in amyloid deposits *in vivo*.[1] It was initially used to describe starchy constituents of plants, and the term was adopted to replace the terms "lardaceous" or "waxy" changes found in several pathological human conditions as these changes were considered to be composed of starch-like substances (Fig. 1). Later analyses revealed that it was proteins that form the major component of "amyloid," not sugars.[7]

In present-day, about 30 amyloid proteins have been identified and each associated with a specific disease condition. Most of the amyloid proteins (or peptides) are derived from corresponding precursor proteins.[1,8] A protein defined as amyloid must satisfy two basic criteria: (1) the protein is present in an insoluble fibrillar form that deposits in tissues, ultimately associated with a disease condition; and (2) the deposited proteins are stained by Congo red with concomitant apple-green birefringence under cross-polarized light (Fig. 2).

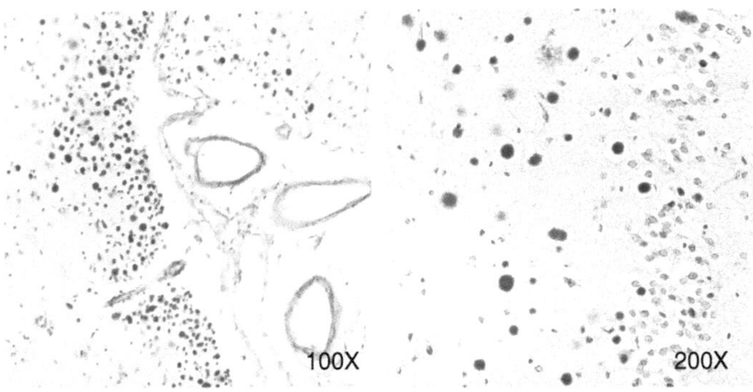

FIG. 1. Sulfated Alcian blue staining of cerebral corpora amylacea in human brain (courtesy of Paul O'Callaghan, Department of Public Health and Caring Sciences, Molecular Geriatrics, University of Uppsala, Uppsala Sweden). Corpora amylacea accumulate in the brain during normal aging, and can also be found in other organs, for example, heart, liver, prostate, lung, thyroid, and muscle.[167] In 1853, Rudolph Virchow demonstrated the reaction of cerebral corpora amylacea of the brain to the iodine test on the assumption that they were cellulose derivatives. He also applied the test to what appeared to be similar to corpuscles in a "waxy" spleen. These experiments convinced him that the components of cerebral corpora amylacea and waxy spleen could be identical to starch; he preferred "amyloid" to the then commonly used terms "lardaceous" or "waxy" changes.[7] A few years later, the presence of protein was demonstrated in waxy spleen; this finding switched attention to the protein moiety in amyloid research thereafter. The term "amyloid" has become the common name for the protein components in amyloid deposits (For review see Puchtler et al.[168]). (See Color Insert.)

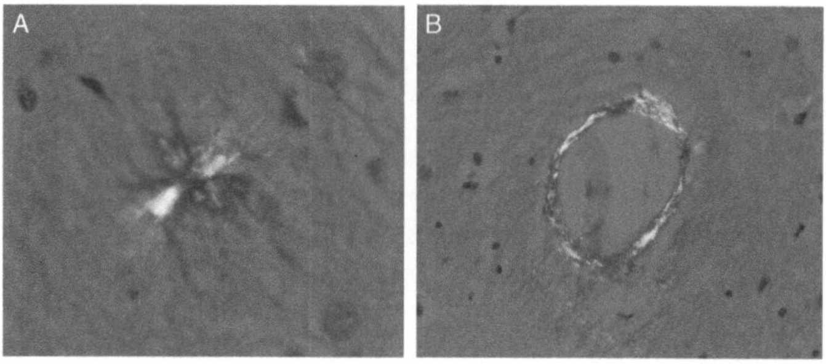

FIG. 2. Congo red staining of Aβ deposits in the brain of patient with Alzheimer's disease (courtesy of Paul O'Callaghan, Department of Public Health and Caring Sciences, Molecular Geriatrics, University of Uppsala, Uppsala Sweden). (A) A neuritic plaque stained with Congo red showing typical apple-green birefringence with a "Maltese cross" morphology as viewed under polarized light. (B) Apple-green birefringence of Congo red staining of a blood vessel with Aβ deposits. (See Color Insert.)

All forms of amyloid appear homogeneous when viewed under light microscope. With electromicroscopy, Cohen and Calkins demonstrated a nonbranching fibrillar structure of amyloid independent of their diverse origins.[9]

The fibrillar structure of amyloid is attributed to defective posttranslational conformation changes of an otherwise soluble protein, that is, "protein misfolding." Normally, following synthesis of linear polypeptides on the ribosome, the polypeptides undergo an energy-consuming process in the endoplasmic reticulum and become soluble proteins with a native conformation that is specific for its biological function. These proteins are mainly composed of α-helical and random coil structures. Protein misfolding leads to amyloid by converting proteins from their soluble forms into insoluble highly ordered fibrillar aggregates which are rigid and linear with ~7.5–10 nm in width.[10] The molecular property of amyloid is characterized by its secondary cross β-pleated sheet structure (β-strands that are perpendicular to the fiber axis), which was first discovered by Geddes et al.[11] The structure of the β-pleated sheet allows the binding to Congo red dye.

Fibril formation of amyloid proteins is often preceded by the formation of various intermediate oligomer states lacking fibrillar appearance, or protofibrils with β-sheet structure.[12] Although fibrillar amyloid deposits in tissues give rise to pathological conditions of different amyloidosis,[8,13,14] oligomers and protofibrils are believed, at least in some cases, to have pathological activities.

B. Amyloidosis

Amyloidosis is associated with or caused by amyloid deposits.[15] About 30 unrelated proteins have been found to form amyloid fibrils *in vivo*, and deposition of the different proteins shares a common pathognomonic structure in tissues,[2,15] although each of them is associated with a distinct pathological conditions. Clinically, the diseases can be classified into systemic and localized amyloidosis. Systemic amyloidosis is derived from plasma proteins and characterized by amyloid deposition in many organs including heart, blood vessels, lungs, skin, tongue, kidneys, adrenals, liver, intestine, lymph nodes, and spleen. Whereas in localized amyloidosis, the amyloid proteins are expressed by cells in proximity of the deposition site and the amyloid deposition is limited to a single organ, for example, the pancreas or the brain.[1,2,15–18]

C. Major Systemic Amyloidosis

Types of systemic amyloidosis are defined based on the species of amyloid precursor protein (APP). The nomenclature of amyloidosis uses the letter A to designate amyloid, followed by an abbreviation of the chemical identity of the fibril protein name.[15] Accordingly, the major types of systemic amyloidosis are amyloid A (AA) amyloidosis, immunoglobulin light chain (AL) amyloidosis, transthyretin-related (ATTR) amyloidosis and beta 2 microglobulin (Aβ2M) amyloidosis. There are several rare systemic amyloidoses caused by diverse amyloid proteins,[19] that will not be discussed in this chapter, because there is no information available about the involvement of HSPG or GAG in these rare amyloidoses.

1. AA Amyloidosis

The disease is associated mainly with many inflammatory and persistently infectious diseases such as rheumatoid arthritis, tuberculosis, and leprosy.[20] The protein fibrils of AA originate from fragments cleaved from the N-terminus of serum amyloid A protein (SAA), an apolipoprotein constituent of high-density lipoprotein.[21] SAA is an acute-phase reactant synthesized by hepatocytes regulated transcriptionally by cytokines.[22,23] During the acute-phase of inflammation, SAA plasma levels are dramatically increased, often by several thousand-folds.[24] In AA amyloidosis, the kidney, liver, and spleen are typically affected. AA amyloidosis presents with proteinuria, followed by progressive renal dysfunction often accompanied by nephrotic syndrome,[19,25,26] leading to nonreversible degeneration of the organ.

2. AL Amyloidosis

Systemic AL amyloidosis includes primary and myeloma-associated amyloidosis.[19] AL fibrils are derived from monoclonal immunoglobulin light chain (κ or λ),[27,28] and AL amyloidosis is the most common form of monoclonal B cell dyscrasia.[29] In AL amyloidosis, the circulating immunoglobulin light chains form aggregates that deposit in almost every organ with the exception of the brain. Nearly all organs involved are severely affected; the patients often die of renal, heart, and liver failure or autonomic neuropathy, with a median survival of only 6–15 months if not treated.[30,31] AL amyloidosis may occur, although rarely, as a localized, sometimes tumorous lesion without any systemic engagement.[15]

3. ATTR Amyloidosis

ATTR amyloidosis is the most common form of autosomal-dominant hereditary disease. TTR is a transport protein of thyroid hormone and retinol-binding protein. Until now, more than 100 variations in the TTR gene have been reported, of which at least 95 are linked to ATTR amyloidosis.[32–34] TTR is a tetramer of four identical subunits of a 127-amino acid (aa) peptide/protein and is almost entirely produced in the liver (also in choroid plexus). However, ATTR fibrils are not likely to be made up of the native tetrameric form,[35] and a transition occurs to dissociate the tetramer, which produces monomers that are modified into a misfolded conformation.[36] Major manifestations of hereditary ATTR amyloidosis are familial amyloid polyneuropathy (FAP) and familial amyloid cardiomyopathy (FAC). FAP is commonly caused by accumulation of ATTR around peripheral nerves, and FAC is characterized by deposition of ATTR in cardiac tissues.[37] Nonmutated TTR protein can also deposit as amyloid in the heart of elderly people, a condition termed sporadic senile systemic amyloidosis (SSA).[38,39] Notably, SSA affects almost 25% of the population > 80 years of age and is the most common systemic amyloidosis.[40]

4. Aβ2M Amyloidosis

Aβ2M amyloidosis is associated with long-term hemodialysis, a complication of end-stage renal disease. It is caused by deposition of amyloid fibrils of Aβ2M mainly in the periarticular tissue and bone.[41,42] The typical clinical manifestations are carpal tunnel syndrome, arthralgia, spondyloarthropathy, subchondral bone cysts, and fractures.[43] β2M, the light chain component of the major histocompatibility complex class I, is a low molecular weight normal serum protein.[44] Following repeated hemodialysis, serum levels of β2M can increase by five to ten times above normal level as β2M does not pass through the conventional dialysis membrane.[45,46] Using advanced high-flux dialyzer membranes in hemodialysis, β2M can be efficiently removed, and the occurrence of Aβ2M amyloidosis is becoming a vanishing complication.[47,48]

D. Major Localized Amyloidosis

Although amyloid deposits are most commonly found in extracellular tissues, increasing evidence from recent studies suggest that some misfolded proteins may initially occur intracellularly with typical amyloid structure. Accordingly, it is accepted that all deposits, regardless of an extra- or intracellular location, fulfilling the amyloid protein definition should be considered amyloid. The pathological hallmarks of Alzheimer's disease (AD) are extracellular deposits of amyloid-β (Aβ) and intracellular deposits of amyloid Tau (ATau) in the brain. In type 2 diabetes, islet amyloid polypeptide (AIAPP) is deposited in the islets of Langerhans of pancreas. Prion disease is another localized amyloidosis, which is caused by amyloid prion protein (APrP).

1. AIAPP Amyloidosis

Islet amyloid polypeptide (IAPP), also known as amylin, is a 37-aa peptide[49–51] derived from a 67 aa precursor protein (human pro-IAPP).[52] IAPP is produced by β cells in the islets of Langerhans in pancreas and stored in the secretory granules along with insulin. Deposits of IAPP in the pancreas are associated with two disease conditions: insulinomas, rare neuroendocrine tumors, and type 2 diabetes. In most patients with type 2 diabetes (90%), amyloid is deposited in the islets (predominantly on β-cells), constituting a pathogenic hallmark of the islet in type 2 diabetes.[53] It seems that the extent of the amyloid deposition correlates with the severity of the disease.[54,55]

2. Aβ Associated Diseases

AD is a major central nervous system disease characterized by a progressive neurodegeneration with a clinical phenotype of cognitive impairment. The histopathological hallmarks of AD are extracellular Aβ deposits in plaques[56,57] (Fig. 2), and aberrant phosphorylated tau deposits, in cell bodies and apical dendrites as neurofibrillary tangles (NFTs).[58] Aβ is a product of an enzyme-catalyzed processing of the amyloid-β precursor protein (AβPP), a transmembrane glycoprotein. Genetically, AD can be classified into sporadic and familial forms. About 90% of AD cases are sporadic in which a normal amino acid sequence of AβPP is present. In some familial AD cases, different mutants of AβPP are thought to be responsible for the pathogenesis of the disease.[57] Moreover, Down syndrome (DS) is also associated with Aβ and ATau deposiiton. DS is a congenital condition resulting from an extra copy of chromosome 21 on which the AβPP gene is located and all DS patients show cognitive impairment.[59,60]

3. APrP Amyloidosis

Prion diseases or transmissible spongiform encephalopathies (TSEs) are APrP amyloidosis. The agent that causes TSEs is a small proteinaceous infectious particle termed "prions" that are resistant to inactivation by most procedures.[61] The major amyloid in prion deposits is PrPSc, that, when comprising 45% β-sheet structure, is formed by conversion of normal cellular PrP (PrPC) through a process whereby a portion α-helical and coil structure of PrPC is refolded into β-sheet.[62] TSEs are fatal mammalian neurodegenerative disorders that encompass Creutzfeldt–Jakob disease (CJD), Gerstmann–Sträussler–Scheinker syndrome (GSS), and fatal familial insomnia in humans, as well as bovine spongiform encephalopathy (BSE) in cattle, scrapie in sheep, and chronic wasting disease in cervids.[63,64] The brains of affected animals or humans show characteristic spongiform degeneration, astrogliosis, and deposits of misfolded prion proteins.[62,65,66]

III. Heparan Sulfate Proteoglycan

HSPG is a family of glycoconjugates commonly composed of a core protein to which a number of HS are attached. The HS side chains, linear polysaccharides, are a member of the glycosaminoglycan (GAG) family. HS is synthesized in most cells through a process catalyzed by multiple enzymes (see previous chapters of this book). HS is heterogeneous in both molecular structure (tissue/cell specific) and size (30–60 kDa). The polysaccharide chains interact with a multitude of proteins, mainly through charge interactions, and thus modulate important processes in development and homeostasis, as well as in some pathological processes.[4–6] The HS chains are degraded primarily by an endoglycosidase, heparanase which cleaves HS to an average size of 3–6 kDa.[67] On the basis of their cellular distribution, HSPG can be classified into two major categories: (1) the plasma membrane associated syndecans and glypicans; (2) the soluble forms of perlecan, agrin, and collagen XVIII.[68]

There are four mammalian syndecans (syndecan 1–4). All isoforms of syndecan are type I transmembrane proteins.[69–71] Glypicans are located on the outer surface of the plasma membrane through binding of its core protein to a glycosylphosphatidylinositol (GPI) anchor. Six mammalian glypican isoforms (1–6) have been indentified.[72–74] Perlecan and agrin are the two major secreted HSPGs.[75,76] Perlecan is a ubiquitous component of the ECM and basement membrane (BM).[77,78] Agrin, recognized as a HSPG in a study on chick brain development,[79] is also a major component of BM along with perlecan and collagen XVIII,[80] but agrin also has been considered a neuronal proteoglycan.[76]

IV. Codeposition of HSPG and GAG with Amyloid

The bulk of amyloid deposits in tissues is the disease-specific component, that is, amyloid proteins or peptides. In addition, a set of common structural constitutes has been identified in most amyloid deposits no matter what kind of amyloidogenic protein is involved; these include serum amyloid P (SAP), apolipoprotein E (ApoE), GAGs (mainly HS), and proteoglycans (mainly HSPG).[3,81,82] Coexistence of GAGs with amyloid deposits has been known for almost a century.[83] The potential significance of the GAG components in amyloidogenesis was first suggested by the observation that accumulation of GAGs in an animal model of AA amyloidosis was coincident with AA amyloid deposits both spatially and temporally.[84] The pioneer studies by Kisilevsky and Snow have contributed significantly to establishing the correlation of GAGs (mainly HS) with amyloidosis.[84–88] The distinct investigations on different amyloidosis point to a common, yet to some extent specific, mechanism of GAGs, particularly HS, in amyloidosis. Nonetheless, a systematic study for the roles of GAGs in the development of amyloidogenesis has not been reported, and available fragmented information will be discussed in the following sections.

A. HSPG and GAG in Major Systemic Amyloidosis

Codeposits of HS/HSPG with amyloid have been detected in most of the systemic and localized amyloidosis diseases described above. GAGs (especially HS) and PGs (mainly HSPG) have been identified by histological and immunohistochemical methods in a number of affected organs with deposition of different amyloid peptides. An early study using immunohistochemistry with different antisera against core proteins of various proteoglycans revealed that HSPG was the only proteoglycan found specifically localized to the amyloid deposits in the kidney of human AA amyloidosis.[89] Whereas, analysis of AA fibrils purified from the liver of human AA type secondary amyloidosis identified different GAGs including HS, chondroitin sulfate, and dermatan sulfate.[90] A further study confirmed the presence of GAGs in purified AA and AL amyloid fibrils with a method that specifically precipitates amyloid fibrils, but not proteoglycans or free GAG chains.[91] HS and HSPG in amyloid fibrils of ATTR and Aβ2M have also been reported.[92–94]

A series of comparative studies of *in situ* ultrastructural characteristics of AA, AL, and ATTR amyloid fibrils found there were some differences among them with respect to their association with HSPG. Amyloid fibrils from experimental AA amyloidosis in mice induced by amyloid enhancing factor (AEF) and an inflammatory stimulus (subcutaneous injection of $AgNO_3$) were examined by high resolution ultrastructural analysis. The results revealed that the *in situ* AA fibrils resemble a structure similar to connective tissue microfibrils. The core of the AA fibril was wrapped in a thick layer of randomly assembled ~5 nm wide double tracks that were composed of BM type HSPG. On its surface, AA amyloid protein composed 1 nm-wide filaments.[95]

Another study using the same method found that the main body of ATTR amyloid fibrils in FAP had an ultra structure similar to that of fibrils in experimental AA amyloidosis and connective tissue microfibrils.[96] However, Aβ2M fibrils, in contrast to AA and ATTR fibrils, in hemodialysis-associated amyloidosis have curved conformations, although the ultrastructural and immunohistochemical features are in common with the core of connective tissue microfibrils and the amyloid fibrils of experimental AA amyloidosis and FAP (one type of ATTR amyloidosis). The external association of the amyloid protein Aβ2M to the core is similar as that seen in AA and FAP. One unusual feature of the fibrils of hemodialysis-associated amyloid Aβ2M, however, is the presence of a peripheral layer composed of chondroitin sulfate proteoglycan rather than HSPG as seen in AA and FAP amyloid structure.[97]

B. HSPG and GAG in Major Localized Amyloidosis

1. AIAPP Amyloidosis

Perlecan, a BM form of HSPG, has been identified as a component of AIAPP deposits in islets of type 2 diabetes patients using histochemical and immunohistochemical methods.[98] Accumulation of perlecan in islet amyloid deposits has also been demonstrated in transgenic mice that over express amylin and develop hyperglycemia.[99–101] However, reports on the correlation of GAGs with AIAPP deposition are rare. With the development of specific antibodies against different types of GAGs and proteoglycans, immunohistochemical examination of specimens from type 2 diabetes patients will assist to characterize the GAGs in AIAPP deposits in more detail.

2. APrP Amyloidosis

Examination of brain sections from human CJD, GSS, kuru and experimental scrapie of hamsters by histochemical staining with sulfated Alcian blue revealed the presence of sulfated GAGs as a part of prion deposits in these diseases.[85,102] It has also been demonstrated that both the protein core and the GAG chains of perlecan are immunolocalized to the prion protein amyloid plaques in each of these diseases.[103] In a scrapie mouse model which exhibits amyloid deposition of abnormal PrP protein, abnormally high levels of HSPG were associated with most types of PrP pathology.[104] Moreover, specific HS structures were identified in the codeposition of PrPSc of the earliest detectable brain lesions of scrapie-infected mice.[105]

3. Aβ Associated Amyloidosis

Association of HS and HSPG with the amyloid pathology of AD has been studied in more detail than that of other localized amyloidoses. Presence of GAGs in Aβ deposits was first identified using Congo red staining for Aβ and Alcian blue dye for sulfated GAGs in brain sections from AD patients.[87] The presence of perlecan in the plaques was later found by immunohistochemical staining of Aβ and HSPG.[106–108] HSPG has also been histochemically and immunohistochemically identified in NFT.[109,110] With development of more specific antibodies against different HSPG, subtypes of HSPG isoforms including syndecan 1–3, glypican 1, and agrin have been immunolocalized in the Aβ deposits in plaques and vasculatures, as well as in NFT of AD brains and transgenic animal models of Aβ amyloidosis.[111–117] Immunostaining has also demonstrated codeposition of HS with Aβ and tau in AD and DS brains.[118,119] Recent studies using advanced anti-HS antibodies which differentially recognize certain structures of the polysaccharide chains have shown pattern-selective colocalizations of HS with different Aβ deposits, for example, neuritic plaques with dense cores and diffuse plaques (Fig. 3).[112,120]

V. Potential Functions of HS/HSPG in Amyloidosis

The constitutive occurrence of HS/HSPG in amyloid deposits, irrespective of the specific amyloid protein identity, has raised the question: are HS/HSPG codeposits biologically inert or functionally involved in amyloid pathogenesis? Evidence obtained from both *in vitro* and *in vivo* experiments imply functional roles of HS/HSPG in amyloid pathogenesis. It has been proposed that HS may facilitate formation of the nidus and/or protofilament around which amyloid fibrillogenesis takes place and impart stability to the amyloid fibril *in vivo*.[87] This notion was supported by the marked increase in fibrillogenesis *in vitro* exerted by HS or its analogue heparin (a mammalian polysaccharide sharing similar structure with HS) on various amyloidogenic polypeptides including Aβ,[121,122] tau,[123,124] PrP,[125] and β2M.[126,127] HS has also been found to affect processing of APPs by acting on proteolytic enzymes and to alter cytotoxicity of amyloid proteins. So, HS and HSPG may be functionally engaged in amyloid deposition through common mechanisms, but they can also exert distinct functions in other aspects of amyloid pathology, for example, cell toxicity.

A. Interaction of HS/HSPG with Amyloid Proteins *In Vitro*

HS, largely owing to its negatively charged property, has been shown to interact with a multitude of molecules including amyloid proteins and amyloid-associated proteins, for example, ApoE. On the other side, some amyloidogenic proteins contain a sequence that appears to favor binding to HS or heparin.[81] Aiming to determine the structures of HS that are necessary for binding amyloid proteins, different approaches have been employed to investigate the interactions between HS/heparin and synthetic or authentic amyloid peptides. The data available point to a common mechanism that the binding of amyloid

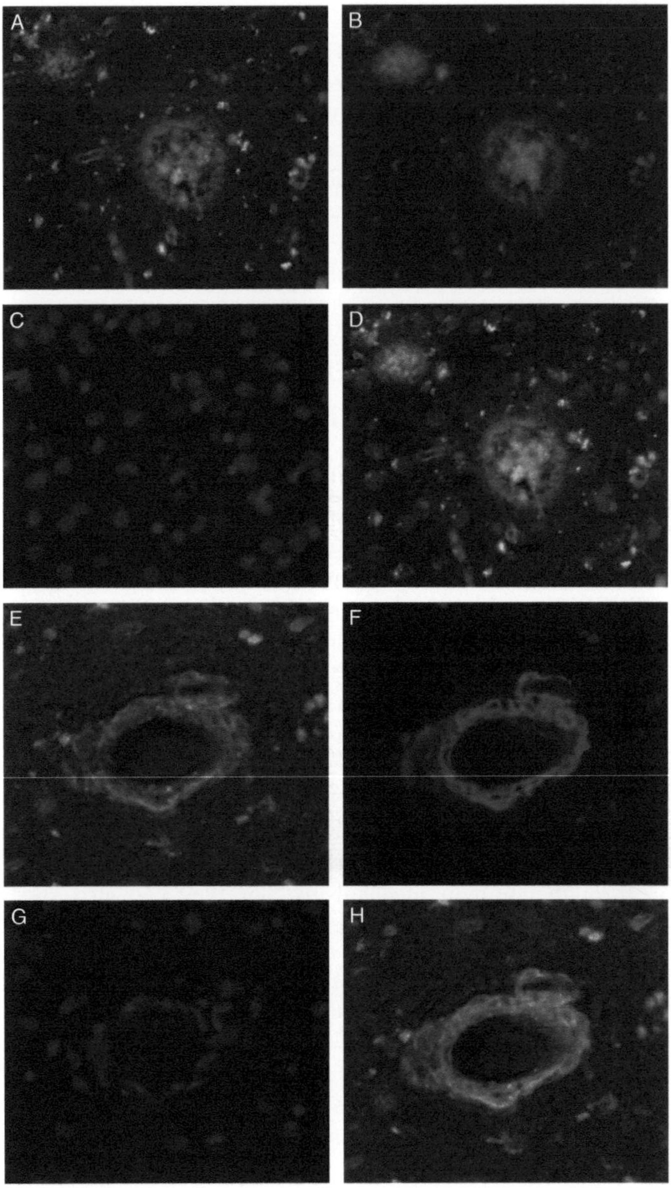

FIG. 3. Codeposition of amyloid-β (Aβ) and heparan sulfate (HS) in the brain of a patient with sporadic Alzheimer's disease. Double immnostaining was performed with an anti-HS antibody in combination with anti-Aβ42 antibody for plaques and Aβ40 antibody for blood vessels. Codeposition of HS with Aβ in neuritic plaques: (A) HS; (B) neuritic plaques; (C) DAPI counterstaining for nuclei; (D) merge of (A)–(C). Codeposition of HS with Aβ in blood vessel wall: (E) HS deposits; (F) Ab deposits; (G) DAPI counterstaining for nuclei; (H) merge of (E)–(G). (See Color Insert.)

proteins to HS is affected by the length of the chain and charge (sulfation degree) of the polysaccharide; however, the possibility that specific HS chain structures are important for interactions cannot be ruled out.

AβPP and Aβ were found to bind to both the core protein of glypican 1 and free HS chains.[128–130] Heparin actually binds AβPP with a higher affinity than HS,[131] likely due to its higher sulfation degree. Further studies showed that HS could initiate Aβ fibril nucleation and promote its growth,[121] suggesting a role for HS in Aβ aggregation. HS has also been demonstrated to induce the formation of paired helical filaments of tau protein *in vitro*.[123] Analysis of the Aβ sequence revealed a consensus heparin-binding domain at residues 12–17 (VHHQKL), and this motif contains two histidine residues that may be involved in the interaction with HS.[131–133] Efforts have been made to illustrate the fine structure of HS that interacts with Aβ by incubation of HS/heparin oligosaccharides with the amyloid peptide. The results showed that a hexasaccharide domain of HS identified in human cerebral cortex containing critical 2-*O*-sulfated iduronic acid residue binds to fibrillar Aβ.[130] Thus, experimental data reported by independent studies suggest that the interaction of HS with Aβ has a proamyloidogenic effect.

Synthetic human IAPP has been shown to bind to heparin and perlecan resulting in the acceleration of IAPP fibril formation *in vitro*.[132,134] A heparin-binding domain was first identified in the N-terminal cleavage site of pro-IAPP.[135] Further investigation found that Arg-22 and His-29 in the N-terminal fragment of pro-IAPP are critical for the interaction with HS, triggering amyloid formation of the N-terminal fragment.[136] As the N-terminal flanking regions of pro-IAPP and HSPG are found in islet amyloid deposits, they are assumed to play roles together in AIAPP amyloidogenesis.[98,137] Studies using cell culture models have shown that HSPG synthesized in β cells are capable of binding amyloidogenic human AIAPP.[138] Inhibition of synthesis of HSPG by WAS-406 or HS by Azaserine in explant culture of islets isolated from transgenic mouse overexpressing human IAPP was found to similarly reduce amyloid formation in the cultured islets.[139]

In contrast to the extensive study on the binding of Aβ and IAPP to HS, information about the interactions of HS with other amyloid peptides is limited. SAA has been shown to be retained in a heparin-sephorose column,[140] however, direct evidence for SAA–HS/heparin interactions has not been reported. A peptide corresponding to the proposed HS binding site of AA was demonstrated to block amyloid deposition in a cell AA amyloidosis model.[141] Moreover, analogues of *N*-acetylglucosamine, one of the sugar units occurring in HS and heparin, exhibit inhibitory effects on AA amyloidosis in the cell and the mouse models.[142] HS was isolated from the liver and spleen of patients with AA amyloidosis and healthy controls to investigate structural properties of the HS chains codeposited with SAA. The results showed that the disaccharide composition of HS from AA amyloidosis organs was different from that of healthy control organs, suggesting an association of a specifically modified HS structure with the amyloid deposits in the affected organs.[143]

Reports regarding to HS/HSPG in Aβ2M and ATTR amyloidosis are circumstantial; nonetheless, fibril formation of Aβ2M was found to involve GAGs.[144] This finding is strengthened by recent report that heparin had a strong enhancing effect on Aβ2M fibril formation in the presence of collagen.[127] As for ATTR amyloidosis, although an early study reported codeposition of ATTR with unspecified GAGs in extracts of cardiac amyloid fibrils from FAC patient, no substantial progress has been made up-to-date. Our ongoing study on the interactions of recombinant wild-type TTR with HS/heparin shows that the GAGs promote fibril formation of ATTR and this promoting effect is dependent on the length and total charge of the oligosaccharides.

Collectively, although further studies are required before reaching more conclusive findings, based on the fact that HS/heparin appears to interact with most of the amyloid peptides, thereby promoting fibrillogenesis, it is encouraging for a hypothesis that HS may function as a scaffold "collecting" misfolded amyloid peptides during amylodogenesis.

B. Regulatory Roles of HS in Processing of Amyloid Precursor Proteins

Most of the amyloid peptides are derived from their precursors; an example is Aβ, a proteolytic product of AβPP. AβPP is a type I transmembrane protein with one form of 695 aa predominantly expressed in neurons, and the forms of 751 and 770 aa expressed in other cells. AβPP is sequentially processed essentially by three enzymes namely α-, β-, and γ-secretase. Cleavage of AβPP by β- and γ-secretase produces amyloidogenic Aβ species, mainly Aβ1–40 and Aβ1–42.[145,146] β-Secretase, also called β-site APP cleaving enzyme 1(BACE1), is the rate-limiting factor for Aβ production.[147] BACE1 seems to be essential for Aβ pathogenesis in AβPP transgenic mice; partial reduction in BACE1 activity dramatically reduced Aβ burden.[148] HS has been reported to exert regulatory functions for the activity of BACE1. In a human neuroblastoma cell line, heparin was found to induce the synthesis of AβPP and stimulate AβPP processing by BACE1, resulting in an increased secretion of cleaved products.[149] Contrarily, HS and heparin have also been shown to interact with BACE1 and to inhibit processing of the AβPP in an *in vitro* system.[150] This inhibitory activity was dependent on the size and structure of HS that may function by masking the active site of the enzyme. The results of this study led to the conclusion that HS is a natural regulator of BACE1 activity for the control of AβPP processing. The inhibitory effect of heparin on BACE1 activity was further demonstrated by other studies.[151,152] A recent study, using cortical neurons isolated from an AβPP transgenic mouse (Tg2576), demonstrated the inhibitory effect of heparin on Aβ secretion, but concluded that the effect was not mediated via inhibition of BACE1.[153] Despite these apparently contradictory results, a common conclusion can be derived, that is, HS is involved in AβPP processing through either modulating the activity of BACE1 or mechanisms unrelated to the cleavage enzymes. Considering that BACE1 is currently a strategic target for treatment of AD, clarifying the mechanism by which HS modulates BACE1 activity may open a new dimension for AD treatment.

Incompletely processed proIAPP is proposed to be a factor in AIAPP deposition. Although there is no report about the involvement of HS in proIAPP processing, it was found that the incompletely processed N-terminal pro region of the peptide binds to HSPG of the BM, thereby acting as a seed for amyloid formation.[154] PrPC is associated with lipid rafts through GPI anchor and binds to HSPG. A recent study reported that glypican 1 acts as a scaffold facilitating the transversion of PrPC to PrPSc within cholesterol-rich lipid rafts; consequently, glypican 1 is regarded to be intimately involved in the misfolding of the prion protein.[155]

C. Effect of HS/HSPG on Amyloid Deposition *In Vivo*

In vivo effects of HSPG on amyloid deposition have been studied in a number of animal models. For instance, infusion of synthetic human Aβ1–40 peptide only or Aβ1–40 along with perlecan into the hippocampus in rats resulted in Aβ deposits with fibrillar content as demonstrated by Congo red staining in areas adjacent to the infusion site. A significant enhancement of amyloid deposition was observed in the rats receiving both Aβ and perlecan in comparison to rats receiving Aβ only as judged by Congo red staining. The morphology of the Congo red positive material resembled amyloid plaques seen in AD patients. It is proposed that perlecan stabilized the plaque-like Aβ deposits and protected the Aβ aggregate from proteolytic degradation of the Aβ aggregate.[156] However, the core protein of perlecan alone was not sufficient to lead to the AD-like pathology, suggesting the important role of HS in the complex proteoglycan molecule.[157]

To address the role of HSPG in the amyloid pathology of prion diseases, a scrapie mouse model has been used. The experimentally infected mice displayed both amyloid and nonamyloid deposition of abnormal PrP. HSPG was found in both amyloid plaques and early diffused forms of PrP. Involvement of HSPG with formation of early diffused PrP deposits suggests a fundamental role of HSPG in scrapie pathogenesis.[104]

Effects of low molecular weight heparin (LMWH) have been studied in animal models of AD and AA amyloidosis. Enoxaparin, an LMWH, was given to transgenic mice overexpressing human AβPP intraperitoneally, and the treatment significantly reduced the Aβ burden in the brains of the transgenic mice.[158] In another experiment, subcutaneous administration of enoxaparin and dalteparin (both are LMWH preparations) resulted in a reduction of AA amyloid deposition using a mouse model of AA amyloidosis.[159]

Solid *in vivo* evidence for functional roles of HS in amyloidogenesis has been obtained from experiments performed with a transgenic mouse model overexpressing human heparanase,[160] an endo-β-D-glucuronidase that catalyzes the hydrolytic cleavage of HS.[161] In the transgenic mice, heparanase was differentially overexpressed with the highest level in the liver resulting in substantially shortened HS chains. Experimental AA amyloidosis was induced in these transgenic mice. After induction, the spleen that expressed low level of heparanase, accordingly produced normal HS, developed amyloidosis, which was indistinguishable from the control spleen. In contrast, the liver, expressing high level of heparanase, of the transgenic mice became resistant to the systemic AA amyloidosis. Therefore, it is proposed that HS may function as a scaffold to facilitate amyloid deposition.[162]

However, the effect of HS or HSPG on amyloid pathogenesis *in vivo* might be disputable because complete association of HS/HSPG with amyloid is not always found. For example, in bovine renal AA amyloidosis, amyloid deposits were detected in the areas without HS/HSPG.[163] In accordance, by applying specific antibodies that recognize the different forms of Aβ species (Aβ1–40 or Aβ1–42), we have observed that HS and HSPG were only detectable in congophilic neuritic Aβ plaques, but not in some diffuse Aβ deposits of sporadic AD.[112]

D. Implications of HS in Amyloid Toxicity

The amyloid plaques formed in the organs may impair their functions, causing the clinical phenotypes of different amyloidosis. However, in AD the extent of Aβ plaque accumulation does not always correlate well with severity scores of the disease and a significant number of nondemented individuals have been found to have considerable Aβ plaque loads in their brains.[164] Similarly, in some AβPP transgenic animal models, clinical phenotypes are frequently observed prior to the onset of Aβ deposition. It has become clear that intermediate states, rather than mature fibrils, of the amyloid peptide, was proposed to represent the cytotoxic species, because the levels of soluble Aβ fibrils correlated better with degree of dementia than insoluble fibrillar deposits.[165] Distinct mechanisms for the toxic effect of soluble oligomers have been suggested.

Our recent study demonstrated that HS is an important cell surface factor mediating uptake of Aβ1–40, and the uptake event coincides with Aβ cytotoxicity.[166] In this study, wild-type Chinese hamster ovary (CHO-WT) cells showed reduced viability following exposure to Aβ1–40, whereas the HS-deficient cell line, pgsD-677, was essentially resistant to the Aβ1–40 treatment. Immunocytochemistry revealed that uptake of Aβ1–40 by CHO-WT was significantly greater than that by pgsD-677 cells. Aβ1–40 toxicity was also attenuated in human embryonic kidney cells overexpressing heparanase. Finally, addition of heparin to human umbilical vein endothelial cells (HUVEC) prevented internalization of added Aβ1–40 and protected against its toxicity. Taken together, these findings suggest that cell surface HS mediates Aβ internalization and toxicity in a chain length dependent manner, as the human embryonic kidney cells overexpressing heparanase produced shorter HS chains.

VI. Concluding Remarks

The functional importance of HS in development and homeostasis is ascribed to its interactions with a variety of proteins, including cytokines with essential biological activities. Interaction of HS with different amyloid proteins has been experimentally evidenced *in vitro*. These *in vitro* data are not sufficient for understanding the mechanisms regarding the impacts of HS in different amyloidosis; however, some scarce *in vivo* experimental results do support a functional role for HS in amyloidosis. Collectively, current knowledge in the literature, as summarized in this chapter, points out that HS/HSPG are implicated at a number of steps/phases of amyloidosis as illustrated in Fig. 4; some of these proposed mechanisms may be common, and some may be distinct to different amyloidoses. Considering that HS structures are differentially expressed in different organs as a result of the controlled biosynthesis process, the structural heterogeneity of HS may be relevant to the characterized pathological patterns of an individual amyloidosis through a typical interaction with a given amyloid protein. Evidence of the molecular mechanism of HS-amyloid interaction *in vivo* will help us to understand the onset and progress of amyloid deposition. Therefore, further efforts are needed to characterize HS structures or other GAGs isolated from disease specific amyloid lesions. On the other hand, to unveil the roles of HS in the pathogenesis of amyloidosis *in vivo*, will also be achieved by applying different animal models carrying mutant forms of the enzymes involved in biosynthesis of the glycosaminoglycans as described in previous chapters.

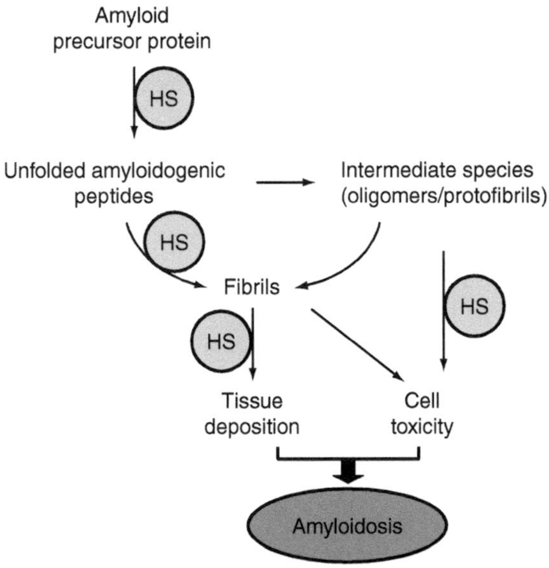

FIG. 4. Potential implications of HS in amyloidosis. Unfolded amyloidogenic peptides are typically derived from their precursor protein through a process involving enzymes or other factors. The unfolded amyloidogenic peptides tend to misfold generating insoluble fibrils as well as soluble intermediate species that can further aggregate to insoluble fibrils. Both the fibril and intermediate forms of amyloid peptide may impair the normal functions of cell/organs, resulting in amyloidodic diseases. HS is evidenced to be involved in a number of steps/phases of amyloidogenesis.

REFERENCES

1. Westermark P, Benson MD, Buxbaum JN, Cohen AS, Frangione B, Ikeda S, et al. Amyloid: toward terminology clarification. Report from the nomenclature committee of the international society of amyloidosis. *Amyloid* 2005;**12**:1–4.
2. Sideras K, Gertz MA. Amyloidosis. *Adv Clin Chem* 2009;**47**:1–44.
3. van Horssen J, Wesseling P, van den Heuvel LP, de Waal RM, Verbeek MM. Heparan sulphate proteoglycans in Alzheimer's disease and amyloid-related disorders. *Lancet Neurol* 2003;**2**:482–92.
4. Bernfield M, Gotte M, Park PW, Reizes O, Fitzgerald ML, Lincecum J, et al. Functions of cell surface heparan sulfate proteoglycans. *Annu Rev Biochem* 1999;**68**:729–77.
5. Lindahl U, Li JP. Interactions between heparan sulfate and proteins-design and functional implications. *Int Rev Cell Mol Biol* 2009;**276**:105–59.
6. Bishop JR, Schuksz M, Esko JD. Heparan sulphate proteoglycans fine-tune mammalian physiology. *Nature* 2007;**446**:1030–7.
7. Kyle RA. Amyloidosis: a convoluted story. *Br J Haematol* 2001;**114**:529–38.
8. Harrison RS, Sharpe PC, Singh Y, Fairlie DP. Amyloid peptides and proteins in review. *Rev Physiol Biochem Pharmacol* 2007;**159**:1–77.
9. Cohen AS, Calkins E. Electron microscopic observations on a fibrous component in amyloid of diverse origins. *Nature* 1959;**183**:1202–3.

10. Sipe JD, Cohen AS. Review: history of the amyloid fibril. *J Struct Biol* 2000;**130**:88–98.
11. Geddes AJ, Parker KD, Atkins ED, Beighton E. "Cross-beta" conformation in proteins. *J Mol Biol* 1968;**32**:343–58.
12. Chiti F, Dobson CM. Protein misfolding, functional amyloid, and human disease. *Annu Rev Biochem* 2006;**75**:333–66.
13. Tompa P. Structural disorder in amyloid fibrils: its implication in dynamic interactions of proteins. *FEBS J* 2009;**276**:5406–15.
14. Soto C, Estrada LD. Protein misfolding and neurodegeneration. *Arch Neurol* 2008;**65**:184–9.
15. Westermark P, Benson MD, Buxbaum JN, Cohen AS, Frangione B, Ikeda S, et al. A primer of amyloid nomenclature. *Amyloid* 2007;**14**:179–83.
16. Hawkins PN. Amyloidosis. *Blood Rev* 1995;**9**:135–42.
17. Pepys MB. Amyloidosis. *Annu Rev Med* 2006;**57**:223–41.
18. Georgiades CS, Neyman EG, Barish MA, Fishman EK. Amyloidosis: review and CT manifestations. *Radiographics* 2004;**24**:405–16.
19. Lachmann HJ, Hawkins PN. Systemic amyloidosis. *Curr Opin Pharmacol* 2006;**6**:214–20.
20. Merlini G, Westermark P. The systemic amyloidoses: clearer understanding of the molecular mechanisms offers hope for more effective therapies. *J Intern Med* 2004;**255**:159–78.
21. Cunnane G. Amyloid precursors and amyloidosis in inflammatory arthritis. *Curr Opin Rheumatol* 2001;**13**:67–73.
22. Yamada T. Serum amyloid A (SAA): a concise review of biology, assay methods and clinical usefulness. *Clin Chem Lab Med* 1999;**37**:381–8.
23. Urieli-Shoval S, Linke RP, Matzner Y. Expression and function of serum amyloid A, a major acute-phase protein, in normal and disease states. *Curr Opin Hematol* 2000;**7**:64–9.
24. Ledue TB, Weiner DL, Sipe JD, Poulin SE, Collins MF, Rifai N. Analytical evaluation of particle-enhanced immunonephelometric assays for C-reactive protein, serum amyloid A and mannose-binding protein in human serum. *Ann Clin Biochem* 1998;**35**(Pt 6):745–53.
25. Lachmann HJ, Goodman HJ, Gilbertson JA, Gallimore JR, Sabin CA, Gillmore JD, et al. Natural history and outcome in systemic AA amyloidosis. *N Engl J Med* 2007;**356**:2361–71.
26. Gertz MA, Kyle RA. Secondary systemic amyloidosis: response and survival in 64 patients. *Medicine (Baltimore)* 1991;**70**:246–56.
27. Glenner GG, Ein D, Eanes ED, Bladen HA, Terry W, Page DL. Creation of "amyloid" fibrils from Bence Jones proteins in vitro. *Science* 1971;**174**:712–4.
28. Glenner GG, Terry W, Harada M, Isersky C, Page D. Amyloid fibril proteins: proof of homology with immunoglobulin light chains by sequence analyses. *Science* 1971;**172**:1150–1.
29. Sirohi B, Powles R. Epidemiology and outcomes research for MGUS, myeloma and amyloidosis. *Eur J Cancer* 2006;**42**:1671–83.
30. Gertz MA, Lacy MQ, Dispenzieri A, Hayman SR. Amyloidosis. *Best Pract Res Clin Haematol* 2005;**18**:709–27.
31. Kyle RA, Gertz MA. Primary systemic amyloidosis: clinical and laboratory features in 474 cases. *Semin Hematol* 1995;**32**:45–59.
32. Eriksson M, Buttner J, Todorov T, Yumlu S, Schonland S, Hegenbart U, et al. Prevalence of germline mutations in the TTR gene in a consecutive series of surgical pathology specimens with ATTR amyloid. *Am J Surg Pathol* 2009;**33**:58–65.
33. Benson MD, Kincaid JC. The molecular biology and clinical features of amyloid neuropathy. *Muscle Nerve* 2007;**36**:411–23.
34. Connors LH, Lim A, Prokaeva T, Roskens VA, Costello CE. Tabulation of human transthyretin (TTR) variants, 2003. *Amyloid* 2003;**10**:160–84.
35. Kelly JW, Lansbury Jr PT. Transthyretin-derived senile systemic amyloidosis: clinicopathologic and structural considerations. *Amyloid* 1994;**1**:186–205.

36. Serpell LC, Sunde M, Fraser PE, Luther PK, Morris EP, Sangren O, et al. Examination of the structure of the transthyretin amyloid fibril by image reconstruction from electron micrographs. *J Mol Biol* 1995;**254**:113–8.
37. Hou X, Aguilar MI, Small DH. Transthyretin and familial amyloidotic polyneuropathy. Recent progress in understanding the molecular mechanism of neurodegeneration. *FEBS J* 2007;**274**:1637–50.
38. Westermark P, Bergstrom J, Solomon A, Murphy C, Sletten K. Transthyretin-derived senile systemic amyloidosis: clinicopathologic and structural considerations. *Amyloid* 2003;**10** (Suppl. 1):48–54.
39. Bergstrom J, Gustavsson A, Hellman U, Sletten K, Murphy CL, Weiss DT, et al. Amyloid deposits in transthyretin-derived amyloidosis: cleaved transthyretin is associated with distinct amyloid morphology. *J Pathol* 2005;**206**:224–32.
40. Cornwell 3rd GG, Murdoch WL, Kyle RA, Westermark P, Pitkanen P. Frequency and distribution of senile cardiovascular amyloid. A clinicopathologic correlation. *Am J Med* 1983;**75**:618–23.
41. Gejyo F, Odani S, Yamada T, Honma N, Saito H, Suzuki Y, et al. Beta 2-microglobulin: a new form of amyloid protein associated with chronic hemodialysis. *Kidney Int* 1986;**30**:385–90.
42. Gejyo F, Homma N, Arakawa M. Long-term complications of dialysis: pathogenic factors with special reference to amyloidosis. *Kidney Int Suppl* 1993;**41**:S78–82.
43. Dember LM, Jaber BL. Dialysis-related amyloidosis: late finding or hidden epidemic? *Semin Dial* 2006;**19**:105–9.
44. Berggard I, Bearn AG. Isolation and properties of a low molecular weight beta-2-globulin occurring in human biological fluids. *J Biol Chem* 1968;**243**:4095–103.
45. Wibell L, Evrin PE, Berggard I. Serum 2-microglobulin in renal disease. *Nephron* 1973;**10**:320–31.
46. Vincent C, Revillard JP,˙Galland M, Traeger J. Serum beta2-microglobulin in hemodialyzed patients. *Nephron* 1978;**21**:260–8.
47. Schwalbe S, Holzhauer M, Schaeffer J, Galanski M, Koch KM, Floege J. Beta 2-microglobulin associated amyloidosis: a vanishing complication of long-term hemodialysis? *Kidney Int* 1997;**52**:1077–83.
48. Traut M, Haufe CC, Eismann U, Deppisch RM, Stein G, Wolf G. Increased binding of beta-2-microglobulin to blood cells in dialysis patients treated with high-flux dialyzers compared with low-flux membranes contributed to reduced beta-2-microglobulin concentrations. Results of a cross-over study. *Blood Purif* 2007;**25**:432–40.
49. Westermark P, Wernstedt C, O'Brien TD, Hayden DW, Johnson KH. Islet amyloid in type 2 human diabetes mellitus and adult diabetic cats contains a novel putative polypeptide hormone. *Am J Pathol* 1987;**127**:414–7.
50. Westermark P, Wernstedt C, Wilander E, Hayden DW, O'Brien TD, Johnson KH. Amyloid fibrils in human insulinoma and islets of Langerhans of the diabetic cat are derived from a neuropeptide-like protein also present in normal islet cells. *Proc Natl Acad Sci USA* 1987;**84**:3881–5.
51. Cooper GJ, Willis AC, Clark A, Turner RC, Sim RB, Reid KB. Purification and characterization of a peptide from amyloid-rich pancreases of type 2 diabetic patients. *Proc Natl Acad Sci USA* 1987;**84**:8628–32.
52. Sanke T, Bell GI, Sample C, Rubenstein AH, Steiner DF. An islet amyloid peptide is derived from an 89-amino acid precursor by proteolytic processing. *J Biol Chem* 1988;**263**:17243–6.
53. Westermark P. Quantitative studies on amyloid in the islets of Langerhans. *Ups J Med Sci* 1972;**77**:91–4.
54. Butler AE, Janson J, Bonner-Weir S, Ritzel R, Rizza RA, Butler PC. Beta-cell deficit and increased beta-cell apoptosis in humans with type 2 diabetes. *Diabetes* 2003;**52**:102–10.

55. Hull RL, Westermark GT, Westermark P, Kahn SE. Islet amyloid: a critical entity in the pathogenesis of type 2 diabetes. *J Clin Endocrinol Metab* 2004;**89**:3629–43.
56. Hardy J, Selkoe DJ. The amyloid hypothesis of Alzheimer's disease: progress and problems on the road to therapeutics. *Science* 2002;**297**:353–6.
57. Querfurth HW, LaFerla FM. Alzheimer's disease. *N Engl J Med* 2010;**362**:329–44.
58. Iqbal K, Liu F, Gong CX, Alonso Adel C, Grundke-Iqbal I. Mechanisms of tau-induced neurodegeneration. *Acta Neuropathol* 2009;**118**:53–69.
59. Isacson O, Seo H, Lin L, Albeck D, Granholm AC. Alzheimer's disease and Down's syndrome: roles of APP, trophic factors and ACh. *Trends Neurosci* 2002;**25**:79–84.
60. Antonarakis SE, Lyle R, Dermitzakis ET, Reymond A, Deutsch S. Chromosome 21 and down syndrome: from genomics to pathophysiology. *Nat Rev Genet* 2004;**5**:725–38.
61. Prusiner SB. Novel proteinaceous infectious particles cause scrapie. *Science* 1982;**216**:136–44.
62. Prusiner SB. Prions. *Proc Natl Acad Sci USA* 1998;**95**:13363–83.
63. Cobb NJ, Surewicz WK. Prion diseases and their biochemical mechanisms. *Biochemistry* 2009;**48**:2574–85.
64. Aguzzi A, Calella AM. Prions: protein aggregation and infectious diseases. *Physiol Rev* 2009;**89**:1105–52.
65. Collinge J. Prion diseases of humans and animals: their causes and molecular basis. *Annu Rev Neurosci* 2001;**24**:519–50.
66. Aguzzi A, Polymenidou M. Mammalian prion biology: one century of evolving concepts. *Cell* 2004;**116**:313–27.
67. Li JP, Vlodavsky I. Heparin, heparan sulfate and heparanase in inflammatory reactions. *Thromb Haemost* 2009;**102**:823–8.
68. Bulow HE, Hobert O. The molecular diversity of glycosaminoglycans shapes animal development. *Annu Rev Cell Dev Biol* 2006;**22**:375–407.
69. Carey DJ. Syndecans: multifunctional cell-surface co-receptors. *Biochem J* 1997;**327** (Pt 1):1–16.
70. Oh ES, Couchman JR. Syndecans-2 and -4; close cousins, but not identical twins. *Mol Cells* 2004;**17**:181–7.
71. Woods A. Syndecans: transmembrane modulators of adhesion and matrix assembly. *J Clin Invest* 2001;**107**:935–41.
72. Filmus J, Capurro M, Rast J. Glypicans. *Genome Biol* 2008;**9**:224.
73. Fransson LA. Glypicans. *Int J Biochem Cell Biol* 2003;**35**:125–9.
74. Filmus J, Selleck SB. Glypicans: proteoglycans with a surprise. *J Clin Invest* 2001;**108**:497–501.
75. Knox SM, Whitelock JM. Perlecan: how does one molecule do so many things? *Cell Mol Life Sci* 2006;**63**:2435–45.
76. Bezakova G, Ruegg MA. New insights into the roles of agrin. *Nat Rev Mol Cell Biol* 2003;**4**:295–308.
77. Farach-Carson MC, Carson DD. Perlecan—a multifunctional extracellular proteoglycan scaffold. *Glycobiology* 2007;**17**:897–905.
78. Handler M, Yurchenco PD, Iozzo RV. Developmental expression of perlecan during murine embryogenesis. *Dev Dyn* 1997;**210**:130–45.
79. Tsen G, Halfter W, Kroger S, Cole GJ. Agrin is a heparan sulfate proteoglycan. *J Biol Chem* 1995;**270**:3392–9.
80. Iozzo RV. Basement membrane proteoglycans: from cellar to ceiling. *Nat Rev Mol Cell Biol* 2005;**6**:646–56.
81. Kisilevsky R, Fraser PE. A beta amyloidogenesis: unique, or variation on a systemic theme? *Crit Rev Biochem Mol Biol* 1997;**32**:361–404.

82. Kisilevsky R. Review: amyloidogenesis-unquestioned answers and unanswered questions. *J Struct Biol* 2000;**130**:99–108.
83. Boyd WA. *Textbook of pathology: an introduction to medicine.* Philadelphia: Lee and Febiger; 1932 pp. 105–159.
84. Snow AD, Kisilevsky R. Temporal relationship between glycosaminoglycan accumulation and amyloid deposition during experimental amyloidosis. A histochemical study. *Lab Invest* 1985;**53**:37–44.
85. Snow AD, Kisilevsky R, Willmer J, Prusiner SB, DeArmond SJ. Sulfated glycosaminoglycans in amyloid plaques of prion diseases. *Acta Neuropathol* 1989;**77**:337–42.
86. Snow AD, Willmer J, Kisilevsky R. A close ultrastructural relationship between sulfated proteoglycans and AA amyloid fibrils. *Lab Invest* 1987;**57**:687–98.
87. Snow AD, Willmer J, Kisilevsky R. Sulfated glycosaminoglycans: a common constituent of all amyloids? *Lab Invest* 1987;**56**:120–3.
88. Snow AD, Willmer JP, Kisilevsky R. Sulfated glycosaminoglycans in Alzheimer's disease. *Hum Pathol* 1987;**18**:506–10.
89. Norling B, Westermark GT, Westermark P. Immunohistochemical identification of heparan sulphate proteoglycan in secondary systemic amyloidosis. *Clin Exp Immunol* 1988;**73**:333–7.
90. Magnus JH, Husby G, Kolset SO. Presence of glycosaminoglycans in purified AA type amyloid fibrils associated with juvenile rheumatoid arthritis. *Ann Rheum Dis* 1989;**48**:215–9.
91. Nelson SR, Lyon M, Gallagher JT, Johnson EA, Pepys MB. Isolation and characterization of the integral glycosaminoglycan constituents of human amyloid A and monoclonal light-chain amyloid fibrils. *Biochem J* 1991;**275**(Pt 1):67–73.
92. Ohishi H, Skinner M, Sato-Araki N, Okuyama T, Gejyo F, Kimura A, et al. Glycosaminoglycans of the hemodialysis-associated carpal synovial amyloid and of amyloid-rich tissues and fibrils of heart, liver, and spleen. *Clin Chem* 1990;**36**:88–91.
93. Husby G, Stenstad T, Magnus JH, Sletten K, Nordvag BY, Marhaug G. Interaction between circulating amyloid fibril protein precursors and extracellular tissue matrix components in the pathogenesis of systemic amyloidosis. *Clin Immunol Immunopathol* 1994;**70**:2–9.
94. Magnus JH, Stenstad T, Kolset SO, Husby G. Glycosaminoglycans in extracts of cardiac amyloid fibrils from familial amyloid cardiomyopathy of Danish origin related to variant transthyretin Met 111. *Scand J Immunol* 1991;**34**:63–9.
95. Inoue S, Kisilevsky R. A high resolution ultrastructural study of experimental murine AA amyloid. *Lab Invest* 1996;**74**:670–83.
96. Inoue S, Kuroiwa M, Saraiva MJ, Guimaraes A, Kisilevsky R. Ultrastructure of familial amyloid polyneuropathy amyloid fibrils: examination with high-resolution electron microscopy. *J Struct Biol* 1998;**124**:1–12.
97. Inoue S, Kuroiwa M, Ohashi K, Hara M, Kisilevsky R. Ultrastructural organization of hemodialysis-associated beta 2-microglobulin amyloid fibrils. *Kidney Int* 1997;**52**:1543–9.
98. Young ID, Ailles L, Narindrasorasak S, Tan R, Kisilevsky R. Localization of the basement membrane heparan sulfate proteoglycan in islet amyloid deposits in type II diabetes mellitus. *Arch Pathol Lab Med* 1992;**116**:951–4.
99. Verchere CB, D'Alessio DA, Palmiter RD, Weir GC, Bonner-Weir S, Baskin DG, et al. Islet amyloid formation associated with hyperglycemia in transgenic mice with pancreatic beta cell expression of human islet amyloid polypeptide. *Proc Natl Acad Sci USA* 1996;**93**:3492–6.
100. Kahn SE, Andrikopoulos S, Verchere CB. Islet amyloid: a long-recognized but underappreciated pathological feature of type 2 diabetes. *Diabetes* 1999;**48**:241–53.
101. Verchere CB, D'Alessio DA, Palmiter RD, Baskin DG, Snow AD, Bonner- Weir S, et al. Characterization of islet amyloid in a transgenic mouse model of NIDDM. *Diabetes* 1996; **45**(Suppl. 2):313A.

102. Guiroy DC, Yanagihara R, Gajdusek DC. Localization of amyloidogenic proteins and sulfated glycosaminoglycans in nontransmissible and transmissible cerebral amyloidoses. *Acta Neuropathol* 1991;**82**:87–92.
103. Snow AD, Wight TN, Nochlin D, Koike Y, Kimata K, DeArmond SJ, et al. Immunolocalization of heparan sulfate proteoglycans to the prion protein amyloid plaques of Gerstmann–Straussler syndrome, Creutzfeldt–Jakob disease and scrapie. *Lab Invest* 1990;**63**:601–11.
104. McBride PA, Wilson MI, Eikelenboom P, Tunstall A, Bruce ME. Heparan sulfate proteoglycan is associated with amyloid plaques and neuroanatomically targeted PrP pathology throughout the incubation period of scrapie-infected mice. *Exp Neurol* 1998;**149**:447–54.
105. Leteux C, Chai W, Nagai K, Herbert CG, Lawson AM, Feizi T. 10E4 antigen of Scrapie lesions contains an unusual nonsulfated heparan motif. *J Biol Chem* 2001;**276**:12539–45.
106. Snow AD, Mar H, Nochlin D, Kimata K, Kato M, Suzuki S, et al. The presence of heparan sulfate proteoglycans in the neuritic plaques and congophilic angiopathy in Alzheimer's disease. *Am J Pathol* 1988;**133**:456–63.
107. Su JH, Cummings BJ, Cotman CW. Localization of heparan sulfate glycosaminoglycan and proteoglycan core protein in aged brain and Alzheimer's disease. *Neuroscience* 1992;**51**:801–13.
108. Perlmutter LS, Chui HC, Saperia D, Athanikar J. Microangiopathy and the colocalization of heparan sulfate proteoglycan with amyloid in senile plaques of Alzheimer's disease. *Brain Res* 1990;**508**:13–9.
109. Snow AD, Lara S, Nochlin D, Wight TN. Cationic dyes reveal proteoglycans structurally integrated within the characteristic lesions of Alzheimer's disease. *Acta Neuropathol* 1989;**78**:113–23.
110. Perry G, Siedlak SL, Richey P, Kawai M, Cras P, Kalaria RN, et al. Association of heparan sulfate proteoglycan with the neurofibrillary tangles of Alzheimer's disease. *J Neurosci* 1991;**11**:3679–83.
111. Donahue JE, Berzin TM, Rafii MS, Glass DJ, Yancopoulos GD, Fallon JR, et al. Agrin in Alzheimer's disease: altered solubility and abnormal distribution within microvasculature and brain parenchyma. *Proc Natl Acad Sci USA* 1999;**96**:6468–72.
112. O'Callaghan P, Sandwall E, Li JP, Yu H, Ravid R, Guan ZZ, et al. Heparan sulfate accumulation with Abeta deposits in Alzheimer's disease and Tg2576 mice is contributed by glial cells. *Brain Pathol* 2008;**18**:548–61.
113. Reilly CE. Crucial role of heparan sulfate proteoglycan (agrin) in beta-amyloid formation in Alzheimer's disease. *J Neurol* 2000;**247**:663–4.
114. van Horssen J, Kleinnijenhuis J, Maass CN, Rensink AA, Otte-Holler I, David G, et al. Accumulation of heparan sulfate proteoglycans in cerebellar senile plaques. *Neurobiol Aging* 2002;**23**:537–45.
115. van Horssen J, Otte-Holler I, David G, Maat-Schieman ML, van den Heuvel LP, Wesseling P, et al. Heparan sulfate proteoglycan expression in cerebrovascular amyloid beta deposits in Alzheimer's disease and hereditary cerebral hemorrhage with amyloidosis (Dutch) brains. *Acta Neuropathol* 2001;**102**:604–14.
116. Verbeek MM, Otte-Holler I, van den Born J, van den Heuvel LP, David G, Wesseling P, et al. Agrin is a major heparan sulfate proteoglycan accumulating in Alzheimer's disease brain. *Am J Pathol* 1999;**155**:2115–25.
117. Watanabe N, Araki W, Chui DH, Makifuchi T, Ihara Y, Tabira T. Glypican-1 as an Abeta binding HSPG in the human brain: its localization in DIG domains and possible roles in the pathogenesis of Alzheimer's disease. *FASEB J* 2004;**18**:1013–5.
118. Verbeek MM, Otte-Holler I, Veerhuis R, Ruiter DJ, De Waal RM. Distribution of A beta-associated proteins in cerebrovascular amyloid of Alzheimer's disease. *Acta Neuropathol* 1998;**96**:628–36.

119. Spillantini MG, Tolnay M, Love S, Goedert M. Microtubule-associated protein tau, heparan sulphate and alpha-synuclein in several neurodegenerative diseases with dementia. *Acta Neuropathol* 1999;**97**:585–94.
120. Bruinsma IB, te Riet L, Gevers T, ten Dam GB, van Kuppevelt TH, David G, et al. Sulfation of heparan sulfate associated with amyloid-beta plaques in patients with Alzheimer's disease. *Acta Neuropathol* 2010;**119**:211–20.
121. McLaurin J, Franklin T, Zhang X, Deng J, Fraser PE. Interactions of Alzheimer amyloid-beta peptides with glycosaminoglycans effects on fibril nucleation and growth. *Eur J Biochem* 1999;**266**:1101–10.
122. Castillo GM, Lukito W, Wight TN, Snow AD. The sulfate moieties of glycosaminoglycans are critical for the enhancement of beta-amyloid protein fibril formation. *J Neurochem* 1999;**72**:1681–7.
123. Goedert M, Jakes R, Spillantini MG, Hasegawa M, Smith MJ, Crowther RA. Assembly of microtubule-associated protein tau into Alzheimer-like filaments induced by sulphated glycosaminoglycans. *Nature* 1996;**383**:550–3.
124. Paudel HK, Li W. Heparin-induced conformational change in microtubule-associated protein Tau as detected by chemical cross-linking and phosphopeptide mapping. *J Biol Chem* 1999;**274**:8029–38.
125. Yu S, Yin S, Pham N, Wong P, Kang SC, Petersen RB, et al. Ligand binding promotes prion protein aggregation—role of the octapeptide repeats. *FEBS J* 2008;**275**:5564–75.
126. Yamamoto S, Yamaguchi I, Hasegawa K, Tsutsumi S, Goto Y, Gejyo F, et al. Glycosaminoglycans enhance the trifluoroethanol-induced extension of beta 2-microglobulin-related amyloid fibrils at a neutral pH. *J Am Soc Nephrol* 2004;**15**:126–33.
127. Relini A, De Stefano S, Torrassa S, Cavalleri O, Rolandi R, Gliozzi A, et al. Heparin strongly enhances the formation of beta2-microglobulin amyloid fibrils in the presence of type I collagen. *J Biol Chem* 2008;**283**:4912–20.
128. Snow AD, Wight TN. Proteoglycans in the pathogenesis of Alzheimer's disease and other amyloidoses. *Neurobiol Aging* 1989;**10**:481–97.
129. Narindrasorasak S, Lowery D, Gonzalez-DeWhitt P, Poorman RA, Greenberg B, Kisilevsky R. High affinity interactions between the Alzheimer's beta-amyloid precursor proteins and the basement membrane form of heparan sulfate proteoglycan. *J Biol Chem* 1991;**266**:12878–83.
130. Lindahl B, Westling C, Gimenez-Gallego G, Lindahl U, Salmivirta M. Common binding sites for beta-amyloid fibrils and fibroblast growth factor-2 in heparan sulfate from human cerebral cortex. *J Biol Chem* 1999;**274**:30631–5.
131. Brunden KR, Richter-Cook NJ, Chaturvedi N, Frederickson RC. pH-dependent binding of synthetic beta-amyloid peptides to glycosaminoglycans. *J Neurochem* 1993;**61**:2147–54.
132. Watson DJ, Lander AD, Selkoe DJ. Heparin-binding properties of the amyloidogenic peptides Abeta and amylin. Dependence on aggregation state and inhibition by Congo red. *J Biol Chem* 1997;**272**:31617–24.
133. Giulian D, Haverkamp LJ, Yu J, Karshin W, Tom D, Li J, et al. The HHQK domain of beta-amyloid provides a structural basis for the immunopathology of Alzheimer's disease. *J Biol Chem* 1998;**273**:29719–26.
134. Castillo GM, Cummings JA, Yang W, Judge ME, Sheardown MJ, Rimvall K, et al. Sulfate content and specific glycosaminoglycan backbone of perlecan are critical for perlecan's enhancement of islet amyloid polypeptide (amylin) fibril formation. *Diabetes* 1998;**47**:612–20.
135. Park K, Verchere CB. Identification of a heparin binding domain in the N-terminal cleavage site of pro-islet amyloid polypeptide. Implications for islet amyloid formation. *J Biol Chem* 2001;**276**:16611–6.
136. Abedini A, Tracz SM, Cho JH, Raleigh DP. Characterization of the heparin binding site in the N-terminus of human pro-islet amyloid polypeptide: implications for amyloid formation. *Biochemistry* 2006;**45**:9228–37.

137. Paulsson JF, Westermark GT. Aberrant processing of human proislet amyloid polypeptide results in increased amyloid formation. *Diabetes* 2005;**54**:2117–25.
138. Potter-Perigo S, Hull RL, Tsoi C, Braun KR, Andrikopoulos S, Teague J, et al. Proteoglycans synthesized and secreted by pancreatic islet beta-cells bind amylin. *Arch Biochem Biophys* 2003;**413**:182–90.
139. Hull RL, Zraika S, Udayasankar J, Kisilevsky R, Szarek WA, Wight TN, et al. Inhibition of glycosaminoglycan synthesis and protein glycosylation with WAS-406 and azaserine result in reduced islet amyloid formation in vitro. *Am J Physiol Cell Physiol* 2007;**293**:C1586–93.
140. Ancsin JB, Kisilevsky R. The heparin/heparan sulfate-binding site on apo-serum amyloid A. Implications for the therapeutic intervention of amyloidosis. *J Biol Chem* 1999;**274**:7172–81.
141. Elimova E, Kisilevsky R, Szarek WA, Ancsin JB. Amyloidogenesis recapitulated in cell culture: a peptide inhibitor provides direct evidence for the role of heparan sulfate and suggests a new treatment strategy. *FASEB J* 2004;**18**:1749–51.
142. Kisilevsky R, Szarek WA, Ancsin J, Vohra R, Li Z, Marone S. Novel glycosaminoglycan precursors as antiamyloid agents: part IV. *J Mol Neurosci* 2004;**24**:167–72.
143. Lindahl B, Lindahl U. Amyloid-specific heparan sulfate from human liver and spleen. *J Biol Chem* 1997;**272**:26091–4.
144. Naiki H, Yamamoto S, Hasegawa K, Yamaguchi I, Goto Y, Gejyo F. Molecular interactions in the formation and deposition of beta2-microglobulin-related amyloid fibrils. *Amyloid* 2005;**12**:15–25.
145. Thinakaran G, Koo EH. Amyloid precursor protein trafficking, processing, and function. *J Biol Chem* 2008;**283**:29615–9.
146. Willem M, Lammich S, Haass C. Function, regulation and therapeutic properties of beta-secretase (BACE1). *Semin Cell Dev Biol* 2009;**20**:175–82.
147. Dingwall C. Spotlight on BACE: the secretases as targets for treatment in Alzheimer disease. *J Clin Invest* 2001;**108**:1243–6.
148. McConlogue L, Buttini M, Anderson JP, Brigham EF, Chen KS, Freedman SB, et al. Partial reduction of BACE1 has dramatic effects on Alzheimer plaque and synaptic pathology in APP transgenic mice. *J Biol Chem* 2007;**282**:26326–34.
149. Leveugle B, Ding W, Durkin JT, Mistretta S, Eisle J, Matic M, et al. Heparin promotes beta-secretase cleavage of the Alzheimer's amyloid precursor protein. *Neurochem Int* 1997;**30**:543–8.
150. Scholefield Z, Yates EA, Wayne G, Amour A, McDowell W, Turnbull JE. Heparan sulfate regulates amyloid precursor protein processing by BACE1, the Alzheimer's beta-secretase. *J Cell Biol* 2003;**163**:97–107.
151. Patey SJ, Edwards EA, Yates EA, Turnbull JE. Heparin derivatives as inhibitors of BACE-1, the Alzheimer's beta-secretase, with reduced activity against factor Xa and other proteases. *J Med Chem* 2006;**49**:6129–32.
152. Patey SJ, Edwards EA, Yates EA, Turnbull JE. Engineered heparins: novel beta-secretase inhibitors as potential Alzheimer's disease therapeutics. *Neurodegener Dis* 2008;**5**:197–9.
153. Klaver D, Hung AC, Gasperini R, Foa L, Aguilar MI, Small DH. Effect of heparin on APP metabolism and Abeta production in cortical neurons. *Neurodegener Dis* 2010;**7**:187–9.
154. Meng F, Abedini A, Song B, Raleigh DP. Amyloid formation by pro-islet amyloid polypeptide processing intermediates: examination of the role of protein heparan sulfate interactions and implications for islet amyloid formation in type 2 diabetes. *Biochemistry* 2007;**46**:12091–9.
155. Taylor DR, Whitehouse IJ, Hooper NM. Glypican-1 mediates both prion protein lipid raft association and disease isoform formation. *PLoS Pathog* 2009;**5**:e1000666.
156. Snow AD, Sekiguchi R, Nochlin D, Fraser P, Kimata K, Mizutani A, et al. An important role of heparan sulfate proteoglycan (Perlecan) in a model system for the deposition and persistence of fibrillar A beta-amyloid in rat brain. *Neuron* 1994;**12**:219–34.

157. Hart M, Li L, Tokunaga T, Lindsey JR, Hassell JR, Snow AD, et al. Overproduction of perlecan core protein in cultured cells and transgenic mice. *J Pathol* 2001;**194**:262–9.
158. Bergamaschini L, Rossi E, Storini C, Pizzimenti S, Distaso M, Perego C, et al. Peripheral treatment with enoxaparin, a low molecular weight heparin, reduces plaques and beta-amyloid accumulation in a mouse model of Alzheimer's disease. *J Neurosci* 2004;**24**:4181–6.
159. Zhu H, Yu J, Kindy MS. Inhibition of amyloidosis using low-molecular-weight heparins. *Mol Med* 2001;**7**:517–22.
160. Zcharia E, Metzger S, Chajek-Shaul T, Aingorn H, Elkin M, Friedmann Y, et al. Transgenic expression of mammalian heparanase uncovers physiological functions of heparan sulfate in tissue morphogenesis, vascularization, and feeding behavior. *FASEB J* 2004;**18**:252–63.
161. Pikas DS, Li JP, Vlodavsky I, Lindahl U. Substrate specificity of heparanases from human hepatoma and platelets. *J Biol Chem* 1998;**273**:18770–7.
162. Li JP, Galvis ML, Gong F, Zhang X, Zcharia E, Metzger S, et al. In vivo fragmentation of heparan sulfate by heparanase overexpression renders mice resistant to amyloid protein A amyloidosis. *Proc Natl Acad Sci USA* 2005;**102**:6473–7.
163. Niewold TA, Flores Landeira JM, van den Heuvel LP, Ultee A, Tooten PC, Veerkamp JH. Characterization of proteoglycans and glycosaminoglycans in bovine renal AA-type amyloidosis. *Virchows Arch B Cell Pathol Incl Mol Pathol* 1991;**60**:321–8.
164. Terry RD. The pathogenesis of Alzheimer disease: an alternative to the amyloid hypothesis. *J Neuropathol Exp Neurol* 1996;**55**:1023–5.
165. Lue LF, Kuo YM, Roher AE, Brachova L, Shen Y, Sue L, et al. Soluble amyloid beta peptide concentration as a predictor of synaptic change in Alzheimer's disease. *Am J Pathol* 1999;**155**:853–62.
166. Sandwall E, O'Callaghan P, Zhang X, Lindahl U, Lannfelt L, Li JP. Heparan sulfate mediates amyloid-beta internalization and cytotoxicity. *Glycobiology* 2010;**20**:533–41.
167. Cavanagh JB. Corpora-amylacea and the family of polyglucosan diseases. *Brain Res Brain Res Rev* 1999;**29**:265–95.
168. Puchtler H, Sweat F. A review of early concepts of amyloid in context with contemporary chemical literature from 1839 to 1859. *J Histochem Cytochem* 1966;**14**:123–34.

Heparin as an Inhibitor of Cancer Progression

LUBOR BORSIG*,†

*Institute of Physiology, University of Zürich, Switzerland

†Zürich Center for Integrative Human Physiology, University of Zürich, Switzerland

Heparin is frequently used in the treatment of cancer-associated thromboembolism. Accumulating clinical evidence indicates that cancer patients treated with unfractionated and low-molecular weight heparin (LMWH) survive longer than patients treated by other anticoagulants, especially patients in the early stage of the disease. Experimental analysis from a number of animal models constantly provides evidence for the ability of heparin to attenuate metastasis. The non-anticoagulant activity of heparin on metastasis includes the ability to inhibit cell–cell-interaction through blocking of P- and L-selectin, to inhibit extracellular matrix protease heparanase, and to inhibit angiogenesis. This chapter summarizes current experimental evidence on the biology of heparin during cancer progression, with the focus on potential mechanism of heparin antimetastatic activity.

I. Heparin Affects Cancer: Clinical Evidence

The close relationship between cancer and venous thromboembolism (VTE) has been recognized for a long time, and multiple risk factors during malignancy have been linked to hypercoagulability. Thromboprophylaxis with

Progress in Molecular Biology
and Translational Science, Vol. 93
DOI: 10.1016/S1877-1173(10)93014-7

unfractionated heparin (UFH) and low-molecular weight heparin (LMWH) has been used in the management of hypercoagulable state in cancer patients, and LMWH is the anticoagulant regimen recommended by international guidelines.[1,2] Many retrospective analyses of clinical data have indicated that heparin treatment affects the survival of cancer patients with various tumors, especially of patients in the early stage of a disease.[3–5] On the basis of these observations several prospective clinical trials have been performed to evaluate heparin for its anticancer potential.[6–14] The CLOT study (Comparison of Low-molecular-weight heparin versus Oral anticoagulant therapy) was designed to study the effect of LMWH or an oral anticoagulant coumarin on prevention of VTE in patients with solid tumors.[6,7] While this study showed no overall effect on cancer progression, a significant increase in patient survival was detected in a subgroup of patients who were without metastasis at the beginning of the trial.[6] Furthermore, the treatment with LMWH nadroparin was significantly more effective than that with the oral anticoagulant.[7] In a trial on patients with small cell lung cancer, the use of LMWH dalteparin parallelly with standard chemotherapy significantly prolonged the median survival of patients when compared to patients treated with chemotherapy alone.[9] However, no significant difference was observed in a study of advanced cancer patients treated with nadroparin for a prolonged time.[12] The Fragmin Advanced Malignancy Outcome Study (FAMOUS) trial evaluated the effect of dalteparin treatment for a year in 374 patients with advanced malignancies.[8] While no statistically significant increase in survival was detected in patients with an originally poor prognosis, in patients with a better prognosis, dalteparin treatment resulted in a significant improvement in patient survival. The Malignancy and LMWH Therapy (MALT) trial on patients with advanced cancer of various origins demonstrated an improvement in survival when compared to a control group.[11] The subgroup analysis of patients with a better prognosis in the MALT and FAMOUS studies further underlines the potential of heparins to affect earlier stages of cancer. Several clinical trials are ongoing with the aim of validating the effect of heparin on cancer progression in specific cancer types. Nevertheless, the available evidence on heparin activity in cancer patients obtained from several completed studies indicates that heparin seems to directly affect cancer progression.

II. Heparin Attenuates Metastasis in Experimental Models

The effect of heparin on cancer progression has been reported by a number of laboratories, most of which used a mouse model of experimental metastasis (reviewed in Refs. 15–17). In the initial studies, application of UFH attenuated metastasis of mammary carcinoma and melanoma cells.[18,19] Despite the many limitations of the experimental metastasis model, the direct application of tumor cells in the circulation allowed a more detailed analysis of heparin's inhibitory potential that is limited to a timely defined presence of heparin in the circulation. The application of heparin around the time of tumor cell injection resulted in the attenuation of metastasis that was observed in the majority of studies using different tumor cell lines, a wide range of heparin doses, and a variety of heparin preparations (see Refs. 15–17). A single dose of heparin effectively attenuated metastasis of human and mouse colon carcinoma and melanoma, mouse breast carcinoma, and lung carcinoma. Application of heparin either 24 h before or after the tumor cell injection, had no effect.[19,20] Attenuation of metastasis was generally independent of the route of heparin application. While intravenous application of heparin around the time of tumor cell injection always led to a reduction of metastasis, subcutaneous or intraperitoneal application was mostly effective.[16,21–24] The observed limited response to heparin has been likely because of variations in time and/or route of heparin application, amounts of applied heparins, and differences in the heparin preparations used.[21–24] In the majority of studies, the amount of heparin applied significantly exceeded the clinically used therapeutic dose. Nevertheless, recent studies have provided evidence that heparin attenuated metastasis in two different mouse models at clinically relevant concentrations also.[25,26] Taken together, a number of animal experiments convincingly support the ability of heparins to attenuate metastasis. The mouse model of experimental metastasis proved to be instrumental in narrowing down the potential mechanism underlying the heparin activity during cancer progression (see below). However, this model had several limitations, which did not allow of the analysis of the various stages of metastasis or the root of metastasis (lymphatic versus hematogenous). Heparin has been shown to only attenuate metastasis in one study with a spontaneously metastatic mouse model, while tumor growth remained largely unaffected.[27] Further studies in spontaneously metastatic mouse models are required to confirm the findings from experimental models, the primary focus being the identification of the potential mechanism of heparin activity in this process.

III. Diverse Biological Activities of Heparin

Heparin is a complex mixture of natural glycosaminoglycans and contains a variety of biological activities. Heparin molecules are long unbranched polymers built of disaccharide repeats consisting of glucosamine and iduronic/galacturonic acid with a high degree of sulfation. Heparin is found in granules of mast cells that line blood vessels and are also present in mucosal tissues. Clinical preparations of heparin are of porcine and bovine origins that are adjusted according to their anticoagulant activity. The molecular weight of an UFH ranges between 12,000 and 14,000 Da. The partially depolymerized and fractionated LMWH consists of fragments with an average size of 5000 Da. Although LMWH preparations differ from each other depending on the preparation method, chemical or enzymatical depolymerization, their pharmacokinetical parameters are comparable.[28] The anticoagulant activity of heparin is defined by its ability to bind antithrombin III. A distinct pentasaccharide structure has been identified as the active site of heparin.[29] But this pentasaccharide represents only a small fraction of heparin preparations. Heparin binding to antithrombin alters its conformation and thereby accelerates its antithrombotic activity. The heparin–antithrombin complex binds to an active factor Xa with a high affinity, thereby inactivating the factor Xa. While UFH and LMWH have comparable anti Xa activities, they differ in their ability to inhibit factor IIa.[30] Binding of factor IIa requires the formation of a complex around a longer chain of the heparin fragment than is usually not present in LMWH preparations.

Heparin preparations are mixtures of heterogeneous polysaccharides containing a large variety of biological activities; this makes the identification of the role of heparin during cancer progression rather complicated. There is accumulating evidence that heparin, apart from its anticoagulant activity, can block P- and L-selectin, affect activity of growth factors and cytokines, release tissue factor pathway inhibitor (TFPI), inhibit heparanase and angiogenesis, alter interactions with integrins, and modulate protease activity and thereby the composition of extracellular matrix.[15,29,31–35] While any of these activities may affect cancer progression (Fig. 1), the experimental evidence from a number of studies strongly indicates that heparin affects early events in the metastatic cascade.

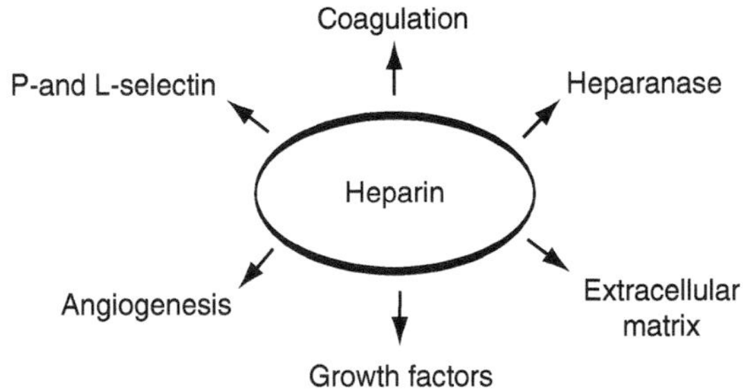

FIG. 1. Diverse biological activities of heparin potentially affecting cancer progression. Heparin may affect any of these activities and thereby inhibit cancer progression. Many of these activities have been shown *in vitro* and the individual contribution to attenuation of cancer progression remains to be determined *in vivo*.

IV. Potential Mechanisms of Heparin Affecting Cancer Progression

Metastasis is a process during which invasive tumor cells enter the circulation, avoid immune responses, attach to the endothelium of distant organs, extravasate from the vasculature, and establish metastatic foci. The common observation in most animal studies that a single dose of heparin before tumor cell injections attenuates metastasis strongly indicates that heparin affects the initial phase of hematogenous metastasis. This observation is further supported by the fact that the half-life of heparin in circulation ranges between 4 and 6 h.[36,37] Despite the relatively short time of heparin availability in the circulation, heparin significantly modulates the metastatic capacity of tumor cells while they are still in the circulation. Thus, heparin is affecting immediate processes upon tumor cell injection into the circulation like inhibition of P- and L-selectin-mediated interactions and affecting coagulation. Consequently, the inhibitory effect of heparin on angiogenesis or extracellular matrix remodeling is unlikely to be affected during this period.

V. Anticoagulant Activity of Heparin

Cancer patients are prone to thrombotic complications, and LMWH and UFH are used as an effective therapeutic anticoagulant treatment or as a prophylaxis.[38] A retrospective analysis of a number of clinical trials, where heparin treatment was compared with other anticoagulant regimen (e.g., coumarin), have indicated that heparin exerts activity beyond its anticoagulation.[29,39,40] Heparin effect on cancer progression has been confirmed in animal models and has been found to be additional to its anticoagulant activity.[15,17,41] The use of modified heparins with minimal or no anticoagulant activity led to attenuation of metastasis in a number of various mouse models using different cancer cell lines, indicating that heparin affects metastasis largely in an anticoagulant independent manner.[19,21,22,27,37,42–44] The use of an antithrombin inhibitor, hirudin, has been shown to reduce metastasis but the amount of hirudin used greatly exceeded clinically relevant concentrations.[45,46] The identification of a pentasaccharide reponsible for the antithrombin inhibitory activity in the heparin sequence enabled the evaluation of the anticancer activity of heparin.[47] While the synthetic pentasaccharide fondaparinux did not affect cancer progression, LMWH effectively attenuated metastasis at clinically tolerable levels.[25,26] However, the antimetastatic activity of LMWH has been found to vary among different preparations.

Heparins also have the ability to release TFPI from the blood vessel walls and thereby affect metastasis.[35] Application of LMWH tinzaparin led to a reduction of metastasis. Intravenous injection of recombinant TFPI at the time of tumor cell injection significantly attenuated metastasis of melanoma.[35,48] The extent of heparin mediated TFPI release and its contribution to inhibition of cancer progression require further *in vivo* investigation in the context of other non-anticoagulant activities of heparin.

VI. Inhibition of Heparanase

Tumor cell invasion is an essential step of cancer progression that is associated with an enhanced capability of tumor cells to degrade extracellular matrix components, including heparin sulfate proteoglycans, collagens, and fibronectin.[49] Tumor cells produce several hydrolytic enzymes including matrix metalloproteinases and heparanase, and their expression correlates with their metastatic potential. Heparanase is an endoglycosidase that cleaves heparin sulfate. Elevated heparanase expression has been detected in association with cancer progression of several carcinomas including colon, liver, pancreas, bladder, breast, and prostate carcinoma, as well as leukemia and multiple myeloma.[50–52] Heparanase overexpression in human tumors confers accelerated tumor growth and invasive phenotype in experimental animals.[52] Silencing of the heparanase gene in cancer cells resulted in a reduced angiogenesis, tumor growth, and metastasis, thereby directly linking heparanase activity to cancer progression.[53] Several studies have provided evidence that heparin inhibits heparanase activity *in vitro* and that modified heparins without anticoagulant activity but with heparanase inhibitory activity attenuated metastasis of melanomas and multiple myelomas.[19,54,55] Furthermore, sulfated oligosaccharides and laminarin sulfate with heparanase inhibitory activity attenuated experimental metastasis of melanomas and breast carcinomas.[56,57] Heparanase has been shown to regulate the function of heparan sulfate and thereby affect tumor growth and metastasis in multiple myeloma.[54,58] Shedding of heparan sulfate proteoglycan syndecan-1 from myeloma cells into their microenvironment is mediated by heparanase and is critical for the cancer progression.[54] The use of a heparanase-specific modified heparin NA-ROH (100%-*N*-acetylated, 25% glycol-split) attenuated primary tumor growth and metastasis.[54,55] Recently, the effect of NA-ROH heparin has been tested also in colon carcinomas, where no attenuation of metastasis was observed.[37] Therefore, inhibition of heparanase leads to attenuation of metastasis only in tumor cells expressing this enzyme.

VII. Selectins as Potential Targets of Heparin

UFH has been found to be an efficient ligand for P- and L-selectin.[59,60] Selectins are vascular cell adhesion molecules involved in adhesive interactions of platelets, leukocytes, and endothelial cells within the blood circulation. The physiological role of selectins has been described in hemostasis, inflammation, and immune response.[61,62] Selectins mediate the initial interactions of leukocytes with the vascular endothelium. The rapid and reversible interactions between selectins and their carbohydrate ligands enable leukocyte rolling that may result in firm adhesion. There are three members of the selectin family: L-, P-, and E-selectin. While L-selectin is constitutively expressed on almost all the leukocyte populations, P-selectin is constitutively expressed in the secretory granules of platelets and endothelial cells. Upon activation, P-selectin is rapidly expressed on the cell surface of both endothelial cells and platelets, thereby initiating cell adhesion. E-selectin expression on endothelial cell surfaces requires *de novo* transcription and therefore occurs several hours after activation.[63] Selectin ligands usually carry sialylated, fucosylated lactosamine oligosaccharide structures containing the terminal tetrasaccharide sialyl Lewis^{x} or its isomer sialyl Lewis^{a} (sLe$^{x/a}$).[63,64] Additional sulfation of the glycan itself or of the protein backbone in close proximity to the oligosaccharide further enhances the specific recognition by P- and L-selectin.[65] In addition to this common selectin ligand, P- and L-selectin can efficiently bind to heparin and sulfated galactosylceramide sulfatides.[59,60,66] The ability of selectins to recognize a variety of different carbohydrate structures indicates that the lectin domain of selectins binds to a certain carbohydrate "patch."[67]

VIII. Carcinomas, Heparin, and Hematogenous Metastasis

Hematogenous metastasis is the common route of cancer spread for epithelial cancer carcinoma. Epithelial cells are lining up the lumen of hollow organs that are covered by mucins. Cell-surface-mucins as well as soluble mucins secreted on apical sites of the epithelium contribute to protection of the organs from pathogens. Mucins are high molecular weight molecules that contain a large portion of O-linked glycan structures.[68,69] Carcinoma cells show altered cell surface glycosylation with an enhanced presence of sLe^{x}, sLe^{a}, Tn, and sialyl-Tn oligosaccharide structures on mucins.[68,69] The positive correlation between sLe$^{x/a}$ expression and poor prognosis, due to metastasis, has been demonstrated in colon, gastric, lung, prostate, renal, and breast cancers.[70–76] The correlation between selectin ligand expression and poor prognosis for carcinoma patients indicates the potential involvement of selectins in cancer progression. Moreover, the presence of carcinomas carrying selectin ligands in blood circulation makes it possible to mediate interactions with selectins on platelet, leukocytes, and endothelium thereby leading to metastasis.[16]

IX. P- and L-Selectins Facilitate Metastasis

Hematogenous metastasis is a process consisting of several events enabling the tumor cell to leave the primary tumor, enter the blood circulation, evade immune responses, adhere to the vascular cells of distant organs, exit from circulation, and form the new metastatic lesion. Of these events, cell–cell interactions leading to tumor cell adhesion to the endothelium of distant organs appear to be critical for this process.[77,78] The rapid expression of P-selectin on platelets and endothelium upon activation and the constant presence of L-selectin on leukocytes further supported the hypothesis that selectins may facilitate metastatic initiation. The recent evidence obtained in P- and/or L-selectin deficient mice confirms the contribution of these early response receptors to metastasis.[20,36,79–82] Attenuation of metastasis was observed in the absence of P-selectin both with carcinomas and melanoma cells expressing selectin ligands.[20,36,37,80] The absence of P-selectin led to a reduction of tumor seeding to the lung vasculature that was associated with a reduced platelet-tumor cell emboli formation.[36] Similarly, the reduction of selectin ligands on tumor cells caused a decrease in platelet-tumor cell emboli formation and attenuation of metastasis.[36,80,83] The platelet-tumor cell emboli formation seems to protect tumor cells from elimination by NK cells and this process is largely P-selectin dependent.[84] Bone marrow reconstitution of lethally irradiated P-selectin deficient mice has shown that the endothelial P-selectin expression also contributes to metastasis.[20]

Attenuation of metastasis was observed also in L-selectin deficient mice, strongly implicating leukocytes to be active facilitators of this process.[80,81] L-selectin-mediated recruitment of leukocytes to vascular tumor cells was associated with an enhanced expression of L-selectin ligands surrounding tumor emboli.[81] Intravenous injection of a function blocking L-selectin antibody resulted in attenuation of metastasis.[81] Although the mechanism of leukocyte contribution to metastasis requires further investigations, the current evidence indicates that leukocytes may potentiate tumor cell extravasation.[85–88] The individual absence of P- or L-selectin significantly attenuated metastasis of carcinoma cells, but in the P- and L-selectin double deficient mice virtually no metastasis was observed.[80] These observations strongly indicate a synergistic effect of both selectins during initiating steps of metastasis.

The overexpression of E-selectin in the liver was shown to divert metastasis to this organ.[89] In another study, experimental liver metastasis has been shown to be blocked by E-selectin blocking antibody.[90,91] In an experimental metastasis model, where mice injected with IL-1 were coinjected with tumor cells together with soluble E-selectin, reduction of metastasis to the lungs has been observed.[92] However, the cytokine-induced E-selectin expression may not reflect the natural process of metastasis. Interestingly, in the absence of E-selectin, no attenuation of experimental metastasis to the lungs was observed.[82] Because of the delayed cell-surface expression, E-selectin may have a limited effect on the very early stages of a naturally occurring metastatic process. Therefore, the presence of early response receptors P- and L-selectin in the blood circulation suggests the implication of endothelium, platelets, and leukocytes in metastasis that can be inhibited by heparins.

X. Heparin Inhibits P- and L-Selectin-Mediated Interactions

Heparin was shown to attenuate metastasis in a number of different animal models when applied at the time of tumor cells injection (reviewed in Refs. 16,17). To address the potential of heparin in the process to affect tumor cell invasion and migration into the blood vessels, spontaneous metastatic models are necessary. Nevertheless, the widely used experimental metastasis model enables to characterize the molecular mechanisms underlying the effect of heparin on the initiation phase of metastasis. The antimetastatic activity of heparin was analyzed in P- and/or L-selectin deficient mice.[20,36,37,80,81,93] Injection of heparin shortly before tumor cell injection attenuated metastasis in wild-type (wt) mice to similar levels as observed in P-selectin deficient mice (Psel−/−).[20,36,80] Heparin injection shortly before tumor cell injection further attenuated metastasis in L-selectin deficient mice (Lsel−/−).[80,81] This heparin administration efficiently inhibited platelet-tumor cell emboli formation that was shown to be largely P-selectin dependent.[36] No additional reduction of metastasis was observed by heparin in Psel−/− mice.[20,36,80] Attenuation of metastasis was also achieved with heparin derivatives without anti-coagulant activity. Together with the finding that heparin showed no effect in Psel−/− mice; heparin treatment at the time when tumor cell are in circulation likely influences P-selectin mediated interactions during the early stages of metastasis.[37]

The additional attenuation of metastasis achieved by heparin treatment at the time of tumor cell injection in Lsel−/− mice indicates that P-selectin-mediated platelet aggregation precedes the involvement of L-selectin and thereby, leukocytes in this process.[81] Accordingly, heparin treatment several hours post-tumor cell injection—late heparin (6–12 h)— further attenuated metastasis in Psel−/− mice, while having no further effect in Lsel−/− mice.[81] The late heparin treatment in wt mice resulted in attenuation of metastasis to a similar extent as observed in Lsel−/− mice. The final evidence that heparin affects primarily P- and L-selectin mediated interactions came from the P- and L-selectin double deficient mice.[80] Heparin treatment, either shortly before or soon after tumor cell injection, had no further effect on metastasis. Interestingly, prolonged treatment of P- and L-selectin double deficient mice with high doses of UFH further reduced metastasis.[93] Thus heparin likely affects also other mechanisms during metastatic progression. Recently, different clinically used LMWH preparations have been shown to attenuate metastasis by inhibition of selectin.[25,26]

XI. Mechanism of Heparin Action During Metastasis

Hematogenous metastasis is directly or indirectly responsible for most cancer-related deaths. Accumulating evidence from a number of laboratories indicates that heparin attenuates experimental metastasis of various cancer cells, as long as dose of heparin or its derivative reaches the clinically used concentration and is applied around the time when tumor cells are still in the circulation (reviewed in Refs. 16,17). There have been only two exceptions, where heparin did not affect metastasis applied around the time of tumor cell injection.[22,23] Heparin treatment of mice either 1 day before or 1 day after tumor cell injection did not affect metastasis.[19,20] Despite the large variation of heparin doses applied, attenuation of metastasis has been observed. Human P-selectin sensitivity is about 10-fold higher when compared to mouse P-selectin sensitivity.[36] The affectivity of non-anticoagulant heparins as metastatic inhibitors further excludes the possibility that the inhibition of prothrombotic activities is the main mechanism of heparin anticancer activity.[19,27,37] Basically, all published studies using heparin or its derivatives in experimental metastasis models are in agreement with the potential inhibition of P- and/or L-selectin function. While other biological activities can clearly contribute to anti-metastatic activity of heparin, the relatively short time of heparin presence in the circulation makes them less likely to be effective in this context. Further analyses of heparin as an inhibitor of cancer progression are required and should validate heparin as an inhibitor of cancer progression in spontaneously metastatic mouse models. Analysis of P- and/or L-selectin inhibitors during metastasis might complement the knowledge about this process. Yet, the very nature of heparin, carrying not only selectin-inhibitory activity but also the activity toward heparanase, binding of cytokines, TFPI release, or modulatory activities on extracellular matrix, might prove beneficial over a single targeting of selectins. The currently available clinical evidence, from already performed clinical trials as well as from ongoing trials, supports the future evaluation of heparin as a potential first treatment specifically for cancer metastasis.

References

1. Lyman GH, Khorana AA, Falanga A, Clarke-Pearson D, Flowers C, Jahanzeb M, et al. American Society of Clinical Oncology guideline: recommendations for venous thromboembolism prophylaxis and treatment in patients with cancer. *J Clin Oncol* 2007;**25**:5490–505.
2. Prandoni P. How I treat venous thromboembolism in patients with cancer. *Blood* 2005;**106**:4027–33.
3. Ornstein DL, Zacharski LR. The use of heparin for treating human malignancies. *Haemostasis* 1999;**29**(Suppl S1):48–60.
4. Hettiarachchi RJ, Smorenburg SM, Ginsberg J, Levine M, Prins MH, Buller HR. Do heparins do more than just treat thrombosis? The influence of heparins on cancer spread. *Thromb Haemost* 1999;**82**:947–52.
5. Kakkar AK, Hedges AR, Williamson RCN, Kakkar VV. Perioperative heparin therapy inhibits late death from metastatic cancer. *Int J Oncol* 1995;**6**:885–8.
6. Lee AY, Rickles FR, Julian JA, Gent M, Baker RI, Bowden C, et al. Randomized comparison of low molecular weight heparin and coumarin derivatives on the survival of patients with cancer and venous thromboembolism. *J Clin Oncol* 2005;**23**:2123–9.
7. Lee AY, Levine MN, Baker RI, Bowden C, Kakkar AK, Prins M, et al. Low-molecular-weight heparin versus a coumarin for the prevention of recurrent venous thromboembolism in patients with cancer. *N Engl J Med* 2003;**349**:146–53.
8. Kakkar AK, Levine MN, Kadziola Z, Lemoine NR, Low V, Patel HK, et al. Low molecular weight heparin, therapy with dalteparin, and survival in advanced cancer: the fragmin advanced malignancy outcome study (FAMOUS). *J Clin Oncol* 2004;**22**:1944–8.
9. Lebeau B, Chastang C, Brechot JM, Capron F, Dautzenberg B, Delaisements C, et al. Subcutaneous heparin treatment increases survival in small cell lung cancer. "Petites Cellules" group. *Cancer* 1994;**74**:38–45.
10. Altinbas M, Coskun HS, Er O, Ozkan M, Eser B, Unal A, et al. A randomized clinical trial of combination chemotherapy with and without low-molecular-weight heparin in small cell lung cancer. *J Thromb Haemost* 2004;**2**:1266–71.
11. Klerk CP, Smorenburg SM, Otten HM, Lensing AW, Prins MH, Piovella F, et al. The effect of low molecular weight heparin on survival in patients with advanced malignancy. *J Clin Oncol* 2005;**23**:2130–5.
12. Sideras K, Schaefer PL, Okuno SH, Sloan JA, Kutteh L, Fitch TR, et al. Low-molecular-weight heparin in patients with advanced cancer: a phase 3 clinical trial. *Mayo Clin Proc* 2006;**81**:758–67.
13. Lazo-Langner A, Goss GD, Spaans JN, Rodger MA. The effect of low-molecular-weight heparin on cancer survival. A systematic review and meta-analysis of randomized trials. *J Thromb Haemost* 2007;**5**:729–37.
14. Kuderer NM, Ortel TL, Francis CW. Impact of venous thromboembolism and anticoagulation on cancer and cancer survival. *J Clin Oncol* 2009;**27**:4902–11.
15. Smorenburg SM, Van Noorden CJ. The complex effects of heparins on cancer progression and metastasis in experimental studies. *Pharmacol Rev* 2001;**53**:93–105.

16. Borsig L, Stevenson JL, Varki A. Heparin in cancer: role of selectin interactions. In: Khorana AA, Francis CW, editors. *Cancer-associated thrombosis*. New York: Informa Healthcare; 2007. p. 97–113.
17. Niers TM, Klerk CP, DiNisio M, Van Noorden CJ, Buller HR, Reitsma PH, et al. Mechanisms of heparin induced anti-cancer activity in experimental cancer models. *Crit Rev Oncol Hematol* 2007;**61**:195–207.
18. Lee AE, Rogers LA, Jeffery RE, Longcroft JM. Comparison of metastatic cell lines derived from a murine mammary tumour, and reduction of metastasis by heparin. *Clin Exp Metastasis* 1988;**6**:463–71.
19. Vlodavsky I, Mohsen M, Lider O, Svahn CM, Ekre HP, Vigoda M, et al. Inhibition of tumor metastasis by heparanase inhibiting species of heparin. *Invasion Metastasis* 1994;**14**:290–302.
20. Ludwig RJ, Boehme B, Podda M, Henschler R, Jager E, Tandi C, et al. Endothelial P-selectin as a target of heparin action in experimental melanoma lung metastasis. *Cancer Res* 2004;**64**:2743–50.
21. Lapierre F, Holme K, Lam L, Tressler RJ, Storm N, Wee J, et al. Chemical modifications of heparin that diminish its anticoagulant but preserve its heparanase-inhibitory, angiostatic, anti-tumor and anti- metastatic properties. *Glycobiology* 1996;**6**:355–66.
22. Yoshitomi Y, Nakanishi H, Kusano Y, Munesue S, Oguri K, Tatematsu M, et al. Inhibition of experimental lung metastases of Lewis lung carcinoma cells by chemically modified heparin with reduced anticoagulant activity. *Cancer Lett* 2004;**207**:165–74.
23. Szende B, Paku S, Racz G, Kopper L. Effect of Fraxiparine and heparin on experimental tumor metastasis in mice. *Anticancer Res* 2005;**25**:2869–72.
24. Niers TM, Bruggemann LW, Klerk CP, Muller FJ, Buckle T, Reitsma PH, et al. Differential effects of anticoagulants on tumor development of mouse cancer cell lines B16, K1735 and CT26 in lung. *Clin Exp Metastasis* 2009;**26**:171–8.
25. Stevenson JL, Choi SH, Varki A. Differential metastasis inhibition by clinically relevant levels of heparins–correlation with selectin inhibition, not antithrombotic activity. *Clin Cancer Res* 2005;**11**:7003–11.
26. Ludwig RJ, Alban S, Bistrian R, Boehncke WH, Kaufmann R, Henschler R, et al. The ability of different forms of heparins to suppress P-selectin function *in vitro* correlates to their inhibitory capacity on bloodborne metastasis *in vivo*. *Thromb Haemost* 2006;**95**:535–40.
27. Kragh M, Binderup L, Vig Hjarnaa PJ, Bramm E, Johansen KB, Frimundt Petersen C. Non-anti-coagulant heparin inhibits metastasis but not primary tumor growth. *Oncol Rep* 2005;**14**:99–104.
28. Fareed J, Hoppensteadt DA, Bick RL. An update on heparins at the beginning of the new millennium. *Semin Thromb Hemost* 2000;**26**(Suppl 1):5–21.
29. Engelberg H. Actions of heparin that may affect the malignant process. *Cancer* 1999;**85**:257–72.
30. Fareed J, Leong WL, Hoppensteadt DA, Jeske WP, Walenga J, Wahi R, et al. Generic low-molecular-weight heparins: some practical considerations. *Semin Thromb Hemost* 2004;**30**:703–13.
31. Lever R, Page CP. Novel drug development opportunities for heparin. *Nat Rev Drug Discov* 2002;**1**:140–8.
32. Varki NM, Varki A. Heparin inhibition of selectin-mediated interactions during the hematogenous phase of carcinoma metastasis: rationale for clinical studies in humans. *Semin Thromb Hemost* 2002;**28**:53–66.
33. Folkman J, Shing Y. Control of angiogenesis by heparin and other sulfated polysaccharides. *Adv Exp Med Biol* 1992;**313**:355–64.
34. Mellor P, Harvey JR, Murphy KJ, Pye D, O'Boyle G, Lennard TW, et al. Modulatory effects of heparin and short-length oligosaccharides of heparin on the metastasis and growth of LMD MDA-MB 231 breast cancer cells *in vivo*. *Br J Cancer* 2007;**97**:761–8.

35. Amirkhosravi A, Mousa SA, Amaya M, Francis JL. Antimetastatic effect of tinzaparin, a low-molecular-weight heparin. *J Thromb Haemost* 2003;**1**:1972–6.
36. Borsig L, Wong R, Feramisco J, Nadeau DR, Varki NM, Varki A. Heparin and cancer revisited: mechanistic connections involving platelets, P-selectin, carcinoma mucins, and tumor metastasis. *Proc Natl Acad Sci USA* 2001;**98**:3352–7.
37. Hostettler N, Naggi A, Torri G, Casu B, Vlodavsky I, Borsig L. P-selectin- and heparanase-dependent antimetastatic activity of non-anticoagulant heparins. *FASEB J* 2007;**21**:3562–72.
38. Zacharski LR, Ornstein DL, Mamourian AC. Low-molecular-weight heparin and cancer. *Semin Thromb Hemost* 2000;**26**(Suppl. 1):69–77.
39. Zacharski LR, Ornstein DL. Heparin and cancer. *Thromb Haemost* 1998;**80**:10–23.
40. Kuderer NM, Khorana AA, Lyman GH, Francis CW. A meta-analysis and systematic review of the efficacy and safety of anticoagulants as cancer treatment: impact on survival and bleeding complications. *Cancer* 2007;**110**:1149–61.
41. Borsig L. Antimetastatic activities of modified heparins: selectin inhibition by heparin attenuates metastasis. *Semin Thromb Hemost* 2007;**33**:540–6.
42. Sciumbata T, Caretto P, Pirovano P, Pozzi P, Cremonesi P, Galimberti G, et al. Treatment with modified heparins inhibits experimental metastasis formation and leads, in some animals, to long-term survival. *Invasion Metastasis* 1996;**16**:132–43.
43. Ono K, Ishihara M, Ishikawa K, Ozeki Y, Deguchi H, Sato M, et al. Periodate-treated, non-anticoagulant heparin-carrying polystyrene (NAC-HCPS) affects angiogenesis and inhibits subcutaneous induced tumour growth and metastasis to the lung. *Br J Cancer* 2002;**86**:1803–12.
44. Mousa SA, Linhardt R, Francis JL, Amirkhosravi A. Anti-metastatic effect of a non-anticoagulant low-molecular-weight heparin versus the standard low-molecular-weight heparin, enoxaparin. *Thromb Haemost* 2006;**96**:816–21.
45. Hu L, Lee M, Campbell W, Perez-Soler R, Karpatkin S. Role of endogenous thrombin in tumor implantation, seeding, and spontaneous metastasis. *Blood* 2004;**104**:2746–51.
46. Im JH, Fu W, Wang H, Bhatia SK, Hammer DA, Kowalska MA, et al. Coagulation facilitates tumor cell spreading in the pulmonary vasculature during early metastatic colony formation. *Cancer Res* 2004;**64**:8613–9.
47. Petitou M, Duchaussoy P, Herbert JM, Duc G, El Hajji M, Branellec JF, et al. The synthetic pentasaccharide fondaparinux: first in the class of antithrombotic agents that selectively inhibit coagulation factor Xa. *Semin Thromb Hemost* 2002;**28**:393–402.
48. Amirkhosravi A, Meyer T, Chang JY, Amaya M, Siddiqui F, Desai H, et al. Tissue factor pathway inhibitor reduces experimental lung metastasis of B16 melanoma. *Thromb Haemost* 2002;**87**:930–6.
49. Vlodavsky I, Abboud-Jarrous G, Elkin M, Naggi A, Casu B, Sasisekharan R, et al. The impact of heparanase and heparin on cancer metastasis and angiogenesis. *Pathophysiol Haemost Thromb* 2006;**35**:116–27.
50. Friedmann Y, Vlodavsky I, Aingorn H, Aviv A, Peretz T, Pecker I, et al. Expression of heparanase in normal, dysplastic, and neoplastic human colonic mucosa and stroma. Evidence for its role in colonic tumorigenesis. *Am J Pathol* 2000;**157**:1167–75.
51. Ilan N, Elkin M, Vlodavsky I. Regulation, function and clinical significance of heparanase in cancer metastasis and angiogenesis. *Int J Biochem Cell Biol* 2006;**38**:2018–39.
52. Vlodavsky I, Ilan N, Nadir Y, Brenner B, Katz BZ, Naggi A, et al. Heparanase, heparin and the coagulation system in cancer progression. *Thromb Res* 2007;**120**(Suppl 2):S112–20.
53. Edovitsky E, Elkin M, Zcharia E, Peretz T, Vlodavsky I. Heparanase gene silencing, tumor invasiveness, angiogenesis, and metastasis. *J Natl Cancer Inst* 2004;**96**:1219–30.
54. Yang Y, MacLeod V, Dai Y, Khotskaya-Sample Y, Shriver Z, Venkataraman G, et al. The syndecan-1 heparan sulfate proteoglycan is a viable target for myeloma therapy. *Blood* 2007;**110**:2041–8.

55. Casu B, Vlodavsky I, Sanderson RD. Non-anticoagulant heparins and inhibition of cancer. *Pathophysiol Haemost Thromb* 2008;**36**:195–203.
56. Miao HQ, Elkin M, Aingorn E, Ishai-Michaeli R, Stein CA, Vlodavsky I. Inhibition of heparanase activity and tumor metastasis by laminarin sulfate and synthetic phosphorothioate oligodeoxynucleotides. *Int J Cancer* 1999;**83**:424–31.
57. Parish CR, Freeman C, Brown KJ, Francis DJ, Cowden WB. Identification of sulfated oligosaccharide-based inhibitors of tumor growth and metastasis using novel *in vitro* assays for angiogenesis and heparanase activity. *Cancer Res* 1999;**59**:3433–41.
58. Yang Y, Macleod V, Bendre M, Huang Y, Theus AM, Miao HQ, et al. Heparanase promotes the spontaneous metastasis of myeloma cells to bone. *Blood* 2005;**105**:1303–9.
59. Nelson RM, Cecconi O, Roberts WG, Aruffo A, Linhardt RJ, Bevilacqua MP. Heparin oligosaccharides bind L- and P-selectin and inhibit acute inflammation. *Blood* 1993;**82**:3253–8.
60. Koenig A, Norgard-Sumnicht K, Linhardt R, Varki A. Differential interactions of heparin and heparan sulfate glycosaminoglycans with the selectins. Implications for the use of unfractionated and low molecular weight heparins as therapeutic agents. *J Clin Invest* 1998;**101**:877–89.
61. McEver RP. Selectin-carbohydrate interactions during inflammation and metastasis. *Glycoconj J* 1997;**14**:585–91.
62. Ley K. The role of selectins in inflammation and disease. *Trends Mol Med* 2003;**9**:263–8.
63. Kansas GS. Selectins and their ligands: current concepts and controversies. *Blood* 1996;**88**:3259–87.
64. Lowe JB. Selectin ligands, leukocyte trafficking, and fucosyltransferase genes. *Kidney Int* 1997;**51**:1418–26.
65. McEver RP, Cummings RD. Role of PSGL-1 binding to selectins in leukocyte recruitment. *J Clin Invest* 1997;**100**:S97–S103.
66. Aruffo A, Kolanus W, Walz G, Fredman P, Seed B. CD62/P-selectin recognition of myeloid and tumor cell sulfatides. *Cell* 1991;**67**:35–44.
67. Varki A. Selectin ligands: will the real ones please stand up? *J Clin Invest* 1997;**99**:158–62.
68. Kim YJ, Varki A. Perspectives on the significance of altered glycosylation of glycoproteins in cancer. *Glycoconj J* 1997;**14**:569–76.
69. Kim YS, Gum Jr. J, Brockhausen I. Mucin glycoproteins in neoplasia. *Glycoconj J* 1996;**13**:693–707.
70. Nakamori S, Kameyama M, Imaoka S, Furukawa H, Ishikawa O, Sasaki Y, et al. Increased expression of sialyl Lewis x antigen correlates with poor survival in patients with colorectal carcinoma: clinicopathological and immunohistochemical study. *Cancer Res* 1993;**53**:3632–7.
71. Tatsumi M, Watanabe A, Sawada H, Yamada Y, Shino Y, Nakano H. Immunohistochemical expression of the sialyl Lewis x antigen on gastric cancer cells correlates with the presence of liver metastasis. *Clin Exp Metastasis* 1998;**16**:743–50.
72. Ogawa J, Tsurumi T, Yamada S, Koide S, Shohtsu A. Blood vessel invasion and expression of sialyl Lewis x and proliferating cell nuclear antigen in stage I non-small cell lung cancer. Relation to postoperative recurrence. *Cancer* 1994;**73**:1177–83.
73. Jorgensen T, Berner A, Kaalhus O, Tveter KJ, Danielsen HE, Bryne M. Up-regulation of the oligosaccharide sialyl LewisX: a new prognostic parameter in metastatic prostate cancer. *Cancer Res* 1995;**55**:1817–9.
74. Idikio HA. Sialyl-Lewis-X, Gleason grade and stage in non-metastatic human prostate cancer. *Glycoconj J* 1997;**14**:875–7.
75. Tozawa K, Okamoto T, Kawai N, Hashimoto Y, Hayashi Y, Kohri K. Positive correlation between sialyl Lewis X expression and pathologic findings in renal cell carcinoma. *Kidney Int* 2005;**67**:1391–6.
76. Renkonen J, Paavonen T, Renkonen R. Endothelial and epithelial expression of sialyl Lewis(x) and sialyl Lewis(a) in lesions of breast carcinoma. *Int J Cancer* 1997;**74**:296–300.

77. Steeg PS. Tumor metastasis: mechanistic insights and clinical challenges. *Nat Med* 2006;**12**:895–904.
78. Gupta GP, Massague J. Cancer metastasis: building a framework. *Cell* 2006;**127**:679–95.
79. Kim YJ, Borsig L, Varki NM, Varki A. P-selectin deficiency attenuates tumor growth and metastasis. *Proc Natl Acad Sci USA* 1998;**95**:9325–30.
80. Borsig L, Wong R, Hynes RO, Varki NM, Varki A. Synergistic effects of L- and P-selectin in facilitating tumor metastasis can involve non-mucin ligands and implicate leukocytes as enhancers of metastasis. *Proc Natl Acad Sci USA* 2002;**99**:2193–8.
81. Läubli H, Stevenson JL, Varki A, Varki NM, Borsig L. L-selectin facilitation of metastasis involves temporal induction of fut7-dependent ligands at sites of tumor cell arrest. *Cancer Res* 2006;**66**:1536–42.
82. Läubli H, Borsig L. Selectins as mediators of lung metastasis. *Cancer Microenviron* 2010;**3**: [in press] DOI 10.1007/s12307-010-0043-6.
83. Fuster MM, Brown JR, Wang L, Esko JD. A disaccharide precursor of sialyl Lewis X inhibits metastatic potential of tumor cells. *Cancer Res* 2003;**63**:2775–81.
84. Nieswandt B, Hafner M, Echtenacher B, Mannel DN. Lysis of tumor cells by natural killer cells in mice is impeded by platelets. *Cancer Res* 1999;**59**:1295–300.
85. Bevilacqua MP, Nelson RM. Endothelial-leukocyte adhesion molecules in inflammation and metastasis. *Thromb Haemost* 1993;**70**:152–4.
86. Wu QD, Wang JH, Condron C, Bouchier-Hayes D, Redmond HP. Human neutrophils facilitate tumor cell transendothelial migration. *Am J Physiol Cell Physiol* 2001;**280**:C814–22.
87. Liang S, Sharma A, Peng HH, Robertson G, Dong C. Targeting mutant (V600E) B-Raf in melanoma interrupts immunoediting of leukocyte functions and melanoma extravasation. *Cancer Res* 2007;**67**:5814–20.
88. Qian B, Deng Y, Im JH, Muschel RJ, Zou Y, Li J, et al. A distinct macrophage population mediates metastatic breast cancer cell extravasation, establishment and growth. *PLoS ONE* 2009;**4**:e6562.
89. Biancone L, Araki M, Araki K, Vassalli P, Stamenkovic I. Redirection of tumor metastasis by expression of E-selectin *in vivo*. *J Exp Med* 1996;**183**:581–7.
90. Brodt P, Fallavollita L, Bresalier RS, Meterissian S, Norton CR, Wolitzky BA. Liver endothelial E-selectin mediates carcinoma cell adhesion and promotes liver metastasis. *Int J Cancer* 1997;**71**:612–9.
91. Khatib AM, Kontogiannea M, Fallavollita L, Jamison B, Meterissian S, Brodt P. Rapid induction of cytokine and E-selectin expression in the liver in response to metastatic tumor cells. *Cancer Res* 1999;**59**:1356–61.
92. Mannori G, Santoro D, Carter L, Corless C, Nelson RM, Bevilacqua MP. Inhibition of colon carcinoma cell lung colony formation by a soluble form of E-selectin. *Am J Pathol* 1997;**151**:233–43.
93. Stevenson JL, Varki A, Borsig L. Heparin attenuates metastasis mainly due to inhibition of P- and L-selectin, but non-anticoagulant heparins can have additional effects. *Thromb Res* 2007;**120**(Suppl 2):S107–11.

Vascular Dermatan Sulfate and Heparin Cofactor II

Douglas M. Tollefsen

Hematology Division, Washington University Medical School, St. Louis, Missouri, USA

Heparin cofactor II (HCII) is a plasma protease inhibitor of the serpin family that inactivates thrombin by forming a covalent 1:1 complex. The rate of complex formation increases more than 1000-fold in the presence of dermatan sulfate (DS). Endothelial injury allows circulating HCII to enter the vessel wall, where it binds to DS and presumably becomes activated. Mice that lack HCII develop carotid artery thrombosis more rapidly than wild-type mice after oxidative damage to the endothelium. These mice also have increased arterial neointima formation following mechanical injury and develop more extensive atherosclerotic lesions when made hypercholesterolemic. Similarly, low plasma HCII levels appear to be a risk factor for atherosclerosis and in-stent restenosis in human subjects. These observations suggest that a major function of the HCII–DS system is to regulate the physiologic response to arterial injury.

Progress in Molecular Biology and Translational Science, Vol. 93
DOI: 10.1016/S1877-1173(10)93015-9

Dermatan sulfate (DS) is an abundant glycosaminoglycan in the extracellular matrix of the skin as well as in many other tissues.[1] It is a component of the small leucine-rich proteoglycan decorin, which binds to collagen and other matrix molecules and plays an important role in the structural integrity of the skin. DS is also present in bone, cartilage, cornea, and blood vessels, where it is incorporated into several other types of proteoglycans. A large variety of molecules interact with DS proteoglycans, including growth factors, cytokines, lipoproteins, matrix proteins, coagulation factors, and cell surface receptors. Thus, DS has been implicated in physiologic and pathologic processes as diverse as oncogenesis, susceptibility to pathogens, inflammation, and wound repair.

This chapter focuses on the interaction between DS and the plasma protein heparin cofactor II (HCII).[2] Binding to DS "activates" HCII, enabling it to rapidly inhibit the blood coagulation protease thrombin. In some respects, this reaction resembles the well described activation of antithrombin (AT) by heparin or heparan sulfate (HS). In both cases, protease inhibition is accelerated by a combination of allosteric and template effects mediated by the glycosaminoglycan, although the mechanisms differ in important details.[3] Physiologically, AT is thought to bind to HS synthesized by vascular endothelial cells and to inhibit thrombosis within the lumen of blood vessels lined by an intact endothelium. By contrast, HCII appears to interact with DS in deeper layers of the vessel wall only after endothelial disruption and to modulate the response to vascular injury (Fig. 1).

I. Biochemistry of HCII

Brinkhous and coworkers reported in 1939 that the anticoagulant activity of heparin is mediated by a substance present in normal human plasma.[4] The nature of this substance remained unknown until 1968, when Abildgaard purified the protein AT (formerly called AT III) and demonstrated that it possesses "heparin cofactor" activity.[5] Soon thereafter, AT was shown to inhibit the coagulation proteases thrombin, factor Xa and factor IXa by formation of a 1:1 stoichiometric complex that is stable during SDS-PAGE.[6,7] Initially, AT was thought to account for all of the heparin cofactor activity in plasma.

Briginshaw and Shanberge were among the first to note that human plasma contains a second heparin-dependent thrombin inhibitor, which they separated from AT by ion-exchange and gel filtration chromatography.[8,9] Although this inhibitor did not react with factor Xa and reacted poorly with thrombin in the absence of heparin, it comprised roughly one-third of the total heparin-dependent thrombin inhibitory activity of plasma. In the early 1980s, Tollefsen and Blank reported that ^{125}I-thrombin forms two SDS-stable complexes in plasma containing heparin,[10] and other investigators found that the

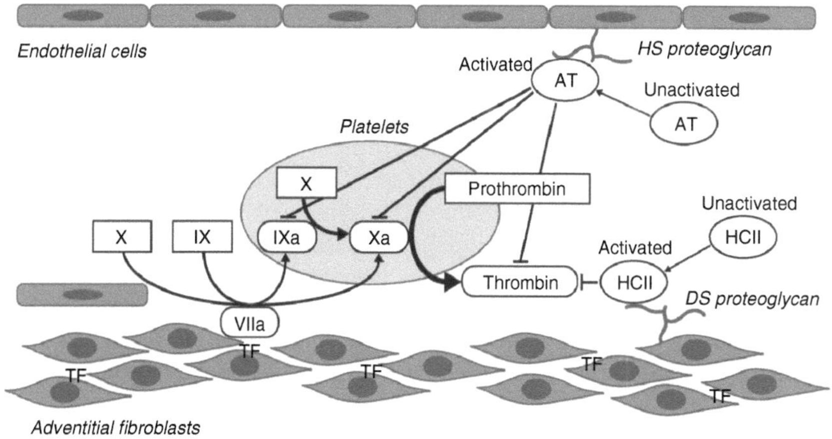

FIG. 1. Model for activation of HCII and AT by vascular glycosaminoglycans. Endothelial injury initiates blood coagulation by allowing circulating factor VIIa to bind to tissue factor (TF) expressed by adventitial fibroblasts. The VIIa-TF complex then converts factors IX and X to their activated forms (IXa and Xa). Factors IXa and Xa assemble with nonenzymatic cofactors on the surface of aggregating platelets to generate larger amounts of Xa and thrombin. Thrombin then cleaves fibrinogen to form the fibrin clot (not shown). AT becomes activated by binding to HS proteoglycans synthesized by normal endothelial cells and inhibits IXa, Xa, and thrombin, thus preventing thrombi from forming within normal blood vessels. HCII is activated by DS proteoglycans deeper in the vessel wall and regulates thrombin activity following vascular injury.

heparin cofactor activity of plasma exceeds the amount predicted from the AT antigen concentration.[11,12] The existence of the second heparin cofactor was firmly established when HCII was purified to homogeneity from human plasma and shown to be distinct from AT.[13]

Human HCII is a single-chain glycoprotein containing 480 amino acid residues.[14,15] It is encoded by the SERPIND1 gene on human chromosome 22q11. HCII mRNA is highly expressed in the liver, which appears to be its major site of biosynthesis.[16] Although low levels of HCII mRNA are also detectable in other tissues, the significance of extrahepatic expression of HCII is unclear.[17] HCII is secreted into the bloodstream, where it circulates at a concentration of approximately 1.2 ± 0.4 μM (mean $\pm$ 2 S.D.).[18] The thrombin–HCII complex is cleared within minutes by the LDL receptor-related protein on hepatocytes.[19]

A. HCII Inhibits Thrombin by the Typical Serpin Mechanism

HCII and AT are members of the serpin family, a large set of homologous proteins that are ~30% identical in amino acid sequence.[20] The serpins have a common tertiary structure characterized by three β-sheets and 8–9 α-helices and most, but not all, function as inhibitors of serine proteases. In the first step of the inhibition reaction, the serpin docks with the catalytic site of its target protease to form a noncovalent encounter complex. Once the encounter complex has formed, the reactive site (P1–P1′) peptide bond of the serpin undergoes nucleophilic attack by the catalytic serine hydroxyl group of the protease. This interaction results in the cleavage of the reactive site with the formation of an acyl bond linking the hydroxyl group to the carbonyl group of the P1 residue. Proteolytic cleavage triggers a rapid conformational change, during which the entire reactive site loop becomes inserted into β-sheet A of the serpin.[21] Consequently, the P1 residue—with the protease still covalently attached—is translocated to the opposite pole of the serpin. This molecular rearrangement deforms the catalytic triad, preventing deacylation and release of the protease from the complex.[22]

The protease specificity of a serpin is determined in part by the amino acid present at the P1 position of the reactive site. The P1 residue of AT is Arg^{393}, and the major target proteases for AT (thrombin, factor Xa and factor IXa) all preferentially attack Arg-X peptide bonds in their natural substrates. By contrast, the P1 residue of HCII is Leu^{444}.[23] In the absence of a glycosaminoglycan, HCII inhibits thrombin at about one-tenth the rate of AT and is virtually inactive against several other proteases involved in coagulation or fibrinolysis.[24] Moreover, HCII inhibits chymotrypsin several times faster than thrombin, although this reaction is probably not physiological and is not accelerated by glycosaminoglycans.[25,26] Substitution of arginine in place of Leu^{444} increases the rate of inhibition of thrombin approximately 100-fold in the absence of a glycosaminoglycan and abolishes chymotrypsin inhibition.[26] These observations emphasize the importance of the P1 residue in restricting the protease specificity of HCII and suggest that the ability of HCII to inhibit thrombin may involve regions of the protein apart from the reactive site.

B. Glycosaminoglycans Activate HCII Allosterically

HCII inhibits thrombin at a very slow basal rate, with a second-order rate constant (k_2) on the order of 10^4–10^5 M^{-1} min^{-1}. The rate increases dramatically in the presence of heparin, HS, or DS ($k_2 \sim 10^8$ M^{-1} min^{-1}).[27] Expressed in physiologic terms, the half-life for thrombin inhibition by 1 μM HCII would decrease from ~5 min in the absence of a glycosaminoglycan to ~50 ms in the presence of heparin or DS.

The rate enhancement largely depends on a cluster of acidic amino acids in HCII that resembles the leech polypeptide hirudin, which binds tightly to anion-binding exosite I of thrombin.[28] This cluster is contained within an extended N-terminal domain of ~80 residues that is not present in AT or other serpins. Deletion of the acidic domain in recombinant HCII greatly diminishes the ability of DS or heparin to stimulate inhibition of thrombin but has little effect on the rate of inhibition in the absence of a glycosaminoglycan.[29–31] Furthermore, N-terminal deletion mutants bind to heparin-agarose more tightly than full-length HCII, suggesting a model in which intramolecular interactions occur between the acidic domain and the glycosaminoglycan-binding site of HCII. According to this model, binding of a glycosaminoglycan to HCII would competitively displace the N-terminal acidic domain, allowing it to interact with exosite I of thrombin and thus facilitate the formation of the thrombin–HCII encounter complex (Fig. 2). Mutation of amino acid residues in exosite I impair the ability of HCII to inhibit thrombin in the presence of a glycosaminoglycan,[32–34] as does introduction of novel disulfide bonds that tether the N-terminal acidic domain to the body of HCII.[35] Direct evidence for interaction of the N-terminal acidic domain of HCII with exosite I of thrombin was recently obtained by X-ray crystallography.[36]

The thrombin–HCII reaction may also be stimulated to some degree by a bridging mechanism in which both proteins bind to a single glycosaminoglycan chain.[37] Thus, both heparin and DS increase the reaction rate of HCII N-terminal

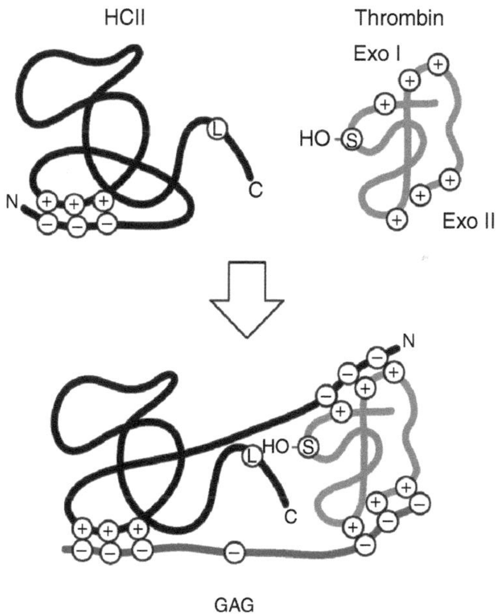

FIG. 2. Mechanism of activation of HCII. According to this model, glycosaminoglycan (GAG) binding displaces the N-terminal acidic domain of HCII, allowing it to interact with anion-binding exosite I (Exo I) of thrombin. If the GAG chain is of sufficient length, it may also bind to exosite II (Exo II) of thrombin, forming a bridge between the two proteins. These interactions bring the active site serine (HO-S) of thrombin into proper alignment with the reactive site leucine (L) residue of HCII and lead to rapid formation of the covalent thrombin–HCII complex.

deletion mutants by a factor of 10–100, although this is far less than the 10,000-fold increase observed with full-length HCII. However, mutations in exosite II of thrombin that abolish binding to heparin or DS have little effect on the ability of these glycosaminoglycans to stimulate the thrombin–HCII reaction.[33] In addition, kinetic studies with heparin fragments as small as 10-mers suggest that simultaneous binding of both proteins to a glycosaminoglycan chain is not required for stimulation; thus, a 2000-fold increase in the rate of thrombin inhibition can be achieved if sufficiently high concentrations of these fragments are used.[38] These observations suggest that glycosaminoglycans accelerate the thrombin–HCII reaction primarily by allosteric effects rather than by a bridging mechanism. DS and heparin preparations generally used for *in vitro* studies bind more tightly to thrombin than to HCII. Therefore, the preferred pathway for allosteric activation of HCII may involve formation of a thrombin–glycosaminoglycan complex which then binds to HCII.[37]

C. The Binding Sites for Heparin and DS in HCII Are Distinct

Analysis of the natural variant $HCII^{Oslo}$ ($Arg^{189} \rightarrow$ His) established that heparin and DS interact with different amino acid residues on the surface of the protein.[39] This mutation results in a large decrease ($\sim$60-fold) in the affinity of HCII for DS but does not significantly alter the affinity for heparin. Arg^{189} occurs within a cluster of basic amino acid residues that can be aligned with the heparin-binding site in AT but are poorly conserved in other serpins. Mutations of Lys^{173}, Arg^{184}, and Arg^{185} in recombinant HCII affect heparin binding, whereas mutations of Arg^{184}, Arg^{185}, Arg^{189}, Arg^{192}, and Arg^{193} affect DS binding.[29,30,39–41] These results indicate that the binding sites in HCII for heparin and DS overlap but are not identical.

D. HCII Binds to Specific DS Oligosaccharides

HCII binds to heparin with a lower affinity in comparison with AT; consequently, higher concentrations of heparin or HS are necessary to activate HCII.[27] When heparin is present in plasma at typical therapeutic concentrations, it preferentially stimulates the thrombin–AT reaction.[10] HCII appears to bind nonspecifically to heparin oligosaccharides that contain at least four sugar residues, regardless of their composition.[42] In addition, HCII is activated by heparin chains that lack the 3-*O*-sulfated glucosamine residue required for high-affinity binding to AT.[43] Certain other natural, synthetic, or semisynthetic polyanions, some of which are of pharmaceutical interest, also stimulate HCII activity.[44]

HCII is unique among serpins in its ability to be activated by DS. DS is synthesized as a repeating polymer of D-glucuronic acid (GlcA) and *N*-acetyl-D-galactosamine (GalNAc), which is then modified by epimerization of GlcA to L-iduronic acid (IdoA), sulfation of the 2-OH group of IdoA, and sulfation of the 4- and/or 6-OH groups of GalNAc.[1] Incomplete epimerization and *O*-sulfation lead to microheterogeneity within the DS polymers.

A minor subpopulation of oligosaccharides obtained by partial depolymerization of DS binds to HCII. The smallest fragment of porcine skin DS that binds to HCII at a physiologic salt concentration is the hexasaccharide containing three $IdoA2SO_3 \rightarrow GalNAc4SO_3$ disaccharide subunits, which comprise only 5% of the total disaccharides present in the intact polymer.[45] Hexasaccharides that contain only one or two of these disulfated disaccharides also bind to HCII, albeit with lower affinity, while hexasaccharides composed entirely of monosulfated uronic acid $(UA) \rightarrow GalNAc4SO_3$ subunits do not bind. Pavão and coworkers showed that the ability of DS to bind and activate HCII is not simply a function of its overall charge density but appears to require the presence of $GalNAc4SO_3$ residues.[46,47] Thus, invertebrate DS polymers composed mainly of $IdoA2SO_3 \rightarrow GalNAc4SO_3$ activate HCII at low concentrations, while polymers composed mainly of $IdoA2SO_3 \rightarrow GalNAc6SO_3$ are 1000 times less active. HCII-binding sites in DS from other tissues, such as porcine intestinal mucosa, contain one or more $UA \rightarrow GalNAc4,6SO_3$ disaccharide subunits.[48] Therefore, modification of $UA \rightarrow GalNAc4SO_3$ subunits in the DS polymer by either 2-*O*-sulfation of IdoA or 6-*O*-sulfation of GalNAc can generate HCII-binding sites. The extent of these biosynthetic modifications varies in different tissues and may determine the anatomical sites in which HCII becomes activated.

E. Little is Known About the Tissue Distribution of HCII

Clearance studies of HCII in humans, baboons, and rabbits suggest that approximately 40–60% of the protein distributes into a noncirculating compartment within hours after intravenous injection and that the circulating pool is cleared with a half-life of 1.6–2.5 days.[49–51] Although HCII levels have not been measured in interstitial fluid, lymph collected from rabbit hindlimbs contains other hemostatic proteins at substantial concentrations, which are roughly in inverse proportion to the proteins' molecular weights.[52] Since AT, which is similar to HCII in molecular weight, is present in hindlimb lymph at approximately 40% of its plasma concentration, it is likely that HCII would also be found in lymph. These and other plasma proteins appear to cross the endothelium by transcytosis mainly at the level of the capillary or postcapillary venule.[53] Although relatively little transcytosis is thought to occur across the endothelia of large blood vessels, investigators have reported that HCII is present in the intima of normal human arteries[54] and that radiolabeled HCII is taken up into the intima or media of the intact rabbit aorta.[49]

F. Vascular Glycosaminoglycans Activate HCII *In Vitro*

Glycosaminoglycans extracted from blood vessels prolong the clotting time of plasma *in vitro*. In one study, Tovar *et al.*[55] reported that although similar amounts of DS and HS are present in the human thoracic aorta (1.0–1.3 μg hexuronic acid per mg dry weight), aortic DS is several times more potent than HS in its ability to prolong the activated partial thromboplastin time. Similar results were obtained with glycosaminoglycans extracted from the human saphenous vein. Therefore, DS appears to be the major anticoagulant glycosaminoglycan in the walls of large blood vessels. Since DS does not prolong the clotting time of HCII-depleted plasma,[56] its anticoagulant effect depends upon the presence of HCII.

The major cellular components of the vessel wall include endothelial cells, smooth muscle cells and fibroblasts. McGuire and Tollefsen[57] found that fibroblast monolayers stimulate the thrombin–HCII reaction. Analysis of extracts from these cells indicated that a small DS-containing proteoglycan is primarily responsible for activation of HCII. Although the fibroblast proteoglycan was not identified, biglycan and decorin are possible candidates, since both of these proteoglycans isolated from other tissues have been shown to activate HCII.[58] Vascular smooth muscle cells in culture also stimulate the thrombin–HCII reaction, while umbilical vein endothelial cells do not.[57] These observations have led to the hypothesis that disruption of the endothelial barrier allows HCII in the bloodstream to enter the vessel wall, become activated by binding to DS, and regulate the activity of thrombin produced at the site of vascular injury (Fig. 1).

II. Studies with HCII Knockout Mice

Experiments with HCII-deficient mice generated independently in St. Louis and Tokushima provide evidence that interaction of HCII with vascular DS modulates the response to arterial injury. He *et al.*[59] replaced the entire first coding exon of the HCII gene with the *neo* gene by homologous recombination in embryonic stem cells. The deleted exon encodes the N-terminal half of the protein, which contains several structural elements required for activity—namely, the acidic cluster, the glycosaminoglycan-binding site, and three of the five strands of β-sheet A. Chimeric $HCII^{+/-}$ (St. Louis) mice derived from these embryonic stem cells produced $HCII^{-/-}$ offspring at close to the expected Mendelian frequency (21.6%). Plasma from $HCII^{-/-}$ mice contains no HCII antigen detectable by Western blotting or ELISA using polyclonal anti-HCII antibodies and lacks HCII activity assayed as DS-dependent thrombin inhibition. However, normal levels of plasma AT measured as heparin-dependent Xa inhibitory activity are present in $HCII^{-/-}$ mice. $HCII^{-/-}$ (St. Louis) mice grow and survive normally, are fertile, and have normal hematopoietic, renal and hepatic function.

Aihara *et al.*[60] inactivated the HCII gene by replacement of a portion of the first coding exon with the *neo* gene. The resulting $HCII^{+/-}$ (Tokushima) mice were studied after being backcrossed for ten generations with C57BL/6J mice. In contrast to the viability of $HCII^{-/-}$ (St. Louis) mice, the $HCII^{-/-}$ (Tokushima) genotype is lethal before day E8.5 of gestation. Aihara *et al.* used fluorescence *in situ* hybridization to demonstrate correct targeting of the *neo* cassette to the HCII locus on chromosome 16 and real-time PCR to show that three neighboring genes (*Ube2l3*, *Pik4ca*, and *Snap29*) are expressed at normal levels in the livers of $HCII^{+/-}$ (Tokushima) mice. They concluded that the more severe phenotype of their knockout could be due to minor strain-dependent differences in genetic background.

A. Thrombosis

Unchallenged $HCII^{-/-}$ (St. Louis) mice have no evidence of spontaneous thrombosis on necropsy. However, they form occlusive thrombi in the carotid artery more quickly than $HCII^{+/+}$ mice after disruption of the endothelium.[59] In these experiments, the photoactive dye Rose Bengal is injected intravenously, and the carotid artery is transilluminated with a green laser to produce singlet oxygen within the lumen of the vessel. Electron microscopic studies indicate that this procedure causes rapid detachment of the endothelial cells, leaving the elastic lamina and underlying smooth muscle cells intact.[61] Disruption of the endothelium allows factor VIIa in the plasma to come in contact with tissue factor (TF) in the arterial adventitia (Fig. 1).[62] The factor VIIa/TF complex then converts factor X to factor Xa, which in combination with factor Va converts prothrombin to thrombin.[63] Thrombin not only converts fibrinogen to fibrin monomers, which polymerize to form a clot, but it also interacts with G-protein-coupled protease activated receptors (specifically, PAR3 and PAR4) on the platelet membrane to stimulate platelet aggregation and degranulation.[64]

In the C57BL/6 genetic background, complete and stable occlusion of the carotid artery of $HCII^{+/+}$ mice occurs in 60 ± 12 min (mean $\pm$ S.D.), while thrombosis occurs in 34 ± 4 min in $HCII^{-/-}$ mice[65] (Fig. 3). The occlusion times are shorter for both genotypes in a mixed C57/129 genetic background (24 ± 5 min for $HCII^{+/+}$ vs. 15 ± 3 min for $HCII^{-/-}$ mice), indicating that genes outside the HCII locus modify the thrombotic phenotype.[59] The shortened occlusion time of $HCII^{-/-}$ mice is normalized by intravenous injection of purified HCII prior to induction of endothelial injury. These observations provide direct evidence that HCII has antithrombotic activity *in vivo* and focus attention on the arterial system as a potential site of action of HCII.

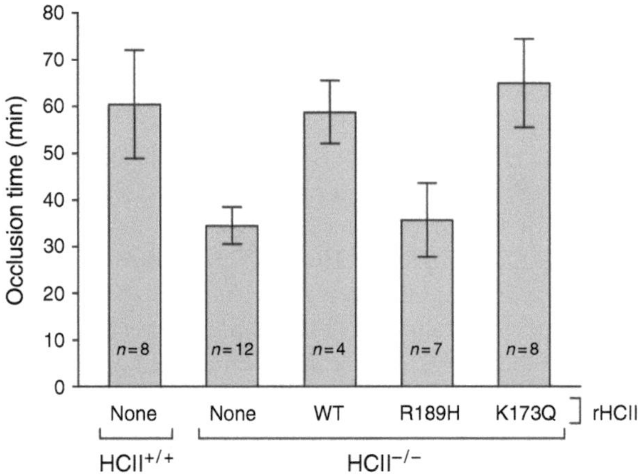

FIG. 3. Thrombotic occlusion times after photochemical injury of the mouse carotid artery. Wild-type ($HCII^{+/+}$) or knockout ($HCII^{-/-}$) mice were injected intravenously with saline (none) or purified recombinant HCII (rHCII) before induction of endothelial injury. Recombinant proteins included wild-type (WT) HCII and variants with decreased affinity for either DS (R189H) or heparin (K173Q). The abnormally rapid occlusion time of $HCII^{-/-}$ mice was corrected by $HCII^{WT}$ or $HCII^{K173Q}$ but not by $HCII^{R189H}$, indicating that the DS-binding site of HCII is required for antithrombotic activity *in vivo*. Data from He *et al.*[65]

Intravenous administration of porcine skin DS to $HCII^{+/+}$ mice prolongs the carotid artery occlusion time in a dose-dependent manner, reaching a maximum of approximately twice the baseline value at a dose of 10 μg DS per g body weight.[66] Unexpectedly, the onset of antithrombotic activity is delayed for 5–15 min after the injection of DS at this dose, during which time more than 80% of the DS is cleared from the circulation. Full antithrombotic activity is still observed even if injury is induced 60 min after the injection of DS, when only 7% of the injected DS remains in the circulation. These pharmacokinetic observations suggest that DS expresses its antithrombotic activity after being transferred from the plasma to sites in the vessel wall.

DS administration does not prolong the occlusion time of $HCII^{-/-}$ mice,[66] which is consistent with the observation that prolongation of the *in vitro* clotting time of plasma by DS depends on the presence of HCII.[56] By contrast, heparin prolongs the occlusion time to an equal degree in $HCII^{+/+}$ and $HCII^{-/-}$ mice by activation of AT, which is present at similar levels in both types of mice. Proposed HCII-independent mechanisms of action of DS, such as release of tissue plasminogen activator,[67,68] do not appear to be as important as activation of HCII in the carotid artery thrombosis model.

B. Neointima Formation

Experiments with both $HCII^{-/-}$ (St. Louis) and $HCII^{+/-}$ (Tokushima) mice suggest that HCII downregulates the accumulation of smooth muscle cells in the arterial intima after mechanical injury.[60,69] Neointima formation was induced in $HCII^{-/-}$ (St. Louis) or $HCII^{+/+}$ mice by dilation of the common carotid artery to approximately twice its normal diameter using a wire probe tipped with an epoxy bead,[70] a technique that mimics balloon angioplasty. Three weeks after injury, serial cross-sections of the artery were prepared and the neointimal area determined. The median lesion area was 2–3 times greater in $HCII^{-/-}$ mice in comparison with $HCII^{+/+}$ mice.[69]

Femoral arteries of $HCII^{+/-}$ (Tokushima) mice were injured by placement of a polyethylene cuff around the vessel or by insertion of a wire into the vessel lumen.[60] The arteries were removed for examination 4 weeks later. After both forms of injury, neointimal areas were two to three times greater in $HCII^{+/-}$ mice in comparison with $HCII^{+/+}$ mice. This is a striking result, considering the fact that the plasma HCII activity in the heterozygous mice was decreased by only 40%. In addition, the neointima formed in $HCII^{+/-}$ mice showed evidence of increased proliferation of vascular mesenchymal cells by bromodeoxyuridine staining, increased expression of several inflammatory cytokines and chemokines, and a greater number of cells that expressed the thrombin receptor, protease-activated receptor-1 (PAR-1). Intraperitoneal administration of purified human HCII to the $HCII^{+/-}$ mice decreased neointimal areas to the level observed in $HCII^{+/+}$ mice, confirming that the degree of neointima formation is related to the level of HCII.

HCII most likely functions in these models by inhibiting thrombin. In support of this hypothesis, the amount of thrombin amidolytic activity in rinsed carotid artery segments harvested from $HCII^{-/-}$ (St. Louis) mice 30 min after mechanical dilation was approximately twice that of injured arteries from $HCII^{+/+}$ mice.[69] Furthermore, staining of frozen sections with the thrombin-specific inhibitor hirudin indicated that more thrombin was present in the injured carotid arteries obtained from $HCII^{-/-}$ mice than from $HCII^{+/+}$ mice.

Several independent lines of investigation suggest that thrombin promotes formation of the neointima. In humans, the earliest histologic response to angioplasty and stent placement includes local deposition of fibrin and platelets, providing good evidence for the generation of thrombin in this setting.[71] Thrombin can activate PAR-1 on nearby endothelial cells.[64] In response to this stimulus, the endothelial cells express adhesion molecules on their surface and release a variety of chemokines and other mediators that recruit platelets and leukocytes. Thus, thrombin could play a role in the infiltration of neutrophils, lymphocytes and macrophages that is observed to occur within the first few days after angioplasty and stent placement. Over the next 2–4 weeks, the fibrin and platelets disappear, and restenosis occurs as a result of accumulation of smooth muscle cells and deposition of extracellular matrix in the neointima.[71] Thrombin may induce smooth muscle cell proliferation both directly by activation of PAR-1 on these cells and indirectly by causing platelets to secrete platelet-derived growth factor. Therefore, thrombin could have multiple effects in both the early and late stages of in-stent restenosis.

Neointima formation in response to mechanical injury of the carotid artery is less intense in PAR-1-null mice than in wild-type mice.[72] This difference appears to reflect defective thrombin signaling in smooth muscle cells or endothelial cells, since the platelets of PAR-1-null mice remain responsive to thrombin (in contrast to human platelets, mouse platelets express PAR-3 and PAR-4, but not PAR-1). In addition, a synthetic peptide analog that selectively antagonizes PAR-1 reduces neointima formation in rats.[73] Various thrombin-specific inhibitors (e.g., hirudin and derivatives thereof) also diminish neointima formation in experimental animals,[74] but other anticoagulants such as heparin are ineffective.

Vicente *et al.*[69] reported that intravenous administration of DS during the first 48 h after injury markedly inhibits neointima formation in $HCII^{+/+}$ mice but not in $HCII^{-/-}$ (St. Louis) mice, while heparin has no effect on neointima formation in $HCII^{+/+}$ mice. These results are consistent with a previous report that DS, but not heparin, attenuates smooth muscle cell proliferation after carotid injury in rabbits[75] and indicate that the anti-proliferative effect is HCII-dependent. Some of the thrombin generated after vascular injury may remain bound to fibrin or to components of the vessel wall in an active form that is protected from inhibition by circulating AT-heparin complexes but is susceptible to inhibition by HCII–DS.[76] The striking inhibitory effect of DS on neointima formation may have therapeutic implications in the setting of vascular surgery, angioplasty or stent placement.

C. Atherogenesis

Many of the thrombin-mediated cellular events that accompany neointima formation (e.g., platelet activation, stimulation of endothelial cells, expression of inflammatory mediators, and proliferation of smooth muscle cells) also occur during development of an atherosclerotic plaque,[77] and active thrombin has been detected in atherosclerotic lesions with hirudin and chromogenic substrates as probes.[78] TF is abundant in atherosclerotic plaques and provides a strong stimulus for thrombin generation during episodes of limited endothelial desquamation or disruption of the microvessels present within the plaque; such episodes are thought to initiate rapid expansion of the atheroma. Heterozygous deficiency of TF pathway inhibitor, which blocks the procoagulant activity of factor VIIa bound to TF, promotes atherosclerosis in apolipoprotein E (apoE)-null mice[79]; this type of observation supports the hypothesis that unregulated thrombin activity plays an important role in atherogenesis.

HCII has been detected in atherosclerotic plaques, where it would be in a position to inhibit thrombin.[80] Although DS is more abundant in atherosclerotic plaques than in normal arteries, its structure is altered such that its ability to stimulate HCII is reduced.[81] Hence, HCII's role in atherogenesis is unclear.

Studies with HCII-deficient mice provide experimental evidence that HCII regulates atherogenesis. Atherogenesis was induced in $HCII^{-/-}$ (St. Louis) or $HCII^{+/+}$ mice bred into an apoE-null background by feeding them a high-cholesterol, high-fat diet for 12 weeks. This diet caused marked elevation of serum cholesterol, triglycerides, free fatty acids and glucose in the apoE-null mice independent of the HCII genotype. Aortas were harvested at the end of the feeding period, and the percentage of the vessel surface occupied by atherosclerotic plaques was determined. $HCII^{-/-}$ mice had significantly larger plaque areas than $HCII^{+/+}$ mice, particularly in the aortic arch ($\sim$18% in $HCII^{-/-}$ mice vs. $\sim$11% $HCII^{+/+}$ mice).[69] No plaques were observed in $HCII^{-/-}$ or $HCII^{+/+}$ mice with wild-type apoE. Therefore, HCII deficiency by itself is insufficient to trigger atherogenesis in this model. A similar increase in plaque formation was observed with $HCII^{+/-}$ (Tokushima) mice bred into the apoE-null background.[60]

D. Vascular DS Interacts with HCII *In Vivo*

Recent experiments by He *et al.*[65] demonstrate that the antithrombotic activity of HCII depends on its ability to interact with DS *in vivo*. Prior to photochemical injury of the carotid artery, $HCII^{-/-}$ mice were injected intravenously with recombinant wild-type human HCII or with variants having single amino acid substitutions that reduce the affinity for DS ($Arg^{189} \rightarrow His$) or heparin ($Lys^{173} \rightarrow Gln$) (Fig. 3). Neither mutation alters the kinetics of thrombin inhibition in the absence of a glycosaminoglycan. $HCII^{-/-}$ mice reconstituted with either wild-type HCII or the Lys^{173} variant had occlusion times (59 ± 7 and 65 ± 9 min, respectively) that were similar to those of $HCII^{+/+}$ mice. By contrast, $HCII^{-/-}$ mice reconstituted with the Arg^{189} variant had occlusion times (36 ± 8 min) that were indistinguishable from those of $HCII^{-/-}$ mice given saline. These results indicate that the DS-binding site in HCII is required for antithrombotic activity *in vivo* and raise the possibility that HCII interacts with DS present in the arterial wall. The results of this experiment also suggest that interaction with heparin-like molecules (e.g., HS) may not be required for the antithrombotic activity of HCII.

The locations of DS and HS in the mouse carotid artery were established using immunochemical methods.[65] Frozen sections of the artery were first treated with chondroitin B-lyase, which cleaves DS to generate oligosaccharide "stubs" with unsaturated $\Delta^{4,5}$-hexuronic acid residues at their nonreducing termini, and then stained with a monoclonal antibody (ΔDi-4S) that recognizes the new epitope.[82] DS was readily detected in the adventitia of the carotid artery but not in the intima or media (Fig. 4). Additional sections were treated with *Flavobacterium* heparitinase, which cleaves HS, and stained with an appropriate antibody (ΔHS).[83] In contrast to DS, HS was detected mainly in the intima and media.

To identify HCII-binding sites in the arterial wall, frozen sections of carotid arteries from $HCII^{-/-}$ mice were incubated with purified recombinant HCII, rinsed to remove excess unbound protein, and stained with an anti-HCII

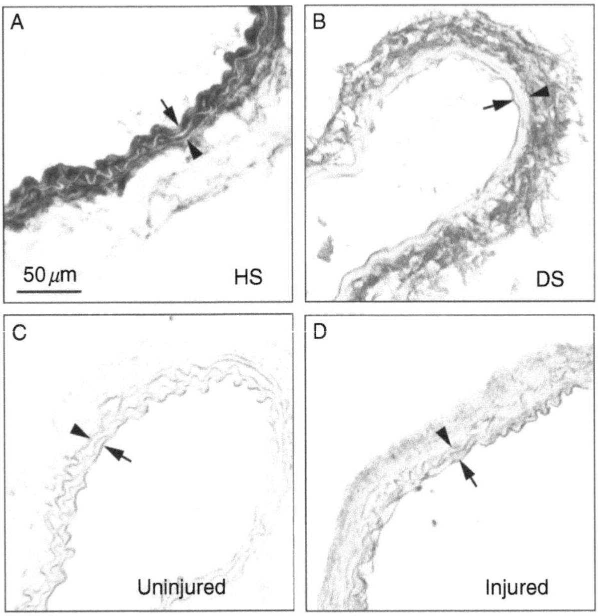

FIG. 4. Colocalization of HCII and DS in the arterial adventitia after endothelial injury. In the upper panels, HS and DS were localized by treating carotid artery sections with *Flavobacterium* heparitinase (panel A) or chondroitin B-lyase (panel B) and then staining with monoclonal antibodies specific for the degradation products. HS was found mainly in the intimal and medial layers between in internal elastic lamina (arrow) and the external elastic lamina (arrowhead), while DS was detected mainly in the adventitial layer outside the external elastic lamina. In the lower panels, carotid arteries were harvested from wild-type mice before (panel C) or 30 min after (panel D) the onset of photochemical injury. HCII antigen was detected with a polyclonal IgG. Although little or no HCII was present in the wall of the uninjured artery, HCII was detected in the adventitia of the injured artery. Data from He *et al.*[65]

antibody.[65] Binding of wild-type HCII was restricted to the adventitia. The binding sites appeared to be composed predominantly of DS, since binding of the Arg^{189} variant was much weaker in comparison with wild-type HCII or the Lys^{173} variant. Furthermore, pre-treatment of frozen sections with chondroitin ABC-lyase, which degrades DS, greatly reduces the amount of wild-type HCII bound.

Little or no endogenous HCII antigen could be detected in carotid arteries harvested from $HCII^{+/+}$ mice prior to photochemical injury. By contrast, HCII antigen was clearly present in the adventitia 30 min after the onset of photochemical injury (i.e., before formation of an occlusive thrombus, Fig. 4). In $HCII^{-/-}$ mice reconstituted with recombinant HCII, uptake into the adventitia correlated with the ability of HCII to bind DS; thus, wild-type HCII and the Lys^{173} variant were present in the adventitia at much higher levels than the Arg^{189} variant 30 min after the onset of injury.

The findings of He *et al.*[65] support a model in which circulating HCII enters the vessel wall after disruption of the endothelium and becomes activated by binding to DS (Fig. 1). Thrombin generation may be initiated in the adventitia, which is the predominant site of expression of TF in normal blood vessels,[62] and HCII bound to DS in the adventitia would be in a good position to regulate thrombin in the injured vessel. Hence, disorders of HCII or DS might contribute to vascular pathology through aberrant control of thrombin activity.

III. Human Studies

The concentration of HCII in plasma varies within fairly narrow limits in normal human subjects, averaging $\sim 1.2\ \mu M$.[18] In one study of 4000 unselected individuals 18–65 years of age, HCII activity ranged from 77% to 109% (mean ± S.D.).[84] Decreased HCII levels as low as 10% are found in some patients with liver failure or disseminated intravascular coagulation, and less severe abnormalities have been reported in patients with a variety of other conditions, including chronic renal insufficiency, diabetes mellitus, and chronic hemolytic anemias.[85] It is uncertain whether the abnormal HCII levels contribute to the pathogenesis of these disorders.

HCII appears to inhibit thrombin *in vivo* in the absence of disease, since thrombin–HCII complexes are detectable at nanomolar concentrations in normal human plasma.[86] Several clinical conditions in which HCII appears to play a role have received particular attention.

A. Pregnancy and Severe Preeclampsia

Elevated plasma concentrations of HCII have been reported in women who are pregnant or use oral contraceptives,[87] and thrombin–HCII complexes are increased approximately fourfold over baseline at term.[86] Conversely, decreased HCII levels (~40–50% of normal) have been reported in patients with severe preeclampsia.[88] During pregnancy, both the maternal and fetal plasma contain trace amounts of a DS proteoglycan that stimulates inhibition of thrombin by HCII.[89] The placenta is rich in DS and may be the source of this proteoglycan. Whereas placental HS has little or no anticoagulant activity when assayed with either HCII or AT, DS appears to be the major anticoagulant glycosaminoglycan in this organ.[90] Immunohistochemical studies have shown that DS is associated with fetal blood vessels and stromal regions of placental villi but is notably absent from the syncytiotrophoblast cells in contact with the maternal circulation. HCII colocalizes with DS in the walls of fetal blood vessels and is also associated with syncytiotrophoblasts.[90] These data suggest that DS is in a position to activate HCII in the fetal blood vessels or in the stroma of placental villi after injury to the syncytiotrophoblast layer.

B. Venous Thrombosis

HCII deficiency does not appear to be an important risk factor for venous thromboembolism. Although there are anecdotal reports associating inherited deficiency of HCII with venous thromboembolic disease, heterozygous HCII deficiency (~50% of normal) appears to be equally prevalent in individuals with or without venous thromboembolism.[85] Two sisters have been reported to have HCII activities only 10–15% of normal; one of these individuals was also heterozygous for AT deficiency and had recurrent venous thromboembolism, but the other was asymptomatic.[91] Individuals who completely lack HCII have not been identified.

C. Atherosclerosis and In-Stent Restenosis

Several clinical studies suggest that the concentration of HCII in plasma is inversely correlated with atherosclerotic cardiovascular disease in certain patient populations. In a group of 306 elderly patients in whom carotid artery atherosclerosis was assessed by ultrasonography, a negative correlation between plasma HCII activity and plaque thickness was observed.[92] A separate study of 110 patients with acute myocardial infarction found that those with the highest HCII levels, comprising ~30% of the total group, had a trend toward decreased major adverse cardiovascular events during 2–6 years of follow-up.[93] Finally, peripheral artery disease (PAD) was diagnosed in 62 of 494 elderly patients with cardiovascular risk factors by measurement of the ankle-brachial pressure index; the mean HCII level was slightly but significantly higher in the non-PAD patients than in the PAD patients ($95 \pm 18\%$ vs. $88 \pm 20\%$, $P = 0.009$).[94] By contrast, in a prospective study of 378 initially asymptomatic middle-aged subjects, no relationship was found between baseline levels of HCII and development of symptomatic coronary heart disease during an average follow-up period of 11.7 years.[95]

An inverse relationship between HCII levels and restenosis (neointima formation) after angioplasty has also been reported. In a study that examined 134 patients 6 months after coronary angioplasty and stent placement, the rate of in-stent restenosis was 6.7% in 36 patients with plasma HCII levels > 110% but was 4.5-fold higher in 32 patients with levels < 80%.[96] Similarly, a prospective study that measured HCII levels in 63 patients prior to femoropopliteal stent placement found that the rate of in-stent restenosis after 12 months of follow-up was 28% in patients with levels > 100% but was 2.3-fold higher in patients with levels $\leq$ 100%.[97] In general, these preliminary clinical findings are consistent with experimental observations that HCII deficiency enhances atherosclerosis and neointima formation in mice.[60,69]

D. Mucopolysaccharidoses

The mucopolysaccharidoses (MPSs) comprise a heterogeneous group of disorders that result from deficiencies of lysosomal enzymes, leading to organ damage caused by accumulation of HS, DS, or other glycosaminoglycans. Recent studies have shown that high levels of thrombin–HCII complexes form *ex vivo* during clotting of blood collected from patients or mice with MPS type I (α-L-iduronidase deficiency) or MPS type II (iduronate sulfatase deficiency) and correlate with disease severity.[98,99] Formation of these complexes appears to depend on the presence of excess DS in the blood. It is not known whether activation of HCII *in vivo* by undegraded glycosaminoglycans plays a role in the clinical manifestations of MPSs.

References

1. Trowbridge JM, Gallo RL. Dermatan sulfate: new functions from an old glycosaminoglycan. *Glycobiology* 2002;**12**:117R–25R.
2. Tollefsen DM. Heparin cofactor II modulates the response to vascular injury. *Arterioscler Thromb Vasc Biol* 2007;**27**:454–60.
3. Huntington JA. Mechanisms of glycosaminoglycan activation of the serpins in hemostasis. *J Thromb Haemost* 2003;**1**:1535–49.
4. Brinkhous KM, Smith HP, Warner ED, Seegers WH. The inhibition of blood clotting: an unidentified substance which acts in conjunction with heparin to prevent the conversion of prothrombin into thrombin. *Am J Physiol* 1939;**125**:683.
5. Abildgaard U. Highly purified antithrombin 3 with heparin cofactor activity prepared by disc electrophoresis. *Scand J Clin Lab Invest* 1968;**21**:89–91.
6. Rosenberg JS, McKenna PW, Rosenberg RD. Inhibition of human factor IXa by human antithrombin. *J Biol Chem* 1975;**250**:8883–8.
7. Rosenberg RD, Damus PS. The purification and mechanism of action of human antithrombin-heparin cofactor. *J Biol Chem* 1973;**248**:6490–505.
8. Briginshaw GF, Shanberge JN. Identification of two distinct heparin cofactors in human plasma. Separation and partial purification. *Arch Biochem Biophys* 1974;**161**:683–90.

9. Briginshaw GF, Shanberge JN. Identification of two distinct heparin cofactors in human plasma II. Inhibition of thrombin and activated factor X. *Thromb Res* 1974;**4**:463–77.
10. Tollefsen DM, Blank MK. Detection of a new heparin-dependent inhibitor of thrombin in human plasma. *J Clin Invest* 1981;**68**:589–96.
11. Friberger P, Egberg N, Holmer E, Hellgren M, Blombäck M. Antithrombin assay—the use of human or bovine thrombin and the observation of a "second" heparin cofactor. *Thromb Res* 1982;**25**:433–6.
12. Griffith MJ, Carraway T, White GC, Dombrose FA. Heparin cofactor activities in a family with hereditary antithrombin III deficiency: evidence for a second heparin cofactor in human plasma. *Blood* 1983;**61**:111–8.
13. Tollefsen DM, Majerus DW, Blank MK. Heparin cofactor II. Purification and properties of a heparin-dependent inhibitor of thrombin in human plasma. *J Biol Chem* 1982;**257**:2162–9.
14. Blinder MA, Marasa JC, Reynolds CH, Deaven LL, Tollefsen DM. Heparin cofactor II: cDNA sequence, chromosome localization, restriction fragment length polymorphism, and expression in *Escherichia coli*. *Biochemistry* 1988;**27**:752–9.
15. Ragg H. A new member of the plasma protease inhibitor gene family. *Nucleic Acids Res* 1986;**14**:1073–88.
16. Zhang GS, Mehringer JH, Van Deerlin VM, Kozak CA, Tollefsen DM. Murine heparin cofactor II: purification, cDNA sequence, expression, and gene structure. *Biochemistry* 1994;**33**:3632–42.
17. Kamp P, Strathmann A, Ragg H. Heparin cofactor II, antithrombin-beta and their complexes with thrombin in human tissues. *Thromb Res* 2001;**101**:483–91.
18. Tollefsen DM, Pestka CA. Heparin cofactor II activity in patients with disseminated intravascular coagulation and hepatic failure. *Blood* 1985;**66**:769–74.
19. Kounnas MZ, Church FC, Argraves WS, Strickland DK. Cellular internalization and degradation of antithrombin III-thrombin, heparin cofactor II-thrombin, and alpha 1-antitrypsin-trypsin complexes is mediated by the low density lipoprotein receptor-related protein. *J Biol Chem* 1996;**271**:6523–9.
20. Gettins PG. Serpin structure, mechanism, and function. *Chem Rev* 2002;**102**:4751–804.
21. Huntington JA. Shape-shifting serpins—advantages of a mobile mechanism. *Trends Biochem Sci* 2006;**31**:427–35.
22. Huntington JA, Read RJ, Carrell RW. Structure of a serpin-protease complex shows inhibition by deformation. *Nature* 2000;**407**:923–6.
23. Griffith MJ, Noyes CM, Tyndall JA, Church FC. Structural evidence for leucine at the reactive site of heparin cofactor II. *Biochemistry* 1985;**24**:6777–82.
24. Parker KA, Tollefsen DM. The protease specificity of heparin cofactor II. Inhibition of thrombin generated during coagulation. *J Biol Chem* 1985;**260**:3501–5.
25. Church FC, Noyes CM, Griffith MJ. Inhibition of chymotrypsin by heparin cofactor II. *Proc Natl Acad Sci USA* 1985;**82**:6431–4.
26. Derechin VM, Blinder MA, Tollefsen DM. Substitution of arginine for Leu444 in the reactive site of heparin cofactor II enhances the rate of thrombin inhibition. *J Biol Chem* 1990;**265**:5623–8.
27. Tollefsen DM, Pestka CA, Monafo WJ. Activation of heparin cofactor II by dermatan sulfate. *J Biol Chem* 1983;**258**:6713–6.
28. Hortin GL, Tollefsen DM, Benutto BM. Antithrombin activity of a peptide corresponding to residues 54–75 of heparin cofactor II. *J Biol Chem* 1989;**264**:13979–82.
29. Ragg H, Ulshöfer T, Gerewitz J. On the activation of human leuserpin-2, a thrombin inhibitor, by glycosaminoglycans. *J Biol Chem* 1990;**265**:5211–8.
30. Ragg H, Ulshöfer T, Gerewitz J. Glycosaminoglycan-mediated leuserpin-2/thrombin interaction. Structure–function relationships. *J Biol Chem* 1990;**265**:22386–91.

31. Van Deerlin VM, Tollefsen DM. The N-terminal acidic domain of heparin cofactor II mediates the inhibition of alpha-thrombin in the presence of glycosaminoglycans. *J Biol Chem* 1991;**266**:20223–31.
32. Fortenberry YM, Whinna HC, Gentry HR, Myles T, Leung LL, Church FC. Molecular mapping of the thrombin-heparin cofactor II complex. *J Biol Chem* 2004;**279**:43237–44.
33. Sheehan JP, Tollefsen DM, Sadler JE. Heparin cofactor II is regulated allosterically and not primarily by template effects. Studies with mutant thrombins and glycosaminoglycans. *J Biol Chem* 1994;**269**:32747–51.
34. Sheehan JP, Wu Q, Tollefsen DM, Sadler JE. Mutagenesis of thrombin selectively modulates inhibition by serpins heparin cofactor II and antithrombin III. Interaction with the anion-binding exosite determines heparin cofactor II specificity. *J Biol Chem* 1993;**268**:3639–45.
35. Brinkmeyer S, Eckert R, Ragg H. Reformable intramolecular cross-linking of the N-terminal domain of heparin cofactor II: effects on enzyme inhibition. *Eur J Biochem* 2004;**271**:4275–83.
36. Baglin TP, Carrell RW, Church FC, Esmon CT, Huntington JA. Crystal structures of native and thrombin-complexed heparin cofactor II reveal a multistep allosteric mechanism. *Proc Natl Acad Sci USA* 2002;**99**:11079–84.
37. Verhamme IM, Bock PE, Jackson CM. The preferred pathway of glycosaminoglycan-accelerated inactivation of thrombin by heparin cofactor II. *J Biol Chem* 2004;**279**:9785–95.
38. O'Keeffe D, Olson ST, Gasiunas N, Gallagher J, Baglin TP, Huntington JA. The heparin binding properties of heparin cofactor II suggest an antithrombin-like activation mechanism. *J Biol Chem* 2004;**279**:50267–73.
39. Blinder MA, Andersson TR, Abildgaard U, Tollefsen DM. Heparin cofactor IIOslo. Mutation of Arg-189 to His decreases the affinity for dermatan sulfate. *J Biol Chem* 1989;**264**:5128–33.
40. Blinder MA, Tollefsen DM. Site-directed mutagenesis of arginine 103 and lysine 185 in the proposed glycosaminoglycan-binding site of heparin cofactor II. *J Biol Chem* 1990;**265**:286–91.
41. Whinna HC, Blinder MA, Szewczyk M, Tollefsen DM, Church FC. Role of lysine 173 in heparin binding to heparin cofactor II. *J Biol Chem* 1991;**266**:8129–35.
42. Maimone MM. Characterization of heparin and dermatan sulfate molecules that bind and activate heparin cofactor II. St. Louis, MO: Washington University; 1990. Ph.D. thesis.
43. Maimone MM, Tollefsen DM. Activation of heparin cofactor II by heparin oligosaccharides. *Biochem Biophys Res Commun* 1988;**152**:1056–61.
44. Colwell NS, Grupe MJ, Tollefsen DM. Amino acid residues of heparin cofactor II required for stimulation of thrombin inhibition by sulphated polyanions. *Biochim Biophys Acta* 1999;**1431**:148–56.
45. Maimone MM, Tollefsen DM. Structure of a dermatan sulfate hexasaccharide that binds to heparin cofactor II with high affinity. *J Biol Chem* 1990;**265**:18263–71.
46. Pavão MSG, Aiello KR, Werneck CC, Silva LC, Valente A-P, Mulloy B, et al. Highly sulfated dermatan sulfates from Ascidians. Structure versus anticoagulant activity of these glycosaminoglycans. *J Biol Chem* 1998;**273**:27848–57.
47. Pavão MSG, Mourão PAS, Mulloy B, Tollefsen DM. A unique dermatan sulfate-like glycosaminoglycan from ascidian. Its structure and the effect of its unusual sulfation pattern on anticoagulant activity. *J Biol Chem* 1995;**270**:31027–36.
48. Halldórsdóttir AM, Zhang L, Tollefsen DM. *N*-Acetylgalactosamine 4, 6-*O*-sulfate residues mediate binding and activation of heparin cofactor II by porcine mucosal dermatan sulfate. *Glycobiology* 2006;**16**:693 701.
49. Hatton MW, Hoogendoorn H, Southward SM, Ross B, Blajchman MA. Comparative metabolism and distribution of rabbit heparin cofactor II and rabbit antithrombin in rabbits. *Am J Physiol* 1997;**272**:E824–31.
50. Sié P, Dupouy D, Pichon J, Boneu B. Turnover study of heparin cofactor II in healthy man. *Thromb Haemost* 1985;**54**:635–8.

51. Sié P, Lansen J, Lacheretz F, Verschuere B, Boneu B. Comparative turn-over of heparin cofactor II and antithrombin III in baboons. Influence of heparin and pentosan polysulfate administration. *Thromb Haemost* 1986;**56**:302–7.
52. Le DT, Borgs P, Toneff TW, Witte MH, Rapaport SI. Hemostatic factors in rabbit limb lymph: relationship to mechanisms regulating extravascular coagulation. *Am J Physiol* 1998;**274**: H769–76.
53. Tuma PL, Hubbard AL. Transcytosis: crossing cellular barriers. *Physiol Rev* 2003;**83**:871–932.
54. Cooper ST, Neese LL, DiCuccio MN, Liles DK, Hoffman M, Church FC. Vascular localization of the heparin-binding serpins antithrombin, heparin cofactor II, and protein C inhibitor. *Clin Appl Thromb Hemost* 1996;**2**:185–91.
55. Tovar AMF, de Mattos DA, Stelling MP, Sarcinelli-Luz BSL, Nazareth RA, Mourão PAS. Dermatan sulfate is the predominant antithrombotic glycosaminoglycan in vessel walls: implications for a possible physiological function of heparin cofactor II. *Biochim Biophys Acta* 2005;**1740**:45–53.
56. Sié P, Ofosu F, Fernandez F, Buchanan MR, Petitou M, Boneu B. Respective role of antithrombin III and heparin cofactor II in the *in vitro* anticoagulant effect of heparin and of various sulphated polysaccharides. *Br J Haematol* 1986;**64**:707–14.
57. McGuire EA, Tollefsen DM. Activation of heparin cofactor II by fibroblasts and vascular smooth muscle cells. *J Biol Chem* 1987;**262**:169–75.
58. Whinna HC, Choi HU, Rosenberg LC, Church FC. Interaction of heparin cofactor II with biglycan and decorin. *J Biol Chem* 1993;**268**:3920–4.
59. He L, Vicente CP, Westrick RJ, Eitzman DT, Tollefsen DM. Heparin cofactor II inhibits arterial thrombosis after endothelial injury. *J Clin Invest* 2002;**109**:213–9.
60. Aihara K, Azuma H, Akaike M, Ikeda Y, Sata M, Takamori N, et al. Strain-dependent embryonic lethality and exaggerated vascular remodeling in heparin cofactor II-deficient mice. *J Clin Invest* 2007;**117**:1514–26.
61. Saniabadi AR, Umemura K, Matsumoto N, Sakuma S, Nakashima M. Vessel wall injury and arterial thrombosis induced by a photochemical reaction. *Thromb Haemost* 1995;**73**:868–72.
62. Wilcox JN, Smith KM, Schwartz SM, Gordon D. Localization of tissue factor in the normal vessel wall and in the atherosclerotic plaque. *Proc Natl Acad Sci USA* 1989;**86**:2839–43.
63. Mann KG, Butenas S, Brummel K. The dynamics of thrombin formation. *Arterioscler Thromb Vasc Biol* 2003;**23**:17–25.
64. Coughlin SR. Protease-activated receptors in the cardiovascular system. *Cold Spring Harb Symp Quant Biol* 2002;**67**:197–208.
65. He L, Giri TK, Vicente CP, Tollefsen DM. Vascular dermatan sulfate regulates the antithrombotic activity of heparin cofactor II. *Blood* 2008;**111**:4118–25.
66. Vicente CP, He L, Pavao MS, Tollefsen DM. Antithrombotic activity of dermatan sulfate in heparin cofactor II-deficient mice. *Blood* 2004;**104**:3965–70.
67. Abbadini M, Zhu GJ, Maggi A, Pangrazzi J, Donati MB, Mussoni L. Dermatan sulphate induces plasminogen activator release in the perfused rat hindquarters. *Blood* 1987;**70**:1858–60.
68. Barbanti M, Calanni F, Marchi E, Semeraro N, Colucci M. Desmin 370, a low molecular weight dermatan sulfate, reduces the weight of preformed thrombi in rats made afibrinogenemic by ancrod. *Thromb Haemost* 1995;**73**:287–90.
69. Vicente CP, He L, Tollefsen DM. Accelerated atherogenesis and neointima formation in heparin cofactor II deficient mice. *Blood* 2007;**110**:4261–7.
70. Leidenfrost JE, Khan MF, Boc KP, Villa BR, Collins ET, Parks WC, et al. A model of primary atherosclerosis and post-angioplasty restenosis in mice. *Am J Pathol* 2003;**163**:773–8.
71. Farb A, Sangiorgi G, Carter AJ, Walley VM, Edwards WD, Schwartz RS, et al. Pathology of acute and chronic coronary stenting in humans. *Circulation* 1999;**99**:44–52.

72. Cheung WM, D'Andrea MR, Andrade-Gordon P, Damiano BP. Altered vascular injury responses in mice deficient in protease-activated receptor-1. *Arterioscler Thromb Vasc Biol* 1999;**19**:3014–24.
73. Andrade-Gordon P, Derian CK, Maryanoff BE, Zhang HC, Addo MF, Cheung W, et al. Administration of a potent antagonist of protease-activated receptor-1 (PAR-1) attenuates vascular restenosis following balloon angioplasty in rats. *J Pharmacol Exp Ther* 2001;**298**:34–42.
74. Chen X, Ren S, Ma MG, Dharmalingam S, Lu L, Xue M, et al. Hirulog-like peptide reduces restenosis and expression of tissue factor and transforming growth factor-beta in carotid artery of atherosclerotic rabbits. *Atherosclerosis* 2003;**169**:31–40.
75. Buchanan MR, Brister SJ. Inhibition of chronic vessel wall intimal hyperplasia following acute anticoagulant treatment: relative effects of heparin and dermatan sulphate. *Thromb Res* 1998;**91**:157–67.
76. Liaw PC, Becker DL, Stafford AR, Fredenburgh JC, Weitz JI. Molecular basis for the susceptibility of fibrin-bound thrombin to inactivation by heparin cofactor ii in the presence of dermatan sulfate but not heparin. *J Biol Chem* 2001;**276**:20959–65.
77. Borissoff JI, Spronk HM, Heeneman S, ten Cate H. Is thrombin a key player in the 'coagulation-atherogenesis' maze? *Cardiovasc Res* 2009;**82**:392–403.
78. Stoop AA, Lupu F, Pannekoek H. Colocalization of thrombin, PAI-1, and vitronectin in the atherosclerotic vessel wall: a potential regulatory mechanism of thrombin activity by PAI-1/vitronectin complexes. *Arterioscler Thromb Vasc Biol* 2000;**20**:1143–9.
79. Westrick RJ, Bodary PF, Xu Z, Shen YC, Broze GJ, Eitzman DT. Deficiency of tissue factor pathway inhibitor promotes atherosclerosis and thrombosis in mice. *Circulation* 2001;**103**:3044–6.
80. Rau JC, Deans C, Hoffman MR, Thomas DB, Malcom GT, Zieske AW, et al. Heparin cofactor II in atherosclerotic lesions from the Pathobiological Determinants of Atherosclerosis in Youth (PDAY) study. *Exp Mol Pathol* 2009;**87**:178–83.
81. Shirk RA, Parthasarathy N, San Antonio JD, Church FC, Wagner WD. Altered dermatan sulfate structure and reduced heparin cofactor II-stimulating activity of biglycan and decorin from human atherosclerotic plaque. *J Biol Chem* 2000;**275**:18085–92.
82. Couchman JR, Caterson B, Christner JE, Baker JR. Mapping by monoclonal antibody detection of glycosaminoglycans in connective tissues. *Nature* 1984;**307**:650–2.
83. David G, Bai XM, Van der Schueren B, Cassiman JJ, Van den Berghe H. Developmental changes in heparan sulfate expression: *in situ* detection with mAbs. *J Cell Biol* 1992;**119**:961–75.
84. Rodeghiero F, Tosetto A. The VITA Project: population-based distributions of protein C, antithrombin III, heparin-cofactor II and plasminogen-relationship with physiological variables and establishment of reference ranges. *Thromb Haemost* 1996;**76**:226–33.
85. Tollefsen DM. Heparin cofactor II deficiency. *Arch Pathol Lab Med* 2002;**126**:1394–400.
86. Liu L, Dewar L, Song Y, Kulczycky M, Blajchman MA, Fenton 2nd JW, et al. Inhibition of thrombin by antithrombin III and heparin cofactor II *in vivo*. *Thromb Haemost* 1995;**73**:405–12.
87. Massouh M, Jatoi A, Gordon EM, Ratnoff OD. Heparin cofactor II activity in plasma during pregnancy and oral contraceptive use. *J Lab Clin Med* 1989;**114**:697–9.
88. Bellart J, Gilabert R, Cabero L, Fontcuberta J, Monasterio J, Miralles RM. Heparin cofactor II: a new marker for pre-eclampsia. *Blood Coagul Fibrinolysis* 1998;**9**:205–8.
89. Andrew M, Mitchell L, Berry L, Paes B, Delorme M, Ofosu F, et al. An anticoagulant dermatan sulfate proteoglycan circulates in the pregnant woman and her fetus. *J Clin Invest* 1992;**89**:321–6.
90. Giri TK, Tollefsen DM. Placental dermatan sulfate: isolation, anticoagulant activity, and association with heparin cofactor II. *Blood* 2006;**107**:2753–8.

91. Villa P, Aznar J, Vaya A, España F, Ferrando F, Mira Y, et al. Hereditary homozygous heparin cofactor II deficiency and the risk of developing venous thrombosis. *Thromb Haemost* 1999;**82**:1011–4.
92. Aihara K, Azuma H, Takamori N, Kanagawa Y, Akaike M, Fujimura M, et al. Heparin cofactor II is a novel protective factor against carotid atherosclerosis in elderly individuals. *Circulation* 2004;**109**:2761–5.
93. Huang SS, Huang PH, Chen YH, Sung SH, Chiang KH, Chen JW, et al. Plasma heparin cofactor II activity is an independent predictor of future cardiovascular events in patients after acute myocardial infarction. *Coron Artery Dis* 2008;**19**:597–602.
94. Aihara K, Azuma H, Akaike M, Kurobe H, Takamori N, Ikeda Y, et al. Heparin cofactor II is an independent protective factor against peripheral arterial disease in elderly subjects with cardiovascular risk factors. *J Atheroscler Thromb* 2009;**16**:127–34.
95. Giri TK, Ahn CW, Wu KK, Tollefsen DM. Heparin cofactor II levels do not predict the development of coronary heart disease: the Atherosclerosis Risk in Communities (ARIC) study. *Arterioscler Thromb Vasc Biol* 2005;**25**:2689–90.
96. Takamori N, Azuma H, Kato M, Hashizume S, Aihara K, Akaike M, et al. High plasma heparin cofactor II activity is associated with reduced incidence of in-stent restenosis after percutaneous coronary intervention. *Circulation* 2004;**109**:481–6.
97. Schillinger M, Exner M, Sabeti S, Mlekusch W, Amighi J, Handler S, et al. High plasma heparin cofactor II activity protects from restenosis after femoropopliteal stenting. *Thromb Haemost* 2004;**92**:1108–13.
98. Langford-Smith K, Arasaradnam M, Wraith JE, Wynn R, Bigger BW. Evaluation of heparin cofactor II-thrombin complex as a biomarker on blood spots from mucopolysaccharidosis I, IIIA and IIIB mice. *Mol Genet Metab* 2010;**99**:269–74.
99. Randall DR, Colobong KE, Hemmelgarn H, Sinclair GB, Hetty E, Thomas A, et al. Heparin cofactor II-thrombin complex: a biomarker of MPS disease. *Mol Genet Metab* 2008;**94**:456–61.

Diverse Functions of Glycosaminoglycans in Infectious Diseases

Rafael S. Aquino, Eui Seung Lee, and Pyong Woo Park

Division of Respiratory Diseases, Children's Hospital, Harvard Medical School, Boston, Massachusetts, USA

Glycosaminoglycans (GAGs) are complex carbohydrates that are expressed ubiquitously and abundantly on the cell surface and in the extracellular matrix (ECM). The extraordinary structural diversity of GAGs enables them to interact with a wide variety of biological molecules. Through these interactions, GAGs modulate various biological processes, such as cell adhesion, proliferation and migration, ECM assembly, tissue repair, coagulation, and immune responses, among many others. Studies during the last several decades have indicated that GAGs also interact with microbial pathogens. GAG–pathogen interactions affect most, if not all, the key steps of microbial pathogenesis, including host cell attachment and invasion, cell–cell transmission, systemic dissemination and infection of secondary organs, and evasion of host defense mechanisms. These observations indicate that GAG–pathogen interactions serve diverse functions that affect the pathogenesis of infectious diseases.

I. Introduction

Infectious diseases are a major global health problem. They kill over 10 million people annually, accounting for approximately 25% of all deaths in the world. Microbial pathogens perform various functions to promote their survival in the host environment, which is generally hostile for the pathogens because

Progress in Molecular Biology
and Translational Science, Vol. 93
DOI: 10.1016/S1877-1173(10)93016-0

the host mounts effective defense mechanisms to eradicate them. Infection occurs when the balance of host–pathogen interactions shifts to favor the pathogen. Hygienic, prophylactic, and therapeutic interventions are continuing to reduce the incidence and mortality of infectious diseases. However, infectious diseases continue to be a major public health threat in both developed and developing countries, and the emergence of new pathogens and drug-resistant strains, globalization of travel and trade, and bioterrorism are adding to this threat.

Microbial pathogens elaborate a multitude of virulence factors, which enable the pathogens to attach to and invade host cells, damage host tissues, disseminate and cause secondary infections, and evade host defense mechanisms, among other virulence activities. Virulence factors frequently subvert host components to promote pathogenesis. In essence, the outcome of an infection is largely governed by the ability of pathogens to exploit host components and their activities. For example, many pathogens express cell surface molecules called adhesins that specifically bind to host tissue components to facilitate their attachment. Pathogens also secrete enzymes that digest host components to generate nutrients and to inactivate host defense factors. Some pathogens synthesize toxins that not only kill host cells and cause tissue damage, but also modify the host cell's cytoskeleton to facilitate their internalization. Further, several pathogens express factors that inhibit specific arms of host defense (e.g., complement-mediated killing, phagocytosis) or dysregulate the host inflammatory response to their advantage. The genetic variability that microbial pathogens can generate allows variant pathogens to exploit host components for their growth, survival, and propagation.

Among the many host components subverted by microbial pathogens, glycosaminoglycans (GAGs) have been implicated in various steps during the course of infection. GAGs are expressed ubiquitously on the cell surface, in the ECM, and in intracellular compartments. Further, they play critical roles in various host defense mechanisms, allowing pathogens abundant opportunities to exploit them for infectious purposes. Consistent with this notion, studies during the last several decades have demonstrated that a wide variety of pathogens subvert GAGs for their attachment and invasion, cellular transmission, systemic dissemination, and evasion of host defense mechanisms.[1–5] The list includes viruses, bacteria, parasites, and fungi, such as herpes simplex virus (HSV), dengue virus, *Neisseria gonorrhoeae*, *Plasmodium falciparum*, and *Candida albicans*, among many others.

GAGs are linear polysaccharides comprised of repeating disaccharide units of hexosamine and uronic acid or galactose with various substitutions. The list of GAGs includes heparan sulfate (HS)/heparin, chondroitin sulfate (CS), dermatan sulfate (DS), keratan sulfate (KS), and hyaluronan (HA) as described in other chapters of this book. Except for HA, all other GAGs are sulfated to various degrees and covalently complexed to proteoglycan core proteins.

The biosynthesis of GAGs is a highly complex, template-free process catalyzed by specific enzymes. HS synthesis, for example, is initiated by the action of four glycosyltransferases assembling the initiation tetrasaccharide linkage sequence (-glucuronic acid-galactose-galactose-xylose) on certain serine residues of HS proteoglycan (HSPG) core proteins, such as syndecans and glypicans. The heparan (or heparosan) polysaccharide chain is elongated on the tetrasaccharide linkage region by adding alternate units of *N*-acetylglucosamine and glucuronic acid. The heparan precursor chain is then sequentially modified by *N*-deacetylase/*N*-sulfotransferases (NDSTs), C5 epimerase; and 2-*O*-sulfotransferase, 6-*O*-sulfotransferases, and 3-*O* sulfotransferases in the Golgi.[6,7] The mature HS chain can be further modified in the Golgi or extracellular compartments by the action of 6-*O*-sulfatases[8] or heparanases.[9] An important feature of HS biosynthesis is that the polymerization and modification reactions do not go to completion. Thus, the biosynthetic process generates an incredibly diverse array of HS structures. By one estimate, 10^{36} HS structures can potentially be synthesized.[10] Perhaps this enormous structural variability explains, in part, why HS can interact with a diverse group of microbial pathogens.

GAG–pathogen interactions impinge on multiple steps of microbial pathogenesis. Many viral, bacterial, parasitic, and fungal pathogens use GAGs as low affinity, initial attachment sites to facilitate their interaction with respective secondary internalization receptors.[2,4,11,12] In some cases, pathogens bind to GAGs and use them as bridging molecules to bind to GAG-binding host components on the cell surface and in the ECM. Some pathogens also use GAGs as scaffolds that induce conformational changes in virulence factors.[13,14] To efficiently invade host cells, pathogens not only use GAGs as low affinity coreceptors or high affinity internalization receptors, but in some cases, also use certain GAGs to modulate endogenous processes, such as caveola-mediated endocytosis[15] and filopodia formation and filopodia-mediated transport of viral particles to cell bodies.[16] GAGs can also influence the cell–cell transmission of pathogens, such as in the transfer of viruses between host cells.[17–19] In addition, recent studies suggest that certain pathogens subvert GAGs to inhibit host defense mechanisms to promote their pathogenesis.[20–24] This chapter reviews these diverse functions of GAG–pathogen interactions in infectious diseases using select examples from recent studies.

II. GAGs in Microbial Pathogenesis

A. GAGs in Pathogen Attachment

The initial attachment of pathogens to host tissues is a critical step in the pathogenesis of most infectious diseases. Pathogens that lack the capacity to rapidly and firmly attach to host tissues are effectively removed by nonspecific mechanical defenses, such as ciliary motion, intestinal peristalsis, and lung reflexes (coughing and sneezing). Many pathogens express GAG-binding adhesins that mediate their attachment to host tissues.[1,4,5] In most cases, GAGs serve as coreceptors that allow pathogens access to the cell surface compartment, enabling pathogens to interact more efficiently with their respective secondary internalization receptors (Fig. 1A). There is abundant evidence indicating that the capacity to attach to GAGs is biologically significant. Several studies have shown that the attachment of HS-binding pathogens is inhibited by exogenous addition of excess HS or heparin, and HS-binding pathogens show significantly reduced attachment to host cells whose cell surface HS levels have been decreased by enzymatic treatment or chemical mutagenesis of HS biosynthetic enzyme genes. Further, GAG antagonists inhibit attachment of GAG-binding pathogens to host cells *in vitro* and reduce virulence in animals, suggesting that microbial binding to GAGs is an important first step in pathogenesis. Some of the recently described GAG attachment mechanisms are described below.

1. Viral Attachment

The majority of HS-binding viral pathogens utilize cell surface HS as a coreceptor. Because HS is ubiquitously expressed, the secondary internalization receptors are thought to largely mediate the tissue tropism of HS-binding viruses. Several hepatitis viruses (hepatitis B virus (HBV), HCV, HEV), which are etiologic agents of acute and chronic liver infections, have been shown to use cell surface HS for their initial attachment to hepatocytes.[25–28] HEV is a nonenveloped, positive-sense, single-stranded RNA virus that causes hepatitis E, a recently described liver inflammatory disease of humans. HEV is transmitted via the fecal–oral transmission route. HEV binds to cell surface HS through the major capsid protein pORF2.[25] HEV pORF2 binding is inhibited when Huh-7 human hepatoma cells are treated with sodium chlorate, an inhibitor of GAG sulfation that blocks the formation of the universal sulfate donor PAPS (3′-phosphoadenosine-5′-phosphosulfate). HEV pORF2 binding to Huh-7 cells is also markedly reduced by addition of HS or heparin, and by treatment with heparinase I or III, but not with chondroitinase ABC. These data indicate that HEV binds selectively to cell surface HS.

Syndecan-1 appears to be the primary target HSPG of HEV as pORF2 binds specifically to syndecan-1 on Huh-7 cells.[25] Knockdown of syndecan-1, but not syndecan-4, by RNA interference, significantly decreased pORF2 binding to Huh-7 cells. However, treatment of Huh-7 cells with phosphatidylinositol-specific phospholipase C had no effect on pORF2 binding to cells, suggesting that glypicans anchored to the cell surface via GPI-linkages do not mediate HEV binding. In addition, experiments with chemically modified heparin showed that pORF2 binding to Huh-7 cells is inhibited by 2-*O*-desulfated heparin, but

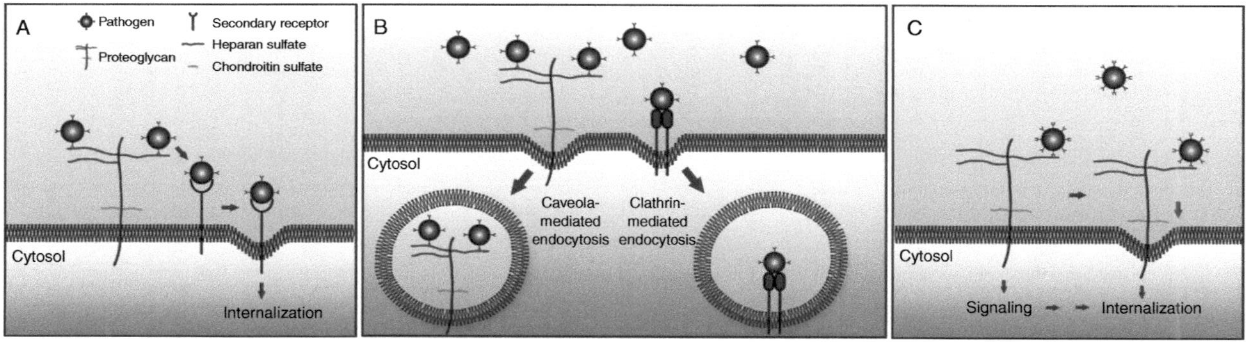

FIG. 1. GAGs in pathogen attachment and invasion. (A) GAGs as cell surface coreceptors: pathogens bind to cell surface GAGs to increase their concentration on the cell surface and facilitate their interaction with secondary internalization receptors. (B) Viral internalization by endocytic mechanisms of host cells: the initial interaction of FMDV with GAGs or other receptors specifies whether FMDV virions will be internalized via a caveola-mediated or clathrin-mediated endocytic pathway. (C) Bacterial internalization by activation of host cell signaling: binding of *N. gonorrhoeae* to cell surface HS activates an intracellular mechanism that leads to its internalization. (See Color Insert.)

not 6-*O*-desulfated heparin, suggesting that 6-*O*-sulfated glucosamine residues play a critical role in mediating the binding of HEV to target hepatocytes. However, precisely how 6-*O*-sulfation facilitates HEV binding is not known. Moreover, it is not clear if 6-*O*-sulfated domains in HS chains of other hepatocyte HSPGs (e.g., syndecan-2, glypicans) can also mediate HEV attachment.

Human papillomavirus (HPV) is the most common sexually transmitted virus in the United States. It is a small, nonenveloped DNA virus that infects skin and mucosal epithelial cells and alters the cell cycle to cause both benign and malignant epithelial tumors. There are over 100 HPV types and, among them, HPV types 5 and 11,[29,30] and the high risk sexually transmitted types 16, 31, 33, and 39[29,31] have been shown to use cell surface HS as low affinity coreceptors. For example, HPV16, which preferentially infects the genital mucosal epithelium and causes cervical cancers, binds to HS through the viral capsid protein L1. Site-directed mutagenesis studies showed that discontiguous Lys residues at positions 278, 356, and 361 in L1 are critical in HS binding,[14] suggesting that structural features of L1 mediate its binding to HS. Interestingly, binding of L1 to HS induces a conformational change in the minor capsid protein L2, which exposes residues in L2 that bind to a secondary internalization receptor whose identity remains to be determined. A recent study showed that cyclophilin B in association with HSPGs on the host cell facilitates the conformational change of L2.[32] These data suggest that L1 binding to cell surface HS brings together L2 and cyclophilin B, facilitating the cyclophilin B-mediated conformational change of L2 that is required for efficient host cell invasion.

Using a recently developed murine cervicovaginal challenge model,[33] the importance of HS in infection of the female genital tract by HPV5, 16, and 31 was evaluated.[29] Luciferase-expressing pseudoviruses of all HPV types infected the genital tract, although HPV5 was the least efficient. Heparinase III treatment of the genital tract significantly inhibited virion attachment to the basement membrane and cell surfaces and inhibited infection of all three types by greater than 90%, demonstrating that HS serves as the primary attachment site for these HPV types *in vivo*. However, various chemically modified heparin compounds differentially inhibited HPV infection of the genital tract. A highly sulfated form of heparin inhibited infections by HPV16 and HPV31, but not HPV5. Instead, HPV5 infection was inhibited by *N*-desulfated, *N*-deacetylated heparin, although HPV5 bound to both highly sulfated and *N*-desulfated, *N*-deacetylated heparin compounds. Because inhibition of infection was closely associated with the relative ability of the variants to inhibit basement membrane and cell surface binding, these findings may suggest that HPV5, a cutaneous HPV, and HPV16 and HPV31, genital mucosal HPVs, may have adapted or evolved to recognize different forms of HS at different tissue sites to enable them to infect their preferred site of infection.[29] These findings suggest that structural features of HS may also influence the tissue tropism of HPV and other HS-binding pathogens.

Clinical data also suggest that the HS–HPV interaction is important *in vivo*. Because HPV16 is highly immunogenic, antibodies against HPV16 capsid proteins are frequently observed in infected women.[34] Importantly, IgG and IgA antibodies against HPV16 extracted from low-grade squamous intraepithelial lesions were found to inhibit the binding of HPV16 virus-like particles (VLPs) to immobilized HS. However, antibodies against HPV16 extracted from cervical cancer patients had no effect on HPV16 VLP binding to HS.[35] Whether successful generation of antibodies that inhibit HS–HPV interactions correlates with the attenuation of HPV disease progression remains to be determined.

2. Bacterial attachment

Several bacterial pathogens have also been shown to subvert GAGs for its attachment. *Staphylococcus aureus* is a major Gram-positive bacterial pathogen that can cause a wide range of infections and toxinoses, such as food poisoning, toxic shock syndrome, fasciitis, sepsis, endocarditis, osteomyelitis, and pneumonia.[36–38] Its impressive pathogenic capacity is mediated by the large arsenal of virulence factors, including adhesins (MSCRAMMs) that recognize various host components. *S. aureus* binding to the human colon carcinoma cell lines Caco-2 and HT-29 was inhibited by heparin or HS.[39] Further, *S. aureus* bound to the ARH-77 human B lymphoblastic cell line expressing syndecan-1, but not to ARH-77 cells devoid of syndecan-1,[39,40] suggesting that *S. aureus* binds selectively to the HS moiety of syndecan-1.

A recent study implicated the pore-forming cytotoxin Panton-Valentine leukocidin (PVL) as the HS adhesin of *S. aureus*.[41] PVL is widely expressed by community-associated, methicillin-resistant isolates,[42,43] and has been proposed as a virulence factor in necrotizing pneumonia,[44] a recently described staphylococcal disease associated with high morbidity and mortality. PVL-positive *S. aureus* strains displayed a fourfold or higher binding capacity to surfaces coated with various matrix components (e.g., type I and IV collagens, laminin) compared to PVL-negative *S. aureus* strains.[41] The expression of recombinant fragments of the PVL subunits (LukS-PV and LukF-PV) and the addition of synthetic peptides indicated that the processed LukS-PV signal peptide (LukS-PV SP) enhances the ability of *S. aureus* to attach to the ECM. Interestingly, adhesion of *S. aureus* expressing the LukS-PV signal peptide to the ECM was inhibited by HS or heparin, suggesting that HS may function as a molecular bridge between the ECM and PVL-positive *S. aureus*. Further, the LukS-PV signal peptide was expressed on the *S. aureus* surface, bound to HS in solid phase assay, and enhanced *S. aureus* adhesion onto the ECM.[41] Collectively, these data suggest a unique adhesion mechanism where the cell surface-associated PVL mediates the attachment of *S. aureus* to matrix components exposed at damaged tissue sites by binding to HS and using it to interact with various HS-binding matrix components, such as collagens and laminins.

A similar HS-bridging model was suggested in the attachment of *Helicobacter pylori* to host tissues. *H. pylori* is a spiral-shaped Gram-negative bacterium that colonizes the gastric mucosa of approximately half of the human population. *H. pylori* infection has been strongly linked to the development of duodenal and gastric ulcers and some gastric cancers.[45] *H. pylori* expresses several HS-binding proteins[46] that have been shown to be important in the attachment to Kato III human gastric carcinoma cells and HeLa human cervical carcinoma cells.[47] Rabbit polyclonal antibodies against *H. pylori* HS-binding proteins inhibited *H. pylori* binding to Kato III and HeLa cells. However, addition of soluble *H. pylori* HS-binding proteins did not inhibit, but instead, enhanced *H. pylori* adherence, suggesting that these HS-binding proteins connect *H. pylori* with HS on the host cell surface. Further, *H. pylori* infection up-regulated the expression of syndecan-4 in host cells through a Toll-like receptor (TLR)-mediated mechanism,[48,49] suggesting that *H. pylori* instructs host cells to increase cell surface HS expression and attaches to HS via the bridge created by its HS-binding proteins. However, it is not clear if the *H. pylori* HS-binding proteins interact selectively with the HS moiety on syndecan-4 or if *H. pylori* can induce other HSPGs and interact with them in a similar manner.

Enterococcus faecalis, a common nosocomial pathogen of the gastrointestinal tract, also exploits GAGs to attach to host cells.[50] *E. faecalis* attachment to Caco2 human gastric cells was inhibited by HS or heparin, but not DS. Further, pretreatment of Caco2 cells with heparinase I or sodium chlorate significantly reduced *E. faecalis* adhesion.[51] Interestingly, pretreatment of *E. faecalis* with sodium meta-periodate, but not pronase, inhibited its attachment to Caco2 cells, suggesting that the *E. faecalis* HS-binding adhesin is a glycoconjugate and not a protein. Consistent with these findings, competition assays with *E. faecalis* diglucosyldiacylglycerol (DGlcDAG) and monoglucosyldiacylglycerol (MGlcDAG) showed that DGlcDAG, but not MGlcDAG, inhibits *E. faecalis* attachment to host cells. Further, surface plasmon resonance analyses confirmed that DGlcDAG is a strong HS-binding molecule.[51] Together, these observations suggest that DGlcDAG, either as itself or as a component of lipoteichoic acid, mediates the attachment of *E. faecalis* to HS on gastric host epithelial cells, facilitating its colonization of the colonic epithelium and potentially its translocation from the intestinal tract.

3. Parasitic Attachment

Few parasites, such as *Plasmodium* spp.,[52] *Leishmania* spp.,[53] and *Trypanosoma cruzi*,[54] also interact with cell surface GAGs for their attachment to host tissues. Recent studies suggest that GAG binding may be more widespread among parasites than anticipated. *Encephalitozoon intestinalis* is an obligate intracellular parasite that infects intestinal epithelial cells. *E. intestinalis* infection can cause severe diarrhea and sometimes even systemic infections in immunocompromised hosts. Adhesion of *E. intestinalis* spores to Caco2 cells was inhibited by HS, heparin, CS, and DS, but not by the nonsulfated HA,[55] suggesting that sulfated GAGs mediate its attachment. However, adhesion of *E. intestinalis* to mutant CHO cells overexpressing CS, but devoid of HS (psgD-677), was markedly reduced compared to wild-type CHO cells expressing both GAGs.[55] Further, the *E. intestinalis* spore wall protein EnP1 was shown to bind to HS on host cell surfaces[56] and excess recombinant EnP1 or anti-EnP1 antibody inhibited *E. intestinalis* attachment to monkey kidney Vero cells. These data suggest that the interaction between the spore wall protein EnP1 and cell surface HS mediates the adhesion of *E. intestinalis* to host epithelial cells.

B. GAGs in Pathogen Invasion

Studies so far have indicated that the capacity to bind to GAGs is most prominent among intracellular pathogens. Consistent with this observation, most, if not all, viruses bind to GAGs and many intracellular bacteria, parasites, and fungi also interact with GAGs.[2,4,12,57] Cell surface GAGs are thought to enhance host cell invasion by serving as low affinity receptors that facilitate the interaction of pathogens with their respective secondary internalization receptors. However, several data indicate that GAGs can also function as direct internalization receptors for some pathogens. For example, the alphaherpesviruses HSV serotypes 1 and 2 have been shown to subvert cell surface HS as both attachment coreceptors and direct internalization receptors. In the attachment coreceptor model, HSV binds to cell surface HS via glycoproteins gB and gC.[58–60] This concentrates HSV virions at the surface and facilitates the interaction of HSV gD with secondary internalization receptors, such as herpes virus entry mediator (HVEM) and nectin-1 and -2.[58–60] Engagement of entry receptors by gD induces fusion of the viral envelope with the host plasma membrane, leading to internalization of HSV virions. In contrast, as an internalization receptor, HS modified by 3-*O*-sulfation can directly bind to gD and mediate HSV internalization.[61–63] Although 3-*O*-sulfation is rarely seen in HS, these findings nevertheless suggest that cell surface HS may serve as an important internalization receptor for HSV in some tissues rich in 3-*O*-sulfated HS. Further, HSV may induce the expression of relevant 3-*O*-sulfotransferase isoforms to generate the required 3-*O*-sulfated HS structures, though this potential virulence mechanism has not been examined.

A recent study examining the fine details of the HS–HSV interaction has revealed that HSV may also modulate cellular processes to facilitate its interaction with HS under biological conditions.[16] The formation of filopodia, a cytoplasmic projection extending from the leading edge of migrating cells, is enhanced by HSV-1 infection. Importantly, HSV-1 virions bind to HS on filopodia and are transported to cell bodies in a mechanism resembling surfing of virions on filopodial plasma membrane in an HS-dependent manner. This process is facilitated by the underlying actin cytoskeleton, regulated by the small Rho GTPase Cdc42, and requires HSV gB, but not gD. The transported HSV virions are internalized by secondary entry receptors in cell bodies. Because induction of filopodia formation is widespread among HS-binding viruses,[64] it will be interesting and important to determine if other viruses interact with HS in this manner.

In addition to HS, HSV-1 can also bind to CS and DS.[65] Highly sulfated CS extracted from squid cartilage (CS-E) inhibited HSV-1 binding to GMK AH-1 kidney epithelial-like cells.[66] Further, expression of chondroitin 4-*O*-sulfotransferase-1, an enzyme that generates the 4-*O*-sulfated CS-E disaccharides, in Sog9 cells induced HSV-1 infection of this nonpermissive cell line.[67] Although the biological significance of CS and DS binding has yet to be rigorously tested, these data provide additional evidence that the fine structure of GAGs plays an important role in modulating the attachment and invasion of host cells by HSV.

Foot-and-mouth disease virus (FMDV) is yet another viral pathogen that exploits HS for its host cell entry. FMDV is an aphthovirus, a member of the *Picornaviridae* family, which infects host cells via distinct endocytic pathways. FMDV can use various host receptors for its entry, including α_v integrins and HS.[68,69] FMDV binding to α_v integrins leads to internalization by a clathrin-dependent endocytic mechanism.[70] In contrast, FMDV is internalized by caveola-mediated endocytosis when it binds to HS[15] (Fig. 1B). HS-mediated FMDV internalization is prevented by inhibitors of lipid raft formation (e.g., cytochalasin D, nystatin, nystatin, and progesterone), which is required for caveola-mediated endocytosis, but not inhibited by chlorpromazine, an inhibitor of clathrin-dependent endocytosis. Further, siRNA-mediated knockdown of caveolin-1 inhibited HS-mediated internalization of FMDV in MCF-10A human mammary gland epithelial cells.[15] Thus, binding to HS can also affect the subsequent endogenous cellular mechanism that is hijacked for viral invasion.

GAGs can also serve as receptors for certain secreted virulence factors.[71–75] For example, dengue virus NS1, a secreted glycoprotein that accumulates on the plasma membrane of target host cells, binds to HS and CS-E.[72] Dengue virus is a mosquito-borne RNA virus of the *Flaviviridae* family. Immune recognition of NS1 on endothelial cells has been proposed as a mechanism for vascular injury and leakage during severe dengue virus infection. Studies with CHO cells established that specific HS and CS structures modified by 2-*O*- and 3-*O*-sulfotransferases promote NS1 binding to cell surfaces.[72] Interestingly, NS1 binds preferentially to cultured human microvascular endothelial cells over aortic or umbilical cord vein endothelial cells. Whether this is defined by specific HS or CS structures awaits precise comparison of GAG structures on endothelial cells from different tissues.

Several bacterial pathogens also bind to GAGs to facilitate their entry.[50,76–78] For example, binding of the Gram-negative bacterium *N. gonorrhoeae* to cell surface HS induces host cell signaling, which leads to internalization of the bacterium[79,80] (Fig. 1C). *N. gonorrhoeae* is the etiologic agent of gonorrhoeae, the second most prevalent bacterial sexually transmitted disease in the United States. *N. gonorrhoeae* Opa protein mediates binding to the HS moiety of syndecan-1 and -4 as overexpression of syndecan-1 or -4 in HeLa cells increases *N. gonorrhoeae* infection.[79] However, *N. gonorrhoeae* attaches, but does not enter, HeLa cells expressing mutant syndecan constructs lacking the cytoplasmic domain.[79] Mutant syndecan-4 constructs lacking the dimerization motif in the cytoplasmic domain that binds to protein kinase C (PKC) and phosphatidylinositol 4,5-bisphosphate (PIP2) or lacking the invariant C-terminal Glu-Phe-Tyr-Ala PDZ binding domain also do not support gonococcal invasion.[79] Thus, although *N. gonorrhoeae* binds to the HS moiety of both syndecan-1 and -4, subsequent internalization is regulated by intracellular signaling mechanisms mediated by the cytoplasmic domain of syndecans. Consistent with this mechanism, *N. gonorrhoeae* binding induces signaling involving phosphatidylcholine-specific phospholipase C, diacylglycerol, acidic sphingomyelinase, and ceramide, which are apparently important for its internalization.[80] However, the precise signaling cascade leading to the internalization of *N. gonorrhoeae* remains to be defined.

C. GAGs in Pathogen Transmission and Dissemination

Several data suggest that GAG binding plays an important role in the cellular transmission and systemic dissemination of pathogens. Human immunodeficiency virus (HIV), the causative agent of AIDS, has been shown to bind to HS on macrophages,[81] endothelial cells,[82] dendritic cells (DCs),[83] and spermatozoa.[17] The interaction of HIV with HS on DCs has been shown to play a critical role in the transmission of HIV to T cells. HIV binds to the HS moiety of syndecans on DCs and, along with the C-type lectin DC-SIGN, syndecans prolong the infectivity of HIV-1 and promote its transmission to T cells.[83,84] However, semen is the main vector for HIV, and it is not clear if HIV in semen can directly and efficiently infect the relatively sparse DCs in the vaginal mucosa. A recent study suggested that spermatozoa captures HIV through HS and transmits the virus to DCs, macrophages, and T cells.[17] The precise mechanism has yet to be defined, but available data suggest that the interaction of HIV-bound spermatozoa with DCs leads to the transmission of HIV, internalization of spermatozoa, and the phenotypic maturation of DCs, suggesting that this mechanism affects the early stages of sexually transmitted HIV infection. Interestingly, at low pH values seen in the vaginal mucosa after sexual intercourse, the binding of HIV to spermatozoa HS and transmission of HIV to DCs were markedly increased,[17] providing additional evidence that this mechanism may be physiologically significant. A recent study also suggested that HS on endothelial cells might influence HIV dissemination by facilitating the interaction of infected lymphoid cells with the endothelium through Tat. Here, HS on lymphoid and endothelial cells simultaneously binds to Tat homodimers, leading to the formation of HS-Tat-Tat-HS quaternary complexes that physically link lymphoid cells to the endothelium, promoting their extravasation.[19] Together, these findings suggest that HS on various cells plays an important role in the capture, transmission, and dissemination of HIV.

Results from several *in vivo* studies suggest that HS binding may also affect the dissemination of bacterial and parasitic pathogens. For example, *Mycobacterium tuberculosis*, the causative agent of tuberculosis (TB), expresses a heparin-binding hemagglutinin adhesin (HBHA) that has been determined to be an epithelial adhesin.[85,86] Alveolar epithelial cells are not the primary cellular targets of TB, but their infection apparently plays an important role in the dissemination of TB. Importantly, the capacity to bind to HS on epithelial cells may mediate TB dissemination. *M. tuberculosis* deficient in HBHA shows significantly reduced adhesion to type II alveolar epithelial cells *in vitro*.[87] Moreover, in a mouse model of intranasal TB, HBHA deletion had no effect on the initial pulmonary colonization, but significantly decreased bacterial dissemination.[87] Because the deletion of HBHA did not affect the capacity of TB to colonize or replicate in distant organs, these findings suggest that HBHA functions specifically in the dissemination of primary pulmonary TB. How this is accomplished is incompletely understood, but HBHA binding to epithelial HS leads to the reorganization of the actin cytoskeleton and triggers endocytosis and transport of TB-containing vesicles to the basal surface.[88]

In the pathogenesis of malaria, infected mosquitoes inject *P. falciparum* sporozoites into the skin of a mammalian host, and the sporozoites travel through the bloodstream to the liver where they invade hepatocytes and are transformed into extraerythrocytic forms. Each extraerythrocytic form releases thousands of merozoites, which enter the blood stream and infect erythrocytes, causing the symptoms of malarial disease, such as anemia, fever, arthralgia, and in severe cases, coma and death. To disseminate from the skin to the liver, *P. falciparum* apparently detects the different degree of HS sulfation in tissues that it encounters through circumsporozoite protein (CSP), the major surface protein of circumsporozoites. CSP binds to low sulfated HS on dermal and endothelial cells, but only invades cells with highly sulfated HS, such as those of hepatocytes, allowing *P. falciparum* sporozoites to migrate through tissues and specifically infect the liver.[89]

III. GAGs in Evasion of Host Defense

The host environment is generally hostile for pathogens because it contains various physical barriers, creates inhabitable conditions (e.g., low pH in stomach, limited access to essential nutrients like iron), constantly operates several nonspecific mechanical systems that remove pathogens (e.g., coughing, sneezing, ciliary motion), and also because the host can mount effective innate and adaptive defense mechanisms to eradicate pathogens. Interestingly, several studies suggest that certain pathogens subvert GAGs to evade various host defense mechanisms. For example, GAGs have been shown to bind to and inhibit several antimicrobial peptides (AMPs). AMPs are ubiquitously expressed in many tissues in both vertebrates and invertebrates.[90] Over 1000 AMPs have been identified and the majority of them are highly cationic molecules.[91] Some AMPs are constitutively expressed, whereas others are rapidly induced by infection. In general, AMPs kill bacteria, parasites, and fungi by disrupting the normal arrangement of lipid membranes and either melting or generating holes in the lipid membrane[90–92] (Fig. 2A). As for viruses, AMPs can inactivate them by acting on either the virus or the host cell. For example, human β-defensins can directly inactivate HIV virions or inhibit viral replication in T cells.[93]

Given the highly positive charge of AMPs, several investigators examined the possibility of highly negatively charged GAGs binding to and neutralizing AMPs. Indeed, DS and HS, but not CS, were found to bind to α-defensin and inhibit α-defensin-mediated killing of *Pseudomonas aeruginosa*, *E. faecalis*, and *Streptococcus pyogenes*.[24] Further, *P. aeruginosa*, *E. faecalis*, and *S. pyogenes* proteinases degraded the DS proteoglycan decorin and released soluble DS fragments,[24] suggesting a pathogenic mechanism where several Gram-positive and Gram-negative bacteria exploit the AMP-inhibiting activity of GAGs by enzymatically releasing them from proteoglycans (Fig. 2B).

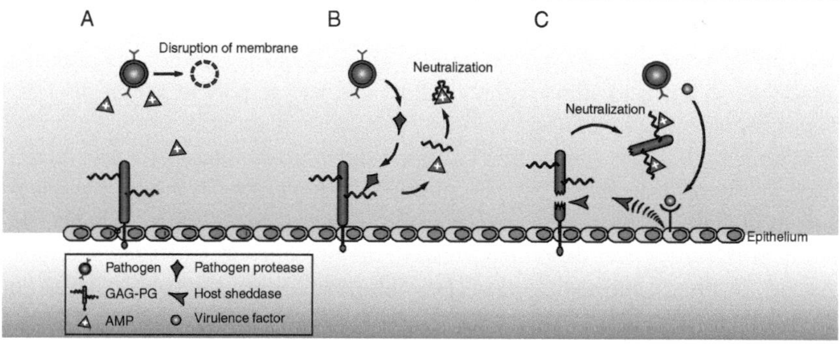

FIG. 2. GAGs in evasion of host defense. (A) AMPs disrupt lipid membranes of pathogens. (B) Pathogens release proteinases that release GAG fragments from proteoglycans. Soluble GAG fragments bind to and inhibit cationic AMPs. (C) Pathogens release virulence factors that trigger the ectodomain shedding of cell surface proteoglycans. Proteoglycan ectodomains, replete with GAG chains, bind to and neutralize cationic AMPs and other GAG-binding host defense factors.

GAGs were also found to bind to and inhibit the antibacterial activity of the human cathelicidin LL-37 in bronchoalveolar lavage (BAL) fluids obtained from patients with cystic fibrosis (CF) against *P. aeruginosa* and *S. aureus.*[21] LL-37 is apparently bound to and inhibited by HS, CS, and HA fragments in the CF airspace because treatment of CF BAL fluids with heparinase II, chondroitinase ABC, hyaluronidase, or high salt increased the antibacterial activity of LL-37 in CF BAL fluids.[21] Similarly, heparin was found to inhibit the antibacterial activity of LL-37 in acute wound fluids against *S. aureus* and *Escherichia coli*.[20] Because both CF BAL fluids and acute wound fluids contain high concentrations of soluble GAG fragments, these observations may explain, in part, why the CF lung and skin wounds are susceptible to bacterial infections despite an abundance of AMPs.

Several bacterial pathogens can also activate the ectodomain shedding mechanism of host cells to generate soluble proteoglycan fragments, replete with AMP-inhibitory GAG chains. *P. aeruginosa*, *S. aureus*, and *Streptococcus pneumoniae* were found to secrete factors that enhance the shedding of syndecan-1 ectodomains.[94–96] Here, bacterial pathogens secrete virulence factors that stimulate the host cell's shedding mechanism after which endogenous or bacterial metalloproteinases cleave the core protein to shed syndecan-1 ectodomains from the cell surface. Importantly, syndecan-1 ectodomain binds to and inhibits AMPs, such as cathelicidins, and potentially other host defense factors (e.g., collectins), in an HS-dependent manner (Fig. 2C). The physiological significance of this mechanism is underscored by the observation that the virulence of *P. aeruginosa* is significantly reduced in syndecan-1 null mice compared to that in wild-type mice in models of intranasal lung infection[23] and burn injury infection.[97] Further, administration of purified syndecan-1 ectodomain or HS increases bacterial virulence in the resistant syndecan-1 null mice, whereas metalloproteinase inhibitors reduce virulence in the susceptible wild-type mice.[23] Whether other bacterial pathogens utilize a similar mechanism to subvert the antihost defense activity of syndecan-1 HS chains remains to be defined.

The HS–*H. pylori* interaction has been shown to facilitate bacterial attachment,[98] but this interaction can also promote pathogenesis by inhibiting phagocytic host defense mechanisms. HS and heparin were found to enhance the ability of *H. pylori* to resist phagocytosis in the presence of untreated and heat-treated serum,[22] suggesting that HS interferes with the phagocytosis of *H. pylori* mediated by the classical and alternative complement pathways. This mechanism may partly explain how *H. pylori* establishes chronic infection because one explanation for the persistence of *H. pylori* in patients is its ability to resist phagocytosis.[98] Again, it remains to be determined whether other bacterial pathogens utilize HS to evade phagocytic host defense mechanisms and cause chronic infections.

Lastly, certain microbial pathogens synthesize their own GAGs as components of the polysaccharide capsule.[99] Bacterial capsules have been recognized as an important virulence structure for over 100 years.[100] Initial studies showed that many highly virulent streptococcal isolates expressed capsules, but in broth cultures, the capsule disappeared within a few hours, suggesting that it is an important virulence structure *in vivo*. Consistent with this notion, bacterial GAGs have been implicated in various steps of pathogenesis, including attachment to host tissues and evasion from host defense mechanisms. Unlike mammalian GAGs, bacterial GAGs are unmodified and not synthesized on Ser residues of core proteins as proteoglycans. *S. pyogenes* and type A *Pasteurella multocida* synthesize HA that is identical or highly similar in structure to vertebrate HA, whereas other types of *P. multocida* can express chondroitin and *N*-acetyl heparosan (unmodified HS), and K5 *E. coli* can also generate *N*-acetyl heparosan.[99]

Streptococcus spp., *Pasteurella* spp., and *E. coli* mutants lacking the genes for GAG synthases, and therefore the capsule, are less virulent than their parent strains expressing GAG synthase genes.[101–103] In most cases, resistance to phagocytosis or complement-mediated killing is reduced in the acapsular mutants. Although the precise mechanism is not understood, GAG capsules are thought to interfere with these host defense mechanisms by inhibiting the binding of host antibodies generated against bacterial cell wall and membrane components. For example, *S. pyogenes* was capable of surviving in human blood containing high levels of antibodies directed against the G-related alpha2-macroglobulin-binding protein (GRAB), a highly conserved *S. pyogenes* surface protein.[104] This was determined to be due to inhibition of binding of anti-GRAB antibodies to the bacterial surface. Anti-GRAB antibodies promoted opsonophagocytosis of a capsule-deficient mutant of *S. pyogenes* but not of the encapsulated parent strain. Protection of *S. pyogenes*, but not acapsular *S. pyogenes*, from opsonophagocytosis in the presence of anti-GRAB antibodies was also observed in a murine model of streptococcal infection.[104] Thus, masking of immunogenic determinants on the bacterial surface by the HA capsule is an important pathogenic mechanism of *S. pyogenes*. Further, GAG capsules are not immunogenic because they are identical or highly similar in structure to the vertebrate GAGs, demonstrating a special case of molecular mimicry where it is used to protect the pathogen from eradication by host defense mechanisms.

IV. Concluding Remarks

The diverse functions of GAGs in the pathogenesis of infectious diseases reflect the multitude of virulence mechanisms elaborated by microbial pathogens. As discussed, GAGs impinge on key steps of pathogenesis, such as adhesion and invasion of host cells, cell–cell transmission, dissemination into the systemic circulation and secondary infection of distant organs, and evasion of host defense mechanisms. Given the ubiquitous nature and versatile capacity of GAGs to modulate many molecular and cellular processes, it is not surprising that many viral, bacterial, parasitic, and fungal pathogens subvert GAGs to facilitate their pathogenesis. However, these features also suggest that GAG–pathogen interactions are potential targets for the development of both highly specific and broadly effective antimicrobial therapy. For example, GAG mimetics that compete with endogenous GAGs and cationic compounds that bind to and inhibit GAGs may serve as therapeutic agents against a broad range of pathogens that use GAGs for their attachment and invasion. Similarly, inhibitors of GAG biosynthesis and enzymes that digest GAGs should interfere with GAG-dependent pathogenic mechanisms. Consistent with this idea, several engineered GAGs, sulfated compounds, and cationic compounds have been shown to effectively inhibit viral, bacterial, and parasitic infections in cell culture-based or animal models of infection.[105–114] However, care must be taken in antagonizing GAGs to combat infectious diseases *in vivo* because GAGs are multifunctional molecules that play essential roles in many endogenous processes. Future studies directed at elucidating the key molecular and cellular mechanisms of GAG–pathogen interactions should enable us to discern the critical functions of GAGs in microbial pathogenesis and endogenous processes. These studies in turn should allow us to design rational GAG-based therapeutic approaches against infectious diseases without compromising the normal functions of GAGs.

References

1. Akhtar J, Shukla D. Viral entry mechanisms: cellular and viral mediators of herpes simplex virus entry. *FEBS J* 2009;**276**:7228–36.
2. Chen Y, Gotte M, Liu J, Park PW. Microbial subversion of heparan sulfate proteoglycans. *Mol Cells* 2008;**26**:415–26.
3. Fears CY, Woods A. The role of syndecans in disease and wound healing. *Matrix Biol* 2006;**25**:443–56.
4. Spillmann D. Heparan sulfate: anchor for viral intruders? *Biochimie* 2001;**83**:811–7.
5. Vives RR, Lortat-Jacob H, Fender P. Heparan sulphate proteoglycans and viral vectors: ally or foe? *Curr Gene Ther* 2006;**6**:35–44.
6. Esko JD, Lindahl U. Molecular diversity of heparan sulfate. *J Clin Invest* 2001;**108**:169–73.
7. Lindahl U, Kjellén L. Biosynthesis of heparin and heparan sulfate. In: Wight TN, Mecham RP, editors. *Biology of proteoglycans*. Orlando, FL: Academic Press Inc.; 1987. p. 59–104.
8. Myette JR, Soundararajan V, Behr J, Shriver Z, Raman R, Sasisekharan R. Heparin/heparan sulfate *N*-sulfamidase from *Flavobacterium heparinum*: structural and biochemical investigation of catalytic nitrogen-sulfur bond cleavage. *J Biol Chem* 2009;**284**:35189–200.
9. Fux L, Ilan N, Sanderson RD, Vlodavsky I. Heparanase: busy at the cell surface. *Trends Biochem Sci* 2009;**34**:511–9.

10. Dietrich CP, Nader HB, Straus AH. Structural differences of heparan sulfates according to the tissue and species of origin. *Biochem Biophys Res Commun* 1983;**111**:865–71.
11. Bernfield M, Götte M, Park PW, Reizes O, Fitzgerald ML, Lincecum J, Zako M. Functions of cell surface heparan sulfate proteoglycans. *Annu Rev Biochem* 1999;**68**:729–77.
12. Rostand KS, Esko JD. Microbial adherence to and invasion through proteoglycans. *Infect Immun* 1997;**65**:1–8.
13. Kines RC, Thompson CD, Lowy DR, Schiller JT, Day PM. The initial steps leading to papillomavirus infection occur on the basement membrane prior to cell surface binding. *Proc Natl Acad Sci USA* 2009;**106**:20458–63.
14. Knappe M, Bodevin S, Selinka HC, Spillmann D, Streeck RE, Chen XS, et al. Surface-exposed amino acid residues of HPV16 L1 protein mediating interaction with cell surface heparan sulfate. *J Biol Chem* 2007;**282**:27913–22.
15. O'Donnell V, Larocco M, Baxt B. Heparan sulfate-binding foot-and-mouth disease virus enters cells via caveola-mediated endocytosis. *J Virol* 2008;**82**:9075–85.
16. Oh MJ, Akhtar J, Desai P, Shukla D. A role for heparan sulfate in viral surfing. *Biochem Biophys Res Commun* 2009;**391**:176–81.
17. Ceballos A, Remes Lenicov F, Sabatte J, Rodriguez Rodrigues C, Cabrini M, et al. Spermatozoa capture HIV-1 through heparan sulfate and efficiently transmit the virus to dendritic cells. *J Exp Med* 2009;**206**:2717–33.
18. Lambert S, Bouttier M, Vassy R, Seigneuret M, Petrow-Sadowski C, et al. HTLV-1 uses HSPG and neuropilin-1 for entry by molecular mimicry of VEGF165. *Blood* 2009;**113**:5176–85.
19. Urbinati C, Nicoli S, Giacca M, David G, Fiorentini S, Caruso A, Alfano M, Cassetta L, Presta M, Rusnati M. HIV-1 Tat and heparan sulfate proteoglycan interaction: a novel mechanism of lymphocyte adhesion and migration across the endothelium. *Blood* 2009;**114**:3335–42.
20. Baranska-Rybak W, Sonesson A, Nowicki R, Schmidtchen A. Glycosaminoglycans inhibit the antibacterial activity of LL-37 in biological fluids. *J Antimicrob Chemother* 2006;**57**:260–5.
21. Bergsson G, Reeves EP, McNally P, Chotirmall SH, Greene CM, Greally P, Murphy P, O'Neill SJ, McElvaney NG. LL-37 complexation with glycosaminoglycans in cystic fibrosis lungs inhibits antimicrobial activity, which can be restored by hypertonic saline. *J Immunol* 2009;**183**:543–51.
22. Dubreuil JD, Ruggiero P, Rappuoli R, Del Giudice G. Effect of heparin binding on *Helicobacter pylori* resistance to serum. *J Med Microbiol* 2004;**53**:9–12.
23. Park PW, Pier GB, Hinkes MT, Bernfield M. Exploitation of syndecan-1 shedding by *Pseudomonas aeruginosa* enhances virulence. *Nature* 2001;**411**:98–102.
24. Schmidtchen A, Frick I, Björck L. Dermatan sulfate is released by proteinases of common pathogenic bacteria and inactivates antibacterial alpha-defensin. *Mol Microbiol* 2001;**39**:708–13.
25. Kalia M, Chandra V, Rahman SA, Sehgal D, Jameel S. Heparan sulfate proteoglycans are required for cellular binding of the hepatitis E virus ORF2 capsid protein and for viral infection. *J Virol* 2009;**83**:12714–24.
26. Leistner CM, Gruen-Bernhard S, Glebe D. Role of glycosaminoglycans for binding and infection of hepatitis B virus. *Cell Microbiol* 2008;**10**:122–33.
27. Barth H, Schnober EK, Zhang F, Linhardt RJ, Depla E, Boson B, Cosset FL, Patel AH, Blum HE, Baumert TF. Viral and cellular determinants of the hepatitis C virus envelope-heparan sulfate interaction. *J Virol* 2006;**80**:10579–90.
28. Schulze A, Gripon P, Urban S. Hepatitis B virus infection initiates with a large surface protein-dependent binding to heparan sulfate proteoglycans. *Hepatology* 2007;**46**:1759–68.
29. Johnson KM, Kines RC, Roberts JN, Lowy DR, Schiller JT, Day PM. Role of heparan sulfate in attachment to and infection of the murine female genital tract by human papillomavirus. *J Virol* 2009;**83**:2067–74.

30. Joyce JG, Tung JS, Przysiecki CT, Cook JC, Lehman ED, Sands JA, Jansen KU, Keller PM. The L1 major capsid protein of human papillomavirus type 11 recombinant virus-like particles interacts with heparin and cell-surface glycosaminoglycans on human keratinocytes. *J Biol Chem* 1999;**274**:5810–22.
31. Giroglou T, Florin L, Schafer F, Streeck RE, Sapp M. Human papillomavirus infection requires cell surface heparan sulfate. *J Virol* 2001;**75**:1565–70.
32. Bienkowska-Haba M, Patel HD, Sapp M. Target cell cyclophilins facilitate human papillomavirus type 16 infection. *PLoS Pathog* 2009;**5**:e1000524.
33. Roberts JN, Buck CB, Thompson CD, Kines R, Bernardo M, Choyke PL, Lowy DR, Schiller JT. Genital transmission of HPV in a mouse model is potentiated by nonoxynol-9 and inhibited by carrageenan. *Nat Med* 2007;**13**:857–61.
34. Rocha-Zavaleta L, Pereira-Suarez AL, Yescas G, Cruz-Mimiaga RM, Garcia-Carranca A, Cruz-Talonia F. Mucosal IgG and IgA responses to human papillomavirus type 16 capsid proteins in HPV16-infected women without visible pathology. *Viral Immunol* 2003;**16**:159–68.
35. Lopez TV, Cancio C, Cruz-Talonia F, Ruiz B, Sapp M, Rocha-Zavaleta L. Binding of human papillomavirus type 16 to heparan sulfate is inhibited by mucosal antibodies from patients with low-grade squamous intraepithelial lesions but not from cervical cancer patients. *FEMS Immunol Med Microbiol* 2008;**54**:167–76.
36. Lowy FD. *Staphylococcus aureus* infections. *N Engl J Med* 1998;**339**:520–32.
37. Sheagren JN. *Staphylococcus aureus*: the persistent pathogen (part II). *N Engl J Med* 1984;**310**:1437–42.
38. Sheagren JN. *Staphylococcus aureus*: the persistent pathogen (part I). *N Engl J Med* 1984;**310**:1368–73.
39. Hess DJ, Henry-Stanley MJ, Erlandsen SL, Wells CJ. Heparan sulfate proteoglycans mediate *Staphylococcus aureus* interactions with intestinal epithelium. *Med Microbiol Immunol* 2006;**195**:133–41.
40. Henry-Stanley MJ, Hess DJ, Erlandsen SL, Wells CL. Ability of the heparan sulfate proteoglycan syndecan-1 to participate in bacterial translocation across the intestinal epithelial barrier. *Shock* 2005;**24**:571–6.
41. Tristan A, Benito Y, Montserret R, Boisset S, Dusserre E, et al. The signal peptide of *Staphylococcus aureus* panton valentine leukocidin LukS component mediates increased adhesion to heparan sulfates. *PLoS One* 2009;**4**:e5042.
42. Carleton HA, Diep BA, Charlebois ED, Sensabaugh GF, Perdreau-Remington F. Community-adapted methicillin-resistant *Staphylococcus aureus* (MRSA): population dynamics of an expanding community reservoir of MRSA. *J Infect Dis* 2004;**190**:1730–8.
43. Gillet Y, Vanhems P, Lina G, Bes M, Vandenesch F, Floret D, Etienne J. Factors predicting mortality in necrotizing community-acquired pneumonia caused by *Staphylococcus aureus* containing Panton-Valentine leukocidin. *Clin Infect Dis* 2007;**45**:315–21.
44. Labandeira-Rey M, Couzon F, Boisset S, Brown EL, Bes M, et al. *Staphylococcus aureus* Panton-Valentine leukocidin causes necrotizing pneumonia. *Science* 2007;**315**:1130–3.
45. Cover TL, Blaser MJ. *Helicobacter pylori* in health and disease. *Gastroenterology* 2009;**136**:1863–73.
46. Ruiz-Bustos E, Ochoa JL, Wadstrom T, Ascencio F. Isolation and characterisation of putative adhesins from *Helicobacter pylori* with affinity for heparan sulphate proteoglycan. *J Med Microbiol* 2001;**50**:215–22.
47. Guzman-Murillo MA, Ruiz-Bustos E, Ho B, Ascencio F. Involvement of the heparan sulphate-binding proteins of *Helicobacter pylori* in its adherence to HeLa S3 and Kato III cell lines. *J Med Microbiol* 2001;**50**:320–9.

48. Smith MF, Novotony J, Carl VS, Comeau LD. *Helicobacter pylori* and Toll-like receptor agonists induce syndecan-4 expression in an NF-kB-dependent manner. *Glycobiology* 2006;**16**:221–9.
49. Magalhaes A, Marcos NT, Carvalho AS, David L, Figueiredo C, Bastos J, David G, Reis CA. Helicobacter pylori cag pathogenicity island-positive strains induce syndecan-4 expression in gastric epithelial cells. *FEMS Immunol Med Microbiol* 2009;**56**:223–32.
50. Baldassarri L, Bertuccini L, Creti R, Filippini P, Ammendolia MG, Koch S, Huebner J, Orefici G. Glycosaminoglycans mediate invasion and survival of *Enterococcus faecalis* into macrophages. *J Infect Dis* 2005;**191**:1253–62.
51. Sava IG, Zhang F, Toma I, Theilacker C, Li B, Baumert TF, Holst O, Linhardt RJ, Huebner J. Novel interactions of glycosaminoglycans and bacterial glycolipids mediate binding of enterococci to human cells. *J Biol Chem* 2009;**284**:18194–201.
52. Rathore D, Hrstka SC, Sacci Jr. JB, De la Vega P, Linhardt RJ, Kumar S, McCutchan TF. Molecular mechanism of host specificity in *Plasmodium falciparum* infection: role of circumsporozoite protein. *J Biol Chem* 2003;**278**:40905–10.
53. Love DC, Esko JD, Mosser DM. A heparin-binding activity on *Leishmania* amastigotes which mediates adhesion to cellular proteoglycans. *J Cell Biol* 1993;**123**:759–66.
54. Lima AP, Almeida PC, Tersariol IL, Schmitz V, Schmaier AH, Juliano L, Hirata IY, Muller-Esterl W, Chagas JR, Scharfstein J. Heparan sulfate modulates kinin release by *Trypanosoma cruzi* through the activity of cruzipain. *J Biol Chem* 2002;**277**:5875–81.
55. Hayman JR, Southern TR, Nash TE. Role of sulfated glycans in adherence of the microsporidian *Encephalitozoon intestinalis* to host cells in vitro. *Infect Immun* 2005;**73**:841–8.
56. Southern TR, Jolly CE, Lester ME, Hayman JR. EnP1, a microsporidian spore wall protein that enables spores to adhere to and infect host cells in vitro. *Eukaryot Cell* 2007;**6**:1354–62.
57. Shukla D, Spear PG. Herpesviruses and heparan sulfate: an intimate relationship in aid of viral entry. *J Clin Invest* 2001;**108**:503–10.
58. Feyzi E, Trybala E, Bergström T, Lindahl U, Spillman D. Structural requirement of heparan sulfate for interaction with herpes simplex virus type I virions and isolated glycoprotein C. *J Biol Chem* 1997;**272**:24850–7.
59. Laquerre S, Argnani R, Anderson DB, Zucchini S, Manservigi R, Glorioso JC. Heparan sulfate proteoglycan binding by herpes simplex virus type 1 glycoproteins B and C, which differ in their contributions to virus attachment, penetration, and cell-to-cell spread. *J Virol* 1998;**72**:6119–30.
60. Spear PG, Shieh MT, Herold BC, WuDunn D, Koshy TI. Heparan sulfate glycosaminoglycans as primary cell surface receptors for herpes simplex virus. *Adv Exp Med Biol* 1992;**313**:341–53.
61. O'Donnell CD, Tiwari V, Oh MJ, Shukla D. A role for heparan sulfate 3-*O*-sulfotransferase isoform 2 in herpes simplex virus type 1 entry and spread. *Virology* 2006;**346**:452–9.
62. Shukla D, Liu J, Blaiklock P, Shworak NW, Bai X, Esko JD, Cohen GH, Eisenberg RJ, Rosenberg RD, Spear PG. A novel role for 3-*O*-sulfated heparan sulfate in herpes simplex virus 1 entry. *Cell* 1999;**99**:13–22.
63. Tiwari V, O'Donnell CD, Oh MJ, Valyi-Nagy T, Shukla D. A role for 3-*O*-sulfotransferase isoform-4 in assisting HSV-1 entry and spread. *Biochem Biophys Res Commun* 2005;**338**:930–7.
64. Lehmann MJ, Sherer NM, Marks CB, Pypaert M, Mothes W. Actin- and myosin-driven movement of viruses along filopodia precedes their entry into cells. *J Cell Biol* 2005;**170**:317–25.
65. Banfield BW, Leduc Y, Esford L, Visalli RJ, Brandt CR, Tufaro F. Evidence for an interaction of herpes simplex virus with chondroitin sulfate proteoglycans during infection. *Virology* 1995;**208**:531–9.

66. Bergefall K, Trybala E, Johansson M, Uyama T, Naito S, Yamada S, Kitagawa H, Sugahara K, Bergstrom T. Chondroitin sulfate characterized by the E-disaccharide unit is a potent inhibitor of herpes simplex virus infectivity and provides the virus binding sites on gro2C cells. *J Biol Chem* 2005;**280**:32193–9.
67. Uyama T, Ishida M, Izumikawa T, Trybala E, Tufaro F, Bergstrom T, Sugahara K, Kitagawa H. Chondroitin 4-*O*-sulfotransferase-1 regulates E disaccharide expression of chondroitin sulfate required for herpes simplex virus infectivity. *J Biol Chem* 2006;**281**:38668–74.
68. Duque H, Baxt B. Foot-and-mouth disease virus receptors: comparison of bovine alpha(V) integrin utilization by type A and O viruses. *J Virol* 2003;**77**:2500–11.
69. Jackson T, Ellard FM, Ghazaleh RA, Brookes SM, Blakemore WE, Corteyn AH, Stuart DI, Newman JW, King AM. Efficient infection of cells in culture by type O foot-and-mouth disease requires binding to cell surface heparan sulfate. *J Virol* 1996;**70**:5282–7.
70. O'Donnell V, LaRocco M, Duque H, Baxt B. Analysis of foot-and-mouth disease virus internalization events in cultured cells. *J Virol* 2005;**79**:8506–18.
71. Argyris EG, Kulkosky J, Meyer ME, Xu Y, Mukhtar M, Pomerantz RJ, Williams KJ. The perlecan heparan sulfate proteoglycan mediates cellular uptake of HIV-1 Tat through a pathway responsible for biological activity. *Virology* 2004;**330**:481–6.
72. Avirutnan P, Zhang L, Punyadee N, Manuyakorn A, Puttikhunt C, Kasinrerk W, Malasit P, Atkinson JP, Diamond MS. Secreted NS1 of dengue virus attaches to the surface of cells via interactions with heparan sulfate and chondroitin sulfate E. *PLoS Pathog* 2007;**3**:1798–812.
73. Rusnati M, Coltrini D, Oreste P, Zoppetti G, Albini A, Noonan D, d'Adda di Fagagna F, Giacca M, Presta M. Interaction of HIV-1 Tat protein with heparin. Role of the backbone structure, sulfation, and size. *J Biol Chem* 1997;**272**:11313–20.
74. Rusnati M, Tulipano G, Spillmann D, Tanghetti E, Oreste P, Zoppetti G, Giacca M, Presta M. Multiple interactions of HIV-1 Tat protein with size-defined heparin oligosaccharides. *J Biol Chem* 1999;**274**:28198–205.
75. Tyagi M, Rusnati M, Presta M, Giacca M. Internalization of HIV-1 tat requires cell surface heparan sulfate proteoglycans. *J Biol Chem* 2001;**276**:3254–61.
76. Banerjee M, Copp J, Vuga D, Marino M, Chapman T, van der Geer P, Ghosh P. GW domains of the Listeria monocytogenes invasion protein InlB are required for potentiation of Met activation. *Mol Microbiol* 2004;**52**:257–71.
77. Baron MJ, Bolduc GR, Goldberg MB, Auperin TC, Madoff LC. Alpha C protein of group B *Streptococcus* binds host cell surface glycosaminoglycan and enters cells by an actin-dependent mechanism. *J Biol Chem* 2004;**279**:24714–23.
78. Jonquieres R, Pizarro-Cerda J, Cossart P. Synergy between the N- and C-terminal domains of InlB for efficient invasion of non-phagocytic cells by Listeria monocytogenes. *Mol Microbiol* 2001;**42**:955–65.
79. Freissler E, Meyer auf der Heyde A, David G, Meyer TF, Dehio C. Syndecan-1 and syndecan-4 can mediate the invasion of Opa_{HSPG}-expressing *Neisseria gonorrhoeae* into epithelial cells. *Cell Microbiol* 2000;**2**:69–82.
80. Grassmé H, Gulbins E, Brenner B, Ferlinz K, Sandhoff K, Harzer K, Lang F, Meyer TF. Acidic sphingomyelinase mediates entry of *N. gonorrhoeae* into nonphagocytic cells. *Cell* 1997;**91**:605–15.
81. Saphire AC, Bobardt MD, Zhang Z, David G, Gallay PA. Syndecans serve as attachment receptors for human immunodeficiency virus type 1 on macrophages. *J Virol* 2001;**75**:9187–200.
82. Argyris EG, Acheampong E, Nunnari G, Mukhtar M, Williams KJ, Pomerantz RJ. Human immunodeficiency virus type 1 enters primary human brain microvascular endothelial cells by a mechanism involving cell surface proteoglycans independent of lipid rafts. *J Virol* 2003;**77**:12140–51.

83. de Witte L, Bobardt M, Chatterji U, Degeest G, David G, Geijtenbeek TB, Gallay P. Syndecan-3 is a dendritic cell-specific attachment receptor for HIV-1. *Proc Natl Acad Sci USA* 2007;**104**:19464–9.
84. Bobardt MD, Saphire AC, Hung HC, Yu X, Van der Schueren B, Zhang Z, David G, Gallay PA. Syndecan captures, protects, and transmits HIV to T lymphocytes. *Immunity* 2003;**18**:27–39.
85. Menozzi FD, Bischoff R, Fort E, Brennan MJ, Locht C. Molecular characterization of the mycobacterial heparin-binding hemagglutinin, a mycobacterial adhesin. *Proc Natl Acad Sci USA* 1998;**95**:12625–30.
86. Menozzi FD, Rouse JH, Alavi M, Laude-Sharp M, Muller J, Bischoff R, Brennan MJ, Locht C. Identification of a heparin-binding hemagglutinin present in mycobacteria. *J Exp Med* 1996;**184**:993–1001.
87. Pethe K, Alonso S, Biet F, Delogu G, Brennan MJ, Locht C, Menozzi FD. The heparin-binding haemagglutinin of *M. tuberculosis* is required for extrapulmonary dissemination. *Nature* 2001;**412**:190–4.
88. Menozzi FD, Reddy VM, Cayet D, Raze D, Debrie AS, Dehouck MP, et al. *Mycobacterium tuberculosis* heparin-binding haemagglutinin adhesin (HBHA) triggers receptor-mediated transcytosis without altering the integrity of tight junctions. *Microbes Infect* 2006;**8**:1–9.
89. Coppi A, Tewari R, Bishop JR, Bennett BL, Lawrence R, Esko JD, Billker O, Sinnis P. Heparan sulfate proteoglycans provide a signal to *Plasmodium* sporozoites to stop migrating and productively invade host cells. *Cell Host Microbe* 2007;**2**:316–27.
90. Brogden KA. Antimicrobial peptides: pore formers or metabolic inhibitors in bacteria? *Nat Rev Microbiol* 2005;**3**:238–50.
91. Lai Y, Gallo RL. AMPed up immunity: how antimicrobial peptides have multiple roles in immune defense. *Trends Immunol* 2009;**30**:131–41.
92. Zasloff M. Antimicrobial peptides of multicellular organisms. *Nature* 2002;**415**:389–95.
93. Klotman ME, Chang TL. Defensins in innate antiviral immunity. *Nat Rev Immunol* 2006;**6**:447–56.
94. Chen Y, Bennett A, Hayashida A, Hollingshead S, Park PW. Streptococcus pneumoniae sheds syndecan-1 ectodomains via ZmpC, a metalloproteinase virulence factor. *J Biol Chem* 2007;**282**:159–67.
95. Park PW, Foster TJ, Nishi E, Duncan SJ, Klagsbrun M, Chen Y. Activation of syndecan-1 ectodomain shedding by *Staphylococcus aureus* alpha-toxin and beta-toxin. *J Biol Chem* 2004;**279**:251–8.
96. Park PW, Pier GB, Preston MJ, Goldberger O, Fitzgerald ML, Bernfield M. Syndecan-1 shedding is enhanced by LasA, a secreted virulence factor of *Pseudomonas aeruginosa*. *J Biol Chem* 2000;**275**:3057–64.
97. Haynes 3rd A, Ruda F, Oliver J, Hamood AN, Griswold JA, Park PW, Rumbaugh KP. Syndecan-1 shedding contributes to *Pseudomonas aeruginosa* sepsis. *Infect Immun* 2005;**73**:7914–21.
98. Chmiela M, Lelwala-Guruge J, Wadstrom T. Interaction of cells of *Helicobacter pylori* with human polymorphonuclear leucocytes: possible role of haemagglutinins. *FEMS Immunol Med Microbiol* 1994;**9**:41–8.
99. DeAngelis PL. Microbial glycosaminoglycan glycosyltransferases. *Glycobiology* 2002;**12**:9R–16R.
100. Stollerman GH, Dale JB. The importance of the group a streptococcus capsule in the pathogenesis of human infections: a historical perspective. *Clin Infect Dis* 2008;**46**:1038–45.
101. Chung JY, Wilkie I, Boyce JD, Townsend KM, Frost AJ, Ghoddusi M, Adler B. Role of capsule in the pathogenesis of fowl cholera caused by *Pasteurella multocida* serogroup A. *Infect Immun* 2001;**69**:2487–92.

102. Svanborg-Eden C, Hagberg L, Hull R, Hull S, Magnusson KE, Ohman L. Bacterial virulence versus host resistance in the urinary tracts of mice. *Infect Immun* 1987;**55**:1224–32.
103. Wessels MR, Goldberg JB, Moses AE, DiCesare TJ. Effects on virulence of mutations in a locus essential for hyaluronic acid capsule expression in group A streptococci. *Infect Immun* 1994;**62**:433–41.
104. Dinkla K, Nitsche-Schmitz DP, Barroso V, Reissmann S, Johansson HM, Frick IM, et al. Identification of a streptococcal octapeptide motif involved in acute rheumatic fever. *J Biol Chem* 2007;**282**:18686–93.
105. Copeland R, Balasubramaniam A, Tiwari V, Zhang F, Bridges A, Linhardt RJ, Shukla D, Liu J. Using a 3-*O*-sulfated heparin octasaccharide to inhibit the entry of herpes simplex virus type 1. *Biochemistry* 2008;**47**:5774–83.
106. Rusnati M, Urbinati C, Caputo A, Possati L, Lortat-Jacob H, Giacca M, Ribatti D, Presta M. Pentosan polysulfate as an inhibitor of extracellular HIV-1 Tat. *J Biol Chem* 2001;**276**:22420–5.
107. Rusnati M, Vicenzi E, Donalisio M, Oreste P, Landolfo S, Lembo D. Sulfated K5 *Escherichia coli* polysaccharide derivatives: a novel class of candidate antiviral microbicides. *Pharmacol Ther* 2009;**123**:310–22.
108. Herold BC, Bourne N, Marcellino D, Kirkpatrick R, Strauss DM, et al. Poly(sodium 4-styrene sulfonate): an effective candidate topical antimicrobial for the prevention of sexually transmitted diseases. *J Infect Dis* 2000;**181**:770–3.
109. Lee E, Pavy M, Young N, Freeman C, Lobigs M. Antiviral effect of the heparan sulfate mimetic, PI-88, against dengue and encephalitic flaviviruses. *Antiviral Res* 2006;**69**:31–8.
110. Lembo D, Donalisio M, Rusnati M, Bugatti A, Cornaglia M, Cappello P, et al. Sulfated K5 Escherichia coli polysaccharide derivatives as wide-range inhibitors of genital types of human papillomavirus. *Antimicrob Agents Chemother* 2008;**52**:1374–81.
111. Roan NR, Sowinski S, Munch J, Kirchhoff F, Greene WC. Aminoquinoline surfen inhibits the action of SEVI (semen-derived enhancer of viral infection). *J Biol Chem* 2010;**285**:1861–9.
112. Schmidtke M, Karger A, Meerbach A, Egerer R, Stelzner A, Makarov V. Binding of a *N*, *N*′-bisheteryl derivative of dispirotripiperazine to heparan sulfate residues on the cell surface specifically prevents infection of viruses from different families. *Virology* 2003;**311**:134–43.
113. Schuksz M, Fuster MM, Brown JR, Crawford BE, Ditto DP, Lawrence R, et al. Surfen, a small molecule antagonist of heparan sulfate. *Proc Natl Acad Sci USA* 2008;**105**:13075–80.
114. Vogt AM, Pettersson F, Moll K, Jonsson C, Normark J, et al. Release of sequestered malaria parasites upon injection of a glycosaminoglycan. *PLoS Pathog* 2006;**2**:e100.

Molecular Mechanism Underlines Heparin-Induced Thrombocytopenia and Thrombosis

Yi Qian, Jing Pan, Xiaodong Zhou, Peter Weiser, Hong Lu, and Lijuan Zhang

Department of Pathology and Immunology, Washington University Medical School, St. Louis, Missouri, USA

Heparin-induced thrombocytopenia (HIT) with thrombosis is the most severe side effect of heparin administration. HIT patients may die or have permanent sequelae, such as a stroke or limb amputation. Contaminated heparin is associated with anaphylactic reactions and deaths by activating the contact system. It is also associated with high incidence of HIT via a yet unknown mechanism. This chapter shows that: (1) the contact system can be activated by a variety of unrelated molecules; (2) kallikrein directly cuts prothrombin to generate functional thrombin through contact system activation; and (3) while heparin contaminants, oversulfated heparin by-product (OS-HB), induce thrombin generation in both normal and HIT patient plasmas

Progress in Molecular Biology and Translational Science, Vol. 93
DOI: 10.1016/S1877-1173(10)93017-2

1877-1173/10 $35.00

through contact system activation, authentic heparin induces thrombin activities only in HIT patient plasmas containing autoantibodies against protein/heparin complex. These data suggest that the negatively charged IgG/protein/heparin or OS-HB complex activate the contact system and produce thrombin in human plasma and thrombin partially activates the platelets allowing subsequent platelet activation through IgG/Fc receptor II signaling. The newly discovered mechanism of heparin-induced thrombin activity could explain the increased incidence of HIT in patients exposed to contaminated heparin. Furthermore, the assays used in these studies would be valuable for HIT diagnosis, prevention, and treatment.

Abbreviation: C1inh, C1 inhibitor; Cephalin, phosphatidylethanolamine; FII, prothrombin; FIX, factor IX; FV, factor V; FVa, activated factor V; FVIII, factor VIII; FX, factor X; FXI, factor XI; FXII, factor XII; HMWK, high molecular weight kininogen; OS-HB, chemically oversulfated heparin by-product; PLs, phospholipids; TAFI, thrombin-activated fibrinolysis inhibitor; TBS, Tris buffered saline; PF4, platelet factor 4

I. Introduction

Heparin-induced thrombocytopenia (HIT) with thrombosis complication (HITT) is the most severe side effect of heparin. Approximately 12 million patients receive heparin treatment annually in the United States. It is estimated that 8% of patients receiving heparin can potentially develop HIT autoantibodies and about 1–5% develop HIT.[1] One-third HIT patients suffer from venous and/or arterial HITT.[2] At least 10–20% of HITT patients die or have permanent sequelae such as stroke or limb amputation.[3]

Heparin-induced IgG antibodies that bind to Platelet factor 4 (PF4)/heparin complexes cause HITT. The IgG/PF4/heparin complex is a procoagulant, which enhances thrombin activities in HIT patient plasmas by an unknown molecular mechanism.[4] Intriguingly, only partially activated platelets allow the IgG/PF4/heparin complex to cross-link Fc receptor II (FcRII) with Fc domain of IgG on human platelet surfaces.[5] The cross-linking initiates FcRII signaling cascade, which leads to platelet activation, aggregation, and subsequent signaling events in blood circulation resulting in thrombosis.[6–8] It is unknown what factors induce partial human platelet activation that serves as a precursor to initiate IgG/FcRII binding and signaling. It is possible that thrombin plays a role due to its ability to activate protease-activated platelet receptor signaling cascade before other secondary platelet activation signaling starts.[7]

Contaminated heparin has resulted in higher incidence of HIT through an as yet unknown molecular mechanism.[9] In the outbreak of events of 2007 and 2008, contaminated heparin was found to cause a myriad of symptoms, including hypotension, anaphylactic reactions, and deaths by activating complement and contact system.[10,11] The contact system was first discovered as an *in vitro* thrombin generation system where negatively charged artificial surfaces induce thrombin generation and clotting.[12] The contact system was thought to have consisted of two serine proteases: prekallikrein and factor XII, a cofactor: high molecular weight kininogen, and a regulator: C1 inhibitor in human plasma.[13] An extension of the roles and players of contact system and a new molecular mechanism for HIT are presented in this chapter.

II. Kallikrein- and Thrombin-Like Activities Is Induced by a Series of Unrelated Negatively Charged Molecules in Normal Human Plasmas

It has been reported that *in vitro* contact system activation occurs on physiologically relevant surfaces such as articular cartilage, skin, sodium urate crystals, and calcium pyrophosphate.[14] *In vivo* contact system activation occurs on developing thrombus, RNA and DNA from degrading cells, phospholipid (PL) enriched polysomes from platelet membranes, β-amyloid, sulfatides, fatty acids, cholesterol sulfate, GAGs, activated platelet surface during platelet transfusion,[15] ionic contrast media,[16] and also under conditions of sepsis, where bacteria lipopolysaccharides provide a negatively charged surface.[17–20] It was recently reported that protein aggregates activate FXII and Prekallikrein. However, FXII activation is greatly enhanced in the presence of dextran sulfate, a chemically sulfated polysaccharide, suggesting that negatively charged-protein aggregates are better contact system activating surfaces.[21] Based upon these established facts,[14] we propose that the contact system is a broad pattern recognition system, which can be activated by unrelated molecules.

To test the idea, we used a variety of negatively charged molecules to induce kallikrein- and thrombin-like activities through contact system activation in pooled normal human plasmas. Since negatively charged molecules can exert both activating and inhibitory effects on kallikrein-like activities in a time-dependent manner,[22] we measured the kinetics instead of a single time point used in previous studies.[10] To this end, the human plasma was incubated with a series of negatively charged molecules at 37 °C for 5 min. The compounds tested include different bacterial LPS, lipid A, yeast zymosan, DNA, PLs, sulfatide, chondroitin sulfate-A, chondroitin sulfate-E, porcine mucosa dermatan sulfate, heparan sulfate, and heparin by-product (a mixture of heparan sulfate, chondroitin sulfate, and dermatan sulfate) oversulfated forms of heparin by-product, heparan sulfate, chondroitin sulfate, dermatan sulfate and five different types of contaminated heparin described previously,[23] including heparin containing oversulfated contaminant (OSCH4), heparin containing both low sulfated and oversulfated contaminants (OSLSCH5), heparin containing

OSHS (lot S1), heparin containing chemically sulfated/desulfated GAGs (G1), heparin containing low-sulfated heparin contaminant (LSCH6), and authentic heparin control purchased from Sigma. Kinetics of kallikrein- and thrombin-like activity induction was measured colorimetrically by kallikrein or thrombin substrate cleavage, respectively.

We found that each negatively charged molecule had distinct characteristics in terms of the lag time, rate, and pattern of induced kallikrein- and thrombin-like activities compared to the control.[24] Since the plasma control exhibited a 7-min delay in kallikrein-activation induced by the plate surface in the presence of 2 mM $CaCl_2$ in the Tris buffered saline (TBS) buffer, we chose optical density (OD) readings at 3 min after adding protease substrate to the activated plasmas as a measure of the rate of kallikrein- and thrombin-like activity generation by tested compounds. We chose OD readings at 40 min as a measure of overall kallikrein- and thrombin-like activity generation by both tested compound and plate surface (Fig. 1).

All forms of oversulfated GAGs induced substantial kallikrein- and thrombin-like activities, which is reflected in the 3 min OD readings shown in Fig. 1A (lanes 11–15) and B (lanes 11–15) after the protease substrate was added to the plasmas that have been activated by different compounds for 5 min. No kallikrein- and thrombin-like activities are induced by LPS, lipid A, or zymosan (Fig. 1A and B, lanes1–4 and 7) at 3 min but a low degree of kallikrein-like activity is induced by herring DNA (lane 6) at 3 min. Interestingly, these compounds enhanced overall kallikrein-like activities by 10–30% compared to the blank control (0 μg/ml) at 40 min (Fig. 1C, lanes 1–4, 6, and 7 compared to lane 22). Sulfatide (lane 5) induces both kallikrein- and thrombin-like activities (Fig. 1A–D). Interestingly, low concentration of sulfatide is preferred for inducing kallikrein-like activities, whereas high concentration of sulfatide is preferred for inducing thrombin-like activities (lane 5).

Natural GAGs, such as dermatan sulfate, and heparin, induced kallikrein-like activities, but induced no or low thrombin-like activities at 3 min (Fig. 1A and B, lanes 9, 11, and 21). Dermatan sulfate, chondroitin sulfate-E, and heparin had overall lower kallikrein-like activities at 40 min, especially at 200 μg/ml concentration, compared to the blank control (Fig. 1C, lanes 9, 11, and 21 compared to lane 22). These results indicate that natural GAGs activate kallikrein-like activities initially due to their charge, but they inhibit kallikrein-like activities once the contact system is activated.

We found that oversulfated GAGs were the most potent kallikrein- and thrombin-like activity inducers as a result of chemical sulfation (Fig. 1A–D, lanes 12–15). The contaminated heparins had decreased level of oversulfated GAG contamination in the order of OSCH4, OSLSCH5, S1, LSCH6, and G1,[23] which correlated with their ability in inducing kallikrein- and thrombin-like activities at 200 μg/ml level (Fig. 1A–D, lanes 16–20).

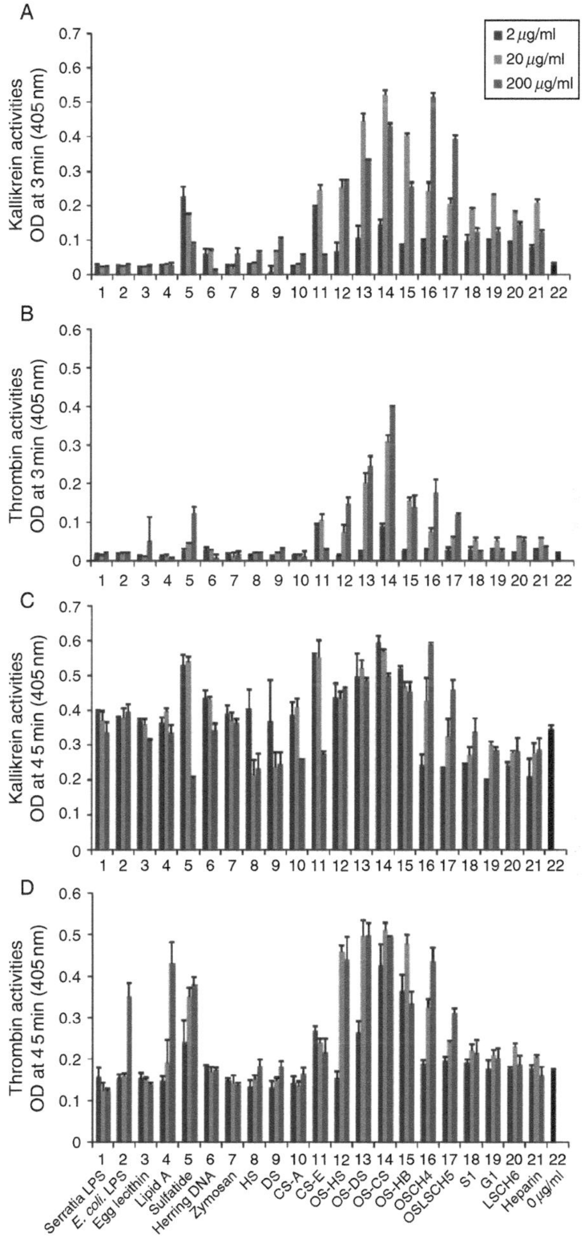

FIG. 1. Negatively charged molecule induced kallikrein- and thrombin-like activities. Normal pooled human plasma (30 μl) was incubated with 20 μl negatively charged molecules at a final concentration of 0, 2, 20, and 200 μg/ml at 37 °C for 5 min. Kallikrein-like activities were assessed by

LPS, lipid A, and zymosan were kallikrein-like activity enhancers once the contact system was activated (Fig. 1C, lanes, 1–4 and 7). Herring DNA was a prekallikrein activator and a kallikrein-like activity enhancer (Fig. 1A and C, lane 6 compared to the control lane 22). *Escherichia coli* LPS and lipid A did not induce thrombin- and kallikrein-like activities at 3 min (Fig. 1B, lanes 2 and 4), however, they enhanced thrombin-like activities at high concentration (200 μg/ml) (Fig. 1D, lanes 2 and 4 compared to the control lane 22).

In conclusion, the kallikrein-like activities are induced by a variety of structure-unrelated compounds in pooled normal human plasmas. Chemically oversulfated GAGs are most potent kallikrein- and thrombin-like activity inducers. Natural GAGs serve as kallikrein- and thrombin-like activity inducers and inhibitors at different times of contact system activation in a structure- and concentration-dependent manner (Fig. 1A–D, lanes 8–11 and 21). Chondroitin sulfate-E (Fig. 1A–D, lane 11) is the most potent kallikrein- and thrombin-like activity inducers among all natural occurring GAGs tested.

III. Thrombin Is Generated Through Contact System Activation

The data in Fig. 1 showed that the contact system is a pattern-dependent signaling system[25] and activated contact system produces both kallikrein- and thrombin-like activities.

The physiological role of the contact system activation-dependent thrombin generation is largely dismissed during the past 50 years for two reasons: (1) It has been assumed that artificial surfaces that are required for contact system activation are not present in the blood circulation; and (2) patients with genetic deficiencies in contact system proteins, such as factor XII or kininogen, are not associated with bleeding phenotypes. However, intravenous injection of some liposomal drugs, diagnostic contrast agents, micelles and other lipid-based nanoparticles can cause acute hypersensitivity reactions similar to contaminated heparin-induced allergic reactions in a high percentage (up to 45%) of patients.[26] It has been known for long that certain nanoparticles-induced thrombosis.[27,28]

---

addition of 150 μl of the kallikrein chromogenic substrate, 0.5 mM *N*-benzoyl-Pro-Phe-Arg-*p*-nitroanilide hydrochloride dissolved in TBS. OD readings were taken every 90 s for 40 mins. Thrombin-like activity was assessed by addition of the thrombin chromogenic substrate, 1 mM Sar-Pro-Arg-*p*-nitroanilide dihydrochloride dissolved in TBS. OD readings were taken every 90 s for 40 min. (A) The OD readings of kallikrein-like activities at 3 min from two independent experiments. (B) The OD readings of thrombin-like activities at 3 min from two independent experiments. (C) The OD readings of kallikrein-like activities at 40 min from two independent experiments. (D) The OD readings of thrombin-like activities at 40 min from two independent experiments. The OD readings of kallikrein activity induced by 20 μg/ml of HS, DS, CS, OSHS, OSDS, and OSCS at 3 and 40 min have been reported previously.[50] This figure was originally published in the *Journal Biological Chemistry*[25]. (See Color Insert.)

We suspected that these particles might have served as the artificial surfaces which have induced thrombin generation through contact system activation. In addition, protein aggregates in human plasma also activate the contact system.[21]

A. Induction of Thrombin Activity from Prothrombin by Kallikrein

While activation of the contact system is known to produce thrombin, the process is thought to proceed primarily via an enzyme cascade that leads to FX activation. It is FXa that cleaves prothrombin to form an active protease[29] in which thrombin is generated from the 72 kDa plasma prothrombin by both thrombin and FXa.[29] Cleavage at positions of 155 and 284 of prothrombin are catalyzed by thrombin whereas the cleavage at positions 271 and 320 are catalyzed by factor Xa to generate A and B chains of the 37 kDa thrombin. The initial cleavage of prothrombin by FXa occurs at position 320, giving rise to meizothrombin. Meizothrombin has the same molecular weight as prothrombin. Both thrombin and meizothrombin have comparable activities toward protein C, factor V, factor XI, and small peptide thrombin substrates.[29]

Recently, Steif showed that incubating purified prothrombin with kallikrein can produce a protease activity that cleaves Sar-Pro-Arg-*p*-nitroanilide, a thrombin substrate.[30] Thus, it is possible that kallikrein may activate prothrombin directly. Since the thrombin substrate can also be cleaved by thrombin, meizothrombin, and other serine proteases, it is unknown if kallikrein can generate meizothrombin or functional thrombin. We used several approaches to examine this possibility further:

(1) We incubated 0.5 μg/ml kallikrein, 5 μg/ml prothrombin, and 0.2 mM of the thrombin substrate (Sar-Pro-Arg-*p*-nitroanilide) in TBS buffer. Thrombin activity was determined by increased OD at 405 nm due to its chromogenic substrate cleavage (Fig. 2A). As reported by Stief,[30] prothrombin and kallikrein incubated together produced an activity that could cleave the thrombin-specific chromogenic substrate.[30] Further analysis indicated that approximately 80% of that production of that activity was inhibited by hirudin, a thrombin-specific inhibitor (Fig. 2A). No activity was obtained when the substrate was incubated with prothrombin alone.

(2) We examined the effects of kallikrein on the prothrombin protein. We incubated kallikrein and prothrombin at 37 °C for 5 min (as above), fractionated the reaction products by SDS-PAGE, and performed a Western blot analysis with antithrombin antibody (Fig. 2B). Incubation of prothrombin with kallikrein resulted in thrombin protein that matched the thrombin standard (Fig. 2B, lanes 2 and 4) and, in addition, produced a protein by its size likely to be des F1-meizothrombin, an intermediate

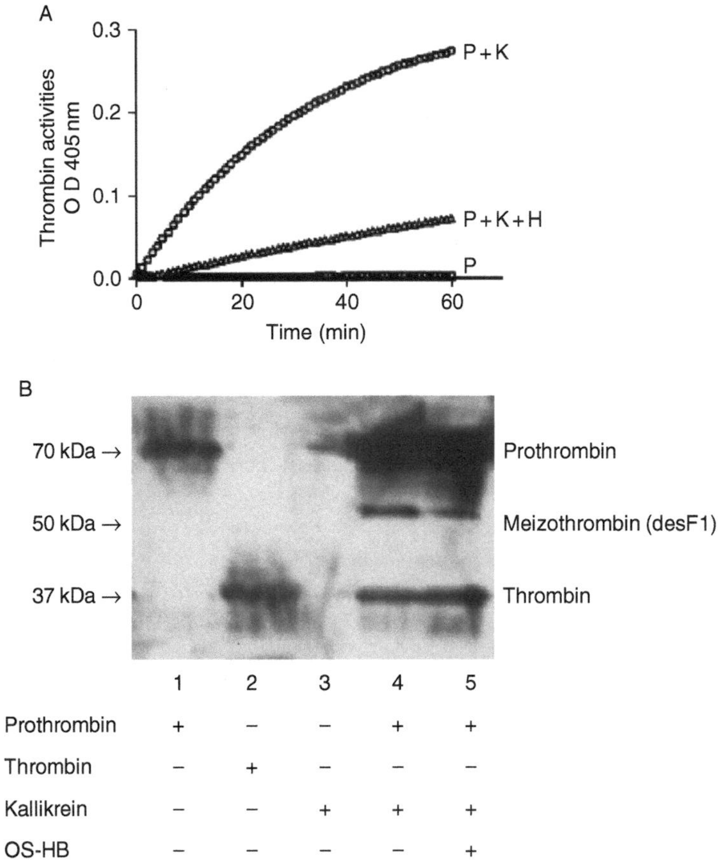

FIG. 2. Thrombin generation by kallikrein. (A) Thrombin activity assay. Prothrombin (1 μg), kallikrein (0.1 μg) + prothrombin (1 μg), or kallikrein (0.1 μg) + prothrombin (1 μg) + hirudin (1 μg) in 10 μl TBS containing 50 mM Tris and 150 mM NaCl pH 7.6 were incubated at 37 °C for 2 h. Thrombin activities was determined by increased OD at 405 nm of the reaction mixture containing 0.1 μg/ml prothrombin alone, 0.1 μg/ml prothrombin + 0.01 μg/ml kallikrein ± 0.1 μg/ml hirudin + 0.2 mM substrate, Sar-Pro-Arg-*p*-nitroanilide dihydrochloride, in TBS. P, prothrombin; P + K, prothrombin + kallikrein, P + K + H, prothrombin + kallikrein + hirudin. (B) Western analysis of thrombin. Prothrombin (0.2 μg), thrombin (0.2 μg), kallikrein (0.2 μg), 0.2 μg kallikrein + 1 μg prothrombin ± 4 μg OS-HB in 10 μl TBS were incubated at 37 °C for 5 min. After adding 10 μl SDS loading buffer, half of sample (10 μl) was used for SDS-PAGE and Western blot analysis. Lanes 1. prothrombin, 2. thrombin, 3. kallikrein, 4. prothrombin + kallikrein, and 5. prothrombin + kallikrein + OS-HB.

in the generation of thrombin from meizothrombin (Fig. 2B, lane 4). Adding OS-HB to kallikrein and prothrombin did not appear to influence the amount of thrombin or des F1-meizothrombin generated (Fig. 2B, lanes 4 and 5).

(3) We examined the capacity of the thrombin activity, generated by kallikrein from prothrombin, to direct fibrinogen clotting. To that end 1 μg of kallikrein and 5 μg of prothrombin were preincubated together (as above) for 5 min, 1, or 2 h, Samples of these reaction mixtures were then incubated in 1 mg/ml fibrinogen solution for up to 500 s. Samples from reactions preincubated for the full 2 h resulted in clotting times of 18.6 ± 0.8 s, similar to the clotting time produced by the thrombin standard (0.33 μg; 16.3 ± 0.4 s). Shorter preincubation times (5 min, 1 h) led to longer clotting times (> 100, 41.2 ± 6 s, respectively). In contrast, no clotting was observed (> 500 s) with 1 or 10 μg kallikrein preincubated alone for up to 2 h.

The data generated from the three independent approaches concluded that functional thrombin can be generated from prothrombin by kallikrein directly.

B. OS-HB, Heparin, and Cephalin (Phosphatidylethanolamine/Silica) Induction of Thrombin- and Kallikrein-Like Activities in Human Plasmas

Given that purified kallikrein can activate thrombin (above), we asked whether this reaction might occur in a more physiological environment. Therefore, we incubated OS-HB in a variety of normal and factor-depleted human plasmas and measured the resulting changes in thrombin- and kallikrein-like activity (Fig. 3). We used heparin as a negative control since we had previously observed minimal effects of heparin on kallikrein- or thrombin-like activity.[25] Also, we used cephalin plus silica as a positive control since this is a clinical clotting reagent that induces thrombin through contact system activation.[13]

OS-HB treatment induced both thrombin-like and kallikrein-like activity in normal human plasma (Fig. 3). Maximum thrombin activity and near maximum kallikrein activity are detected at 2 μg/ml activator. Under those conditions the induction of both thrombin-like and kallikrein-like activities is dependent on the presence of plasma prekallikrein. Moreover, thrombin-like activity but not kallikrein-like activity requires the presence of kininogen, the upstream contact system component. Interestingly, neither activity is dependent on the downstream contact system components factor IX and factor XI nor the factor X, a component common to both the contact system and the tissue factor (extrinsic) coagulation pathways.

Cephalin plus silica treatment also induce both thrombin-like and kallikrein-like activity in control plasma (Fig. 3). In this case maximum thrombin and kallikrein activities are detected at 20 μg/ml of activator. Under those conditions the induction of both thrombin-like and kallikrein-like activities was, like

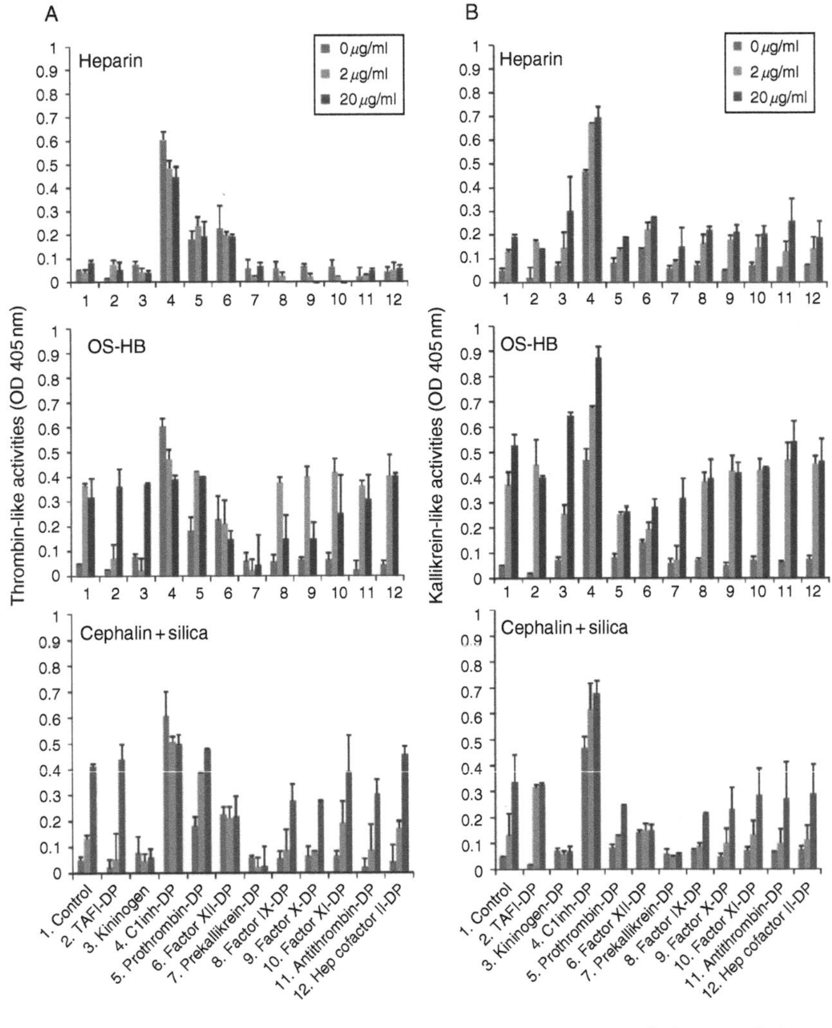

FIG. 3. Induction of thrombin and kallikrein activities. Human control plasma, and plasmas depleted in kallikrein, kininogen, antithrombin, heparin cofactor II, C1 inhibitor, FXII, FXI, FX, FIX, FII (prothrombin), and TAFI were incubated with 0, 2, and 20 μg/ml heparin, OS-HB, or cephalin/silica for 5 min at 37 °C. (A) Kallikrein activity was assessed by addition of 150 μl of the kallikrein chromogenic substrate, 0.5 mM *N*-benzoyl-Pro-Phe-Arg-*p*-nitroanilide hydrochloride dissolved in TBS. OD readings were taken every 90 s for 40 min. The average OD readings of kallikrein activities at 40 min from two independent experiments were presented. (B) Thrombin activity was assessed by addition of the thrombin chromogenic substrate, 1 mM Sar-Pro-Arg-*p*-nitroanilide dihydrochloride dissolved in TBS. OD readings were taken every 90 s for 40 min. The average OD readings of thrombin activities at 40 min from two independent experiments were presented. (See Color Insert.)

the case of OS-HB activator, dependent on the presence of plasma prekallikrein. Both thrombin-like and kallikrein-like activities are dependent on the presence of kininogen, and, as with OS-HB, independent of factor IX, factor X, and factor XI. Of note, maximum thrombin- and kallikrein-like activities are observed in the absence of C1inh, the natural inhibitor of these and other plasma serine proteases, even in the absence of an inducer (Fig. 3A and B).

As expected, of the three potential activators, heparin had the least effect. Induction of thrombin-like activity is relatively modest at 20 μg/ml (Fig. 3A), while, activation of kallikrein-like activity, albeit /although modest, is difficult to interpret due to a relatively high background in the prekallikrein-depleted plasma (Fig. 3B).

C. Prothrombin and TAFI are Integral Members of the Contact System

The data presented in Fig. 3 suggest the both prothrombin and TAFI regulate contact system activation.

Activation of the contact system results in proteolytic cleavage of prekallikrein, FXII, HMWK, and C1inh as well as deactivation of kallikrein and FXII via covalent attachment to C1inh, events that can be systematically monitored and compared to unactivated contact system proteins by Western blot analysis. We decided to test how these events were affected in prothrombin-, TAFI-, and other contact system protein-depleted plasmas.

To this end, normal and factor-depleted human plasmas were activated *in vitro* by adding 20 μl of OS-HB into 30 μl of plasma at 37 °C for 5 min with a final concentration of 200 μg/ml. TBS were used as a nonactivation control. Plasmas were then 1:4 diluted with TBS, and equal volume of 2× Laemmli sample buffer was added. After being boiled at 100 °C for 5 min, 7 μl of each sample were separated by 4–15% SDS-PAGE and transferred onto a nitrocellulose membrane. After blotting with the primary and secondary antibody, protein was visualized using the SuperSignal West Pico Chemiluminescent Substrate from Thermo-Scientific-Pierce.

To define the best contact system activation conditions for Western blot analysis, we first incubated kallikrein-, FXII-, kininogen-, C1inh-, prothrombin-, TAFI-depleted and control human plasmas by varying OS-HB concentration from 20 to 200 μg/ml and by varying incubation time from 1, 5, 10, to 30 min. The Western analysis of the contact proteins kallikrein, FXII, kininogen, C1inh, thrombin, and TAFI indicated that the cleavage of HMWK and C1inh occurred within 1 min incubation time and leveled off at 5 min; and both 20 and 200 μg/ml OS-HB induced the same level of the contact protein changes (data not shown). We decided to use 200 μg/ml of OS-HB to activate the plasma contact system

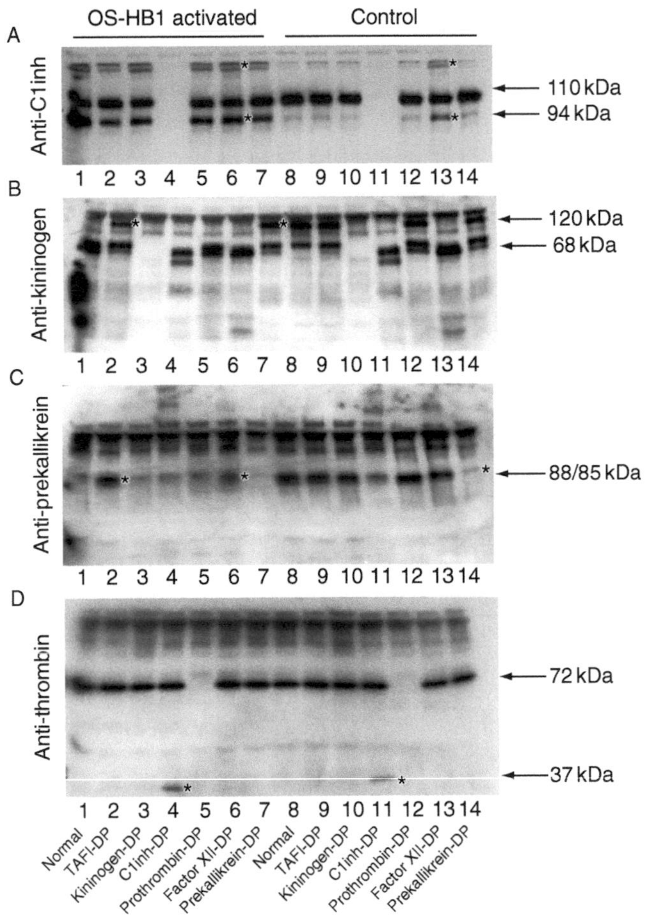

FIG. 4. Western analysis of the contact system proteins in OS-HB activated or control plasmas. Human plasmas depleted of C1inh, prekallikrein, kininogen, prothrombin, TAFI, and FXII were incubated with 200 μg/ml OS-HBs for 5 min at 37 °C. Samples were analyzed by reducing SDS-PAGE and Western blotting using antibodies against C1inh, prekallikrein, HMWK, thrombin, TAFI, and FXII, respectively. Plasma (0.53 μl) was used in each lane. The experiment was performed twice with similar results.

for 5 min followed by Western analysis. As shown in Fig. 4, OS-HB treatment resulted in distinct contact protein changes in a contact protein-dependent manner:

Figure 4A: *C1inh (110 kDa)*. C1inh is a serine protease inhibitor. It circulates in blood at levels around 250 μg/ml. C1inh is the most important physiological inhibitor of plasma kallikrein, factor XIa, factor XIIa, and complement serine proteases by forming a 1:1 stoichiometric complex. Both cleaved 94 kDa C1inh and C1inh/protease complex bands have been reported in clinical C1inh dysfunction patient plasmas.[31] As compared to the 110 kDa band of C1inh in control plasma, the 94 kDa cleaved C1inh band was produced, as well as higher molecular weight bands that may represent complexes with kallikrein, or other serine proteases (indicated by a red mark; Fig. 4A, lanes 1–7). These C1inh/protein complex bands have been reported in clinical C1inh dysfunction patient plasmas (e.g., hereditary angioedema and C1inh-deficient patient plasmas[31]). High and low molecular C1inh protein bands were seen in all OS-HB treated C1inh sufficient plasmas (Fig. 4A, lanes 8–14), which confirms that OS-HB, induced contact system activation.[10]

All contact protein-depleted plasmas were obtained by passing through an antibody column. High and low molecular C1inh protein bands were only observed in the FXII-depleted plasma (Fig. 4A, lane 13), which suggests, that the contact system was largely unaffected during contact protein removal by affinity chromatography. Therefore, FXII deficiency was responsible for the production of high and low molecular C1inh protein bands. This observation suggests that FXII might inhibit the contact system activation in FXII-sufficient plasma.

Figure 4B: *Kininogens (120 kDa)*. The two plasma kininogens, HMWK (120 kDa) and LMWK (70 kDa) are two splicing variants of the same kininogen gene products. They circulate in blood at levels around 80 and 90 μg/ml, respectively. Kininogens are multifunction proteins with antithrombotic, antidiuretic, profibrinolytic, antiadhesive, antiangiogenic, and proinflammation properties.[32] Both HMWK and LMWK at plasma concentrations inhibited thrombin-induced platelet aggregation.[33] Bradykinin and kallidin are released from HMWK and LMWK by plasma kallikrein and tissue kallikreins, respectively. Bradykinin and kallidin modulate the activation, proliferation, migration and effector functions of macrophages, dendritic cells, T and B lymphocytes[34] and are also potent inducers of vasodilation mediated by the secondary release of other signaling molecules including nitric oxide and platelet activating factor through the activation of vascular endothelium bradykinin receptors.[35] HMWK is known to be cleaved by kallikrein to form its 68 kDa cleavage product after the contact system activation.

We found that OS-HB-induced HMWK cleavage required prekallikrein as prekallikrein-depleted plasma did not produce the 68 kDa protein band (Fig. 4B, lane 7). In contrast, the 68 kDa protein band was observed in the C1inh-depleted plasma accompanied by total loss of the 120 kDa band (Fig. 4B, lane 11), which indicates, that C1inh prevents HMWK cleavage in C1inh-sufficient plasmas. Interestingly, HMWK cleavage required TAFI (Fig. 4B, lane 2, the retention of the 120 kDa band). Moreover, FXII-depleted plasma resulted in cleaved HMWK (Fig. 4B, lane 13), suggesting that FXII also inhibit HMWK cleavage. These results indicate that OS-HB-induced HMWK cleavage was regulated by prekallikrein, C1inh, prothrombin, FXII, and TAFI. Different HMWK cleavage products in the contact protein-deleted plasmas suggest that HMWK might be a sensitive indicator of the contact system activation.

Figure 4C: *Prekallikrein (85/88 kDa).* Prekallikrein is the zymogen of kallikrein. It circulates in blood at levels around 42 μg/ml. The conversion of prekallikrein (85/88 kDa) to its active form kallikrein is accomplished by a serine protease cleavage, especially FXIIa, where a single bond (Arg371-Ile372) is split, generating a heavy chain of 371 amino acids and a light chain of 248 amino acids held together by a disulfide bond. Under our Western analysis conditions, kallikrein comigrate with prekallikrein. In plasma, kallikrein is inactivated by two protease inhibitors, C1inh and α2-macroglobulin, each of which form a 1:1 stoichiometric complex with kallikrein. Therefore, an activated contact system might be accompanied by normal or decreased prekallikrein/kallikrein band with or without observable kallikrein/protease inhibitor complexes.

We found that normal plasma showed expected 88/85 kDa band of prekallikrein (Fig. 4C, lane 8). The prekallikrein bands were decreased in C1inh-depleted plasma (Fig. 4C, lane 11). A small amount of prekallikrein was still visible in prekallikrein-depleted plasma (Fig. 4C, lane 14). Greatly reduced prekallikrein 88/85 kDa band was observed in OS-HB-induced control-, kininogen-, and prothrombin-depleted plasmas (Fig. 2C lanes 1, 3, and 5). OS-HB did not induce significant loss of the prekallikrein band in TAFI- and FXII-depleted plasmas (Fig. 4C, lanes 2 and 6). It suggests that FXII might be required for maximum contact system activation and TAFI regulates prekallikrein activation.

Figure 4D: *Thrombin (37 kDa).* Prothrombin (72 kDa) circulates in the blood at levels around 72–144 μg/ml. Cleavage at positions of 155 and 284 of prothrombin are catalyzed by thrombin whereas the cleavage at positions 271 and 320 are catalyzed by FXa to generate A and B chains of the 37 kDa thrombin. Thus, thrombin Western analysis may reveal same or different sizes of cleaved thrombin products reflecting 1–4 cuts of prothrombin by FXa and/or thrombin. In plasma, thrombin is inactivated by two protease inhibitors, antithrombin and heparin cofactor II catalyzed by GAGs, each of which form a 1:1 stoichiometric complex with thrombin. Therefore, an activated contact system might be accompanied by normal or decreased prothrombin/meizothrombin band with or without observable 37 kDa thrombin band. Thrombin/antithrombin and thrombin/heparin cofactor II complexes might be observable as well.

The Western analysis of thrombin and prothrombin (Fig. 4D) shows the presence of prothrombin in all lanes of control and OS-HB treated plasmas with the exception of the prothrombin-depleted plasma (lanes 5 and 12). A 37-kDa thrombin band was seen in both OS-HB-activated and unactivated C1inh-depleted plasma (Fig. 4D, lanes 4 and 11). Thrombin band observed in C1inh-depleted plasma was corelated with abnormal HMWK bands observed in prothrombin-deficient plasma (Fig. 4B, lane 12), which indicates that thrombin was generated through the contact system activation in human plasma in the absence of C1inh. It suggests that C1inh is a major inhibitor of thrombin generation through contact system activation and thrombin is a nature product of the contact system activation.

Interestingly, the unknown high molecular weight species were consistently observed when the contact proteins prekallikrein (Fig. 4C), prothrombin (data not shown), thrombin (Fig. 4D), and TAFI (data not shown) were analyzed by Western in control and contact protein-depleted plasmas. These high molecular weight bands were not observed when using the secondary antibody alone (data not shown), indicating these protein bands might not be artifacts. Since high molecular weight bands were visible (Fig. 4D, lanes 5 and 12), we hypothesized that prothrombin might be present in a complex form with other plasma proteins. To test it, we treated both control and prothrombin-depleted plasmas with or without kallikrein and then analyzed the plasma prothrombin by Western analysis. We found that the intensity of the high molecular weight species was greatly reduced after the kallikrein treatment (data not shown).

We further compared the contact system protein properties of prothrombin-, prekallikrein-, and FXII-depleted plasmas with those of prothrombin-, prekallikrein-, and FXII-deficient plasmas obtained from human patients. We found that both purchased prothrombin- and prekallikrein-deficient human plasmas still retained $\sim$30% prothrombin and prekallikrein protein bands estimated by the Western blot analysis (data not shown).

We conclude based on Fig. 4 that: (1) OS-HB activated the contact system proteins systematically based on the Western analysis; (2) prothrombin and TAFI were part of the established contact system[13]; (3) thrombin was generated in C1inh-depleted human plasma through the contact system activation.

D. A Model of Thrombin Generation Through Contact System Activation

It has been well established that *in vivo* blood coagulation is initiated by exposure of tissue factor that forms a complex with FVIIa, which results in the generation of small quantities of FIXa and FXa. FXa then generates minute amounts of thrombin that result in the activation of FVIII and FV. At this point, the FVIIIa, FIXa, and PL form a "tenase" complex. The "tenase" generates sufficient quantities of FXa to form a "prothrombinase" complex, composed of FVa, FXa, Ca^{2+}, and PL. The prothrombinase then generates sufficient thrombin to induce clot formation[36,37] (Fig. 5).

In addition to the tissue factor or "extrinsic" pathway, an alternate pathway to thrombin generation (the "intrinsic" pathway or contact system) has been described when blood or plasma comes in contact with negatively charged artificial surfaces.[12]

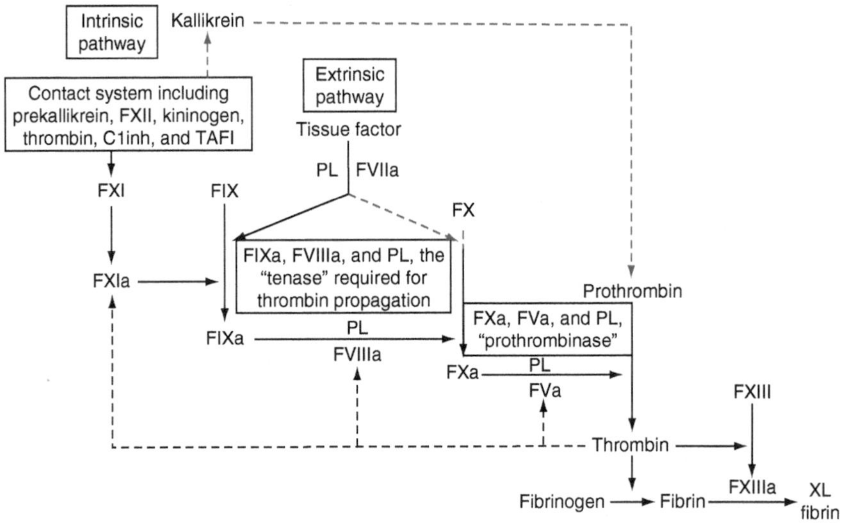

FIG. 5. The coagulation pathways. Either tissue factor or kallikrein (red dotted lines) can generate catalytic amounts of thrombin, which could result in the activation of factor XI, FVIII, and FV. At this point, FIXa generated either by intrinsic or extrinsic pathways forms a "tenase" with FVIIIa and PLs. The tenase generates enough FXa to form a "prothrombinase" complex, composed of FVa, FXa, Ca^{2+}, and PL. The prothrombinase propagates enough thrombin to induce clot formation. The figure is modified from Colman et al.[36] (For interpretation of the references to color in this figure legend, the reader is referred to the Web version of this chapter.)

Here we present evidence for a new route to thrombin formation that begins with the activation of the contact system component kallikrein and leads directly to the cleavage of prothrombin. We showed that purified kallikrein cleaves prothrombin to form an active enzyme that has fibrinogen-dependent clotting activity. Moreover, following our previous observation that treatment of plasma with oversulfated GAGs can induce kallikrein and thrombin activity, we showed that induction of kallikrein, and, most importantly, thrombin occurred in plasma devoid of intrinsic pathway components (FIX and FXI) and one component was common to both the intrinsic and extrinsic pathways (FX) (Fig. 3). In addition, we demonstrated that kallikrein and thrombin activities could also be induced by cephalin plus silica, again independently of factors IX, X, and XI. Kininogen, an upstream component of the contact system, was required for kallikrein activity as induced by cephalin plus silica but not by OS-HB, observations suggesting that OS-HB activates kallikrein by a novel mechanism. Although we are aware of only one report that indicates that kallikrein might induce thrombin formation directly, Asmis et al.[38] showed that clotting occurs in kaolin-treated FXII-depleted plasma in the absence of detectable FXI activation, suggesting the existence of a FXI-independent pathway.

IV. Thrombin Generation Through Contact System Activation in HIT Patient Plasmas

Heparin is responsible for inducing autoantibodies. The autoantibodies are responsible for inducing HIT by activating platelets in affected patients. HIT patient plasmas are associated with high plasma thrombin activities.[4] However, the molecular mechanism of autoantibodies induced thrombin generation in HIT patients is largely unknown.

We obtained 15 suspected HIT patient plasmas (Table I) following the institution's human HIT plasma collecting protocols. After informed consent was given by patients, blood was collected by standard venepuncture into EDTA coated plastic tubes. The platelet-free plasma was aliquoted and stored at -70 °C before use. This study was previously approved by the local IRB, the Human Studies Committee (Washington University School of Medicine).

The suspected HIT patients included 4 females and 11 males whose ages ranged from 38 to 84 years old (Table I). Three of the patients (106, 108, and 113) had platelet counts within the normal range (194, 169, and $246 \times 10^9\ l^{-1}$ compared to $150–400 \times 10^9\ l^{-1}$). Anti-PF4/heparin antibodies were screened

TABLE I
OS-HB-Induced Thrombin Activities

Patients	Gender	Age	Platelets ($\times$ 109 l^{-1})	α-PF4/heparin Ab ELISA (OD)	OS-HB induced thrombin (OD)
101	M	59	8	0.035	*0.9916*
102	M	62	86	0.03	0.4588
103	F	75	118	0.081	*0.5825*
104	F	82	58	0.11	0.3158
105	M	57	83	0.392	0.286
106	M	66	194	*0.506*	*0.6652*
107	M	58	72	0.142	0.2069
108	M	84	168	*1.93*	*0.4842*
109	M	76	58	0.166	0.2343
110	F	52	101	0.284	0.2408
111	M	73	86	0.176	0.4545
112	M	38	84	*2.306*	*0.7705*
113	M	55	246	*0.869*	*0.5168*
116	F	67	108	*2.5*	0.2723
119	M	80	< 100	*0.57*	*0.4911*

by a sensitive ELISA assay. We found that the antibodies were positive in 6 out of 15 patients (106, 108, 112, 113, 116, and 119) including the three patients (106, 108, and 113) with normal platelet counts.

A. OS-HB-Induced Thrombin Activities

OS-HB-induced thrombin activities in control and HIT patient plasmas were measured after 2, 20, and 200 μg/ml OS-HB inductions. OS-HB-induced thrombin activities were concentration-dependent and instant (Fig. 6), which was correlated with instant contact system activation properties of contaminated heparin. By using 0.46 OD at 405 nm as a cut off, 7 out of 15 patient plasmas, including 5 out of 6 (106, 108, 112, 113, and 119) patient plasmas containing autoantibodies against PF4/heparin (Fig. 6; Table I), had higher OS-HB-induced thrombin activities. Patient 116 plasma had a positive antibody test (Table I), but OS-HB did not induce high thrombin activity in this patient plasma (Fig. 6, 116). Conversely, plasma from Patients 101 and 103 were negative for the autoantibody test (Table I) but were positive for the thrombin activity test (Fig. 6, 101 and 103).

B. OS-HB-Induced Kallikrein Activities

The data for the control plasma and plasmas from Patients 101, 102, and 119 are shown in Fig. 7. OS-HB at 200 μg/ml induced comparable kallikrein activities in control and Patents 101 and 102 plasmas (OD 1.52, 1.71, and 1.76, respectively) and less kallikrein activities in Patient 119 plasma (OD 0.56). OS-HB at 20 μg/ml induced higher kallikrein activities only in control and Patient 102 plasmas (Fig. 7A and C, OD 1.39 and 0.92). OS-HB at both 2 and 20 μg/ml induced low kallikrein activities in Patients 101 and 119 plasmas (Fig. 7B and D). Our data in Figs. 6 and 7 indicate that OS-HB-induced kallikrein and thrombin activities did not correlate in the Patient 101 and 119 plasmas.

C. Heparin-Induced Thrombin and Kallikrein Activities

Both thrombin and kallikrein activities in control and patient plasmas were measured after 2, 20, and 200 μg/ml heparin induction (Fig. 8).

Heparin at 200 μg/ml induced high thrombin activities in Patient 101 and 119 plasmas but not in the control and Patient 102 plasmas (Fig. 8 left).

In contrast, heparin at 200 μg/ml induced comparable kallikrein activities in the control and Patient 101 and 102 plasmas (OD 1.00, 0.97, and 0.74 respectively) and less kallikrein activity (OD 0.52) in Patient 119 plasma (Fig. 8 right). Heparin at 2 and 20 μg/ml induced low kallikrein activities in all the plasmas tested (Fig. 8 right).

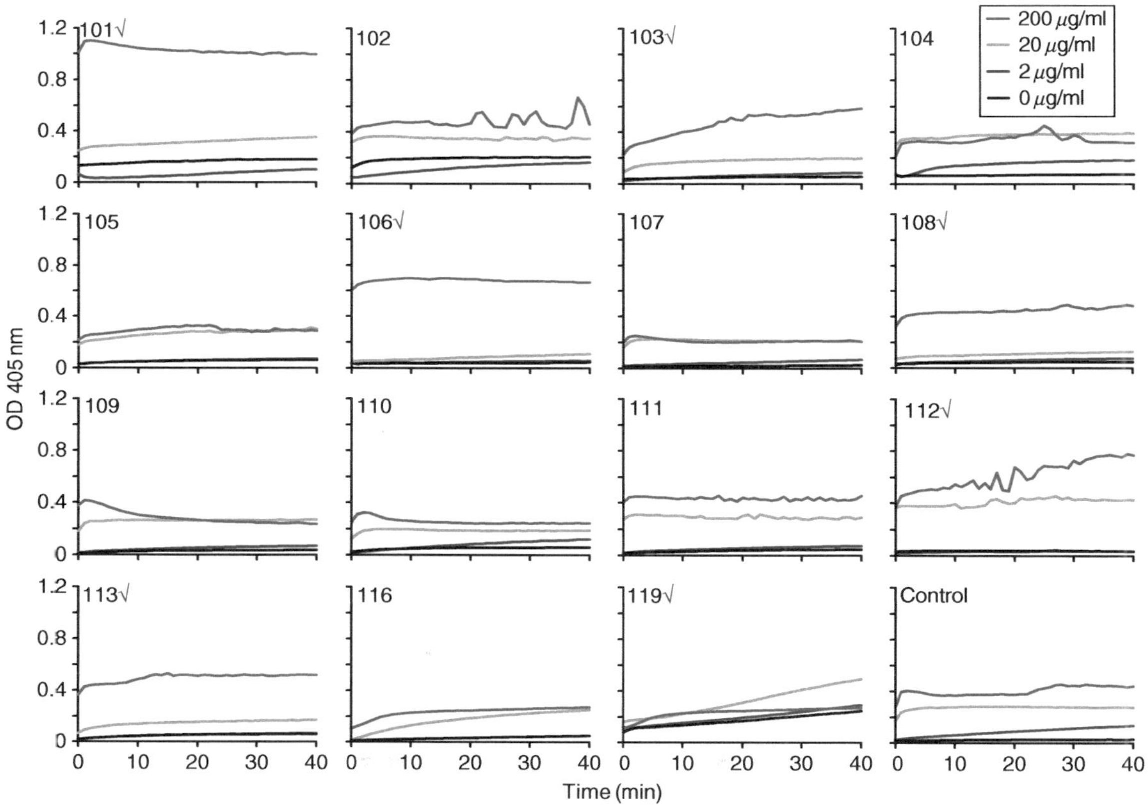

FIG. 6. HIT diagnosis by measuring OS-HB-induced thrombin activities in HIT patient plasmas. Normal pooled human plasma or HIT patient plasma (30 μl) was incubated with 20 μl OS-HB with a final concentration of 0, 2, 20, and 200 μg/ml at 37 °C for 5 min. Thrombin activity was assessed by addition of 150 μl of 100 mM Sar-Pro-Arg-*p*-nitroanilide dihydrochloride chromogenic substrate. OD readings were taken every 90 s for 40 min. (See Color Insert.)

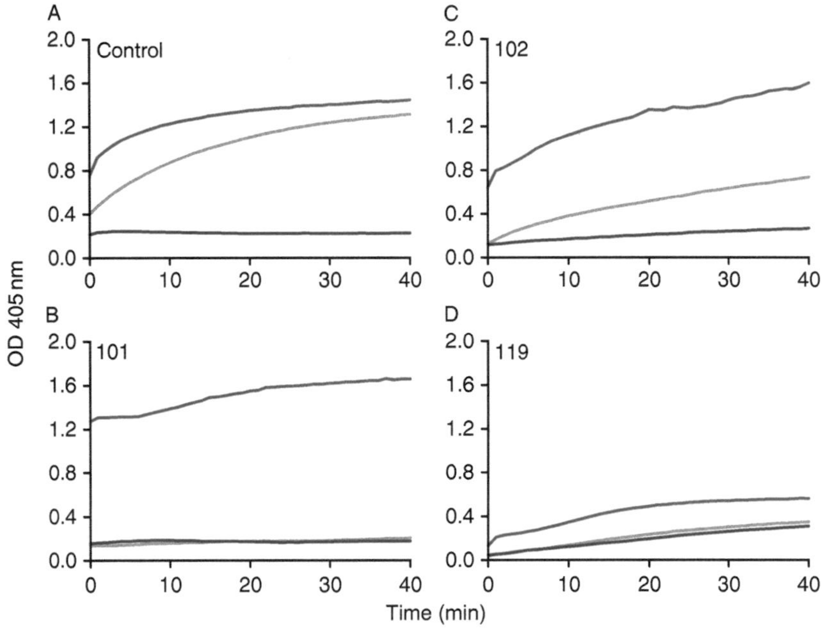

FIG. 7. OS-HB-induced kallikrein activities in control and HIT patient plasmas. Patient or control pooled human plasma (30 μl) was incubated with 20 μl OS-HB, with a final concentration of 0, 2, 20, and 200 μg/ml at 37 °C for 5 min. Kallikrein activity was assessed by addition of 150 μl of the kallikrein chromogenic substrate, 0.5 mM *N*-benzoyl-Pro-Phe-Arg-*p*-nitroanilide hydrochloride dissolved in TBS. OD readings were taken every 90 s for 40 min.

V. Detection HIT Autoantibodies that Do Not Bind to Platelet Factor 4 by Cell Surface Binding Assay

Both OS-HB and heparin-induced thrombin activities (Figs. 6 and 8), were observed in the Patient 101 and 103 plasmas without detectable autoantibodies against PF4/heparin (Table I). Since anti-NAP-2 and anti-IL-8 antibodies are also associated with HITT,[39] it is possible that Patients 101 and 103 might have other HIT-inducing antibodies that were not detectable by the anti-PF4/heparin antibody assay (Table I).

To test this idea, a cell surface binding assay was performed. To this end, three CHO cell lines, 745 (GAG-deficient mutant), K1 (wild-type), and K1.5 (3-O-sulfotransferase-1 expressing K1 cells[40]) were incubated with Patient 101 plasma. The fluorescence labeled anti-human IgG was used to detect IgGs that bound to CHO cell surfaces after extensive wash (Fig. 9). Fluorescence labeled anti-human IgG alone did not bind to any of the three cell surfaces (gray area).

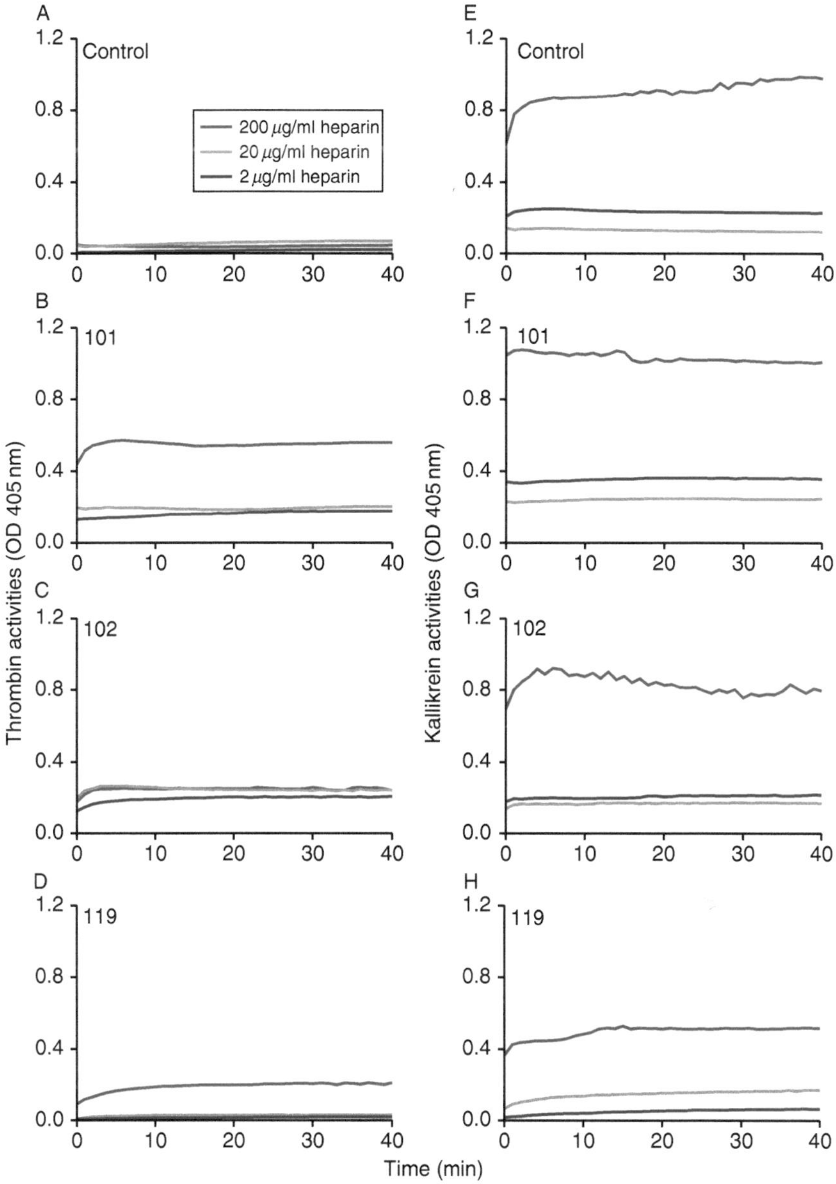

Fig. 8. Heparin-induced thrombin and kallikrein activities in HIT patient plasma. Patient or normal pooled human plasma (30 μl) was incubated with 20 μl heparin, with a final concentration of 0, 2, 20, and 200 μg/ml at 37 °C for 5 min. *Left panel*: Thrombin activity was assessed by addition of the thrombin chromogenic substrate, 1 mM Sar-Pro-Arg-*p*-nitroanilide dihydrochloride dissolved in TBS. OD readings were taken every 90 s for 40 min. The experiment was repeated twice with similar results. *Right panel*: Kallikrein activity was assessed by addition of 150 μl of the kallikrein chromogenic substrate, 0.5 mM *N*-benzoyl-Pro-Phe-Arg-*p*-nitroanilide hydrochloride dissolved in TBS. OD readings were taken every 90 s for 40 min. (See Color Insert.)

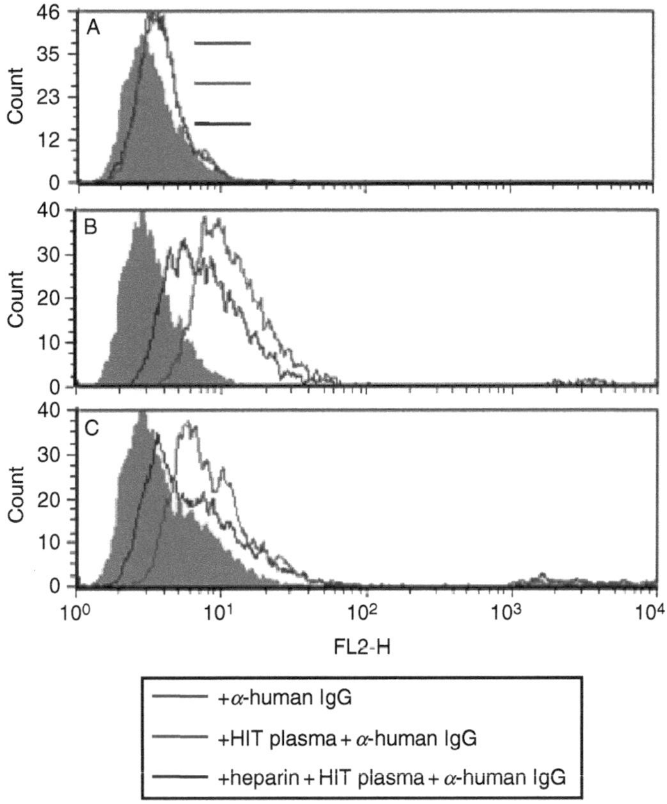

FIG. 9. Flow cytometric analysis of IgG antibody binding to CHO cells. Binding of IgG antibody present in the Patient 101 plasma to CHO-745, CHO-K1, and CHO-K1.5 cells was performed with 2 μl patient plasma on ice in the presence or absence of 1 μg of heparin. IgG binding was detected with PE-labeled anti-human IgG monoclonal antibody. Negative controls include PE-labeled anti-human IgG monoclonal antibody alone (Gray line). The experiments were repeated three times with similar results. (A) CHO-745 the GAG-deficient mutant; (B) CHO-K1; and (C) CHO-K1.5.

The anti-human IgG bound to K1 and K1.5 surfaces but not to the GAG-deficient 745-cell surface. Following heparin addition, IgG-binding was partially inhibited by heparin in both K1 and K1.5 surfaces and unchanged at the 745-cell surface.

VI. Proposed Molecular Mechanism for HIT

The present report indicates that heparin directly induced thrombin activities in HIT patient plasmas free of platelets and other blood cells (Figs. 6–9; Table I). The induction was autoantibody-dependent, which suggests that the negatively charged

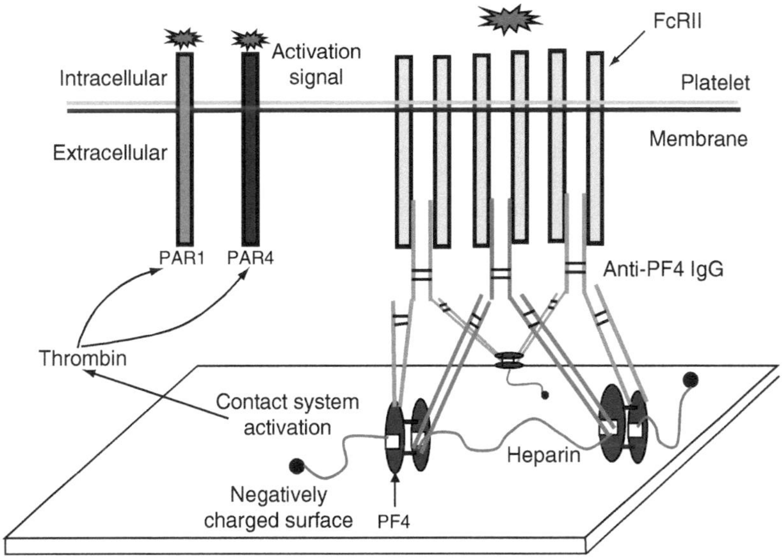

FIG. 10. A model for heparin and IgG antibody-dependent platelet activation. Heparin-induced platelet activation occurs in two phases. First, heparin, PF4 or other heparin-binding protein, and IgG autoantibodies form large complexes in plasma. The complexes serve as a negatively charged surface that induces thrombin generation through the contact system activation. The thrombin then binds to protease-activated receptors 1 and/or 4 (PAR1 and PAR4), which partially activates platelets. Second, partially activated platelets allow FcRII to bind to Fc domain of the IgG in the IgG/PF4/heparin complex. The Fc and FcRII cross-linking propagates platelet activation, aggregation, and blood coagulation through FcRII signaling. The figure is modified from Meyer *et al.*[42]

IgG/protein/heparin complexes activated the plasma contact system and produced the platelet activating thrombin. Based on published and our own data, we propose that HIT occurs in two phases (Fig. 10). First, heparin, PF4, and IgG autoantibodies form large complexes in plasma. The complexes serve as a negatively charged surface that induces thrombin generation through the contact system activation. The thrombin partially activates platelets and the partially activated platelets are required for subsequent IgG/FcRII binding and FcRII intraplatelet signaling.

This model provides a plausible mechanism for autoantibodies against NAP-2 and IL-8 to induce HIT. It also explains why other sulfated polymers, such as vinyl sulfate, but not small heparin pentasaccharide, can substitute for heparin to form the functional large negatively charged complexes.[41] This model also predicts that other IgG antibodies that are capable of forming negatively charged protein/IgG complexes would produce HIT-like symptoms. Indeed, this concept is supported by published studies of IgG antibody-induced thrombosis in unrelated human diseases.[41]

Bevacizumab is the first U.S. FDA approved antibody used for cancer therapy. Surprisingly, Bevacizumab is associated with arterial thromboembolism in colorectal cancer patients.[42] Bevacizumab is shown to activate platelets only in the presence of heparin in an hFcRII transgenic mouse model, originally designed for HIT and other autoimmune disease studies. The thrombotic activity requires the heparin-binding domain of $VEGF_{165}$ because $VEGF_{121}$ lacking heparin-binding domain cannot activate platelets in the mouse model even though Bevacizumab binds to both $VEGF_{121}$ and $VEGF_{165}$.[42] This published report also supports the model presented in Fig. 10.

VII. Concluding Remarks

Thromboembolic disorders are the leading cause of disabilities and death not only in HITT but also in a variety of unrelated human diseases, such as coronary heart disease,[43] cancer,[44] diabetes,[45] kidney failure,[46] and other autoimmune diseases.[47,48] Thrombin is the only known enzyme that causes thrombus formation. Thrombin also plays multiple roles in development, tissue repair, inflammation, and hemostasis.[29] Conditional loss of prothrombin leads to rapid deaths of adult mice,[49] suggesting that thrombin is a key survival factor which is continuously generated in blood circulation. Thus, it is critical to reexam the involvement of abnormal contact system in thrombin generation and in human thromboembolic disorders induced by different molecular identities associated with the disorders. The presence of such molecular identities in different human thromboembolic disorders will be demonstrated in Chondroitin Sulfate and Abnormal Contact System in Rheumatoid Arthritis; Activated Contact System and Abnormal Glycosaminoglycans in Lupus and Other Auto- and Non-Autoimmune Diseasesm; and Glycosaminoglycans and Activated Contact System in Cancer Patient Plasmas.

References

1. Warkentin TE, Levine MN, Hirsh J, Horsewood P, Roberts RS, Gent M, et al. Heparin-induced thrombocytopenia in patients treated with low-molecular-weight heparin or unfractionated heparin. *N Engl J Med* 1995;**332**:1330–5.
2. Comunale ME, Van Cott EM. Heparin-induced thrombocytopenia. *Int Anesthesiol Clin* 2004;**42**:27–43.
3. Greinacher A, Warkentin TE. Contaminated heparin. *N Engl J Med* 2008;**359**:1291–2 (author reply 1293).
4. Tardy-Poncet B, Piot M, Chapelle C, France G, Campos L, Garraud O, et al. Thrombin generation and heparin-induced thrombocytopenia (HIT). *J Thromb Haemost* 2009;**7**:1474–81.

5. Newman PM, Chong BH. Heparin-induced thrombocytopenia: new evidence for the dynamic binding of purified anti-PF4-heparin antibodies to platelets and the resultant platelet activation. *Blood* 2000;**96**:182–7.
6. Cines DB, Rauova L, Arepally G, Reilly MP, McKenzie SE, Sachais BS, et al. Heparin-induced thrombocytopenia: an autoimmune disorder regulated through dynamic autoantigen assembly/disassembly. *J Clin Apher* 2007;**22**:31–6.
7. Coughlin SR. Protease-activated receptors and platelet function. *Thromb Haemost* 1999;**82**:353–6.
8. Kelton JG, Warkentin TE. Heparin-induced thrombocytopenia: a historical perspective. *Blood* 2008;**112**:2607–16.
9. Adiguzel C, Bansal V, Litinas E, Cunanan J, Iqbal O, Nelson K, et al. Increased prevalence of antiheparin platelet factor 4 antibodies in patients may be due to contaminated heparin. *Clin Appl Thromb Hemost* 2009;**15**:145–51.
10. Kishimoto TK, Viswanathan K, Ganguly T, Elankumaran S, Smith S, Pelzer K, et al. Contaminated heparin associated with adverse clinical events and activation of the contact system. *N Engl J Med* 2008;**358**:2457–67.
11. Blossom DB, Kallen AJ, Patel PR, Elward A, Robinson L, Gao G, et al. Outbreak of adverse reactions associated with contaminated heparin. *N Engl J Med* 2008;**359**:2674–84.
12. Waaler BA. Contact activation in the intrinsic blood clotting system; studies on a plasma product formed on contact with glass and similar surfaces. *Scand J Clin Lab Invest* 1959;**11**:1–133.
13. Colman RW. Contact activation (kallikrein-kinin) pathway: multiple physiologic and pathophysiologic activities. In: Colman RW, editor. *Hemostasis and thrombosis*. Philadelphia: Lippincott Williams & Wilkins; 2006. p. 107–30.
14. Schmaier AH. Assembly, activation, and physiologic influence of the plasma kallikrein/kinin system. *Int Immunopharmacol* 2008;**8**:161–5.
15. Gilstad CW. Anaphylactic transfusion reactions. *Curr Opin Hematol* 2003;**10**:419–23.
16. Meth MJ, Maibach HI. Current understanding of contrast media reactions and implications for clinical management. *Drug Saf* 2006;**29**:133–41.
17. Schmaier AH, McCrae KR. The plasma kallikrein–kinin system: its evolution from contact activation. *J Thromb Haemost* 2007;**5**:2323–9.
18. Shimada T, Kato H, Iwanaga S, Iwamori M, Nagai Y. Activation of factor XII and prekallikrein with cholesterol sulfate. *Thromb Res* 1985;**38**:21–31.
19. Shimada T, Sugo T, Kato H, Yoshida K, Iwanaga S. Activation of factor XII and prekallikrein with polysaccharide sulfates and sulfatides: comparison with kaolin-mediated activation. *J Biochem* 1985;**97**:429–39.
20. Hojima Y, Cochrane CG, Wiggins RC, Austen KF, Stevens RL. In vitro activation of the contact (Hageman factor) system of plasma by heparin and chondroitin sulfate E. *Blood* 1984;**63**:1453–9.
21. Maas C, Govers-Riemslag JW, Bouma B, Schiks B, Hazenberg BP, Lokhorst HM, et al. Misfolded proteins activate Factor XII in humans, leading to kallikrein formation without initiating coagulation. *J Clin Invest* 2008;**118**:3208–18.
22. van der Graaf F, Koedam JA, Bouma BN. Inactivation of kallikrein in human plasma. *J Clin Invest* 1983;**71**:149–58.
23. Pan J, Qian Y, Zhou X, Pazandak A, Frazier SB, Weiser P, et al. Identification of chemically sulfated/desulfated glycosaminoglycans in contaminated heparins and development of a simple assay for the detection of most contaminants in heparin. *Glycobiol Insight* 2010;**2**:1–12.
24. Pan J, Qi Y, Zhou X, Lu H, Ramacciotti E, Zhang L. Chemically oversulfated glycosaminoglycans are potent modulators of contact system activation and different cell signaling pathways. *J Biol Chem* 2010;doi:10.1074/jbc.M109.063735.

25. Pan J, Qi Y, Zhou X, Lu H, Ramacciotti E, Zhang L. Chemically oversulfated glycosaminoglycans are potent modulators of contact system activation and different cell signaling pathways. *J Biol Chem* 2010; doi: 10.1074/jbc.M109.063735.
26. Szebeni J, Alving CR, Rosivall L, Bunger R, Baranyi L, Bedocs P, et al. Animal models of complement-mediated hypersensitivity reactions to liposomes and other lipid-based nanoparticles. *J Liposome Res* 2007;**17**:107–17.
27. Radomski A, Jurasz P, Alonso-Escolano D, Drews M, Morandi M, Malinski T, et al. Nanoparticle-induced platelet aggregation and vascular thrombosis. *Br J Pharmacol* 2005;**146**:882–93.
28. Simberg D, Zhang WM, Merkulov S, McCrae K, Park JH, Sailor MJ, et al. Contact activation of kallikrein–kinin system by superparamagnetic iron oxide nanoparticles in vitro and in vivo. *J Control Release* 2009;**140**:301–5.
29. Jenny NS, Lundblad RL, Mann KG. Thrombin. In: Colman RW, editor. *Hemostasis and thrombosis*. Philadelphia: Lippincott Williams & Wilkins; 2006. p. 193–213.
30. Stief TW. Kallikrein activates prothrombin. *Clin Appl Thromb Hemost* 2008;**14**:97–8.
31. Zuraw BL, Curd JG. Demonstration of modified inactive first component of complement (C1) inhibitor in the plasmas of C1 inhibitor-deficient patients. *J Clin Invest* 1986;**78**:567–75.
32. Colman RW, White JV, Scovell S, Stadnicki A, Sartor RB. Kininogens are antithrombotic proteins in vivo. *Arterioscler Thromb Vasc Biol* 1999;**19**:2245–50.
33. Meloni FJ, Schmaier AH. Low molecular weight kininogen binds to platelets to modulate thrombin-induced platelet activation. *J Biol Chem* 1991;**266**:6786–94.
34. Schulze-Topphoff U, Prat A, Bader M, Zipp F, Aktas O. Roles of the kallikrein/kinin system in the adaptive immune system. *Int Immunopharmacol* 2008;**8**:155–60.
35. Ueno A, Oh-ishi S. Roles for the kallikrein–kinin system in inflammatory exudation and pain: lessons from studies on kininogen-deficient rats. *J Pharmacol Sci* 2003;**93**:1–20.
36. Colman RW, Clowes AW, George JN, Goldhaber SZ, Marder VJ. Overview of hemostasis. In: Colman RW, editor. *Hemostasis and thrombosis*. Philadelphia: Lippincott Williams & Wilkins; 2006. p. 3–16.
37. Orfeo T, Brummel-Ziedins KE, Gissel M, Butenas S, Mann KG. The nature of the stable blood clot procoagulant activities. *J Biol Chem* 2008;**283**:9776–86.
38. Asmis LM, Sulzer I, Furlan M, Lammle B. Prekallikrein deficiency: the characteristic normalization of the severely prolonged aPTT following increased preincubation time is due to autoactivation of factor XII. *Thromb Res* 2002;**105**:463–70.
39. Amiral J, Marfaing-Koka A, Wolf M, Alessi MC, Tardy B, Boyer-Neumann C, et al. Presence of autoantibodies to interleukin-8 or neutrophil-activating peptide-2 in patients with heparin-associated thrombocytopenia. *Blood* 1996;**88**:410–6.
40. Zhang L, Beeler DL, Lawrence R, Lech M, Liu J, Davis JC, et al. 6-*O*-sulfotransferase-1 represents a critical enzyme in the anticoagulant heparan sulfate biosynthetic pathway. *J Biol Chem* 2001;**276**:42311–21.
41. Greinacher A, Gopinadhan M, Gunther JU, Omer-Adam MA, Strobel U, Warkentin TE, et al. Close approximation of two platelet factor 4 tetramers by charge neutralization forms the antigens recognized by HIT antibodies. *Arterioscler Thromb Vasc Biol* 2006;**26**:2386–93.
42. Meyer T, Robles-Carrillo L, Robson T, Langer F, Desai H, Davila M, et al. Bevacizumab immune complexes activate platelets and induce thrombosis in FCGR2A transgenic mice. *J Thromb Haemost* 2009;**7**:171–81.
43. Duvall WL. Antithrombotic therapy. *Curr Mol Med* 2006;**6**:603–19.
44. Mousa SA. Role of current and emerging antithrombotics in thrombosis and cancer. *Timely Top Med Cardiovasc Dis* 2006;**10**:E19.
45. Bartnik M, Norhammar A, Ryden L. Hyperglycaemia and cardiovascular disease. *J Intern Med* 2007;**262**:145–56.

46. Grand'Maison A, Charest AF, Geerts WH. Anticoagulant use in patients with chronic renal impairment. *Am J Cardiovasc Drugs* 2005;**5**:291–305.
47. Swigris JJ, Fischer A, Gillis J, Meehan RT, Brown KK. Pulmonary and thrombotic manifestations of systemic lupus erythematosus. *Chest* 2008;**133**:271–80.
48. Linkins L-AMD, Warkentin TMD. The approach to heparin-induced thrombocytopenia. *Semin Respir Crit Care Med* 2008;066–74.
49. Mullins ES, Kombrinck KW, Talmage KE, Shaw MA, Witte DP, Ullman JM, et al. Genetic elimination of prothrombin in adult mice is not compatible with survival and results in spontaneous hemorrhagic events in both heart and brain. *Blood* 2009;**113**:696–704.
50. Pan J, Qian Y, Zhou X, Pazandak A, Frazier SB, Weiser P, et al. Oversulfated chondroitin sulfate is not the sole contaminant in heparin. *Nat Biotechnol* 2010;**28**:203–7.

Chondroitin Sulfate and Abnormal Contact System in Rheumatoid Arthritis

Xiaodong Zhou, Peter Weiser, Jing Pan, Yi Qian, Hong Lu, and Lijuan Zhang

Department of Pathology and Immunology, Washington University Medical School, St. Louis, Missouri, USA

Rheumatoid arthritis (RA) is a heterogeneous autoimmune disease that affects 1% of the population worldwide. In the K/BxN mouse model of RA, autoantibodies specific for glucose-6-phosphate isomerase (GPI) from these mice can transfer joint-specific inflammation to normal mice. The binding of GPI/autoantibody to the cartilage surface is a prerequisite for autoantibody-induced joint-specific inflammation in the mouse model. Chondroitin sulfate (CS) on cartilage surface is the long sought high-affinity receptor for GPI. The binding affinity and structural differences between mouse paw/ankle CS and knee/elbow CS correlate with the distal to proximal disease severity in these joints. The data presented in this chapter indicate that autoantigen/autoantibodies in blood circulation activate contact system to produce vasodilators to

Progress in Molecular Biology and Translational Science, Vol. 93
DOI: 10.1016/S1877-1173(10)93018-4

1877-1173/10 $35.00

allow immune complex, protein aggregates, and other plasma proteins to get into the joints. Cartilage surface CS binds and retains autoantigen/autoantibodies. The CS/autoantigen/autoantibody complexes could induce C3a and C5a production through contact system activation. C3a and C5a trigger degranulation of mast cells, which further recruit plasma contact system and complement proteins, immune cells, and immune activation factors to facilitate joint-specific tissue destruction. Therefore, either reducing autoantibody production or inhibiting autoantibody-induced contact system activation might be effective in RA prevention.

I. Introduction

Rheumatoid arthritis (RA) almost exclusively affects joints of the extremities, the interphalangeal, metacarpophalangeal, and wrists. About 1% of the world's population is afflicted by RA with women three times more often than men. Genetic, environmental, hormonal, microbial factors, autoantibodies, or mechanical insults could trigger joint-specific RA-like symptoms. For some patients, RA develops gradually over several years. For others, RA progresses rapidly. However, for most of patients, RA progresses in an unpredictable cycle of flare and remission.

Different biomarkers are associated with different RA patient plasmas, suggesting RA is a heterogeneous disease triggered by different factors. One of the characteristic features of human RA is the expression of several autoantibodies including joint-specific and joint-irrelevant autoantibodies, such as rheumatoid factors (RFs), which are autoantibodies against the Fc portion of IgG. However, not all the RA patients have RFs. Since the historical description of RFs, numerous additional autoantibodies have been discovered in RA patient plasmas. These autoantibodies are directed to stress proteins, cartilage proteins, enzymes, nuclear proteins, and, most importantly, citrullinated proteins such as fibrinogen or vimentin.[1] Autoantibodies to citrullinated human fibrinogen have better RA diagnostic and prognostic value compared to RFs.[2] Interestingly, the sensitivity of anti-citrulline antibody-based diagnosis increases with RA progression in tested patients,[2] which suggest that ACF might be the consequence rather the cause of RA.

RA is associated with common markers of inflammation, such as elevated erythrocyte sedimentation rate or elevated C-reactive protein levels in blood circulation.[3] RA is also associated with abnormal coagulation and fibrinolysis proteins both in blood circulation and in joint fluid, such as thrombin–antithrombin complexes, thrombin fragment, fibrin D-dimer and thrombin-activated fibrinolysis inhibitor antigenic levels.[3] What causes the system inflammation and how the system inflammation leads to joint-specific tissue destruction in RA is largely unknown.

Dissimilar causes and courses of RA lead to common end-stage effector mechanisms. The commonalities at the end-stage of RA include massive recruitment of neutrophils and macrophages, the implication of inflammatory cytokines like tumor necrosis factor (TNF)-α and interleukin (IL)-1, and the synthesis of a battery of degradative enzymes at joints. Because the molecular cascades that initiate RA remain unknown, current management is focused on treating the consequences but not the causes of RA.

Both C3a and C5a induced by glucose-6-phosphate isomerase (GPI) autoantibodies but not through classical antibody-dependent complement activation pathways are essential for RA initiation and development in the K/BxN mouse model of RA.[4–6] Small joint chondroitin sulfate (CS) is the long sought high-affinity receptor to retain circulating GPI/GPI autoantibodies in the joints.[7,8] However, it was unknown how GPI autoantibodies induce C3a and C5a production through nonclassical complement activation in the K/BxN mouse model of RA.

We recently established that plasma contact system, consisting three serine protease zymogens: prothrombin, prekallikrein, factor XII, one cofactor: high molecular weight kininogen (HMWK), and two regulators: C1 inhibitor and thrombin activatable fibrinolysis inhibitor, is a pattern recognition system.[9] Contact system can be activated by autoantigen/autoantibody/glycosaminoglycan (GAG) complexes[10] or protein aggregates.[11] Once activated, kallikrein produces bradykinin (BK), a potent vasodilator, and thrombin produces C3a[12] and C5a.[13] The facts present in this chapter indicate that autoantibodies/antigen/CS, immune complex, or protein aggregates activate the contact system. Activated contact system regulates endothelial permeability and sets the immune tone whereas cartilage surface CS serve as a specific receptor for autoantigen/autoantibodies to initiate joint-specific disease.

II. The K/BxN Mouse Model of RA

Understanding the etiology and key molecular players in RA has been greatly accelerated by employing the K/BxN mouse model of RA. This model is based on a T-cell receptor transgene, KRN, which has shown to recognize both the foreign antigen bovine RNase, and the ubiquitously expressed self antigen, GPI. K/BxN mice develop polyarthritis by 4–5 weeks of age with many of the hallmarks of human RA, including synovial hyperplasia, cellular infiltration, bone and cartilage erosion and remodeling, and loss of joint function.[14] Naïve mice can also develop polyarthritis by injecting with K/BxN serum, affinity-purified polyclonal anti-GPI antibodies or a combination of two or more anti-GPI monoclonal antibodies.[4] Purified anti-GPI autoantibodies transferred into the mice localized specifically to distal joints in the front and rear limbs within minutes of injection, got saturated within 20 min and remained localized for at least 24 h.[15]

By using transgenic mice, it has been demonstrated[16] that the IgG autoantibody production in the K/BxN mice could be through either T-cell dependent or T-cell independent pathways.[17] The autoantibodies to GPI in K/BxN mice are predominantly of the IgG1 isotype, which do not activate classical and lectin complement pathways, but do activate the alternative complement pathway.[18]

Localization of the autoantibodies to the joints depends partially on mast cells,[19] neutrophils, Fc receptors (FcRs),[20] especially FcRIII.[4] Diseases-relevant and irrelevant immune complexes allow autoantibodies and plasma proteins to enter the joints, where autoantibodies must bind to a target antigen at cartilage surface to induce joint-specific inflammation.[4,6,21] The vasoactive amines histamine and serotonin bypass the need for mast cells, neutrophil, and FcRIII and allow the immune complex to get into the joints.[16] These observations were explained by histamine- and serotonin-receptor mediated vasopermeability.[16] However, vasoactive amines regulated the contact system activation (data are shown in Glycosamino glycans and Activated Contact Sytem in Cancer Patient Plasmas). Further, proper liver function, where all the contact system proteins are made, but not the spleen is required for joint localization of immune complex[16] in the K/BxN mouse model of RA, which suggest that the contact system might play a role in the immune complex-triggered joint vasopermeability.

Circulating C3 is necessary and sufficient for arthritis induction in the mouse RA model[22] and C5a is largely responsible for tissue destruction in RA.[23] Both C3a and C5a have anaphylatoxin activity, directly triggering degranulation of mast cells as well as increasing vascular permeability and smooth muscle contraction. Degranulation of mast cells is appeared within an hour and influx of neutrophils is prominent within 1–2 days; synovial hyperplasia and mononuclear cell infiltration begin within a week. T cells, B cells, macrophages, mast cells, neutrophils and osteoclasts participate in pannus formation, cartilage destruction and erosion of bone at late stage of RA.[15]

In conclusion, RA in the K/BxN mouse model starts with anti-GPI antibody-induced vasopermeability, which allows immune complexes, antibodies, and plasma proteins to enter the joint space. Accumulation of anti-GPI autoantibodies at cartilage surface leads to C3a and C5a generation. C3a and C5a trigger degranulation of mast cells and initiate broad immune responses in the joints. C5a, TNF, and IL-1 are key effectors in the subsequent joint destruction.

III. Contact System-Based LEWIS Rat Model of RA

A. Lewis Rat Model of RA

Colman *et al.* developed contact system-based rat models of acute and chronic inflammatory arthritis by injecting peptidoglycan-polysaccharide (PGPS) into the distal ileum and cecum of Lewis and Buffalo rats. Only the genetically susceptible Lewis rat, but not the Buffalo rat, develops chronic synovitis and joint erosion. In the Lewis rat, but not the Buffalo rat, these changes are accompanied by a decrease in plasma prekallikrein and HMWK, reflecting activation of the contact system.[24] Pretreatment with a specific plasma kallikrein inhibitor modulates both acute and chronic arthritis. To provide direct genetic evidence that the contact system is involved in the induced RA, Sainz *et al.* backcrossed Brown-Norway HMWK-deficient rats with Lewis rats for six generations, two new rat strains were produced, wild-type F6 and HMWK-deficient F6, each with a 98.5% Lewis genome. Inflammatory arthritis was induced by intraperitoneal injection of PGPS. Rats showing HMWK deficiency develop minimal to no disease at all when injected with PGPS compared to disease-laden wild-type F6.[25]

Based on their experimental data, Colman *et al.* proposed that the contact system involves in both acute and chronic inflammatory arthritis in genetically susceptible Lewis rats. Following PGPS exposure, the contact system is initiated by the activation of plasma factor XII to factor XIIa or by prolylcarboxypeptidase (PRCP) on the endothelial cell and leukocyte surface. Factor XIIa/PRCP cleaves plasma prekallikrein generating kallikrein, which in turn cleaves HMWK to yield BK and cleaved HMWK (HMWKa). Kallikrein is chemotactic, induces neutrophil aggregation and stimulates the release of elastase and superoxide (O_2^-) as potent inducers of tissue injury. BK through its receptor BK-2R and cell adhesion molecules stimulate intracellular signaling and the release and synthesis of other mediators generating pain (prostaglandin release), vascular dilation (prostaglandin) or permeability (endothelial nitric oxide system: eNOS). This pathway initiates an acute inflammatory response. In the chronic phase of RA, the coagulation cascade is promoted. HMWKa receptors including urokinase-type plasminogen activator receptor (uPAR) and Mac-1 (M-1) are located on the surface of monocytes and neutrophils. Through these receptors, HMWKa stimulates the release of cytokines TNF-α, IL-1β, chemokines monocyte chemoattractant protein-1 (MCP-), and IL-8 from monocytes; and superoxide (O_2^-) plus elastase from neutrophils. All the contact system elements induce tissue injury and eventually chronic inflammation.[26]

B. Mutation in Kininogen Explains Why Lewis Rats are Susceptible to RA Induction

Lewis rats are one of the most used strains in animal models of RA. Lewis rats are known to have enhanced susceptibility to a number of experimental inflammatory conditions, such as PGPS-induced arthritis, adjuvant-induced arthritis, collagen-induced arthritis, autoimmune encephalitis, autoimmune thyroiditis, and enterocolitis. MHC was accepted to play a dominant role in RA susceptibility. However, Lewis and Fisher share the same rat MHC, but differ in their susceptibility to RA. The high female susceptibility was initially thought to be due to high estrogen levels in the blood and its secondary effects on the reticuloendothelial system. Again the female Fisher rats have comparable levels of serum estrogen to female Lewis rats but do not develop arthritis following PGPS injections. It turns out that the mutation Ser511Asn leads to a new *N*-glycosylation site in the HMWK. The novel *N*-glycosylation of HMWK increases cleavage of HMWK in Lewis rats during the contact system activation. Increased HMWK cleavage leads to increased BK release in Lewis rats compared to Fisher rats, which provides an explanation why Lewis rats have enhanced susceptibility to induced RA.[27]

IV. The Relevance of K/BxN Mouse Model of RA to Human RA

One of the characteristic features of human RA is the expression of several autoantibodies including joint-specific and joint-irrelevant autoantibodies, such as RFs, which are antibodies against the Fc portion of IgG. RF and IgG join to form immune complexes. However, not all the RA patients have RFs and RFs are also present in certain normal patient plasmas. Indeed, published data showed that either anti-GPI autoantibodies or preformed immune complex allow coinjected antibody get into joints. Anti-GPI but not other RA-irrelevant autoantibodies could induce inflammation at joints,[15] which indicates RF-associated immune complexes may only contribute to increased vasopermeability at joints.

Anti-GPI antibody is one of the autoantibodies related to human RA. Anti-GPI antibodies are found in the serum of RA patients ranging from15%[28] to 64%,[29] but they are significantly elevated in patients afflicted with several types of inflammatory arthritis,[28,30] including arthritis caused by trauma and crystal deposits.[28] In synovial fluid, where anti-GPI antibody concentrations are higher than in serum, patients with inflammatory arthritis with an immune origin (e.g., RA, Reiter's syndrome, ankyosing spondylitis, and SLE), have higher anti-GPI levels than patients with nonimmune inflammatory arthritis.[28] Therefore, anti-GPI antibodies are associated with RA. More importantly, IgG from anti-GPI antibody positive patients induces synovitis in cynomolgus monkeys,[31] which resembles the K/BxN serum transfer mouse model of RA. It indicates that normal joints from mouse, monkey, and human have evolutionally conserved their GPI-binding abilities.

It is predictable that other autoantibodies that bind to autoantigens, such as collagen type II, with high affinity to cartilage GAGs or other cartilage proteins may initiate RA.

V. Chondroitin Sulfate and RA

Articular cartilage has high water content, but it forms a meshwork impermeable to molecules as large as antibodies because of its many negatively charged GAG chains.[32,33] Based on histological features, including the number of chondrocytes and their morphology, collagen fiber orientation, and GAG concentration, articular cartilage has four distinct layers, superficial, middle, deep, and calcified. GAGs, including heparan sulfate (HS), CS, and hyaluronan (HA), are linear sugar chains consisting of glycosamine and glucuronic acid (GlcA) disaccharide repeats. They are abundant in each layer of the cartilage. Incomplete epimerization of GlcA to iduronic acid and sulfation at different positions of repeating disaccharides in HS give rise to 48 possible disaccharide structures, of which 23 have been found.[34] GAG sequences are not directly encoded by genes, but are assembled in the *Golgi* by enzymes encoded by over 40 genes. Because of the vast repertoire of expression of the GAG assembly enzymes and existence of GAG structural modification enzymes in extracellular matrix, GAGs display a sulfation pattern, chain length, and fine structure unique to each cell and tissue.[35] Because of building block-disaccharide complexity, GAGs are the most information dense biopolymers found in nature. Transgenic and knockout animal data in the past decade provide compelling evidence that animals use precise GAG sequences in multicellular communications.[34] However, the precise GAG sequences responsible for specific biological functions are largely unknown due to lack of appropriate tools and techniques to study the highly heterogeneous GAGs. CS is a major ($>95\%$) cartilage GAG.

GPI's natural substrate, glucose-6-phosphate, is structurally similar to GlcA in GAGs. Their structure difference is a phosphor group versus a carboxyl group at the carbon 6 of the sugar ring. We surmised that GAGs at the articular cartilage surface might bind to GPI based on their structural similarity. Our data showed that CS is the long sought high-affinity cartilage receptors for GPI.[8] The binding affinity and structural differences between mouse paw/ankle CS and elbows/knee CS correlated with disease severity in these joints. We found that GPI CS-binding domain overlaps with the active site of GPI enzyme. GlcA at the CS nonreducing end is partially involved in CS/GPI interactions. More importantly, we discovered several small molecules, including glucose-6-phosphate, inhibit GPI/CS interactions.[8] Therefore, GAGs at cartilage surface might be involved in RA development.

VI. Coagulation Cascade and Complement Activation in RA

A. Thrombin is Responsible for Fibrin Deposition in Synovium

Accumulation of fibrin in the synovium of RA patients represents one of the most striking pathologic features of RA.[36,37] The fibrin deposition has been considered to be a serious contributor to permanent damage by maintaining a vicious circle of inflammation in RA.[38] Thrombin is the only known enzyme that cleaves fibrinogen to produce fibrin and is directly responsible for synovium fibrin deposition. Activation of the coagulation and fibrinolytic cascades in the joint and in the blood circulation is evident in both inflammatory and degenerative joint diseases.[3,39] Therefore, RA is not only a joint-specific disease but also associated with abnormal thrombin generation in blood circulation.

Thrombin generation is a highly regulated process, which occurs in two phases.[40] During the initiation phase, catalytic amount of thrombin is produced by either factor Xa or kallikrein[41] to activate platelet, factor VIII (FVIII), and factor V (FV), which are required for thrombin propagation. During the propagation phase, thrombin is generated through the same "tenase" consisting of FVIIIa/FIXa/phospholipid (PL) and "prothrombinase" consisting of FVa/FXa/Ca^{2+}/PL shared by two established coagulation pathways, that is, tissue factor-dependent extrinsic pathway and contact system activation-dependent intrinsic pathway. Thrombin can be generated through either tissue factor-dependent extrinsic pathway caused by injury or joint movement-associated injury or through contact system activation-dependent intrinsic pathway by autoantigen/autoantibody/GAG complexes.[10]

B. Complement Activation in RA

Complement proteins consist of over 20 proteins accounting for about 5% of the globulin fraction of blood serum. Almost all the complement proteins bind to GAGs.[42] Three biochemical pathways activate the complement system: the classical complement pathway, the alternative complement pathway, and the lectin pathway. The classical complement pathway typically requires antibodies for activation, whereas the alternative and mannose-binding lectin pathways can be activated by C3 hydrolysis or antigens without the presence of antibodies. Both C3a and C5a generated during complement activation have anaphylatoxin activity, directly triggering degranulation of mast cells as well as increasing vascular permeability and smooth muscle contraction. C3a and C5a also trigger an oxidative burst in macrophages, neutrophils, and eosinophils (meaning that these cells become activated and generate oxygen derived toxins). Thus, C3a and C5a stimulate many pro-inflammatory activities.

Interestingly, antibody-dependent classical pathway of C3a and C5a generation is not required for antibody-induced RA in the K/BxN mouse model.[4] Circulating C3 is necessary and sufficient for arthritis induction in the K/BxN mouse model[22] and C5a is largely responsible for tissue damage in RA.[23] It has been reported that thrombin generates C5a[13] and both plasmin and thrombin generate C3a independent of the three established complement activation pathways.[12] Therefore, an activated contact system generated thrombin also has the capacity to produce C3a and C5a to set off an immune response.

VII. Chondroitin Sulfate and Contact System in Human and K/BxN Mouse Models of RA

Based on the knowledge described above, we suspected that C5a in joint space, produced by contact system activation associated vasopermeability and thrombin generation, might be the missing link in understanding the etiology of RA. To test the idea, we have studied how specific CS structures and plasma contact system proteins from both diseased K/BxN mice and human RA patients are associated with RA.

A. Lost GPI Binding and Abnormal Contact System in Diseased K/BxN Mice

1. Diseased Joint Lost the High-Affinity GPI-Binding GAG Receptor

We isolated paw/ankle GAGs from both 8-week-old (acute) and 15-week-old (chronic) K/BxN mouse paws/ankles. When we performed the GPI-binding gel mobility shift assay, we found that the GAGs from either acute or chronic diseased joints did not cause GPI shift compared to normal paw/ankle GAGs (data not shown). It indicates that diseased paw/ankle GAGs have lost the GPI-binding ability. To confirm the observation, we performed the same *in situ* GPI stain assay published previously.[8] We found that cartilage surface GPI binding was totally lost in the diseased K/BxN ankle joint compared to the normal ankle joint (Fig. 1). We concluded based on these data that GAG structures changed when the joint disease started.

2. Diseased Joint GAGs had High Levels of Heparin

We digested GAGs from normal and diseased K/BxN joint with a combination of heparin lyases and chondroitinase ABC and then analyzed by a combined capillary HPLC coupled MS. We found ~50 times increase in

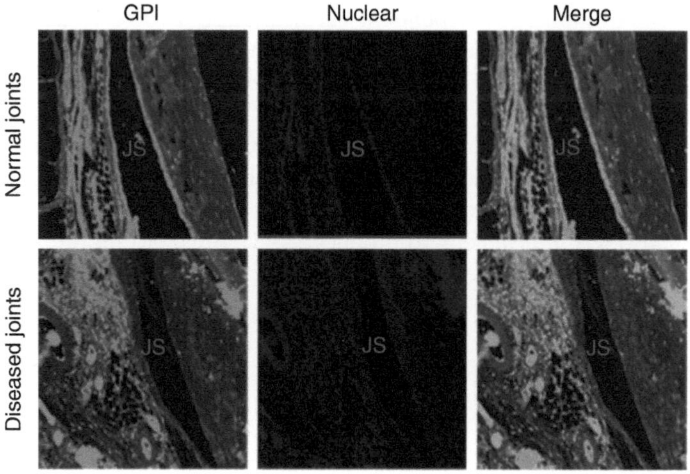

Fig. 1. Diseased cartilage surface did not bind to GPI. Normal and diseased joint sections were deparaffinized and air-dried. Sections were fixed with 50% acetone in PBS, incubated with PBS containing 2 M NaCl to remove GAG binding proteins. Sections were incubated with 20 nM recombinant mouse GPI, followed by incubation with 1:50 diluted sera of arthritic K/BxN mice. After wash, sections were incubated with Alexa-fluor 488 conjugated goat anti-mouse secondary antibody. Nuclear was stained by To-Pro-3. Sections were examined by confocal fluorescence microscopy. The experiments were repeated twice with similar results. (See Color Insert.)

trisulfated HS disaccharide, ΔUA2S-GlcNS6S, in the diseased paw/ankle GAGs and subsequently proved that the disaccharides were derived from heparin (data not shown). The MS data indicate that heparin was significantly upregulated in the diseased joints.

3. Decreased CS Sulfation in the Diseased Joints

To understand why the diseased paw/ankle joints lost the GPI binding, we analyzed the CS structures by capillary HPLC/MS. To overcome the difficulties in quantifying GAG di- and oligosaccharides by capillary HPLC/MS, we adapted Glycan Reductive Isotope Labeling (GRIL) technique.[43] The technique first involves enzymatic digestion of purified GAGs using chondroitinases and/or heparin lyases, followed by reductive amination of the reducing ends of the disaccharides with aniline for one sample and deuterated (D_5)-aniline for another sample to form a secondary amine derivative. The addition of the regular aniline group adds 77 Da to the mass of the ($[M-H]^{-1}$) ion, whereas D_5-aniline group adds 82 Da. Thus, samples tagged with different isotopically labeled anilines can be mixed and simultaneously analyzed to yield quantitative or ratiometric comparisons. To this end, GAGs were isolated from diseased paws/ankles and knees/elbows of the same mouse. After quantification, 20 μg of each were digested by chondroitinase AC, this enzyme does not digest iduronic acid-containing saccharides and leads to the accumulation of tetra- and other larger oligomers containing CS-B (dermatan sulfate) structures. The digested paw/ankle GAG was labeled with aniline and knee/elbow GAG was labeled with D_5-aniline. After the labeling, the two samples were combined and coinjected into capillary HPLC for separation coupled MS for detection (Fig. 2). The nonsulfated CS disaccharide ΔUA-GalNAc (U-a) eluted at 15.9 min with one pair of ions observed (*m/z* 455.2 from paw/ankle CS and 460.2 from knee/elbow CS). The monosulfated nonreducing end GalNAc eluted at 17.5 min (*m/z* 377.1 and 382.1). The monosulfated disaccharide U-a eluted at 18.3 min (*m/z* 535.2 and 540.2). The disulfated tetrasaccharides

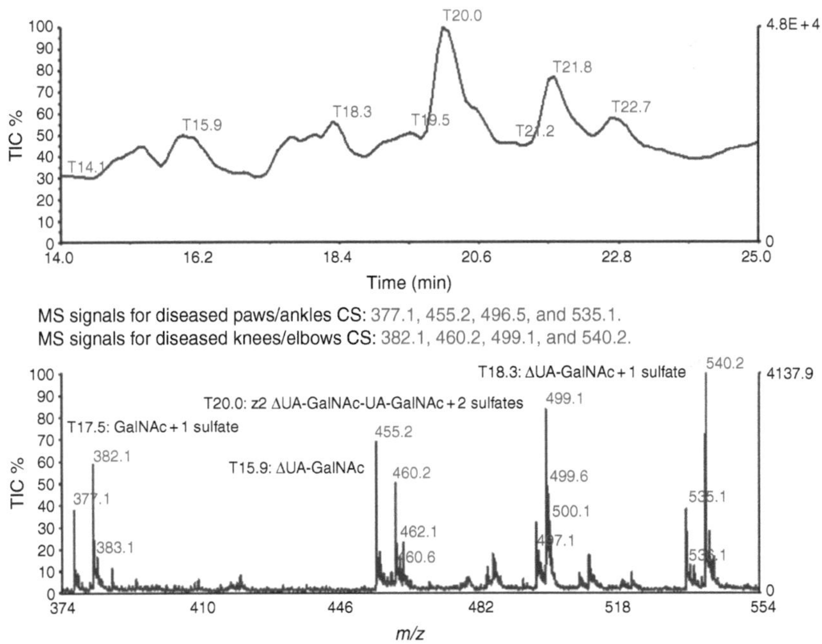

FIG. 2. GRIL-facilitated capillary HPLC coupled MS analysis of CS saccharide compositions of diseased paw/ankle and knee/elbow GAGs. GAGs isolated from either diseased paws/ankles or knees/elbows (20 μg of each) were digested by chondroitinase AC, this enzyme does not digest iduronic acid containing oligosaccharides and lead to the accumulation of tetra- and other larger oligosaccharides. Digested paw/ankle GAG was labeled with aniline and knee/elbow GAG was labeled with D_5-aniline. After labeling reaction, the two samples were combined and coinjected into capillary HPLC for separation coupled MS for detection. (A) Total ion current (TIC) from 14 to 25 min. (B) Accumulated ions observed from 14 to 25 min with *m/z* ranged from 374 to 554.

(U-a-I-a+2 sulfates) eluted at 20.0 min ($[M-2H]^{-2}$ ions *m/z* 496.5 and 499.1). The paw/ankle CS had 1.4 times more U-a than knee/elbow CS, but GalNAc+1 sulfate, U-a+1 sulfate, and U-a-I-a+2 sulfates were 1.5, 2.8, and 2.5 times less than those of the paw/ankle CS. These data plus our published normal paw/ankle GAG structural data showed that the degree of CS sulfation decreased in the paw/ankle CS during the disease progression.[8] In addition, the presence of tetrasaccharides ($[M-2H]^{-2}$ ions, *m/z* 496.5 and 499.1) demonstrated that dermatan sulfate (CS-B)-like structures were present in both paw/ankle and knee/elbow CS GAGs.

B. Abnormal Contact System in K/BxN Mice

We performed Western analysis of the contact system proteins of EDTA plasmas from four age-matched normal mice and five 15-week-old K/BxN mice (data not shown). We found that the amount of prekallikrein, kininogen, and TAFI increased 3.5 ± 1.7, 12.8 ± 6.4, and 3.2 ± 1.4 times in the diseased mice compared to that of the four normal control mice, which indicates that RA development was associated with profound contact system change. By adding 10% K/BxN mouse plasma to 90% normal mouse plasma, we observed that K/BxN mouse plasma induced normal mouse contact system activation (data not shown), which suggests that GPI/anti-GPI/GAG complex in plasma might be the inducer of the contact system activation.

C. Human RA Patient Plasmas Were Associated with Abnormal Contact Systems Proteins

1. Western Analysis of the Contact System Proteins of RA Patient Plasmas

The abnormal contact systems observed in K/BxN mouse plasmas suggest that human RA might also be associated with abnormal contact systems. To test it, we purchased 12 RA patient plasmas from Bioreclamation, Inc., including four males and eight females either on or not on drug treatment (Fig. 3, top). We performed Western analysis on six contact system proteins, that is, kallikrein, FXII, kininogen, C1inb, thrombin, and TAFI on both pooled human plasma of 30+ normal donors and 12 RA patient plasmas. The most profound changes were observed in HMWK and prothrombin (Fig. 3, bottom).

2. HMWK (120 kDa)

HMWK is known to be cleaved by kallikrein to form its 68 kDa cleavage product after the contact system activation. The cleaved 68 kDa kininogen band was observed in all patients except patients 38, 46, and 47. No HMWK was observed in patients 37 and 44, indicating total cleavage of HMWK.

Id no.	Gender	Age	Medications
36	Male	50	None
37	Female	39	None
38	Female	66	Aspirin
39	Female	61	Methotrexate, mucinex, immodium, fexofenadine, folic acid
40	Male	38	Methotrexate, folic acid, prednisone, bextra
41	Female	73	Methotrexate, prednisone, folic acid, fosamax, cymbalta
42	Female	62	ASA, darvocet, mobic, advair, combivent
43	Female	83	Methotrexate, folic acid, prednisone, celebrex, spiriva, albutreol
44	Male	56	Aleve, furosemide, cellcept, enalapril, darcocet, iron supplement
45	Female	40	Methotrexate, celebrex, folic acid, humira, vitamin B12
46	Male	56	Methylprednisone, cellcept, enalapril, furosemide, amlodipine, lescol
47	Female	56	Advair, singulair, diovan, preacid

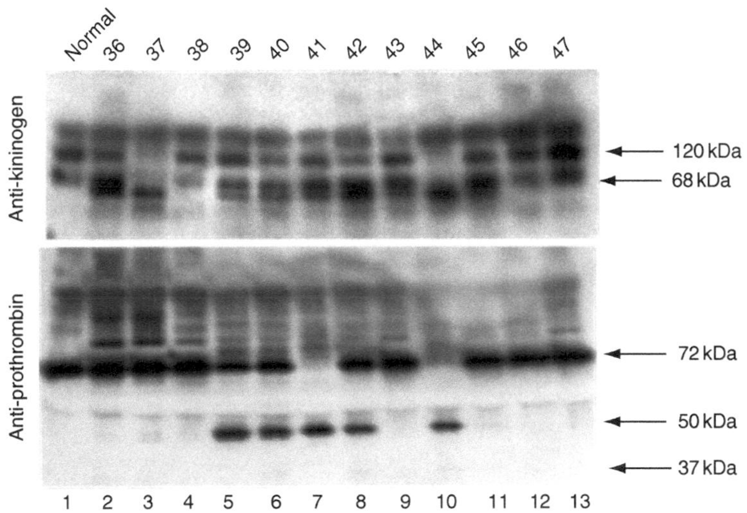

FIG. 3. Western analysis of contact proteins of normal and RA patient plasmas. Pooled normal human plasma and 12 RA patient plasmas were directly analyzed by reducing SDS-PAGE and Western analysis using antibodies to kininogen and prothrombin, respectively. Plasma (0.53 μl) was used in each lane.

Increased amount of HMWK (120 kDa) was observed in patients 46 and 47 compared to the normal control. Thus, most RA patients were associated with changes in HMWK.

3. Prothrombin (72 kDa)

The Western analysis assesses prothrombin (72 kDa), meizothrombin (desF1) (50 kDa), the active intermediate leading to thrombin formation, and thrombin (37 kDa) (Fig. 3 bottom). No prothrombin was observed in patients

41 and 44. Interestingly, meizothrombin (desF) bands (50 kDa) were present in appreciative amount in patients 39, 40, 41, 42, and 44. The abnormal meizothrombin (desF) bands distinquished RA from other autoimmune diseases (Activated Contact System and Abnormal Glycosaminoglycans in Lupus and other Auto- and Non-autoimmune Diseases).

4. Abnormal Thrombin Activities in RA Patient Plasmas

Meizothrombin (desF1) is an intermediate product of thrombin generation. It was accumulated in 5 out of 12 RA patient plasmas. Meizothrombin has protease activity toward fibrinogen and platelets.[44] To test if the observed meizothrombin (desF1) bands correlate with thrombin activity, we measured the kinetics of thrombin activity of normal and RA patient plasmas by using Sar-Pro-Arg-*p*-nitroanilide as a thrombin substrate. We found that the increased thrombin activities in patients 39, 40, 41, 42, and 44 correlated with observed meizothrombin (desF1) bands (Fig. 4). Since patients 39, 40, 41, 42, and 44 were not associated with specific gender, age, or drug treatment (Fig. 3, top), we concluded that meizothrombin (desF1) might represent a new biomarker for a subgroup of RA patients. However, the molecular mechanism that leads to the accumulation of meizothrombin (desF1) in certain RA patient plasmas is unknown.

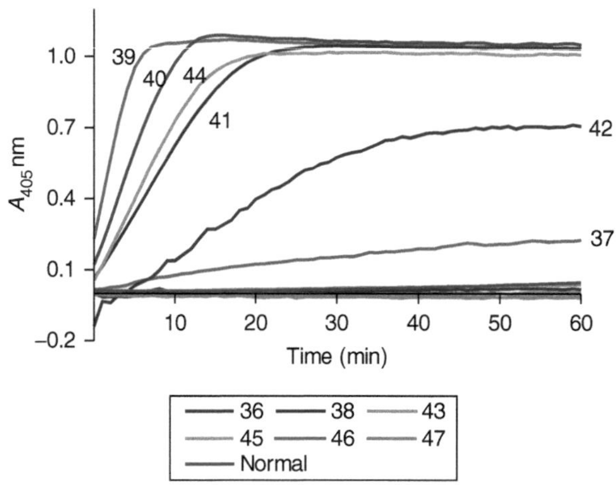

Fig. 4. (A) Kinetics of thrombin activities of normal and RA patient plasmas. The thrombin amidolytic activities of pooled normal human plasma and 12 RA patient plasmas (30 μl) were assessed by addition of 150 μl of 100 mM Sar-Pro-Arg-*p*-nitroanilide dihydrochloride substrate. OD readings at 405 nm were taken every minute for 60 min. The experiments were repeated twice with similar results. (See Color Insert.)

5. RA Patient Plasma Induced the Contact System Activation of Normal Human Plasma

Since K/BxN mouse plasma activated normal mouse plasma probably through GPI/anti-GPI/GAG ternary complex (data not shown), we hypothesized that RA patient plasma might be able to activate normal human plasma if autoantigen/autoantibody were present in the RA patient plasma. We chose patient 37 for the following contact system activation assay since patient 37 was not on drug treatment and was associated with an activated contact system evidenced by losing all HMWK (Fig. 3, bottom). Patient 37 plasma had an intermediate plasma thrombin activities compared to others (Fig. 4).

We incubated pooled human plasma (30 μl), pooled human plasma plus 10%, 50% or 100% patient 37 plasma, at 37 °C for 5 min. We then measured kallikrein amidolytic activity by addition of 150 μl of 50 mM *N*-benzoyl-Pro-Phe-Arg-*p*-nitroanilide. We detected autokallikrein activities in 100% patient 37 plasma with a final OD of 0.38 and no detectable kallikrein activity in normal human plasma at 60 min (Fig. 5). Theoretically, 10% patient plasma + 90% normal plasma should produce an OD of 0.038 and 50% patient plasma + 50% normal plasma should produce an OD of 0.19 if no interaction occur between the two plasmas at 60 min. However, The ODs observed (0.07 and 0.32) were 1.8× and 1.7× higher than the predicted no-plasma interaction values.

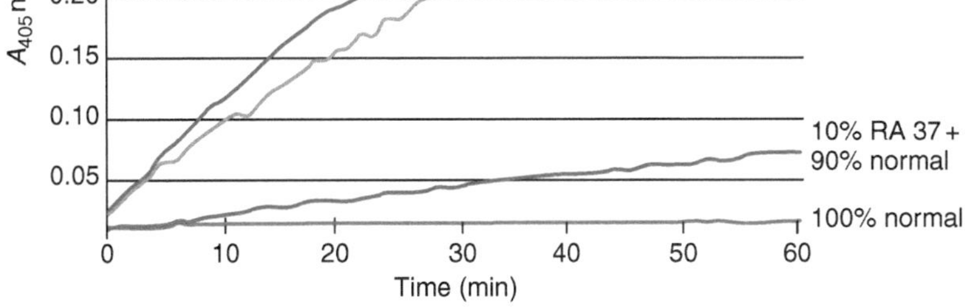

Fig. 5. Kinetics of kallikrein activities of normal human plasma plus 0%, 10%, 50% patient 37 plasma or 100% patient 37 plasma. Pooled human plasma from 30+ normal donors (30 μl) or pooled human plasma plus 10%, 50%, or 100% patient 37 plasma were incubated at 37 °C for 5 min. Kallikrein amidolytic activity was assessed by addition of 150 μl of 50 mM *N*-benzoyl-Pro-Phe-Arg-*p*-nitroanilide hydrochloride substrate. OD readings at 405 nm were taken every min for 60 min. The experiments were repeated twice with similar results.

Mixing 30+ normal human plasmas from different blood type donors did not result in any increase in kallikrein activities (Fig. 5). In contrast, mixing patients 39–42, or 44 plasma with normal human plasma also led to increased kallikrein activities in the mixed plasmas (data not shown). Thus, these results suggest that the factors present in RA patient plasmas have the capacity to activate the contact system of normal human plasmas. Therefore, activated contact system might upregulate immune response both in blood circulation and at cartilage surface by increasing vasopermeability and inflammation through BK, C3a, and C5a generation.

D. Proposed Contact System- and GAG-Involved K/BxN Model of RA

The contact system model of RA establishes the essential role of the contact system in initiating and maintaining RA in the Lewis rats. Our studies showed that GAGs are the GPI-high-affinity acceptor at cartilage surface. Our contact system studies in both human and mouse model suggest that autoantibodies/GPI/GAG complex might be the contact system activator both in the blood circulation and in the joint space. Based on these data, we added possible GAGs and the contact system involvement in the four-step K/BxN model of autoantibody-induced arthritis proposed by Wipke *et al.*[15] (Fig. 6).

VIII. Concluding Remarks

In the K/BxN mouse model of RA, autoantibodies specific for GPI from these mice can transfer joint-specific inflammation to normal mice. The data presented in this chapter indicate that the contact system activation might play a key role in RA initiation and development. Since the contact system is a pattern recognition system,[9] it explains why RF, preformed immune complex, and protein aggregates could enhance vasopermeability that allows autoantibodies to enter the joint space. It predicts that microbial factors could enhance vasopermeability by their innate ability to activate the complement system. It also predicts that thrombin generated by mechanical or joint movement associated injury could accelerate RA development. In summary, autoantibody binds specifically to cartilage surface might be a cause of RA. Therefore, by reducing autoantibody production or by inhibiting autoantibody-induced contact system activation might be effective in RA prevention.

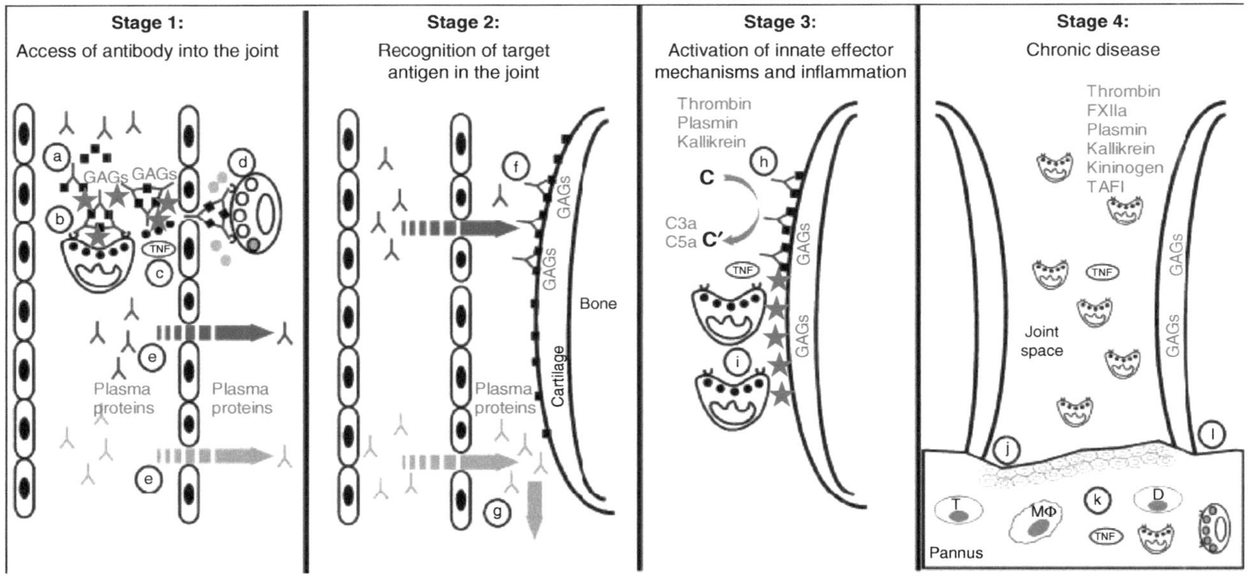

Fig. 6. A four-step K/BxN model of autoantibody-induced arthritis. Stage 1: Access of Ab and plasma proteins into the joint. a, Injected anti-GPI Abs (dark gray) bind to GPI present in the serum (squares) and GAGs form soluble ICs; b, ICs bind to FcRIII on neutrophils or activate the contact system (star-shaped structures); c, Triggering the release of vasoactive mediators such as TNF- (oval) and BK, resulting in a local increase in vascular permeability, thereby allowing the GPI/anti-GPI/ICs to enter the perivascular space where they encounter a mast cell; d, The GPI/anti-GPI/ICs bind to the mast cell via the FcRIII, causing them to degranulate, resulting in a large increase in vascular leakage; e, Allowing anti-GPI Abs (dark gray), nonspecific Abs (light gray), and plasma proteins to enter the joint. The increased vascular permeability is a transient event, and the vascular integrity is restored within 20 min. Stage 2: Recognition of GPI in the joint. f, Anti-GPI Abs bind to GPI that binds to cartilage surface GAGs; g, but nonspecific Abs have no target to bind and are cleared and/or degraded. Stage 3: Activation of innate effector mechanisms and inflammation. h, Anti-GPI Abs bound to the cartilage expose the negatively charged the cartilage surface, which activate the contact system (star-shaped structures). Thrombin and plasmin produce the anaphylatoxins, C3a and C5a; i, C5a then activates multiple cell types (neutrophils, mast cells, macrophages, and endothelial cells) to produce pro-inflammatory molecules (i.e., TNF-, IL-1, and chemokines), thereby causing inflammation. Stage 4: Chronic disease. After prolonged inflammation, chronic changes in the joint occur such as: j, synovial hyperplasia; k, pannus formation with inflammatory infiltrates comprising of T cells, B cells, macrophages, mast cells, and neutrophils; and l, activation of osteoclasts, resulting in bone erosion associated with persistent contact system activation.

References

1. Steiner G. Auto-antibodies and autoreactive T-cells in rheumatoid arthritis: pathogenetic players and diagnostic tools. *Clin Rev Allergy Immunol* 2007;**32**:23–36.
2. Nielen MM, van der Horst AR, van Schaardenburg D, van der Horst-Bruinsma IE, van de Stadt RJ, Aarden L, et al. Antibodies to citrullinated human fibrinogen (ACF) have diagnostic and prognostic value in early arthritis. *Ann Rheum Dis* 2005;**64**:1199–204.
3. So AK, Varisco PA, Kemkes-Matthes B, Herkenne-Morard C, Chobaz-Peclat V, Gerster JC, et al. Arthritis is linked to local and systemic activation of coagulation and fibrinolysis pathways. *J Thromb Haemost* 2003;**1**:2510–5.
4. Ji H, Ohmura K, Mahmood U, Lee DM, Hofhuis FM, Boackle SA, et al. Arthritis critically dependent on innate immune system players. *Immunity* 2002;**16**:157–68.
5. Matsumoto I, Maccioni M, Lee DM, Maurice M, Simmons B, Brenner M, et al. How antibodies to a ubiquitous cytoplasmic enzyme may provoke joint-specific autoimmune disease. *Nat Immunol* 2002;**3**:360–5.
6. Solomon S, Kolb C, Mohanty S, Jeisy-Walder E, Preyer R, Schollhorn V, et al. Transmission of antibody-induced arthritis is independent of complement component 4 (C4) and the complement receptors 1 and 2 (CD21/35). *Eur J Immunol* 2002;**32**:644–51.
7. Wipke BT, Wang Z, Kim J, McCarthy TJ, Allen PM. Dynamic visualization of a joint-specific autoimmune response through positron emission tomography. *Nat Immunol* 2002;**3**:366–72.
8. Studelska DR, Mandik-Nayak L, Zhou X, Pan J, Weiser P, McDowell LM, et al. High affinity glycosaminoglycan and autoantigen interaction explains joint specificity in a mouse model of rheumatoid arthritis. *J Biol Chem* 2009;**284**:2354–62.
9. Pan J, Qi Y, Zhou X, Lu H, Ramacciotti E, Zhang L. Chemically oversulfated glycosaminoglycans are potent modulators of contact system activation and different cell signaling pathways. *J Biol Chem* 2010;doi:10.1074/jbc.M109.063735.
10. Qian Y, Pan J, Zhou X, Weiser P, Lu H, Shih FF, et al. Heparin and oversulfated heparin byproduct induce thrombin generation through contact system in plasma of patients with HIT. *Clin Appl Thromb Hemost* 2010;**16**:251–60.
11. Maas C, Govers-Riemslag JW, Bouma B, Schiks B, Hazenberg BP, Lokhorst HM, et al. Misfolded proteins activate Factor XII in humans, leading to kallikrein formation without initiating coagulation. *J Clin Invest* 2008;**118**:3208–18.
12. Clark A, Weymann A, Hartman E, Turmelle Y, Carroll M, Thurman JM, et al. Evidence for non-traditional activation of complement factor C3 during murine liver regeneration. *Mol Immunol* 2008;**45**:3125–32.
13. Huber-Lang M, Sarma JV, Zetoune FS, Rittirsch D, Neff TA, McGuire SR, et al. Generation of C5a in the absence of C3: a new complement activation pathway. *Nat Med* 2006;**12**:682–7.
14. Kouskoff V, Korganow AS, Duchatelle V, Degott C, Benoist C, Mathis D. Organ-specific disease provoked by systemic autoimmunity. *Cell* 1996;**87**:811–22.
15. Wipke BT, Wang Z, Nagengast W, Reichert DE, Allen PM. Staging the initiation of autoantibody-induced arthritis: a critical role for immune complexes. *J Immunol* 2004;**172**:7694–702.
16. Binstadt BA, Patel PR, Alencar H, Nigrovic PA, Lee DM, Mahmood U, et al. Particularities of the vasculature can promote the organ specificity of autoimmune attack. *Nat Immunol* 2006;**7**:284–92.
17. Mandik-Nayak L, Racz J, Sleckman BP, Allen PM. Autoreactive marginal zone B cells are spontaneously activated but lymph node B cells require T cell help. *J Exp Med* 2006;**203**:1985–98.
18. Nandakumar KS, Holmdahl R. Antibody-induced arthritis: disease mechanisms and genes involved at the effector phase of arthritis. *Arthritis Res Ther* 2006;**8**:223.

19. Lee DM, Friend DS, Gurish MF, Benoist C, Mathis D, Brenner MB. Mast cells: a cellular link between autoantibodies and inflammatory arthritis. *Science* 2002;**297**:1689–92.
20. Maccioni M, Zeder-Lutz G, Huang H, Ebel C, Gerber P, Hergueux J, et al. Arthritogenic monoclonal antibodies from K/BxN mice. *J Exp Med* 2002;**195**:1071–7.
21. Solomon S, Kassahn D, Illges H. The role of the complement and the FcgammaR system in the pathogenesis of arthritis. *Arthritis Res Ther* 2005;**7**:129–35.
22. Monach PA, Verschoor A, Jacobs JP, Carroll MC, Wagers AJ, Benoist C, et al. Circulating C3 is necessary and sufficient for induction of autoantibody-mediated arthritis in a mouse model. *Arthritis Rheum* 2007;**56**:2968–74.
23. Okroj M, Heinegard D, Holmdahl R, Blom AM. Rheumatoid arthritis and the complement system. *Ann Med* 2007;**39**:517–30.
24. Schwab JH, Henderson B, Edwards JCW, Pettipher ER. Bacterial cell-wall induced arthritis: models of chronic recurrent polyarthritis and reactivation of monoarticular arthritis. In: Henderson B, Edwards JCW, Pettipher ER, editors. *Mechanisms and models in rheumatoid arthritis*. London: Academic Press; 1995. p. 431–46.
25. Sainz IM, Isordia-Salas I, Castaneda JL, Agelan A, Liu B, DeLa Cadena RA, et al. Modulation of inflammation by kininogen deficiency in a rat model of inflammatory arthritis. *Arthritis Rheum* 2005;**52**:2549–52.
26. Isordia-Salas I, Pixley RA, Sainz IM, Martinez-Murillo C, Colman RW. The role of plasma high molecular weight kininogen in experimental intestinal and systemic inflammation. *Arch Med Res* 2004;**35**:369–77.
27. Isordia-Salas I, Pixley RA, Parekh H, Kunapuli SP, Li F, Stadnicki A, et al. The mutation Ser511Asn leads to *N*-glycosylation and increases the cleavage of high molecular weight kininogen in rats genetically susceptible to inflammation. *Blood* 2003;**102**:2835–42.
28. Schaller M, Stohl W, Tan SM, Benoit VM, Hilbert DM, Ditzel HJ. Raised levels of anti-glucose-6-phosphate isomerase IgG in serum and synovial fluid from patients with inflammatory arthritis. *Ann Rheum Dis* 2005;**64**:743–9.
29. Schaller M, Burton DR, Ditzel HJ. Autoantibodies to GPI in rheumatoid arthritis: linkage between an animal model and human disease. *Nat Immunol* 2001;**2**:746–53.
30. Matsumoto I, Lee DM, Goldbach-Mansky R, Sumida T, Hitchon CA, Schur PH, et al. Low prevalence of antibodies to glucose-6-phosphate isomerase in patients with rheumatoid arthritis and a spectrum of other chronic autoimmune disorders. *Arthritis Rheum* 2003;**48**:944–54.
31. Matsumoto I, Muraki Y, Yasukochi T, Hua Z, Kori Y, Hayashi T, et al. The exploration of joint-specific immunoreactions on immunoglobulins G of anti-glucose-6-phosphate isomerase antibody-positive patients with rheumatoid arthritis. *Int J Mol Med* 2005;**16**:793–800.
32. Snowden JM, Maroudas A. The distribution of serum albumin in human normal and degenerate articular cartilage. *Biochim Biophys Acta* 1976;**428**:726–40.
33. Maroudas A. Transport of solutes through cartilage: permeability to large molecules. *J Anat* 1976;**122**:335–47.
34. Esko JD, Selleck SB. Order out of chaos: assembly of ligand binding sites in heparan sulfate. *Annu Rev Biochem* 2002;**71**:435–71.
35. Esko JD, Lindahl U. Molecular diversity of heparan sulfate. *J Clin Invest* 2001;**108**:169–73.
36. Carmassi F, De Negri F, Morale M, Puccetti R, Chung SI. Elastase- and plasmin-mediated fibrinolysis in rheumatoid arthritis. *Int J Tissue React* 1994;**16**:89–93.
37. Sanchez-Pernaute O, Largo R, Calvo E, Alvarez-Soria MA, Egido J, Herrero-Beaumont G. A fibrin based model for rheumatoid synovitis. *Ann Rheum Dis* 2003;**62**:1135–8.
38. Busso N, Hamilton JA. Extravascular coagulation and the plasminogen activator/plasmin system in rheumatoid arthritis. *Arthritis Rheum* 2002;**46**:2268–79.
39. Gabazza EC, Osamu T, Yamakami T, Ibata H, Sato T, Sato Y, et al. Correlation between clotting and collagen metabolism markers in rheumatoid arthritis. *Thromb Haemost* 1994;**71**:199–202.

40. Orfeo T, Brummel-Ziedins KE, Gissel M, Butenas S, Mann KG. The nature of the stable blood clot procoagulant activities. *J Biol Chem* 2008;**283**:9776–86.
41. Qian Y, Pan J, Zhou X, Weiser P, Hourcade D, Lizsewski KM, et al. Oversulfated heparin byproducts induce thrombin generation through contact system activation. *Clin Appl Thromb Hemost* 2010;**16**:244–50.
42. Yu H, Munoz EM, Edens RE, Linhardt RJ. Kinetic studies on the interactions of heparin and complement proteins using surface plasmon resonance. *Biochim Biophys Acta* 2005;**1726**:168–76.
43. Lawrence R, Olson SK, Steele RE, Wang L, Warrior R, Cummings RD, et al. Evolutionary differences in glycosaminoglycan fine structure detected by quantitative glycan reductive isotope labeling. *J Biol Chem* 2008;**283**:33674–84.
44. Doyle MF, Mann KG. Multiple active forms of thrombin. IV. Relative activities of meizothrombins. *J Biol Chem* 1990;**265**:10693–701.

Activated Contact System and Abnormal Glycosaminoglycans in Lupus and other Auto- and Non-Autoimmune Diseases

Peter Weiser,* Yi Qian,* Jing Pan,* Xiaodong Zhou,* Hong Lu,* Daniel R. Studelska,* Fei F. Shih,† and Lijuan Zhang*

*Department of Pathology and Immunology, Washington University Medical School, St. Louis, Missouri, USA

†Department of Pediatrics, Washington University Medical School, St. Louis, Missouri, USA

Progress in Molecular Biology and Translational Science, Vol. 93
DOI: 10.1016/S1877-1173(10)93019-6

Systemic lupus erythematosus (SLE), heparin-induced thrombocytopenia (HIT), rheumatoid arthritis (RA) are marked by the presence of autoantibodies against negatively changed DNA, phospholipids, heparin, and chondroitin sulfate, respectively. Heparin/protein complexes induce contact system activation in HIT patient plasmas. The activated contact system generates thrombin. Thrombin is responsible for thrombosis, a common cause of death and disabilities for both HIT and SLE. In this chapter, we analyze plasma contact system proteins, thrombin- and kallikrein-like activities, glucosamine and galactosamine content from SLE-, RA-, osteoarthritis (OA)-, and psoriasis (Ps)-patient plasmas in addition to pooled 30+ healthy patient plasmas. We found that all SLE patient plasmas exhibited abnormal contact systems marked by the absence of high molecular weight kininogen, the presence of processed C1 inhibitor (C1inh), the display of abnormal thrombin- and kallikrein-like activities, and increased levels of plasma glucosamine and galactosamine. Different patterns of contact system activation distinquish SLE, RA, and Ps whereas no contact system activation is observed in normal and OA patient plasmas. The presence of paradoxical "lupus anticoagulants" in certain thrombosis-prone SLE patient plasmas, marked by delayed clotting in clinical plasma test, was explained by the consumption of contact system proteins, especially high molecular weight kininogen. Finally, we discovered that mouse and human SLE autoantibodies bind to cell surface GAGs with structural selectivity. In conclusion, markers of abnormal contact system activation represent potential new targets for autoimmune disease diagnosis, prevention, and treatment. These markers might also be useful in monitoring SLE activity/severity and in pinpointing patients with SLE-associated arthritis and psoriasis.

Abbreviations: SLE, systemic lupus erythematosus; HIT, heparin-induced thrombocytopenia; GAG, glycosaminoglycan; C1inh, C1 inhibitor; TBS, Tris buffered saline; OA, osteoarthritis; RA, rheumatoid arthritis; Ps, psoriasis

I. Introduction

Autoantibodies against negatively charged molecules such as DNA and phospholipids are essential agents of systemic lupus erythematosus (SLE),[1] although they can be present indefinitely without symptoms, or reside for months before the onset of disease.[2] The disease can affect almost any organ in the body; causing blood clotting, arthritis, fatigue, heart disease, osteoporosis, kidney failure, and cognitive loss in a patient-specific manner.[3]

SLE patients with autoantibodies to phospholipid, cardiolipin, or the phospholipid-binding protein, β-2-glycoprotein I, suffer the most severe forms of the disease.[4] However, some SLE autoantibodies react with phospholipids other than cardiolipin and proteins other than β-2-glycoprotein I.[5] Antiphospholipid autoantibodies in certain SLE patient sera also bind to negatively charged heparin/platelet factor 4 (PF4) complexes.[6] Therefore, autoantibodies in SLE may cross-react with a series of negatively charged molecules. SLE is different from a typical antibody-mediated immune disease in that thrombin and complement factors C3a and C5a are major effecters in disease etiology. However, the molecular mechanism of how autoantibodies induce thrombin and complement activation is largely unknown.

Contaminated heparin was associated with at least 149 deaths in 2007 and 2008.[7] Heparin contaminants, that is, oversulfated glycosaminoglycans (OS-GAGs) induce contact system activation and C3a and C5a production in normal human plasmas through an unclassical complement activation pathway.[8] It has been previously established that thrombin produces C3a and C5a *in vivo*.[9,10] We reasoned that thrombin might be responsible for producing C3a and C5a through heparin contaminant-induced contact system activation. Indeed, the contact system was first discovered as an *in vitro* thrombin generation system where negatively charged artificial surfaces induce thrombin generation and clotting.[11] The contact system consists of two serine proteases: prekallikrein and factor XII, a cofactor: high molecular weight kininogen, and a regulator: C1 inhibitor (C1inh) in human plasma.[12]

We found that heparin contaminants, that is chemically oversulfated heparin byproduct (OS-HB),[13,14] induced thrombin generation through contact system activation in human plasmas.[15] We then discovered that the negatively charged IgG/PF4/heparin complex activates the contact system and produces thrombin in heparin-induced thrombocytopenia patient plasmas.[16] Our results suggest that partially activated platelets induced by thrombin/protease activated receptor signaling allow subsequent platelet activation through the IgG/Fc receptor II pathway.[16] We further showed that a variety of endogenous negatively charged molecules can induce contact system activation in normal human plasma.[17] These findings suggest that the contact system is an *in vivo* thrombin generation system activated by negatively charged protein complexes. We reasoned that abnormal thrombin generation induced by an autoantibody/negatively charged molecule complex might initiate SLE and other autoimmune pathogenesis via contact system activation.

Indeed, we found that SLE patient plasmas exhibit elevated thrombin activity and activated contact systems as shown by the absence of high molecular weight kininogen, the presence of processed C1inh, and the display of abnormal kallikrein activity. Abnormal contact system activation explains why "lupus anticoagulants" are present in certain SLE patient plasmas. Because of activated contact systems, lupus patient plasmas are deficient in contact system-dependent clinical clotting assays.[16]

GAGs represent a family of sulfated polysaccharides composed of repetitive disaccharide units consisting of one hexosamine and one uronic acid (reviewed in Ref. 18). GAGs are made by all animal cells and are divided into different subclasses according to their disaccharide composition and sulfation. The four major GAGs are heparin, heparan sulfate, chondroitin sulfate, and dermatan sulfate. Heparin is the most sulfated natural occurring GAG and induces autoantibodies against heparin/protein complex in heparin-induced thrombocytopenia patients. GAGs are modified by multiple enzymes which themselves are regulated by tissue specific, developmental, and environmental factors. Autoantibodies to GAGs have been reported in SLE patients.[19] Heparan sulfate antibodies have also been implicated in SLE nephritis but have not gained wide acceptance as causative factors.[20] Immunization of mice with GAGs results in joint inflammation and arthritis, demonstrating that GAGs other than heparin can initiate autoimmune disease.[21]

GAGs, phospholipids, RNAs, and DNAs are the major highly negatively charged molecules made by animal cells. However, only GAGs cover cell surfaces and are abundantly present in the extracellular matrix. Anti-DNA antibodies are pathognomonic but it is difficult to understand how autoantibodies to DNA initiate disease without ready access to an antigen. Based on the highly patterned and negatively charged structure of DNA, we propose that SLE autoantibodies cross-react with GAGs or GAG/protein complexes, which are also highly patterned and negatively charged structures. Since heparin not only induces autoantibodies but also is recognized by autoantibodies in heparin/protein complexes,[16] we hypothesize that abnormal GAG production in SLE patients precipitates SLE pathogenesis. This mechanism is supported by heparin-induced autoantibodies that initiated HIT, by increased glucosamine and galactosamine levels in SLE patient plasmas, and different affinities of autoantibodies to different cell surface GAG structures with high affinity. In addition, we discovered that different patterns of contact system activation distinquish SLE, RA, and Ps whereas no contact system activation was observed in normal and osteoarthritis (OA) patient plasmas. We conclude that the contact system and the abnormal GAGs in patient plasma represent new targets for autoimmune disease diagnosis, prevention, and treatment.

II. Methodology

A. Materials

Kallikrein substrate, *N*-benzoyl-Pro-Phe-Arg-*p*-nitroanilide hydrochloride, was from Sigma. OS-HB was synthesized in our laboratory.[17] Pooled normal human plasma from 30+ normal donors was purchased from George King Biomedical, Inc. The teenage SLE patient plasma collection protocol was approved by the Human Studies Committee (Washington University School of Medicine). After informed written consent was given by patients, blood was collected by standard venepuncture into EDTA in plastic tubes. The platelet-free plasma was aliquoted and stored at − 70 °C before use. All adult patient plasmas were from 12 SLE-, 12 OA-, 9 rheumatoid arthritis (RA)- (only nine samples were available at the time of purchase), and 12 Ps-patient plasmas were purchased from Bioreclamation, Inc. The plasma was aliquoted and stored at − 70 °C before use.

B. Kallikrein Activity Assay

Plasmas were activated *in vitro* by adding 20 μl OS-HB into 30 μl of plasma at 37 °C for 5 min at a final concentration of 0, 2, 20, or 200 μg/ml, in a 96-well plate. TBS buffer (50 mM Tris, pH 7.6, 50 mM NaCl buffer) was then added to each well to a total volume of 150 μl. Absorbance at 405 nm was recorded as blank by Spectramax M2 plate reader (Molecular Devices). Substrate (50 μl in TBS) was added to each well and absorbance at 405 nm was recorded every 90 s for 40 min

C. Western Analysis

Both Normal and SLE patient plasmas were diluted 1:4 with TBS, and an equal volume of 2× Laemmli sample buffer was added. After boiling at 100 °C for 5 min, 7 μl of each sample was separated by 4–15% SDS-PAGE and transferred onto a nitrocellulose membrane. After blotting with the primary and secondary antibody, protein was visualized using the SuperSignal West Pico Chemiluminescent Substrate from Thermo-Scientific-Pierce.

D. Flow Cytometric Analysis of IgG Antibody Binding to CHO Cells

CHO cells were treated with EDTA to lift them off culture plates at 37 °C for 15 min and washed with PBS once and counted. Each 5×10^5 cell aliquot was incubated with 2 μl plasma in 100 μl FACS buffer on ice. Binding of murine or human antibodies to a panel of CHO cells expressing specific GAG was detected with anti-mouse Ig-FITC or anti-human Ig-Cy2, respectively. The cells were finally washed with FACS buffer once, resuspended in 0.2 ml FACS buffer and analyzed on BD FACSCaliber with 10,000 cells. Mean fluorescence intensity was calculated by comparing with the negative control.

E. Glucosamine and Galactosamine Quantification

Two times 1 μl of plasma or purified GAGs were used for glucosamine and galactosamine quantification. Samples were subjected to acid hydrolysis, sodium borohydride reduction, precolumn derivatization with *o*-phthaldialdehyde and 3-mercaptopropionic acid, and reversed-phase HPLC separation with fluorescence detection of the isoindole derivatives as described in detail elsewhere.[22]

III. An Activated Contact System is Common in Plasmas from SLE Patients

A. Abnormal Kallikrein Activity in SLE Patient Plasmas

A hallmark of SLE is the production of autoantibodies resulting in autoantibody mediated pathology such as thrombosis and fetal loss through complement system activation.[4] We reasoned that if SLE autoantibodies induce contact system activation, SLE patient plasmas should have high basal levels of kallikrein and thrombin activity and abnormal contact system protein profiles. To test this hypothesis, we purchased 12 SLE patient plasmas from Bioreclamation, Inc. (Table I).

Kallikrein activity of plasmas from the 12 SLE patients and plasmas pooled from 30+ healthy individuals were measured by monitoring the colored cleavage product of a kallikrein substrate. The results shown in Fig. 1 were derived

TABLE I
ADULT SLE PATIENT INFORMATION PROVIDED BY BIORECLAMATION, INC.

Patient number	Gender	Age	Medication
SLE1	F	39	Motrin, Hydrocortisone, Coumadin
SLE2	F	56	Prednisone, Hyzaar, Lotrel, Clonidine
SLE3	M	24	Prednisone, Plaquenil, Imuran, Diovan, Metoprolol
SLE4	F	49	Ibuprofen, HCTZ, Requip, Albuterol
SLE5	F	32	Prednisone, Baby Aspirin, Imuran, Plaquenil, Lotensin, Iron
SLE6	F	66	Methotrexate, Cortisol, Triamcinolone, Fish Oil, Folic Acid, Simvastatin, hydrochlorothiazide, Glucovance, Plaquenil, Cardizem, Multivitamin,
SLE7	F	57	Celexa, Ambien, Lipitor
SLE8	F	37	Plaquenil, CellCept, Remeron, Ambien, Phenergan, Effexor, Topamax, Nexium
SLE9	F	65	Methotrexate, Cortisol, Triamcinolone, Fish Oil, Folic Acid, Plaquenil, MVI, Vitamin E
SLE10	F	31	Prednisone, Baby Aspirin, Imuran, Lotensin, Plaquenil, Iron,
SLE11	F	31	Prednisone, Baby Aspirin, Imuran, Lotensin, Plaquenil, Iron,
SLE12	F	41	Plaquenil, Vasoretic

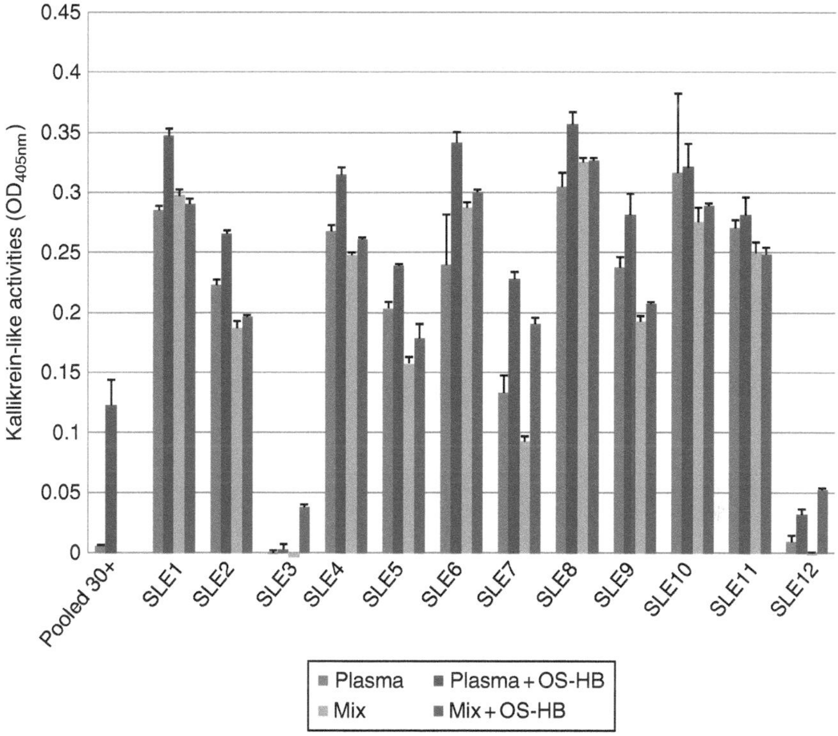

FIG. 1. Abnormal kallikrein activity in SLE patient plasmas. Plasmas were incubated with 20 μl TBS or 20 μl OS-HB at a final concentration of 200 μg/ml at 37 °C for 5 min. Kallikrein activity was assessed by addition of 150 μl of the kallikrein chromogenic substrate, 0.5 mM *N*-benzoyl-Pro-Phe-Arg-*p*-nitroanilide hydrochloride dissolved in TBS. Pooled 30+ indicates the normal plasma control obtained from 30+ healthy individuals. Detailed information about the SLE patients are shown in Table I. Bars represent arithmetic mean of the change of the $OD_{405\ nm}$ at 60 min from starting $OD_{405\ nm}$; each value was established from three independent measurements. The error indicated is the standard deviation. Blue bar, kallikrein activities in normal and SLE patient plasmas. Red bar, OS-HB-induced kallikrein activities in normal and SLE patient plasmas. Green bar, kallikrein activities in a 1:1 mixture of normal and SLE patient plasmas. Purple bar, kallikrein activities in an OS-HB-induced 1:1 mixture of normal and SLE patient plasmas. (See Color Insert.)

from three independent measurements. No kallikrein activity was observed in normal control plasma (OD 0.006) (Fig. 1). In contrast, kallikrein activity was elevated in most of patient plasmas except patients 4 (OD 0.0001), 8 (OD 0.133), and 13 (OD 0.010).

B. Kallikrein Activity in SLE Plasmas is Not Further Induced by OS-HB Treatment

Plasmas were incubated with 200 μg/ml OS-HB at 37 °C for 5 min before exposing them to the kallikrein chromogenic substrate (Fig. 1). OS-HB induced a 21-fold increase in kallikrein activity in normal human plasma (from 0.006 to 0.123), consistent with previous observations.[8] In contrast, we found that OS-HB only induced minimal kallikrein activity in SLE patient plasmas.

C. Mixing Normal and SLE Plasmas Recruits SLE-Like Plasma Kallikrein Activity in Normal Plasma

Because SLE is a chronic inflammatory disease, the plasmas were from patients with established disease. To test whether SLE plasmas could induce kallikrein activity in normal plasma, we mixed 50% SLE plasma with 50% pooled plasmas from 30+ healthy individuals and measured the kallikrein activity. We found that the kallikrein activity was only slightly reduced instead by 50%. This indicates that kallikrein could be induced in normal plasma by mixing with SLE plasma. Therefore, autoantibodies in SLE plasmas might be responsible for this induction in mixed plasmas.

D. Mixed Normal and SLE Plasmas has a SLE Plasma-Like Response to OS-HB-Induced Kallikrein Activation

To further confirm that the abnormal kallikrein activity was induced by the diseased SLE plasmas, OS-HB was added to the 1:1 mixed plasmas. We found that OS-HB application resulted in minimal kallikrein activities in the mixed plasmas in contrast to the 21-fold-induction observed in normal plasma (Fig. 1).

E. Elevated Thrombin-Like Activity in SLE Patient Plasmas

Since thrombin is essential for thrombosis in SLE patients, we measured plasma thrombin-like activities colormetrically by monitoring cleavage of the thrombin substrate, Sar-Pro-Arg-*p*-nitroanilide dihydrochloride. We found that OS-HB strongly induced thrombin-like activity in normal human plasmas but not in SLE patient plasmas (Fig. 2). Interestingly, the plasma from patient 8 had the third lowest kallikrein activity but the highest thrombin-like activity, whereas the plasma from patients 4 and 13 had the lowest kallikrein and thrombin activities among the 12 patient plasmas tested, which distinguished patients 4, 8, and 13 from the rest of the SLE patients.

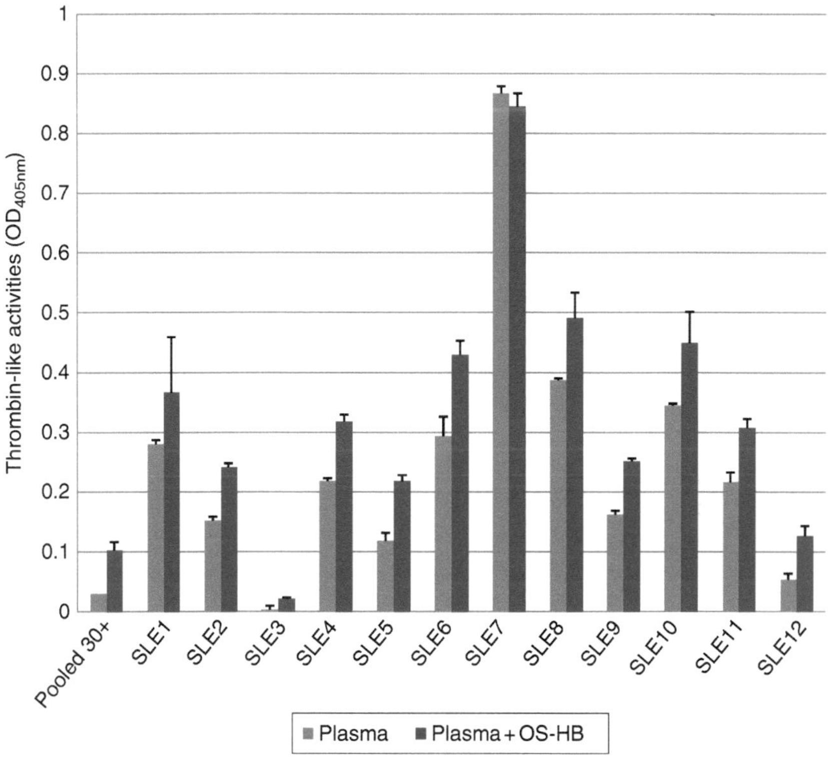

FIG. 2. Elevated thrombin-like activity in SLE patient plasmas. Plasmas were incubated with 20 μl TBS or 20 μl OS-HB with a final concentration of 200 μg/ml at 37 °C for 5 min. Thrombin-like activity was assessed by addition of 150 μl of the thrombin chromogenic substrate, 0.5 mM Sar-Pro-Arg-*p*-nitroanilide dihydrochloride dissolved in TBS. Pooled 30+ indicates the normal plasma control obtained from 30+ healthy individuals. Detailed information about the SLE patients are shown in Table I. Bars represent arithmetic mean of the change of the $OD_{405\ nm}$ at 60 min from starting $OD_{405\ nm}$; value established from three independent measurements with standard deviation. Blue bar, thrombin-like activity in normal and SLE patient plasmas. Red bar, OS-HB-induced thrombin-like activity in normal and SLE patient plasmas. (For interpretation of the references to color in this figure legend, the reader is referred to the Web version of this chapter.)

F. Activated Contact System in Teenage SLE Patient Plasmas

To extend our findings, we obtained the Institutional Review Board (IRB) approval and collected plasmas from 15 additional SLE patients in St. Louis Children's hospital with appropriate consent. Clinical information of these patients is shown in Table II.

TABLE II
TEENAGE SLE PATIENT INFORMATION

Patient number	Gender	Age	αDNA auto antibody	Platelets	Disease course	Arthritis	Rash	Medication
8	F	15	23	331	Stable, anemia,	No	Yes	MMF, Plaquenil, and Prednisone
19	F	16	488	297	Active, worse	No	Yes	Plaquenil
31	F	18	8	384	Active-MMF	No	Yes	MMF, Prednisone, Plaquenil
34	F	18	48	115	Simmering-active	No	No	MMF, Plaquenil
63	F	14	101	254	Controlled	No	No	Prednisone, Plaquenil
107	M	18	316	230	Steady, aPL Ab	No	No	MMF, Prednisone
113	F	17	338	183	Stable	No	No	MMF, Plaquenil
127	F	18	61	318	Active	Yes	No	Leflunomide, Naproxen
138	F	15		314	Active-controlled		Yes	MMF
163	F	18	3	295	Reactivated	No	Yes	
165	F	18	36	287	Stable controlled	No	No	MMF and Plaquenil
167	F	12	2		Not true SLE	No	Yes	
168	F	18	142	221	Flare	Yes	No	Plaquenil
169	F	18	312	332	New SLE	Yes	No	
170	F	15	732	276	Second visit	No	Yes	Plaquenil

We compared the kallikrein activity in normal control and 15 teenage SLE patient plasmas. No kallikrein activity was observed in normal control plasma (Fig. 3, control panel, 0 μg/ml black line). In contrast, kallikrein activity was measurable in all SLE patient plasmas (Fig. 3, 0 μg/ml black line), albeit the level of kallikrein activity varied among the teenage SLE patients, which was consistent with the results observed in the adult SLE plasmas (Fig. 3). To test if kallikrein activity could be further induced by OS-HB, the control and SLE patient plasmas were treated with OS-HB at 2, 20, or 200 μg/ml, and the kallikrein activity was measured. We found that OS-HB-induced kallikrein activities in normal control plasma in a concentration-dependent manner (Fig. 3, control panel), as previously reported.[8] In contrast, only the 200 μg/ml OS-HB treatment increased kallikrein activity in teenage SLE patient plasmas. These results confirm that normal and SLE plasmas have different natural or OS-HB-induced kallikrein activity profiles.

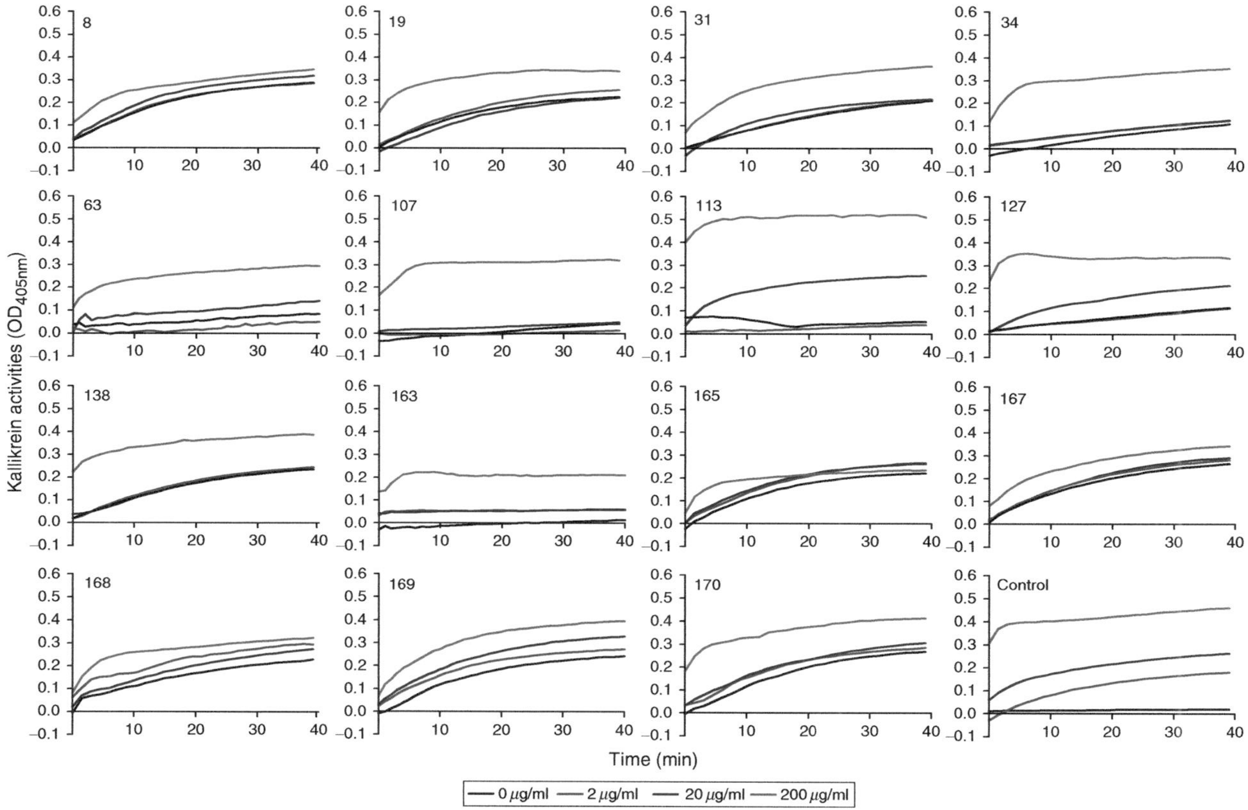

FIG. 3. High naturally occurring and low OS-HB-induced kallikrein activity in teenage SLE patient plasmas. Normal pooled human plasma or 15 SLE patient plasmas were incubated with 20 μl OS-HB with a final concentration of 0, 2, 20, or 200 μg/ml at 37 °C for 5 min. Kallikrein activity was assessed by addition of 150 μl of the kallikrein chromogenic substrate, 0.5 mM *N*-benzoyl-Pro-Phe-Arg-*p*-nitroanilide hydrochloride dissolved in TBS. OD readings were taken every 90 s for 40 min. (See Color Insert.)

G. Activated Contact System Proteins in SLE Patient Plasmas

To further define the relationship between kallikrein activity and other contact system proteins, we performed western analysis on C1inh, kininogen, prekallikrein, and thrombin from both normal and teenage SLE patient plasmas (Fig. 4). Because contact system activation results in proteolytic cleavage

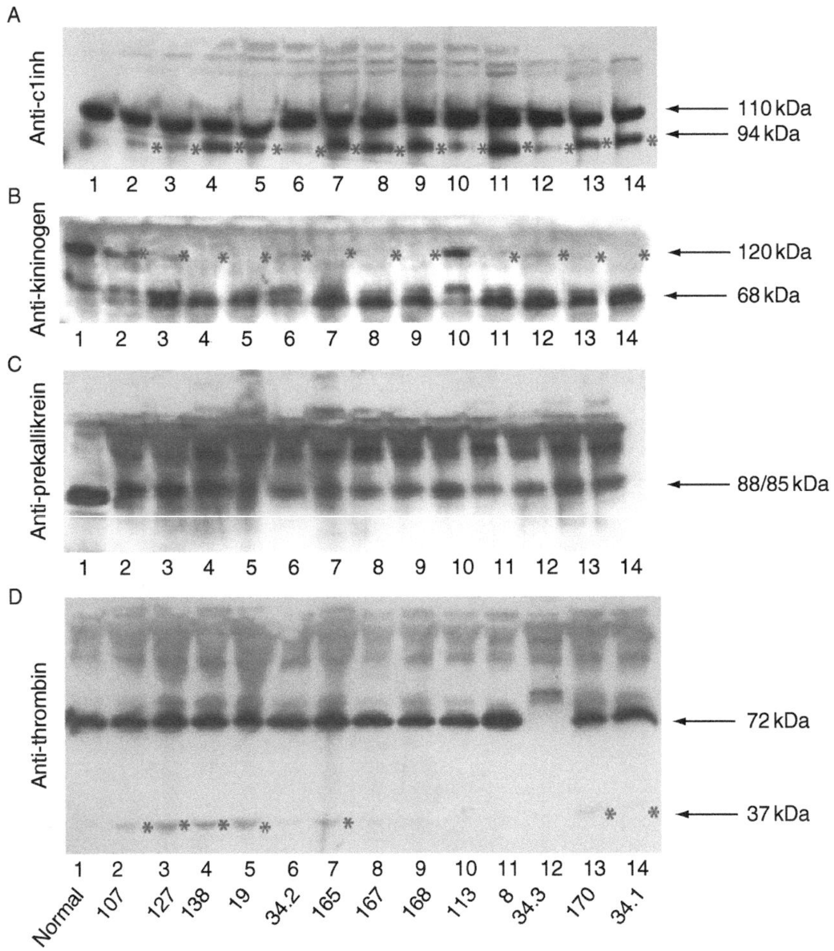

FIG. 4. Western analysis of the contact system proteins of teenage SLE and normal plasmas. Pooled normal human plasma and SLE patient plasmas were analyzed by SDS-PAGE and western blotting using antibodies to C1inh, prekallikrein, HMWK, and thrombin, respectively. Plasma (0.53 μl) was used in each lane. The experiment was performed twice with similar results.

of several of its components as well as the deactivation of serine proteases via covalent attachment of C1inh, an activated contact system can be systematically demonstrated by western analysis (Fig. 4).

A. C1INH

C1inh is a 110 kDa serine protease inhibitor. It circulates in blood at around 250 μg/ml. C1inh is the most important physiological inhibitor of plasma kallikrein, factor XIa, factor XIIa, and several complement serine proteases. It acts by forming a 1:1 inhibitory complex with these proteases. Cleaved C1inh (94 kDa) and C1inh/protease complex bands have been reported in clinical C1inh dysfunction patient plasmas.[23] In contrast to the 110 kDa C1inh band of in normal plasma, we found the 94 kDa cleaved C1inh band and the C1inh/protease complex band (indicated by a red asterick) in all SLE patient plasmas (Fig. 4A, lane 2–11). The western analysis indicated that all SLE patient plasmas tested were associated with C1inh abnormalities.

B. KININOGENS

The two plasma kininogens, high molecular weight kininogen (120 kDa) and low molecular weight kininogen (70 kDa) are spliced variants of the kininogen gene. Kininogens are multifunction proteins with antithrombotic, antidiuretic, profibrinolytic, antiadhesive, antiangiogenic, and proinflammation properties. Bradykinin and kallidin, two antithrombotic, profibrinolytic, and potent vasodilating peptides, are released from high and low molecular weight kininogens by plasma kallikrein and tissue kallikreins, respectively. High molecular weight kininogen is known to be cleaved by plasma kallikrein to form its 68 kDa cleavage product after contact system activation. High molecular weight kininogen was rare in SLE patient plasmas, except for small amounts detected in plasmas from patients 107, 127, 34.2, and 113 (Fig. 4B, lanes 2, 3, 6, and 10). Abnormal C1inh and kininogens in SLE patient plasmas were consistent with the abnormal kallikrein activity observed (Figs. 1 and 3), suggesting kallikrein activities were generated through contact system activation.

C. PREKALLIKREIN

The conversion of prekallikrein (85/88 kDa) to active kallikrein is accomplished by serine protease cleavage, generally by factor XIIa, where Arg371-Ile372 is split, generating a heavy chain of 371 amino acids and a light chain of 248 amino acids held together by a disulfide bond. Under western analysis conditions, kallikrein comigrates with prekallikrein. In plasma, C1inh or α2-macroglobulin, either of which forms a 1:1 inhibitory complex with kallikrein, and inactivates kallikrein. Therefore, an activated contact system might be accompanied by a normal or decreased prekallikrein/kallikrein band with or without kallikrein/protease inhibitor complexes. Decreased prekallikrein bands were observed in most SLE patients (Fig. 4C, lanes 2–11 compared to the control lane 1), which is consistent with the consumption of contact system proteins in SLE patient plasmas.

D. THROMBIN (37 KDA)

Thrombin is normally generated from the 72 kDa plasma prothrombin by both thrombin and factor Xa.[24] Cleavage at positions of 155 and 284 are catalyzed by thrombin whereas cleavage at positions 271 and 320 are catalyzed by factor Xa to generate A and B chains of the 37 kDa thrombin. The initial cleavage of prothrombin by factor Xa occurs at position 320, giving rise to meizothrombin, which has the same molecular weight as prothrombin. Thrombin and meizothrombin have comparable activities toward protein C, factor V, factor XI, and small peptide substrates.[24] Therefore, thrombin western analysis may reveal the same or different sizes of cleaved thrombin products reflecting 1–4 cuts of prothrombin by factor Xa and/or thrombin. In plasma, thrombin is inactivated by antithrombin or heparin cofactor II, each catalyzed by glycosaminoglycans, and each of which form a 1:1 stoichiometric complex with thrombin. Therefore, an activated contact system might be accompanied by a normal or decreased prothrombin/meizothrombin band with or without a 37 kDa thrombin band. Thrombin/antithrombin and thrombin/heparin co-factor complexes might be observed as well. We found that 37 kDa thrombin bands were present in a subset of SLE patient plasmas (107, 127, 138, 19, 165, 170, and 34.1; Fig. 4D, lanes 2, 3, 4, 5, 7, 13, and 14). In summary, the western analysis demonstrated that all SLE patient plasmas contained activated contact systems in that abnormal kallikrein activity (Figs. 1–3) was accompanied by thrombin generation, abnormal C1inh, and kininogen cleavage (Fig. 4).

Since SLE autoantibodies mediate pathology through thrombosis and complement system activation,[4] we included C3a determination in the western analysis of 12 adult SLE patient plasmas. C3a is an indicator of complement activation. The result is shown in Fig. 5. Protein bands present in the normal plasma are depicted green, if the patient plasmas contained less of the same protein band; it is coded with lighter green, if none, the color is white. On the other hand, if bands of activated forms were present in the SLE plasmas, they were represented by two shades of red, the darker indicating more prominent protein presence. We found that all SLE patient plasmas had activated contact systems in that abnormal kallikrein activity (Fig. 1) was accompanied by abnormal C1inh and kininogen cleavage, as well as by kallikrein complex formation. Complement C3a was present in all SLE plasma samples.

SLE is associated with anti-DNA and antiphospholipid autoantibodies. These autoantibodies have been termed lupus anticoagulants because they are thought to prolong bleeding because of interference with cell membrane clotting factors. Our data indicate that the consumption of contact system proteins (Fig. 2), especially lack of high molecular weight kininogen, leads to the observed lupus anticoagulant

	C1 inhibitor			Kininogen			Prekallikrein		Prothrombin		C3a
	Complex	110 kDa	94 kDa	110 kDa	70 kDa	64 kDa	Complex	88/85 kDa	70 kDa	50 kDa	10 kDa
Pooled 30+		x		x	x			x	x		
SLE1		i	i	i	x	x	x	x	x		x
SLE2		i	i	i	x	x	x	x	x		x
SLE3		x		x	x			x	x		i
SLE4			i	i	x	x	x	x	x		x
SLE5		x	i	i	x	i	x	x	x		i
SLE6		i	i	i	x	x	x	x	x		x
SLE7		x	i	x	x			x			x
SLE8		i	x	i	x	x	x	x	x		x
SLE9		x	i	x	x	x	x	x	x		x
SLE10		i	x	i	i	x	x	x	x		x
SLE11		x	i	i	x	x	x	x	x		x
SLE12		x		x	x	x		x	x		i

FIG. 5. Western analysis results of the contact system proteins and C3a in adult SLE and normal plasmas. Pooled normal human plasma and SLE patient plasmas were analyzed by SDS-PAGE and western blotting with antibodies to C1inh, prekallikrein, HMWK, thrombin, and C3a, respectively. Plasma (0.53 μl) was used in each lane for Western analysis. Results from two independent Western analyses were summarized. Protein bands present in the normal plasma are depicted green. If a patient plasma contained less of the same protein band, it was coded with lighter green, if the band was absent, and the color was white. On the other hand, if abnormal protein bands were present in the SLE plasmas, they were represented by different shades of red, the darker indicating more prominent protein presence. (For interpretation of the references to color in this figure legend, the reader is referred to the Web version of this chapter.)

phenotype because the clotting is delayed due to lack of high molecular weight kininogen in SLE patient plasmas detected by the clinical clotting test.[15] Therefore, autoantibody-induced contact system activation underlines SLE pathogenesis.

IV. Different Patterns of Contact System Activation Distinquish SLE, RA, and Ps whereas No Contact System Activation was Observed in Normal and OA Patient Plasmas

To test if abnormal kallikrein- and thrombin-like activites and abnormal contact system protein profiles were restricted to SLE, we performed the same tests on 9 RA- (Fig. 6), 12 OA- (Fig. 7), and 12 Ps- (Fig. 8) patient plasmas along with pooled 30+ normal patient plasma as a negative control.

We have summarized the analysis results in Fig. 9.

	C1 inhibitor			Kininogen			Prekallikrein		Prothrombin		C3a	C5a
	Complex	110 kDa	94 kDa	120 kDa	70 kDa	64 kDa	Complex	88/85 kDa	72 kDa	50 kDa	10 kDa	10 kDa
Pooled 30+		x		i	i			x	x			
RA1	x	x	x	i	i	x		x	x		x	i
RA2	x	x	i	x	x			x	x		i	i
RA3	x	x	i	x	x	x		x	x		x	i
RA4	x	x	i	x	x	x		x	x		i	i
RA5	x	x	x	x	x	x		x	x		i	i
RA6	x	x	x	x	x	x		x	x		i	x
RA7	x	x	i	x	x	x		x	x			i
RA8	x	x	x	x	x	x		x	x			x
RA9	x	x	i	x	x	i		x	x		i	x

FIG. 6. Western analysis results of the contact system proteins, C3a, and C5a in rheumatoid arthritis (RA) and normal plasmas. Pooled normal human plasma and RA patient plasmas were analyzed by SDS-PAGE and western blotting with antibodies to C1inh, prekallikrein, kininogen, prothrombin, C3a, and C5a respectively. Plasma (0.53 μl) was used in each lane for Western analysis. Results from two independent Western analyses were summarized. Protein bands present in the normal plasma are depicted green. If a patient plasma contained less of the same protein band, it was coded with lighter green, if the band was absent, the color was white. On the other hand, if abnormal protein bands were present in the RA patient plasmas, they were represented by two shades of red, the darker indicating more prominent protein presence. (For interpretation of the references to color in this figure legend, the reader is referred to the Web version of this chapter.)

We found that SLE patient plasmas were associated with highest kallikrein- (Fig. 9A) and thrombin-like (Fig. 9B) activities compared to those of RA, OA, and Ps patient plasmas. Interestingly, three Ps patient plasmas (Ps5, Ps8, and Ps9), two OA patient plasmas (OS4 and OS10), and one RA patient plasma (RA1) were also associated with increased kallikrein- and thrombin-like activities.

Since SLE is a heterogeneous disease, we suspect the three Ps (Ps5, Ps8, and Ps9), 2OS (OS4 and OS10), and one RA (RA1) patients might be SLE-related. To test this idea, we performed Western analysis of contact system proteins plus C3a and C5a on all patient plasmas. Figure 9C shows the simplified data summary of the Western analysis.

We found that high molecular weight kininogen was undetectable in 23 out of 27 SLE patient plasmas (Figs. 4 and 5), which distinguished SLE patient plasmas from most of RA (Fig.6) OA (Fig.7) and Ps (Fig.8) patient plasmas tested because only the 3 out of 12 Ps (Ps5, Ps8, and Ps9), the 2 out of 12 OS (OS4 and OS10), and the 1 out of 9 RA (RA1) patient plasmas pinpointed by

	C1 inhibitor			Kininogen			Prekallikrein		Prothrombin		C3a	C5a
	Complex	110 kDa	94 kDa	110 kDa	70 kDa	64 kDa	Complex	88/85 kDa	72 kDa	50 kDa	10 kDa	10 kDa
Pooled 30+		x			x i			x	x			
OA1		x		x	i	x		x	x	x		
OA2		x		x	i	i		x	x	x		
OA3		x		x	i	i		x	x	x		
OA4			x			x	x	x	x	x		x
OA5		x		x	i			x	x	i		
OA6		x		x	i			x	x	x		
OA7		x		x	x			x	x	x		
OA8		x		x	x	x		x	x	x		
OA9		x		x	x	i		x	x	x		
OA10		x	x			x	x	x	x	x		x
OA11		x		x	x	x		x	x	x		
OA12		x		x	x	x		x	x	x		

FIG. 7. Western analysis results of the contact system proteins, C3a, and C5a in osteoarthritis (OA) and normal plasmas. Pooled normal human plasma and OA patient plasmas were analyzed by SDS-PAGE and western blotting with antibodies to C1inh, prekallikrein, kininogen, prothrombin, C3a, and C5a respectively. Plasma (0.53 μl) was used in each lane for Western analysis. Results from two independent Western analyses were summarized. Protein bands present in the normal plasma are depicted green. If a patient plasma contained less of the same protein band, it was coded with lighter green, if the band was absent, the color was white. On the other hand, if abnormal protein bands were present in the RA patient plasmas, they were represented by two shades of red, the darker indicating more prominent protein presence. (For interpretation of the references to color in this figure legend, the reader is referred to the Web version of this chapter.)

kallikrein- and thrombin-activity tests were associated with undetectable high molecular weight kininogen. These observations confirmed that these PS, OA, and RA patients had a SLE-related disease.

However, these Ps, OS, RA patient plasmas were different from those of SLE patient plasmas in that C5a were only detected in the Ps and RA patient plasmas.

In summary, the combined Western analysis of C1inhibitor, kininogen, prekallikrein, prothrombin, C3a, and C5a (Fig. 9C) plus kallikrein- (Fig. 9A) and thrombin-(Fig. 9B) activity measurement provided unique molecular characteristics for SLE, RA, OA, and Ps, respectively. Such molecular characteristics might be useful in monitoring SLE activity/severity and in pinpointing patients with SLE-associated RA, OA, and Ps.

	C1 inhibitor			Kininogen			Prekallikrein		Prothrombin		C3a	C5a
	Complex	110 kDa	94 kDa	110 kDa	70 kDa	64 kDa	Complex	88/85 kDa	70 kDa	50 kDa	10 kDa	10 kDa
Pooled 30+		x		x	x			x	x			
Ps1		x		x	x	i		x	x			x
Ps2		x		x	x	i		x	x		i	x
Ps3		x		x	x	i		x	x		i	x
Ps4		x		x	x	x		x	x			x
Ps5		x			x	x		x	x			x
Ps6		x		i	x	x		x	x		i	x
Ps7		x		x	x	x		x	x		x	x
Ps8		i	x			x	x	x	i		i	x
Ps9		i	x			x	x	x	x		xxx	x
Ps10		x		x	x	x		x	x		xx	x
Ps11		x		x	x	x		x	x		x	x
Ps12		x		x	x	x		x	x			x

FIG. 8. Western analysis results of the contact system proteins, C3a, and C5a in Ps and normal plasmas. Pooled normal human plasma and Ps patient plasmas were analyzed by SDS-PAGE and western blotting with antibodies to C1inh, prekallikrein, kininogen, prothrombin, C3a, and C5a respectively. Plasma (0.53 μl) was used in each lane for Western analysis. Results from two independent Western analyses were summarized. Protein bands present in the normal plasma are depicted green. If a patient plasma contained less of the same protein band, it was coded with lighter green, if the band was absent, the color was white. On the other hand, if abnormal protein bands were present in the RA patient plasmas, they were represented by two shades of red, the darker indicating more prominent protein presence. (For interpretation of the references to color in this figure legend, the reader is referred to the Web version of this chapter.)

An activated contact system generates kallikrein, thrombin, and factor XIIa. All of them are multi-substrate serine proteases. Kallikrein not only cleaves high molecular weight kininogen to produce the potent vasodilator bradykinin, but also directly or indirectly produces a series of potent biological regulators, such as thrombin,[25] TAFI,[26] anaphylatoxins C3a[27] and C5a,[10] and factor H.[28] Thrombin generation could activate the coagulation pathway. C3a and C5a generation could activate the complement pathway. Thrombin, C3a, and C5a production is correlated with both thrombosis and the multiorgan attack events observed in SLE patients. It has been established that activated contact systems can lead to inflammation,[29,30] vasodilatation, platelet activation, blood coagulation/fibrinolysis, complement activation, and other cell signaling events.[12] Hundreds of allergic reactions and at least 149 deaths[7] associated with activated contact system induced by contaminated heparin clearly showed that the contact system is a powerful signaling system that bridges hemostasis and immune response (Tables III–V).[31]

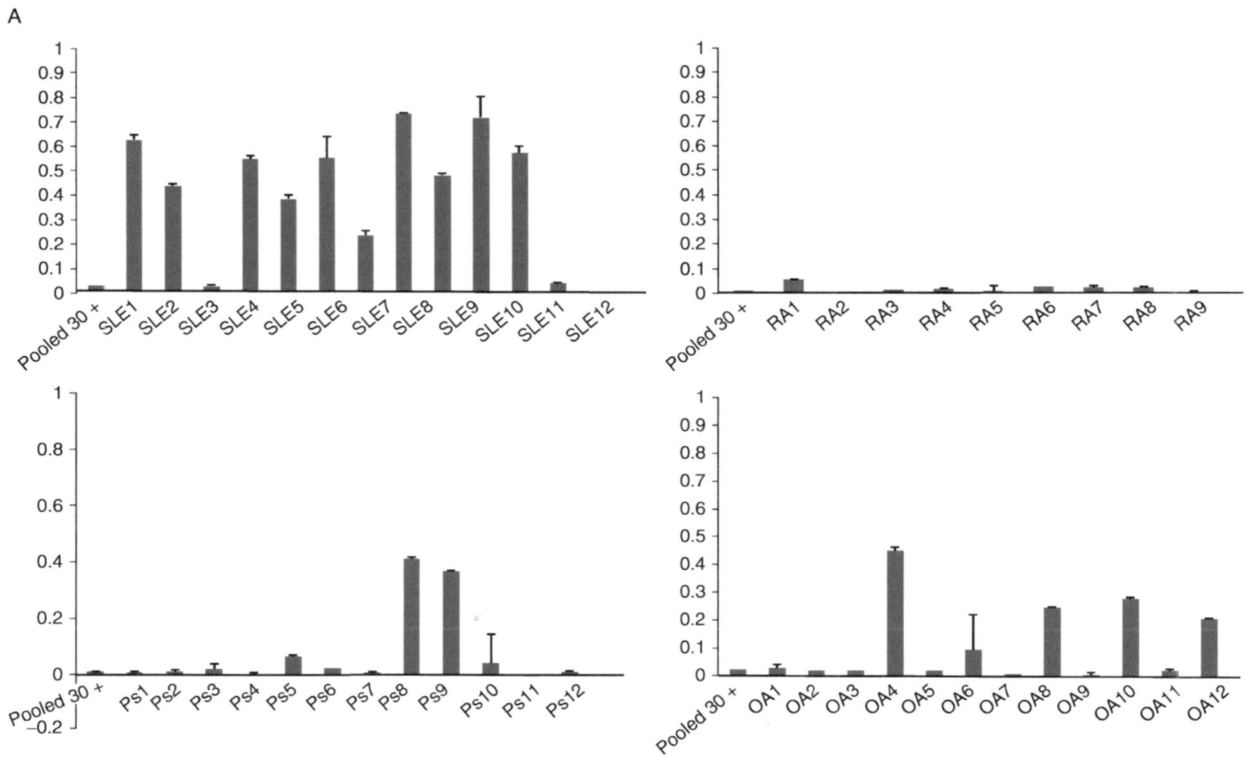

FIG. 9. (Continued)

B

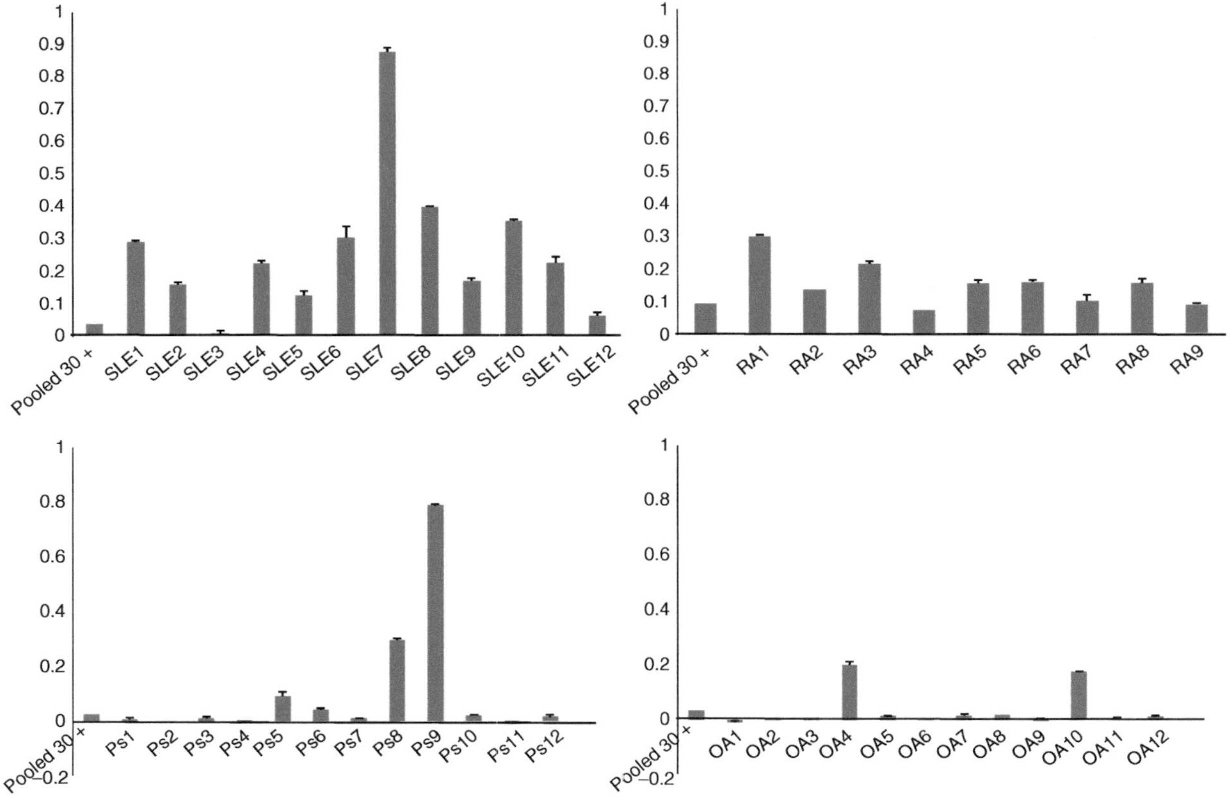
1
0.9
0.8
0.7
0.6
0.5
0.4
0.3
0.2
0.1
0
Pooled 30 +
SLE1
SLE2
SLE3
SLE4
SLE5
SLE6
SLE7
SLE8
SLE9
SLE10
SLE11
SLE12
RA1
RA2
RA3
RA4
RA5
RA6
RA7
RA8
RA9
Ps1
Ps2
Ps3
Ps4
Ps5
Ps6
Ps7
Ps8
Ps9
Ps10
Ps11
Ps12
OA1
OA2
OA3
OA4
OA5
OA6
OA7
OA8
OA9
OA10
OA11
OA12
−0.2

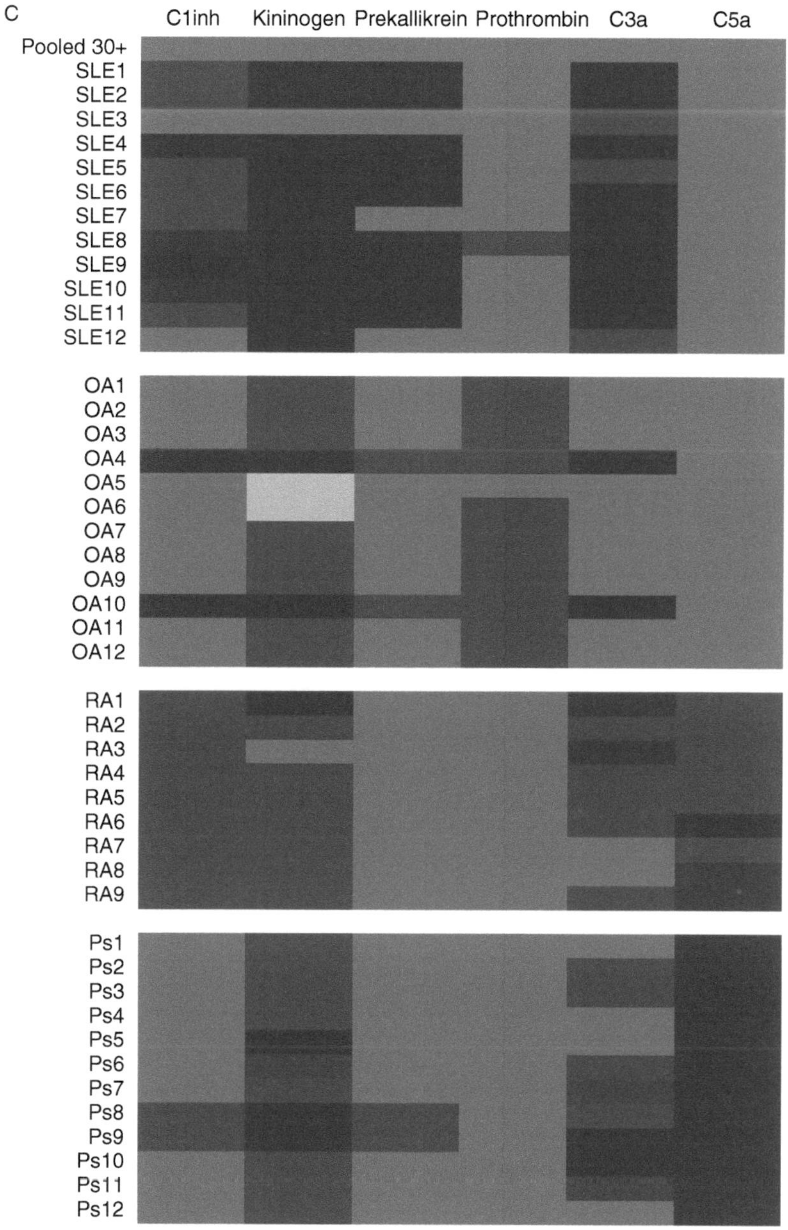

FIG. 9. Systematic compariaon of kallikrein-, thrombin-like activities and contact system proteins of plasmas from 12 SLE, 9 rheuamtoid arthritis, 12 psoresis, and 12 osteoarthritis patients. (A) Kalikrein-like activities. Plasma kallikrein activity was assessed by addition of 150 μl of the

TABLE III
RHEUMATOID ARTHRITIS PATIENT INFORMATION

Patient number	G	Age	Medication
RA1	F	46	Mtx, Ibu, F
RA2	F	61	En, Pred, VitD, Ca
RA3	F	83	Mtx, Pred. Cel, F
RA4	F	40	Mtx, Hum, Cel, F
RA5	F	54	Mtx, Actonel, F, Diovan, Vytorin
RA6	F	55	Flexeril, Clonidine, Diovan
RA7	M	57	Mtx, Pred, Enap, Norvasc, F
RA8	M	54	Mtx, Plq, Asa, Amaryl, Privinil, Toprol
RA9	M	56	MFM, MetPred, Amlo, Enap, Furo, Lescol

V. Mouse and Human SLE Autoantibodies Recognize GAGs on Cell Surfaces

Although DNA reactivity is a predominant feature of SLE autoantibodies, we suspect that GAGs are the physiological targets of the autoantibodies. To test this hypothesis, we made use of a murine model of systemic inflammation with many features in common with SLE. These mice are generated by crossing KRN TCR transgenic mice to G7m mice that express the G7m ligand in the thymus.[32] KRN/G7m double transgenic mice develop systemic inflammation and autoantibodies.[33] These mice uniformly produce anti-DNA antibodies. Half of these mice

kallikrein chromogenic substrate, 0.5 mM *N*-benzoyl-Pro-Phe-Arg-*p*-nitroanilide hydrochloride dissolved in TBS. (Detailed osteoarthritis, rheumatoid arthritis, Ps patient information is presented in supplemental materials.) Bars represent arithmetic mean of the change of the $OD_{405\ nm}$ at 60 min from starting $OD_{405\ nm}$; each value was established from three independent measurements. The error indicated is the standard deviation. (B) Thrombin-like activites. Plasma thrombin-like activity was assessed by addition of 150 μl of the thrombin chromogenic substrate, 0.5 mM Sar-Pro-Arg-*p*-nitroanilide dihydrochloride dissolved in TBS. (Detailed osteoarthritis, rheumatoid arthritis, Ps patient information is presented in supplemental materials.) Bars represent arithmetic mean of the change of the $OD_{405\ nm}$ at 60 min from starting $OD_{405\ nm}$; value established from three independent measurements with standard deviation. (C) Western analysis of the contact system proteins plus C3a and C5a. Pooled normal human plasma and various patient plasmas were analyzed by SDS-PAGE and western blotting with antibodies to C1inh, prekallikrein, HMWK, thrombin, C3a, and C5a, respectively. The results from two independent Western analyses of patient plasmas were summarized. Proteins present in the normal plasma are depicted green. When abnormal protein bands were observed in the patient plasmas, they were represented by different shades of red, the darker indicating more prominent protein presence. If additional protein band was detected, the protein was depicted as yellow. (For interpretation of the references to color in this figure legend, the reader is referred to the Web version of this chapter.)

TABLE IV
Osteoarthritis Patient Information

Patient number	G	Age	Medication
OS1	M	51	–
OS2	M	45	–
OS3	M	65	–
OS4	F	35	–
OS5	M	45	–
OS6	F	58	–
OS7	M	56	–
OS8	F	44	–
OS9	M	59	–
OS10	F	52	–
OS11	F	50	–
OS12	F	67	–

TABLE V
Psoriasis Patient Information

Patient number	G	Age	Medication
Ps1	M	63	–
Ps2	F	43	–
Ps3	M	42	–
Ps4	Ff	76	–
Ps5	M	43	–
Ps6	F	41	–
Ps7	M	52	Enbrel, Lotrel, Lipitor
Ps8	F	65	Mtx, Plq, Cortisol, Triam
Ps9	M	52	Enbrel, Lotrel, Lipitor
Ps10	F	63	Naproxen, Diovan
Ps11	F	65	Naproxen, Synthroid
Ps12	M	52	Enbrel, Lotrel, Lipitor

also develop anti-RBC antibodies and rheumatoid factor. Hematologic abnormalities include hemolytic anemia, thrombocytopenia, and lymphopenia. Histologically, the mice exhibit vasculitis in the lungs, liver, kidneys, and carditis/pericarditis. Many succumb to colitis. A few of these mice develop arthritis. It is important to note that the original KRN (aka K/BxN) mice[34] develop an arthritis resembling human RA. This is mediated by autoantibodies to glucose-6-phosphate isomerase (GPI, the physiologic antigen for KRN). We have previously shown that the autoantigen, glucose-6-phosphate isomerase, binds to mouse paw/ankle joint chondroitin sulfate with high affinity,[35] which again suggests that the negatively charged chondroitin sulfate/autoantigen/autoantibody complex might be responsible for the pathogenesis of the autoimmune disease through activation of the contact system. The arthritis seen in the KRN/G7m mice differs from that seen in K/BxN mice in that it is less penetrated, less severe, and does not involve anti-GPI antibodies. The immune effectors of the KRN/G7m arthritis have not been identified.

To determine whether antibodies from KRN/G7m mice can bind to GAGs, we used a panel of CHO mutants that express a limited repertoire of GAG structures due to specific enzyme mutations. By flow cytometric analysis, we found distinct patterns of GAG recognition among sera from KRN/G7m mice (Fig. 10A) demonstrating that these antibodies can distinguish specific GAGs structures. For example, sera from mouse 880 recognized GAGs from all CHO mutants except those from AT15, which expresses short heparan sulfate chains. Sera from mouse 1070 recognized GAGs from all CHO lines except wild-type CHOK1.

We found that sera from SLE patients with high titer anti-DNA exhibited distinct patterns in GAG binding (Fig. 10B). Of note, patients JN, TH, and NH presented with significant arthritis and nephritis correlating with GAG reactivity. TW presented with vasculitis. In summary, we have evidence of anti-GAG reactivity in a SLE-like mouse model and in SLE patients. Antibodies exhibit exquisite fine specificity. The diverse GAG structures present in different tissues provide plausible targets for autoantibodies in SLE.

VI. Increased Glucosamine and Galactosamine Levels in SLE Patient Plasmas

Heparin induces IgG autoantibody production in HIT patients.[36,37] We previously discovered that autoantibodies against heparin/protein complexes activate the contact system.[16] We suspect that abnormal GAGs in SLE patient plasmas are a cause of SLE. To test this hypothesis, we quantified glucosamine and galactosamine content in control and SLE patient plasmas. Glucosamine is present in heparin and heparin sulfate whereas galactosamine is present in chondroitin and dermatan sulfate. Both glucosamine and galactosamine are present in N-linked and O-linked glycans. It has been previously established that purified carcinoma mucins from patient sera induce contact system activation in human plasma.[38] We reasoned that if GAGs and/or other sulfated glycans in

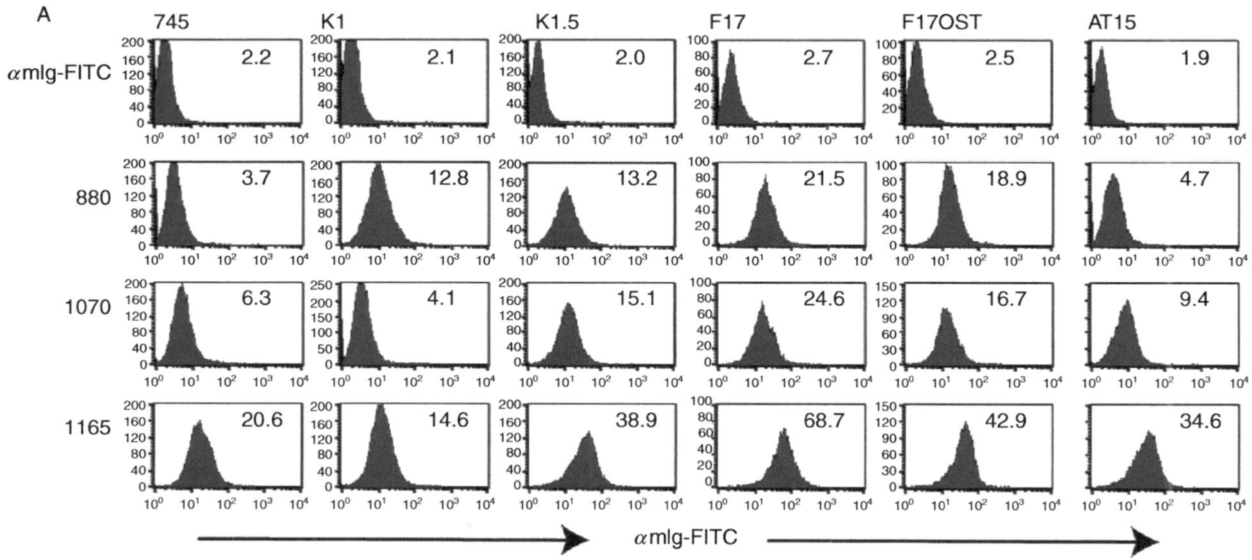

FIG. 10. (Continued)

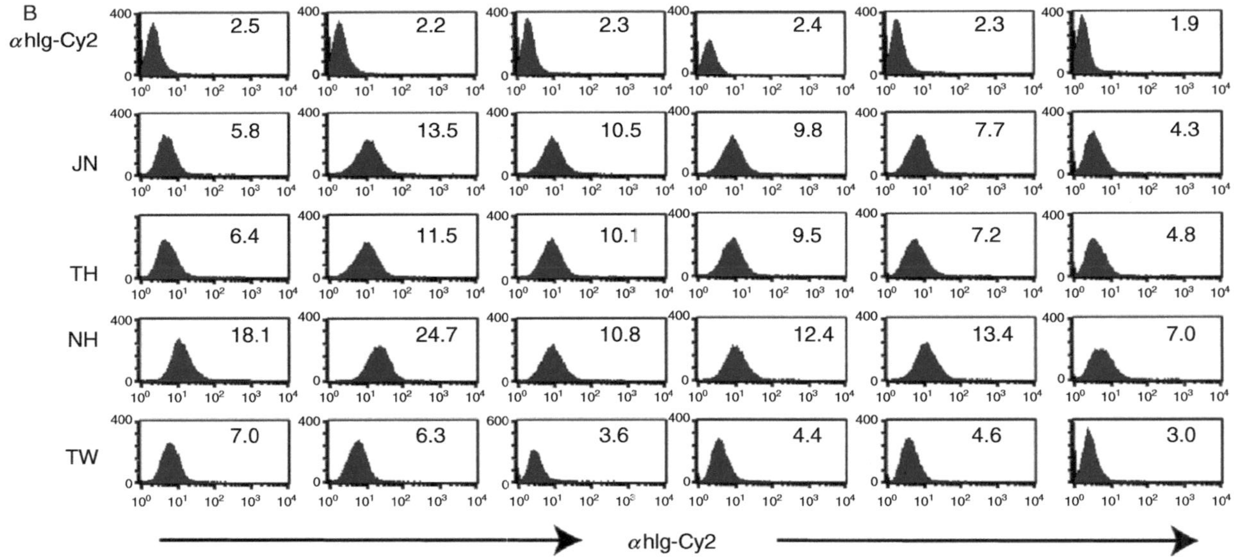

FIG. 10. Sera from KRN/G7m mice and SLE patients recognize cell surface GAGs. (A) Sera from three *KRN/G7m* mice show distinct GAG selectivity. (B) Sera from SLE patients show distinct GAG binding. Sera Ig binding to a panel of CHO cells expressing specific GAG was assessed at 1:50 dilution and detected with anti-mouse Ig-FITC (A) or anti-human Ig-Cy2 (B). CHO745 is a GAG deficient CHO mutant. CHOK1 is wild-type CHO. CHOK1.5 is 3OSTI expressing CHOK1. CHOF17 is defective in 2-*O*-sulfation of heparan sulfate. CHOF173OST is a 3OSTI expressing CHOF17. CHOAT15 has a point mutation in EXT1 (heparan sulfate polymerase) and expresses short heparan sulfate chains. Numbers in the histogram indicate mean fluorescence intensity.

mucins are contact system activators, the glucosamine and galactosamine level in SLE patient plasmas might be elevated. We quantified glucosamine and galactosamine in 1 μl of plasmas from four normal controls, seven SLE patients, and three HIT patients based on our published method.[22] The data are shown in Fig. 11. We found that plasma galactosamine level (Fig. 11, top panel) was increased in SLE patient plasmas (0.55 ± 0.07 mM), but was largely unchanged in HIT patients (0.48 ± 0.02 mM) compared to the four normal controls (0.50 ± 0.11 mM). In contrast, glucosamine levels (Fig. 11, bottom panel) were increased in both SLE patient (0.75 ± 0.07 mM) and HIT patient plasmas (0.65 ± 0.02 mM) compared to normal controls (0.56 ± 0.06 mM). Our data indicate that SLE patient plasmas exhibit increased glucosamine and galactosamine, consistent with increased amounts of the corresponding glycans. Abnormal GAGs or glycans in SLE patient plasmas are therefore candidates for precipitating factors of SLE pathogenesis.

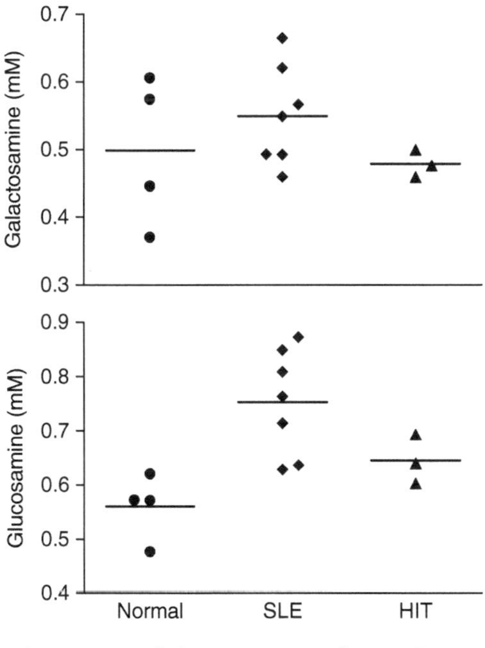

FIG. 11. Plasma galactosamine and glucosamine quantification. One micro liter plasmas from normal controls, HIT patients, or SLE patients were hydrolyzed and derivatized. The fluorescent derivatives of glucosamine and galactosamine are separated by the reversed-phase HPLC and then quantified.[22]

Our data (Figs. 11) suggest that abnormal glycans in SLE patient plasmas could act like heparin in initiating autoantibody production and facilitating autoimmune attack through autoantibody/glycans/protein complexes. Both GAGs and glycans have a tightly controlled circulation concentration due to the presence of multiple hepatic clearance receptors to eliminate them.[39–41] As a result, intravenously injected anticoagulant heparin has a half life of only 1 h in blood circulation. It is unknown whether the high level of blood glycans observed in our studies (Fig. 11) was a result of abnormal glycans biosynthesis or a result of defective GAG/glycan clearance. The relative contribution of glycans and GAGs in activating the contact system is also unknown. The answers to these questions could lead to novel SLE prevention and treatment strategies.

VII. Concluding Remarks

SLE, antiphospholipid syndrome, HIT, and RA[35] are caused by autoantibodies against DNA, phospholipid, phospholipid/protein complex, heparin/protein complex, and chondroitin sulfate/protein/IgG complex,[25] respectively. These diseases share similar characteristics. (1) All the autoantibodies are against negatively charged molecules or negatively charged molecule/protein complexes. (2) The autoantibodies are relatively common in the general population, but only a subpopulation of patients develop disease. (3) The autoimmune disorders are characterized by a similar pathogenesis that includes platelet activation, blood coagulation, complement activation, vasodilatation, and inflammation. (4) Except RA, each disorder is associated with thrombocytopenia, a high risk of venous and arterial thrombosis. (5) Catastrophic thrombotic outcomes, such as multiorgan failure, tissue destruction, amputation, and death, are trigged by an unknown mechanism. Indeed, infection and vaccinations are associated with increased levels of autoantibodies but most infections and virtually all vaccinations in human lack links to autoimmune diseases.[42] This indicates that only certain autoantibodies have potential for the etiology of autoimmune diseases. The data presented suggest that a common trigger of autoimmune disease is contact system activation induced by negatively charged immune complexes.

The kallikrein (Figs. 1 and 3) and thrombin activity assay (Fig. 2), the contact protein western analysis (Figs. 4–8), specific cell surface GAG binding (Fig. 10), and plasma glycans quantification (Fig. 11) assays used in our studies should be useful as SLE diagnostic tools. Abnormal plasma GAGs/glycans, and the consequential abnormal contact system activation in SLE patients represent potential targets for autoimmune disease prevention and treatments.

References

1. Kavanaugh A, Tomar R, Reveille J, Solomon DH, Homburger HA. Guidelines for clinical use of the antinuclear antibody test and tests for specific autoantibodies to nuclear antigens. American College of Pathologists. *Arch Pathol Lab Med* 2000;**124**:71–81.
2. Lyons R, Narain S, Nichols C, Satoh M, Reeves WH. Effective use of autoantibody tests in the diagnosis of systemic autoimmune disease. *Ann NY Acad Sci* 2005;**1050**:217–28.
3. Marshall E. Lupus: mysterious disease holds its secrets tight. *Science* 2002;**296**:689–91.
4. Salmon JE, Girardi G, Theodore E. Woodward Award: antiphospholipid syndrome revisited: a disorder initiated by inflammation. *Trans Am Clin Climatol Assoc* 2007;**118**:99–114.
5. Levine JS, Branch DW, Rauch J. The antiphospholipid syndrome. *N Engl J Med* 2002;**346**:752–63.
6. Martin-Toutain I, Piette JC, Diemert MC, Faucher C, Jobic L, Ankri A. High prevalence of antibodies to platelet factor 4 heparin in patients with antiphospholipid antibodies in absence of heparin-induced thrombocytopenia. *Lupus* 2007;**16**:79–83.
7. *FDA. Information on adverse event reports and heparin.* http://www.fda.gov/Drugs/DrugSafety/PostmarketDrugSafetyInformationforPatientsandProviders/ucm112669.htm; 2009.
8. Kishimoto TK, Viswanathan K, Ganguly T, Elankumaran S, Smith S, Pelzer K, et al. Contaminated heparin associated with adverse clinical events and activation of the contact system. *N Engl J Med* 2008;**358**:2457–67.
9. Clark A, Weymann A, Hartman E, Turmelle Y, Carroll M, Thurman JM, et al. Evidence for non-traditional activation of complement factor C3 during murine liver regeneration. *Mol Immunol* 2008;**45**:3125–32.
10. Huber-Lang M, Sarma JV, Zetoune FS, Rittirsch D, Neff TA, McGuire SR, et al. Generation of C5a in the absence of C3: a new complement activation pathway. *Nat Med* 2006;**12**:682–7.
11. Waaler BA. Contact activation in the intrinsic blood clotting system; studies on a plasma product formed on contact with glass and similar surfaces. *Scand J Clin Lab Invest* 1959;**11**:1–133.
12. Colman RW. Contact activation (kallikrein–kinin) pathway: multiple physiologic and pathophysiologic activities. In: Colman RW, editor. *Hemostasis and Thrombosis*. Philadelphia: Lippincott Williams & Wilkins; 2006.
13. Pan J, Qian Y, Zhou X, Pazandak A, Frazier SB, Weiser P, et al. Oversulfated chondroitin sulfate is not the sole contaminant in heparin. *Nat Biotech* 2010;**28**:203–7.
14. Pan J, Qian Y, Zhou X, Pazandak A, Frazier SB, Weiser P, et al. Identification of chemically sulfated/desulfated glycosaminoglycans in contaminated heparins and development of a simple assay for the detection of most contaminants in heparin. *Glycobiol Insight* 2010;**2**:1–12.
15. Qian Y, Pan J, Zhou X, Weiser P, Hourcade D, Lizsewski KM, et al. Oversulfated heparin byproducts induce thrombin generation through contact system activation. *Clin Appl Thromb Hemost* 2010;**16**:244–250.
16 Qian Y, Pan J, Zhou X, Weiser P, Lu H, Shih FF, et al. Heparin and Oversulfated heparin byproduct induce thrombin generation through contact system activation in plasma of patients with HIT. *Clin Appl Thromb Hemost* 2010;**16**:251–260.
17. Pan J, Qi Y, Zhou X, Lu H, Ramacciotti E, Zhang L. Chemically oversulfated glycosaminoglycans are potent modulators of contact system activation and different cell signaling pathways. *J Biol Chem* 2010;doi:10.1074/jbc.M109.063735.
18. Sasisekharan R, Raman R, Prabhakar V. Glycomics approach to structure–function relationships of glycosaminoglycans. *Annu Rev Biomed Eng* 2006;**8**:181–231.
19. Berden JH, Termaat RM, Brinkman K, Smeenk RJ, Swaak AJ, Faaber P, et al. Binding of anti-DNA antibodies to glomerular heparan sulfate: a new clue for the pathogenesis of SLE nephritis? *Nephrologie* 1989;**10**:127–32.

20. Ravirajan CT, Rowse L, MacGowan JR, Isenberg DA. An analysis of clinical disease activity and nephritis-associated serum autoantibody profiles in patients with systemic lupus erythematosus: a cross-sectional study. *Rheumatology (Oxford)* 2001;**40**:1405–12.
21. Wang JY, Roehrl MH. Glycosaminoglycans are a potential cause of rheumatoid arthritis. *Proc Natl Acad Sci USA* 2002;**99**:14362–7.
22. Studelska DR, Giljum K, McDowell LM, Zhang L. Quantification of glycosaminoglycans by reversed-phase HPLC separation of fluorescent isoindole derivatives. *Glycobiology* 2006;**16**:65–72.
23. Zuraw BL, Curd JG. Demonstration of modified inactive first component of complement (C1) inhibitor in the plasmas of C1 inhibitor-deficient patients. *J Clin Invest* 1986;**78**:567–75.
24. Jenny NS, Lundblad RL, Mann KG. Thrombin. In: Colman RW, editor. *Hemostasis and thrombosis*. Philadelphia: Lippincott Williams & Wilkins; 2006. p. 193–213.
25. Stief TW. Kallikrein activates prothrombin. *Clin Appl Thromb Hemost* 2008;**14**:97–8.
26. Myles T, Nishimura T, Yun TH, Nagashima M, Morser J, Patterson AJ, et al. Thrombin activatable fibrinolysis inhibitor, a potential regulator of vascular inflammation. *J Biol Chem* 2003;**278**:51059–67.
27. DiScipio RG. The activation of the alternative pathway C3 convertase by human plasma kallikrein. *Immunology* 1982;**45**:587–95.
28. Saito A. Plasma kallikrein is activated on dermatan sulfate and cleaves factor H. *Biochem Biophys Res Commun* 2008;**370**:646–50.
29. Sainz IM, Pixley RA, Colman RW. Fifty years of research on the plasma kallikrein–kinin system: from protein structure and function to cell biology and in vivo pathophysiology. *Thromb Haemost* 2007;**98**:77–83.
30. Coughlin SR. Protease-activated receptors in hemostasis, thrombosis and vascular biology. *J Thromb Haemost* 2005;**3**:1800–14.
31. Blossom DB, Kallen AJ, Patel PR, Elward A, Robinson L, Gao G, et al. Outbreak of adverse reactions associated with contaminated heparin. *N Engl J Med* 2008;**359**:2674–84.
32. Shih FF, Allen PM. T cells are not as degenerate as you think, once you get to know them. *Mol Immunol* 2004;**40**:1041–6.
33. Shih FF, Mandik-Nayak L, Wipke BT, Allen PM. Massive thymic deletion results in systemic autoimmunity through elimination of $CD4^+$ $CD25^+$ T regulatory cells. *J Exp Med* 2004;**199**:323–35.
34. Korganow A-S, Ji H, Mangialaio S, Duchatelle V, Pelanda R, Martin T, et al. From systemic T cell self-reactivity to organ-specific autoimmune disease via immunoglobulins. *Immunity* 1999;**10**:451–61.
35. Studelska DR, Mandik-Nayak L, Zhou X, Pan J, Weiser P, McDowell LM, et al. High affinity glycosaminoglycan and autoantigen interaction explains joint specificity in a mouse model of rheumatoid arthritis. *J Biol Chem* 2009;**284**:2354–62.
36. Greinacher A, Kohlmann T, Strobel U, Sheppard JA, Warkentin TE. The temporal profile of the anti-PF4/heparin immune response. *Blood* 2008;**113**:4970–6.
37. Warkentin TE, Sheppard JA, Moore JC, Cook RJ, Kelton JG. Studies of the immune response in heparin-induced thrombocytopenia. *Blood* 2009;**113**:4963–9.
38. Pineo GF, Brain MC, Gallus AS, Hirsh J, Hatton MW, Regoeczi E. Tumors, mucus production, and hypercoagulability. *Ann NY Acad Sci* 1974;**230**:262–70.
39. Wahrenbrock MG, Varki A. Multiple hepatic receptors cooperate to eliminate secretory mucins aberrantly entering the bloodstream: are circulating cancer mucins the "tip of the iceberg"? *Cancer Res* 2006;**66**:2433–41.
40. Harris EN, Weigel JA, Weigel PH. The human hyaluronan receptor for endocytosis (HARE/Stabilin-2) is a systemic clearance receptor for heparin. *J Biol Chem* 2008;**283**:17341–50.
41. Harris EN, Kyosseva SV, Weigel JA, Weigel PH. Expression, processing, and glycosaminoglycan binding activity of the recombinant HUMAN 315-kDa hyaluronic acid receptor for endocytosis (HARE). *J Biol Chem* 2007;**282**:2785–97.
42. Mackay IR, Leskovsek NV, Rose NR. Cell damage and autoimmunity: a critical appraisal. *J Autoimmun* 2008;**30**:5–11.

Glycosaminoglycans and Activated Contact System in Cancer Patient Plasmas

Jing Pan, Yi Qian,
Peter Weiser, Xiaodong Zhou,
Hong Lu, Daniel R.
Studelska, and Lijuan Zhang

*Department of Pathology and Immunology,
Washington University Medical School,
St. Louis, Missouri, USA*

Oncogenic mutations create cancer cells. Cancer cells require thrombin for growth, angiogenesis, and metastasis. All cancer patients display a hypercoagulable state, which includes platelet activation, blood coagulation, complement activation, vasodilatation, and inflammation. This often results in thrombosis, the second leading cause of death in cancer patients. It is established that chemically oversulfated glycosaminoglycans (GAGs) induce thrombin generation through contact system activation in human plasma. Thrombin is responsible for thrombosis. In this chapter, we show that plasmas from lung cancer patients contain activated contact systems apparent by the absence of high molecular weight kininogen and processed C1inh, by abnormal kallikrein and

Progress in Molecular Biology
and Translational Science, Vol. 93
DOI: 10.1016/S1877-1173(10)93020-2

1877-1173/10 $35.00

thrombin activities, and by increased glucosamine, galactosamine, and GAG levels. Activated contact systems were also evident in plasmas from breast, colon, and pancreatic cancer patients. These data suggest that GAGs or other molecules produced by tumors induce abnormal thrombin generation through contact system activation. Therefore, the contact system and glycans represent new targets for cancer diagnosis, prevention, and treatment.

Abbreviations: OS-HB, chemically oversulfated heparin by-product; C1inh, C1 inhibitor; HMWK, high molecular weight kininogen; PL, phospholipids; FII, prothrombin; FV, factor V; FVa, activated factor V; FVIII, factor VIII; FIX, factor IX; FX, factor X; FXI, factor XI; FXII, factor XII; TAFI, thrombin activated fibrinolysis inhibitor; ephalin, phosphatidylethanolamine; TBS, Tris buffered saline

I. Introduction

Lung cancer is the leading cause of cancer related deaths worldwide.[1] Most lung cancers have disseminated at the time of presentation and affected individuals have a mortality rate of nearly 90%. More than 1.4 million people die from lung cancer annually[1] with no early diagnostic, staging, or relapsing biomarkers available.

All cancer patients, including lung cancer patients, display a hypercoagulable state,[2] which includes platelet activation, blood coagulation, complement activation, vasodilatation, and inflammation.[3] This often results in thrombosis, the second leading cause of death in cancer patients.[2] The connection of the hypercoagulable state to cancer was first made by Armand Trousseau in 1865.[4] He noted that unexpected thrombosis events could forewarn an occult tumor. In fact, within 2 years of his discovery, Dr. Trousseau recognized that he had developed the symptoms, "Trousseau's sign," himself, and died of a gastric cancer after his self-diagnosis.

Abnormal blood clot formation can result in thrombosis, the formation of a clot large enough to obstruct blood flow, or in disseminated intravascular coagulation, the formation of small blood clots in blood vessels throughout the body. Abnormal bleeding occurs because the small clots consume circulating coagulation proteins and platelets.[5] Sack and colleagues[6] extended Trousseau's syndrome to include this disseminated coagulopathy.

Thrombin is the only known enzyme that cleaves fibrinogen to produce fibrin, which causes clot formation. Thrombin also plays multiple roles unrelated to coagulation.[7] The loss of prothrombin biosynthesis leads to the death of healthy adult mice in a few days. The mice die from severe hemorrhagic events within multiple tissues, particularly in the heart and brain. This sequel is not due to the constitutive loss of clotting or to the loss of platelet function.[8] These observations suggest that thrombin might be a key survival or vitality factor for animals. A vitality factor should be highly regulated, and thrombin is constantly generated from prothrombin *in vivo*. However, uncontrolled thrombin generation may lead to clot formation or create a hypercoagulable state to produce uncontrolled bleeding.

A hypercoagulable state is required for tumor angiogenesis and for tumor dispersion, or metastasis.[9] Angiogenesis is essential for expanding tumor mass[10]: Thrombin generates a fibrin scaffold, which attracts endothelial cells and activates various protease-activated receptors on endothelial cells to increase expression of VEGF receptors and to activate hypoxia inducible factor 1α, which leads to production of several angiogenic molecules.[9] Metastasis is thrombin dependent: Thrombin activates platelets[11,12] and initiates thrombosis,[7] each required for platelet–tumor aggregation and essential for cancer metastasis.[4] In addition, thrombin induces cancer cell adhesion to the subendothelial matrix, which promotes tumor growth.[9] Thrombin produces anaphylactoxins C5a and osteopontin,[13,14] which not only induce inflammation but also help cancer cells to spread to other organs or tissues. Evidence has also been presented that thrombin serves to preserve and/or activate dormant tumors.[9] Therefore, tumors are dependent on thrombin for angiogenesis, metastasis, growth, and dormancy survival.

Because of its importance, thrombin generation from prothrombin *in vivo* is well regulated. Both tissue factor and the activated contact system protein kallikrein produce small amounts of thrombin.[15–17] Thrombin is required for activating platelets, factor VIII (FVIII), and factor V (FV) for thrombin propagation and other physiological processes. Only thrombin propagation leads to clot formation. Thrombin can be generated and propagated by two established pathways, the tissue factor-dependent extrinsic pathway and the contact system activation-dependent intrinsic pathway.[18]

The tissue factor-dependent extrinsic pathway has been accepted as the only *in vivo* thrombin generation pathway and therefore tissue factor has been taken to be responsible for the hypercoagulable state present in cancer patients.[19,20] However, tissue factor is a transmembrane protein. Its level either in circulation or on blood encountered cell surfaces is not only low but also does not correlate with hypercoagulable state in cancer patients.[19] Even though cancer cells have upregulated tissue factor expression and tissue factor has multiple signaling capacities in cancer cells,[21] tissue factor has no predictive value as a cancer diagnostic biomarker.

The contact system consists of the serine proteases, prekallikrein and factor XII; a high molecular weight kininogen cofactor; and C1inh, a regulatory C1 inhibitor.[22] It was first discovered as a coagulation system, which rapidly induces clotting when blood or plasma interacts with or "contacts" artificial surfaces.[23] Both the *in vivo* thrombin generation and coagulation roles of an activated contact system have been dismissed because the contact system protein-deficiency in humans is not associated with a bleeding phenotype[22] and artificial surfaces are not present in blood circulation. However, the contact system participates in the initiation of inflammation, complement activation, fibrinolysis, angiogenesis, and kinin formation.[22] If it generates thrombin, an activated contact system has the capacity to create a hypercoagulable state.

Contaminated heparin produces anaphylactic reactions and death by activating the contact system.[24] We found that heparin contaminants, such as chemically oversulfated heparin by-product (OS-HB),[25,26] induced thrombin generation through contact system activation in human plasmas.[16] We then discovered that the negatively charged IgG/PF4/heparin complex activates the contact system and produces thrombin in human plasma, and thrombin partially activates the platelets allowing subsequent platelet activation.[12] We further showed that a variety of endogenous negatively charged molecules, including glycosaminoglycans (GAGs), could induce the contact system activation in normal human plasma.[27] These findings suggest that the contact system is an *in vivo* thrombin generation system and that abnormal thrombin generation through contact system activation might be responsible for the hypercoagulable state observed in cancer patients.

In this chapter, we have employed several techniques to study normal and cancer plasma contact system proteins. We have discovered that lung cancer patient plasmas contain activated contact systems, apparent from the absence of high molecular weight kininogen and processed C1inh, from abnormal kallikrein and thrombin activities, and from increased glucosamine, galactosamine, and GAG levels. Activated contact systems were also evident in plasmas from breast, colon, and pancreatic cancer patients. Our data suggest that GAGs or other molecules produced by tumors induce abnormal thrombin generation through contact system activation.

II. Methodology

A. Materials

Heparin (H4784), kallikrein substrate, *N*-benzoyl-Pro-Phe-Arg-*p*-nitroanilide hydrochloride, and the thrombin substrate, Sar-Pro-Arg-*p*-nitroanilide dihydrochloride, were purchased from Sigma. Heparin by-product (HB) was provided by the FDA. It contained ~50% heparin/heparan sulfate and ~50% chondroitin/dermatan sulfate based on glucosamine/galactosamine quantification, heparin lyase digestion, chondroitinase digestion, CE, NMR, and LC/MS analysis.[25,26] Oversulfated heparin by-product (OS-HB) was chemically sulfated and analyzed by both CE and NMR in our lab as described.[27] All antibodies in this study are from Affinity Biologicals, Inc. including sheep anti-human thrombin, sheep anti-human prothrombin, sheep anti-human prekallikrein, sheep anti-human Kininogen, and sheep anti-human C1 inhibitor. Pooled normal human plasma from 30+ normal donors was purchased from George King Biomedical, Inc. All cancer patient plasmas were purchased from Bio-reclamation, Inc. The plasma was aliquoted and stored at − 70 °C before use.

B. Western Analysis

Plasmas were activated *in vitro* by adding 20 μl of OS-HB into 30 μl of plasma at 37 °C for 5 min with a final OS-HB concentration of 200 μg/ml. Tris buffered saline (TBS) was used as a nonactivation control. Plasmas were then 1:4 diluted with TBS, and an equal volume of 2× Laemmli sample buffer was added. After boiling at 100 °C for 5 min, 7 μl of each sample was separated by 4–15% SDS-PAGE and transferred onto a nitrocellulose membrane. After blotting with the primary and secondary antibody, protein was visualized using the SuperSignal West Pico Chemiluminescent Substrate from Thermo-Scientific-Pierce.

C. Kallikrein and Thrombin Activity Assays

Plasmas were activated *in vitro* by adding 20 μl of heparin or OS-HB into 30 μl of plasma at 37 °C for 5 min with a final concentration of 0, 2, and 20 μg/ml, respectively, in a 96-well plate. TBS buffer (50 mM Tris, pH 7.6, 50 mM NaCl buffer) was then added to each well to a total volume of 150 μl. Absorbance at 405 nm was recorded as blank by Spectramax M2 plate reader (Molecular Devices). Substrate (50 μl in TBS) was added to each well and absorbance at 405 nm was recorded every 90 s for 40 min. For kallikrein and thrombin assays, 0.5 mM *N*-benzoyl-Pro-Phe-Arg-*p*-nitroanilide hydrochloride and 1 mM Sar-Pro-Arg-*p*-nitroanilide dihydrochloride were used as substrates, respectively. Both substrates were dissolved in TBS buffer.

D. Glucosamine and Galactosamine Quantification

One microliter of plasma or purified GAG was used in duplicate for glucosamine and galactosamine quantification. Samples were subjected to acid hydrolysis, sodium borohydride reduction, precolumn derivatization with *o*-phthaldialdehyde and 3-mercaptopropionic acid, and reversed-phase HPLC separation with fluorescence detection of the isoindole derivatives as described in detail elsewhere.[28]

III. An Activated Contact System is Common in Plasmas from Cancer Patients

Contact system activation results in proteolytic cleavage of prekallikrein, prothrombin, high molecular weight kininogen, and C1inh as well as deactivation of kallikrein via covalent attachment to C1inh. These events can be systematically monitored in cancer patient plasma and compared to the contact system proteins of normal human plasmas by Western analysis. Ten lung cancer patient plasmas (Table I) were compared to pooled normal human plasma. We performed western analysis on: (A) C1inh, (B) kininogens, (C) prekallikrein, and (D) thrombin, to make this comparison. The results are presented in order below.

A. C1inh

C1inh is the most prominent inhibitor of blood coagulation serine proteases. It is a 110 kDa protein present in blood at levels of around 250 μg/ml. It forms a 1:1 inhibitory complex with kallikrein, factor XIa, factor XIIa, and complement serine proteases. In contrast to normal plasma, where the 110 kDa band of C1inh was intact, we found a 94 kDa cleaved C1inh band and a C1inh/protease complex band (indicated by an asterisk) in all lung cancer patient plasmas (Fig. 1A, lanes 2–11). Cleaved C1inh (94 kDa) and C1inh/protease complex bands have been reported in patient plasmas with clinical C1inh dysfunction.[29]

TABLE I
LUNG CANCER PATIENT INFORMATION

Patient no.	Age	Gender	Treatment	Cancer stage
48	80	M	Taxol, Aloxi, Carboplatin	3
49	55	F	Taxol, Aloxi, Carboplatin	2
50	75	F	None	2
51	58	M	Gemzar, Carboplatin	3
52	88	F	None	2
53	81	F	Carboplatin	2
54	27	F	None	2
55	70	F	None	2
56	84	F	None	2
57	72	F	None	2

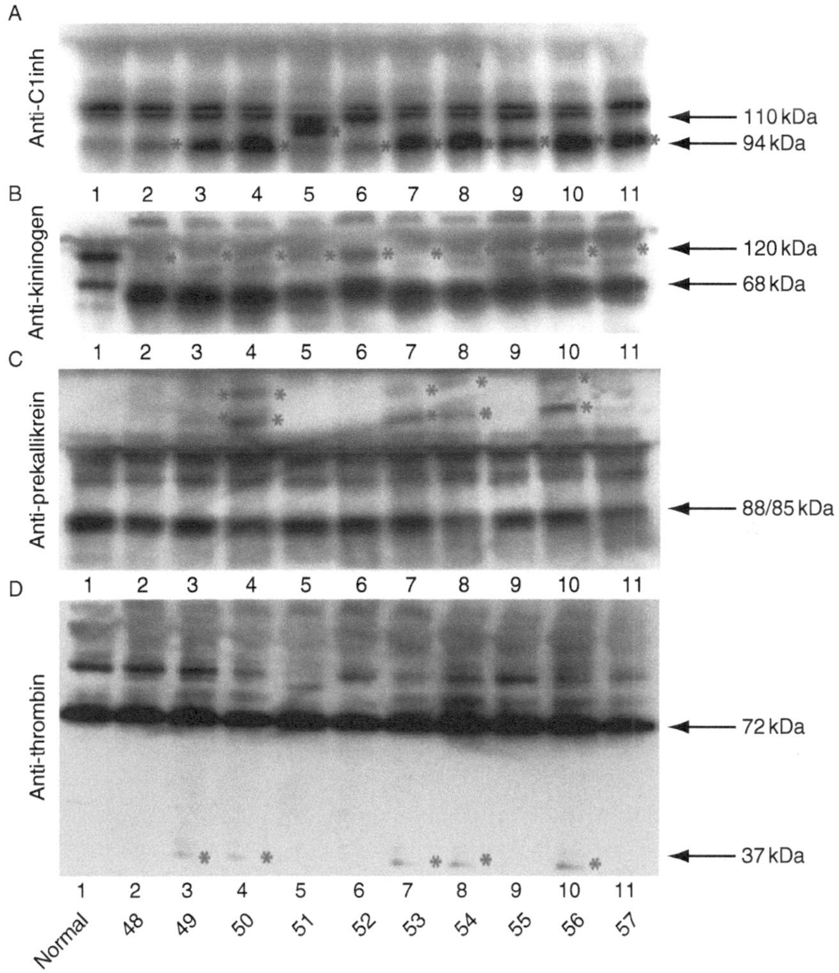

FIG. 1. Western analysis of the contact system proteins in lung cancer and normal patient plasmas. Pooled normal human plasma and cancer patient plasmas were incubated with 200 μg/ml OS-HB or control buffer for 5 min at 37 °C. Samples were analyzed by reducing SDS-PAGE and Western blotting using antibodies to C1inh, prekallikrein, HMWK, and thrombin, respectively. Plasma (0.53 μl) was used in each lane. The experiment was performed three times with similar results. The asterisk markings indicate the protein bands uniquely present in cancer patient plasmas.

B. Kininogens

The two plasma kininogens, high molecular weight kininogen (120 kDa) and low molecular weight kininogen (70 kDa) are splicing products of the kininogen gene. Kininogens are multifunction proteins with antithrombotic, antidiuretic, profibrinolytic, antiadhesive, antiangiogenic, and pro-inflammation properties. Bradykinin and kallidin, two potent antithrombotic and profibrinolytic peptides, are released from high and low molecular weight kininogens by plasma kallikrein and tissue kallikreins, respectively. There are 15 known tissue kallikreins (KLK1–15). Tissue kallikrein 3 (KLK3, prostate cancer-specific antigen or PSA), was the first FDA-approved cancer biomarker. High molecular weight kininogen is known to be cleaved by plasma kallikrein to form a 68 kDa cleavage product after contact system activation. We found that cancer patient plasmas, with the exception of the plasma of patient 52 (Fig. 1A, lanes 1 and 6), had cleaved 68 kDa but not the 120 kDa high molecular weight kininogen bands evident in normal plasma (Fig. 1B). This abnormality provides further evidence of activated contact systems in lung cancer patients.

C. Prekallikrein

The conversion of prekallikrein (85/88 kDa) to active kallikrein is accomplished by a serine protease, usually factor XIIa. A single bond (Arg371-Ile372) is split, generating kallikrein with a heavy chain of 371 amino acids and a light chain of 248 amino acids held together by a disulfide bond. Under our Western analysis conditions, kallikrein comigrates with prekallikrein. In plasma, kallikrein is inactivated by two protease inhibitors, C1inh and α2-macroglobulin, each of which forms a 1:1 stoichiometric complex with kallikrein. Therefore, an activated contact system might be accompanied by normal or decreased prekallikrein/kallikrein band with or without observable kallikrein/protease inhibitor complexes. Decreased amounts of prekallikrein were observed in the plasmas of patients 48, 50, and 54–57 (Fig. 1C, lanes 2, 4, 8–11 compared to the control lane 1). In addition, kallikrein/protease inhibitor complex bands (marked by asterisks in the figure) were observed in plasmas of patients 49, 50, 53, 54, and 56 (Fig. 1C, lanes 3, 4, 7, 8, and 10), suggesting activated contact systems.

D. Thrombin

Thrombin is generated from 72 kDa prothrombin by thrombin, factor Xa,[7] and kallikrein.[16] Cleavage at positions 155 and 284 are catalyzed by thrombin whereas cleavage at positions 271 and 320 are catalyzed by factor Xa. Both proteases generate A and B chains with a combined mass of 37 kDa for thrombin. Meizothrombin arises from a single cleavage at 320; it has the same molecular weight as prothrombin, and the full proteolytic activity of thrombin for protein C, factor V, factor XI, and small peptide substrates.[7] Therefore, thrombin Western analysis may reveal the same or different sizes of cleaved thrombin products reflecting 1–4 cuts of prothrombin by factor Xa and/or thrombin. In plasma, thrombin is inactivated by one of two protease inhibitors, antithrombin or heparin cofactor II. The inhibitory activity of each protein inhibitor is greatly enhanced by GAGs. Each inhibitor forms a 1:1 complex with thrombin. Therefore, an activated contact system might be accompanied by a normal or decreased prothrombin/meizothrombin band with or without an observable 37 kDa thrombin band. Thrombin/antithrombin and thrombin/heparin cofactor complexes might be observable as well. We found that the 37 kDa thrombin bands were present in plasmas of patients 49, 50, 53, 54, and 56 (Fig. 1D, lanes 3, 4, 7, 8, and 10). The thrombin bands were correlated with the kallikrein/protease inhibitor bands observed in the same lung cancer patient plasmas (Fig. 1C, lanes 3, 4, 7, 8, and 10).

The Western data from lung cancer patients (Fig. 1) demonstrated that thrombin generation was correlated with kallikrein activation and C1inh and kininogen abnormalities. To test if abnormal contact system proteins were also present in plasmas from patients with other kinds of cancer, we performed Western analyses on plasmas from 11 colon cancer patients, 9 breast cancer patients, 3 pancreatic cancer patients, and 1 renal cancer patient. Patient information and the Western analysis results are shown in Table II and Figs. 2–4. We found that all cancer patient plasmas were associated with an activated contact system and the degree of the activation varied with the type of cancer and was cancer stage-dependent.

TABLE II
OTHER CANCER PATIENT INFORMATION

Patient no.	Gender	Age	Stage	Conditions
L290	Female	81		Lung adenocarcinoma, COPD, diabetes, high cholesterol, degenerative arthritis
B291	Female	70	2	Breast cancer
B292	Female	33	4	Breast cancer
B293	Female	73	2	Breast cancer
B294	Female	63	1	Breast cancer
B295	Female	80	2	Breast cancer
B296	Female	72	2	Breast cancer
B297	Female	31	1	Breast cancer
B298	Female	70	2	Breast cancer
B299	Female	89	1	Breast cancer
B300	Female	93	2	Breast cancer
B301	Female	69		Breast cancer, diabetes, asthma
C302	Female	65	4	Colon cancer

(*Continues*)

TABLE II (*Continued*)

Patient no.	Gender	Age	Stage	Conditions
C303	Female	59	3	Colon cancer
C304	Male	71	2	Colon cancer
C305	Female	63	2	Colon cancer
C306	Male	79	3	Colon cancer
C307	Female	63	2	Colon cancer
C308	Male	71	1	Colon cancer
C309	Female	72	3	Colon cancer
C310	Female	80	3	Colon cancer
C311	Male	75	3	Colon cancer
C312	Male	56		Crohns disease
B313	Female	64		HTN, breast cancer, high cholesterol, psoriasis
B314	Female	64		HTN, breast cancer, psoriasis
P315	Male	67	3	Pancreatic cancer
P316	Female	85	2	Pancreatic cancer
P317	Female	55	3	Pancreatic cancer
C318	Male	47		Rectal cancer, colon cancer, cirrhosis, low level liver impairment
R319	Male	66		Renal cell carcinoma, osteoarthritis, HTN, high cholesterol

Lung and pancreatic cancers are associated with high fatality rates. We found that 9 out of 10 lung cancer patient plasmas and 3 out of 3 pancreatic cancer patient plasmas had no detectable high molecular weight kininogen. In contrast, only 1 out of 11 colon cancer plasmas had no high molecular weight kininogen, and the negative patient had stage IV disease. These results suggest that the fatality of cancers correlates with the total consumption of high molecular weight kininogen through contact system activation.

Contact system activation can induce thrombin and other proteases to create a hypercoagulable state through established protease activated receptors, and via C3a-, C5a-, kinin-, and osteopontin-dependent signal transduction pathways.[11,13,14,22] The activated pathways can lead to platelet activation, blood coagulation, complement activation, vasodilatation, and inflammation. Our data suggest that tumor-produced GAGs or other molecules can induce contact system activation and upregulate thrombin generation to support the hypercoagulable state needed for tumor angiogenesis and metastasis.

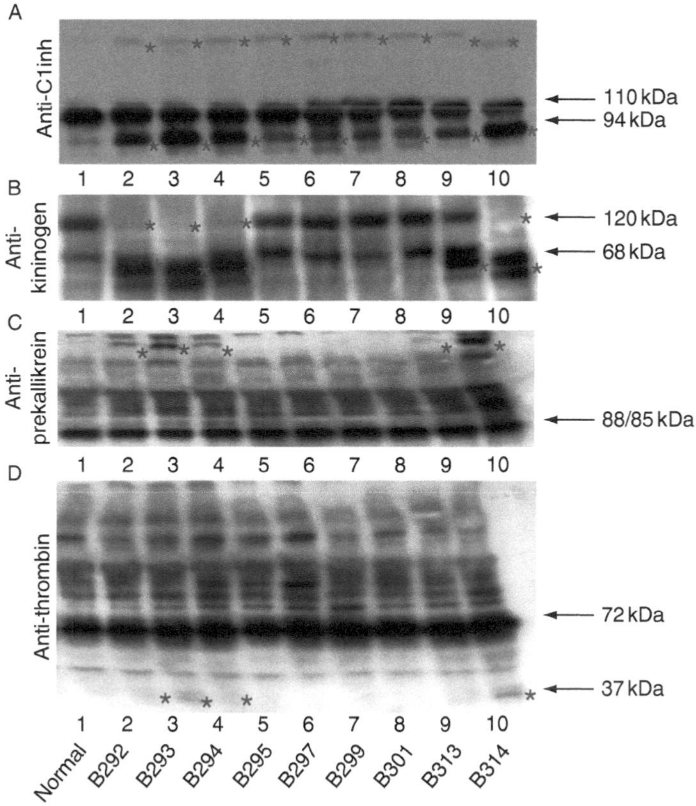

FIG. 2. Western analysis of the contact system proteins in breast cancer and normal patient plasmas. The method is the same as described in Fig. 1.

IV. Activated Contact System but Diminished Kallikrein and Thrombin Activities Were Found in Stage III Cancer Patient Plasmas

Contact system activation results in an increase of multiple active serine proteases. The most critical serine proteases are kallikrein and thrombin.[16] We established previously that the kallikrein and thrombin activities in normal human plasmas could be induced by negatively charged molecules, such as heparin contaminant, oversulfated heparin by-product (OS-HB), or heparin in a concentration-dependent manner.[12,16,27] We expected to observe abnormal kallikrein and thrombin activities in lung cancer patient plasmas based on Western analysis results that show the activated contact system in all cancer

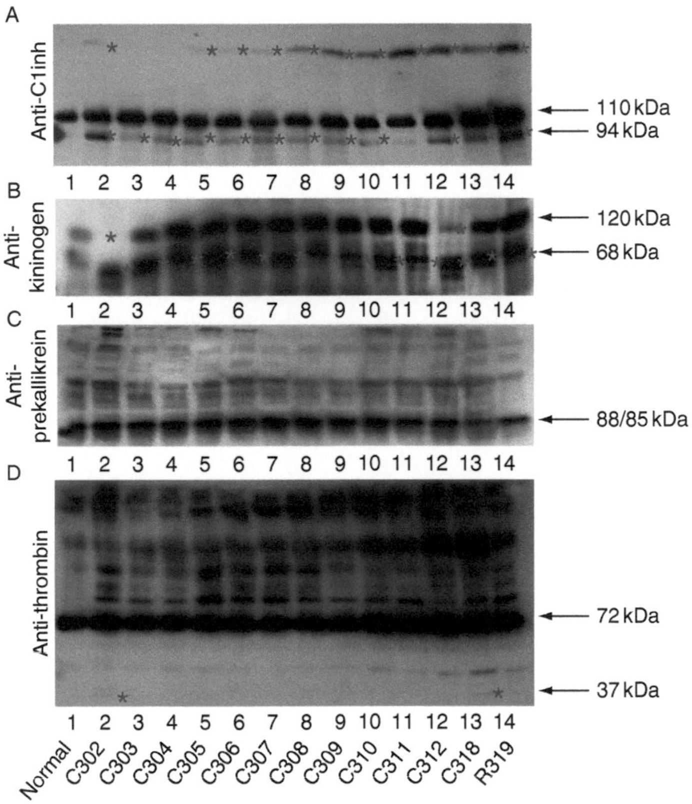

FIG. 3. Western analysis of the contact system proteins in colon cancer, renal cancer, and normal patient plasmas. The method is the same as described in Fig. 1.

patient plasmas (Figs. 1–4). To test this hypothesis, control plasma pooled from 30+ normal donors and lung cancer plasmas were treated with OS-HB or heparin at 0, 2, and 20 μg/ml, respectively. Kallikrein and thrombin activities were measured colormetrically by substrate cleavage.

We found that kallikrein and thrombin activities in normal human plasma were induced by 2 or 20 μg/ml OS-HB or heparin in a concentration-dependent manner compared to the untreated (0 μg/ml) control (Fig. 5). In contrast, 2 μg/ml OS-HB did not increase kallikrein activities in most of the lung cancer patient plasmas, whereas it did increase thrombin activity in most patients. At 20 μg/ml OS-HB induced both kallikrein and thrombin activities in most lung cancer patient plasmas (Fig. 5). Interestingly, two patients (48 and 51) with stage III lung cancer had almost no measurable kallikrein or thrombin activity

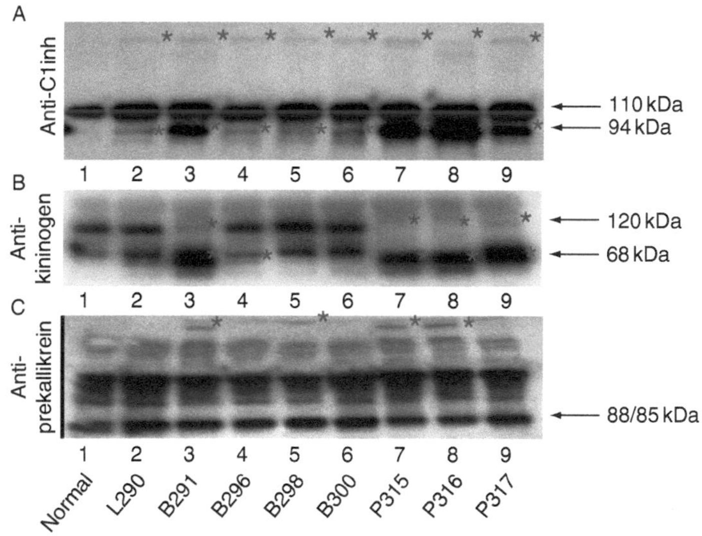

FIG. 4. Western analysis of the contact system proteins in lung cancer, breast cancer, pancreatic cancer, and normal patient plasmas. The method is the same as described in Fig. 1.

even when induced with 20 μg/ml OS-HB (Fig. 5), indicating that kallikrein and thrombin activities were totally inhibited by a yet unknown molecular mechanism in the late stage cancer patient plasmas. Shutting down kallikrein and thrombin activities in cancer patient plasmas might be responsible for loss of vitality in cancer patients considering that prothrombin is required for sustaining life.[8] The kallikrein and thrombin activities measurement used in these studies might be useful as a vitality test for cancer patients.

V. Increased Glucosamine and Galactosamine Levels were Found in Lung Cancer Patient Plasmas

GAG-containing proteoglycans, such as syndecan-1,[30] glypican-3,[30,31] endocan,[32] bamacan (CSPG 6, chondroitin sulfate proteoglycan 6),[33] melanoma associated CSPG,[34,35] and versican,[36] are potential biomarkers for different types of cancers. Up to fourfold increases in heparan and chondroitin sulfate quantities were observed in lung cancer tissues,[37] which indicates that tumor malignancy is associated with increased heparan sulfate and chondroitin sulfate expression. Interestingly, lung cancer associated endothelial cells also contribute to high plasma GAG levels.[38] Endocan, a dermatan sulfate proteoglycan, is

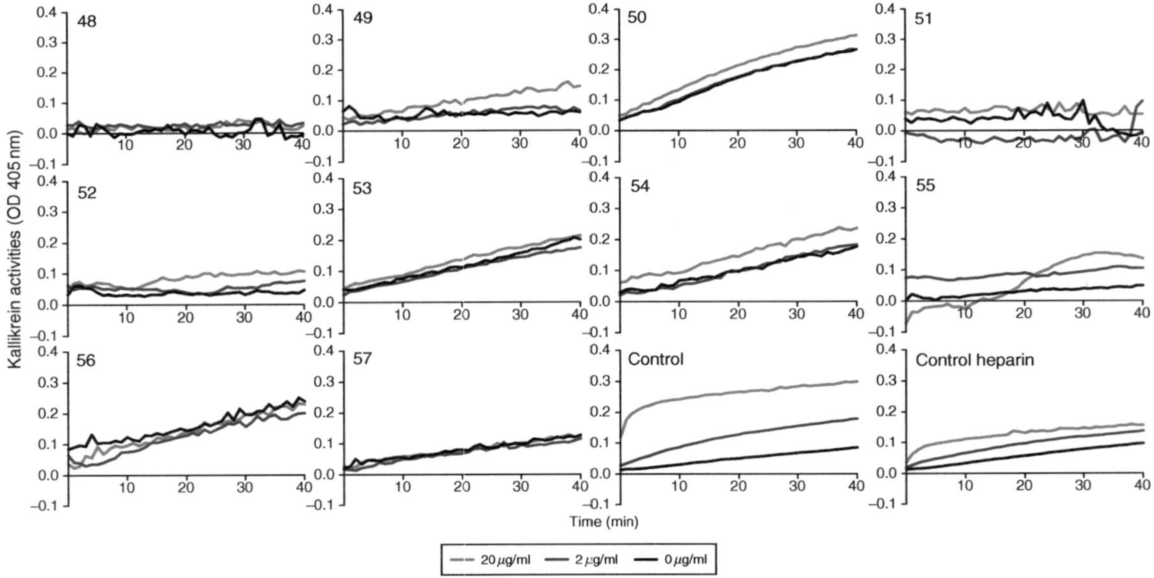
48
49
50
51
52
53
54
55
56
57
Control
Control heparin
Kallikrein activities (OD 405 nm)
Time (min)
20 μg/ml
2 μg/ml
0 μg/ml

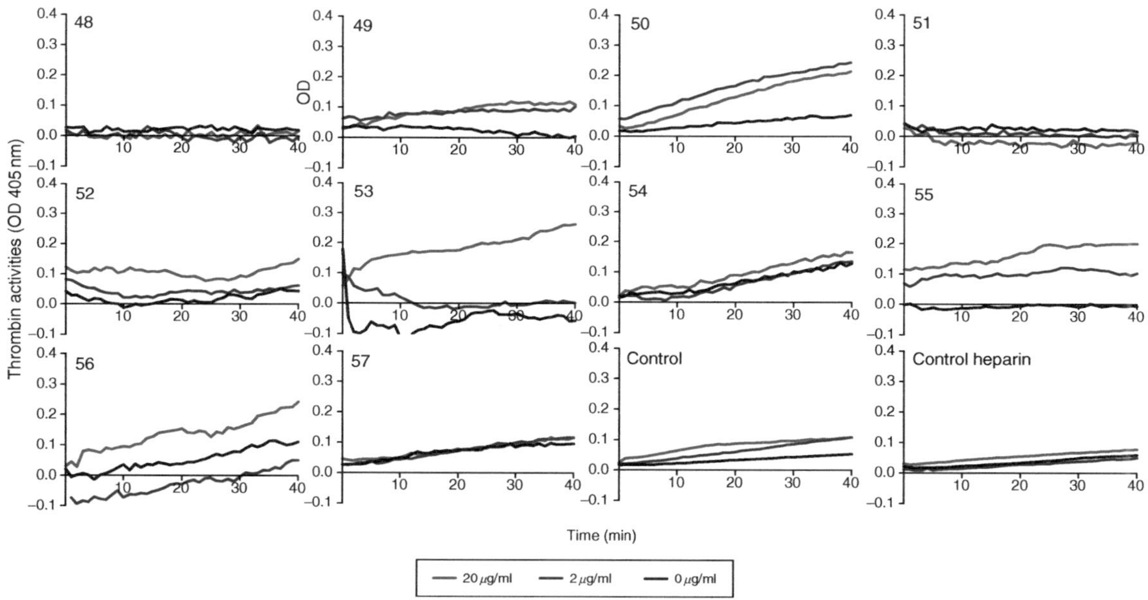

FIG. 5. Abnormal kallikrein and thrombin activities in lung cancer patient plasmas. Normal pooled human plasma (30 μl) was incubated with 20 μl negatively charged molecules at a final concentration of 0, 2, and 20 μg/ml at 37 °C for 5 min. Kallikrein activity was assessed by addition of 150 μl of the kallikrein chromogenic substrate, and 0.5 mM *N*-benzoyl-Pro-Phe-Arg-*p*-nitroanilide hydrochloride dissolved in TBS. Absorbance readings were taken every 90 s for 40 min. Thrombin activity was assessed by addition of the thrombin chromogenic substrate, 1 mM Sar-Pro-Arg-*p*-nitroanilide dihydrochloride dissolved in TBS. Absorbance readings were taken every 90 s for 40 min. (See Color Insert.)

secreted into blood circulation by endothelial cells. Tumor-derived factors, such as VEGF, upregulate endocan expression in tumor-associated endothelial cells.[39] Serum endocan levels in lung cancer are correlated with both survival and time to tumor progression and are being developed as biomarkers for lung cancer prognosis.[38]

In addition, carcinoma mucins, carriers of heavily sialylated, fucosylated, and sulfated glycans, are shed by tumors into the blood circulation and are sometimes measured as cancer prognostic markers.[4] Detection of circulation mucins is used for cancer diagnosis and prognosis. For example, CA19-9 antibody detects sialyl Lewis A (Neu5Ac2-3Gal1-3[Fuc1-4]GlcNAc) presented on a circulating high molecular weight mucin, providing an assay for diagnosis and monitoring of stomach, colon, and pancreatic cancer[40]; and *O*-glycans on circulating mucins recognized by antibody CA-125 provide for diagnosis and monitoring of ovarian cancers.[41] It has been documented that purified carcinoma mucins induce contact system activation in human plasma.[42] These observations suggest that circulating proteoglycans and/or mucins might be the negatively charged molecules that activate the contact system in cancer.

Glucosamine is part of the core structure of heparan sulfate while galactosamine is its counterpart in chondroitin/dermatan sulfate. Both glucosamine and galactosamine are present in O- and N-linked glycans in mucins. We reasoned that if GAGs and/or mucins are the thrombin inducers, the glucosamine and galactosamine level in cancer patient plasmas might be higher than that of normal control. We therefore quantified glucosamine and galactosamine content in 1 μl of plasma sampled from 4 normal human controls, 10 lung cancer patients, 9 breast cancer patients, and 3 pancreatic cancer patients based on the basis of our published method.[28] The data are shown in Fig. 6. We found that plasma galactosamine levels (Fig. 6A) were increased in lung cancer patients (0.63 ± 0.10 mM) and largely unchanged in breast cancer patients (0.48 ± 0.10 mM) and pancreatic cancer patients (0.44 ± 0.08 mM) compared to four normal controls (0.50 ± 0.11 mM). In contrast, glucosamine levels (Fig. 6B) were increased in both lung cancer patients (0.81 ± 0.07 mM) and breast cancer patients (0.60 ± 0.07 mM) and essentially unchanged in pancreatic cancer patients (0.52 ± 0.13 mM) compared to normal controls (0.56 ± 0.06 mM). Our data support the hypothesis that the lung cancer patient plasmas have increased levels of glucosamine- and galactosamine-containing glycans.

Circulatory glucosamine concentration is tightly controlled. Biggee *et al.*[43] demonstrated this by administering 1.5 g glucosamine to 18 fasting patients with osteoarthritis. Afterward, their serum glucosamine concentrations were only marginally higher.[43] Therefore, the increased plasma glucosamine and galactosamine levels of lung patient plasmas might not come from diet, but be a result of cancer. The glucosamine and galactosamine levels did not correlate

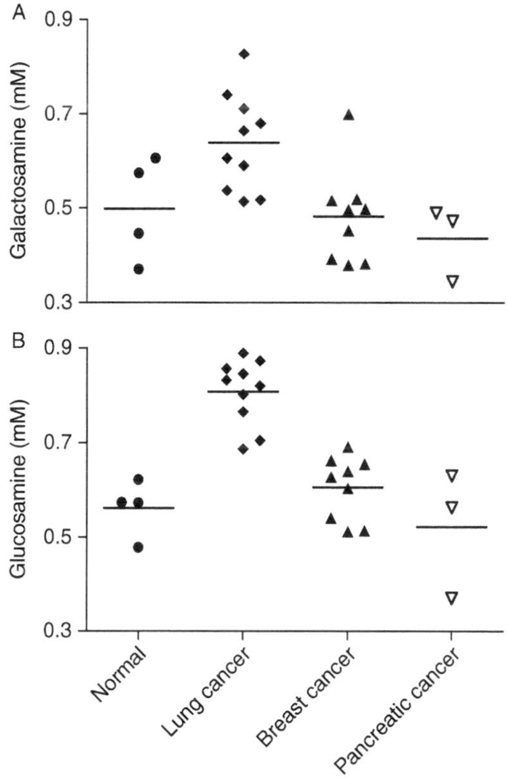

FIG. 6. Plasma galactosamine and glucosamine quantification. One microliter of plasma from control or cancer patients in duplicates were hydrolyzed and derivatized. The fluorescent derivatives are separated by the reversed-phase HPLC and then quantified.[28]

with the stage of cancer of these lung cancer patients (Fig. 6), which suggests that aberrant glucosamine and galactosamine levels might be useful for early detection of lung cancer.

Both lung cancer and pancreatic cancer patient plasmas were associated with a lack of high molecular weight kininogen (Figs. 1 and 4). However, the plasma glucosamine and galactosamine levels of pancreatic cancer patients were not elevated above control values like the plasma levels of lung cancer patients (Fig. 6). Circulating GAGs only contribute $\sim$5% of the glucosamine and galactosamine content in normal human plasma. Therefore, the increase in circulating GAGs in pancreatic cancer patients might not be detectable by whole plasma glucosamine and galactosamine quantification.

VI. Increased GAG Levels in Lung Cancer Patient Plasmas

Circulating GAGs contribute ~5% of the glucosamine and galactosamine present in normal human plasma. Therefore, a significant increase in circulating GAGs in cancer patients might not be detectable by plasma glucosamine and galactosamine quantification. To test this idea, GAGs were isolated from plasmas of four normal controls and four lung cancer patients. The glucosamine and galactosamine content of purified GAGs were quantified by the same HPLC method. We found that the galactosamine level of plasma GAGs (Fig. 7A) increased from 24.6 ± 2.8 μM in normal control to 79.4 ± 45.4 μM in lung cancer patients whereas the glucosamine level of plasma GAGs (Fig. 7B) increased from 29.4 ± 5.1 μM in normal control to 67.9 ± 37.0 μM in lung cancer patients. In conclusion, in lung cancer patients, whole plasma galactosamine and glucosamine levels were increased 126% and 145%, respectively, whereas the galactosamine and glucosamine content of their plasma GAGs were increased 323% and 231%, respectively, with respect to normal controls.

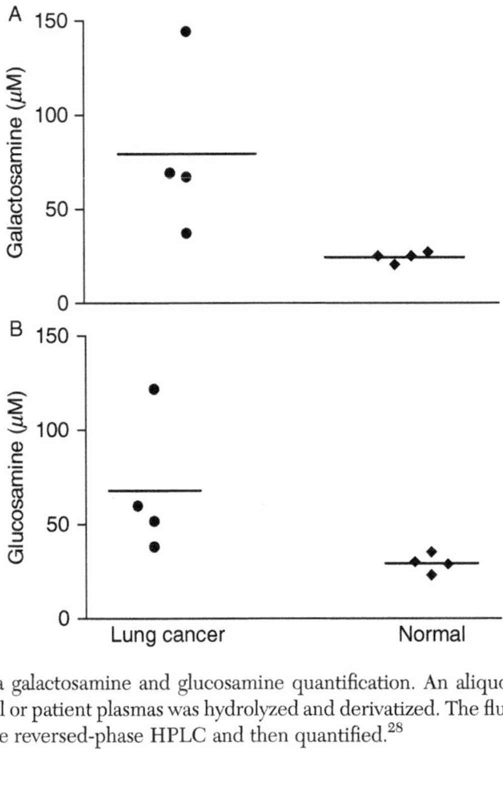

FIG. 7. Plasma galactosamine and glucosamine quantification. An aliquot of GAGs purified from normal control or patient plasmas was hydrolyzed and derivatized. The fluorescent derivatives are separated by the reversed-phase HPLC and then quantified.[28]

These data suggest that galactosamine and glucosamine quantifications either from plasma or from purified plasma GAGs are valid in developing lung cancer diagnosis biomarkers.

Both GAGs and mucins have a tightly controlled circulation concentration due to the presence of multiple hepatic clearance receptors.[44–46] As a result, intravenously injected anticoagulant heparin has a half-life of 1 h in the blood. It is unknown how the high level of GAGs is maintained in lung cancer plasmas or whether dermatan sulfate made by endothelial cells or GAGs made by the cancer cells contribute to the activated contact systems observed in our studies. The relative contributions of mucins or proteoglycan GAGs to the activation of cancer plasma contact systems are also unknown. The answers to these questions may lead to novel cancer prevention and treatment targets.

VII. Amine-Containing Compounds Regulated the Contact System Activation

The data presented in this and previous chapters (Molecular Mechanism Underlines Heparin Induced Thrombocytopenia and Thrombosis, Chondroitin Sulfate and Abnormal Contact System in Rheumatoid Arthritis, Activated Contact System and Abnormal Glycosaminoglycans in Lupus and other Auto- and Non-autoimmune Diseases) indicate that abnormal contact system activation and/or thrombin generation underline unrelated thromboembolic disorder, the most common cause of human deaths. As amine-containing compounds, such as chloroquine, are effective drugs not only for malaria, but also for prion, viral infection, different autoimmune diseases, and cancers, we suspect that such amine-containing compounds might inhibit contact system activation through charge/charge interaction with GAGs.[47]

We tested serotonin, serotonine/creatinine complex, histamine, and chloroquine in similar concentrations reported previously[48] (100–400 μg/ml) for their abilities in inhibiting 200 μg/ml OS-GAG induced contact system activation (Fig. 8). All of them, except serotonin, partially or completely inhibited the normal human plasma contact system activation induced by 200 μg/ml OS-GAG evidenced by lack or reduced 94 kDa and two high molecular weight C1inh bands (indicated by the asterisks). The 120 kDa HMWK bands were only visible in the control plasma (lane 1) and 400 μg/ml serotonin/creatine-treated plasma (lane 8), which indicates that 400 μg/ml serotonin/creatine protected HMWK from being cleaved during the induced contact system activation. Even though chloroquine inhibited high molecular weight C1inh formation at both 200 and 400 μg/ml, it did not prevent the HMWK from being cleaved during the contact system activation. These results suggest that not all

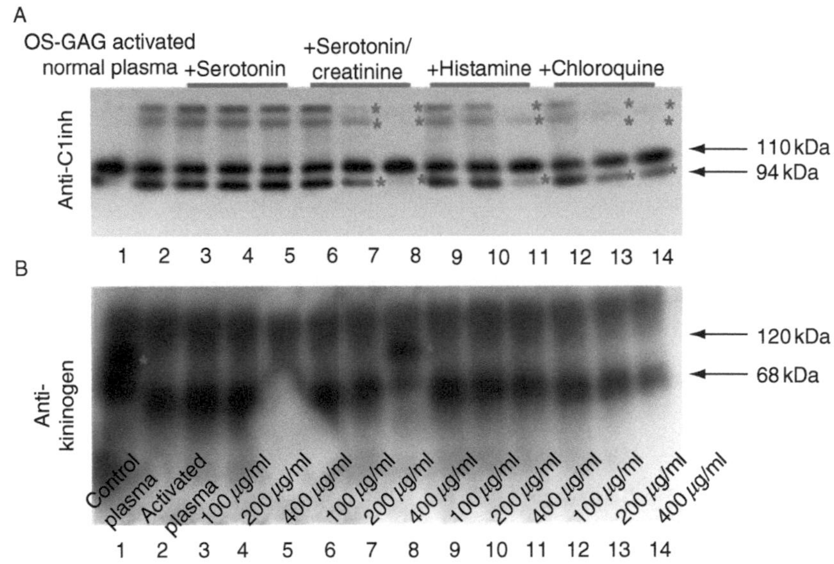

FIG. 8. Western analysis of the contact system activation of normal human plasmas by amine-containing compounds. Pooled normal human plasma was incubated with control buffer (lane 1) or 200 µg/ml OS-CACs plus amine-containing compounds at the concentration indicated for 5 min at 37 °C. Samples were analyzed by reducing SDS-PAGE and Western analysis using antibodies to C1inh and kininogen. Plasma (0.53 µl) was used in each lane.

amine-containing compounds inhibit the contact system equally. The specific structure of amine-containing compounds might add specificity to contact system regulation.

VIII. Concluding Remarks

Contact system activation and/or thrombin generation might be the hemostatic response to a variety of unrelated human diseases, such as coronary heart disease,[49] diabetes,[50] kidney failure,[51] autoimmune diseases,[52] cancers,[53] and heparin induced thrombocytopenia and thrombosis (HITT).[54] Thrombin is well studied, but the biological functions of other contact system activated serine proteases are less known. Proteases and cofactors generated through contact system activation can have opposite roles; for example, kallikrein produces thrombin to initiate clot formation and plasmin to dissolve the clot at the same time. Therefore, the contact system might contribute not only to thrombosis but also to hemostasis in a profound way.[22] It is likely that contact system activated proteases, such as kallikrein, FXIIa, plasmin, TAFIa, FXIa, and FXa, have functions in addition to their critical roles in coagulation. Therefore, understanding the complexity of contact system activation and major molecular players in causing thromboembolic disorders might be the key to cure cancers and other unrelated human diseases in a patient specific-manner.

References

1. Thun MJ, Hannan LM, Adams-Campbell LL, Boffetta P, Buring JE, Feskanich D, et al. Lung cancer occurrence in never-smokers: an analysis of 13 cohorts and 22 cancer registry studies. *PLoS Med* 2008;**5**:e185.
2. Mousa SA. Role of current and emerging antithrombotics in thrombosis and cancer. *Timely Top Med Cardiovasc Dis* 2006;**10**:E19.
3. Sainz IM, Pixley RA, Colman RW. Fifty years of research on the plasma kallikrein–kinin system: from protein structure and function to cell biology and *in vivo* pathophysiology. *Thromb Haemost* 2007;**98**:77–83.
4. Varki A. Trousseau's syndrome: multiple definitions and multiple mechanisms. *Blood* 2007;**110**:1723–9.
5. de la Fouchardiere C, Flechon A, Droz JP. Coagulopathy in prostate cancer. *Neth J Med* 2003;**61**:347–54.
6. Sack Jr GH, Levin J, Bell WR. Trousseau's syndrome and other manifestations of chronic disseminated coagulopathy in patients with neoplasms: clinical, pathophysiologic, and therapeutic features. *Medicine (Baltimore)* 1977;**56**:1–37.
7. Jenny NS, Lundblad RL, Mann KG. Thrombin. In: Colman RW, editor. *Hemostasis and thrombosis*. Philadelphia: Lippincott Williams & Wilkins; 2006. p. 193–213.
8. Mullins ES, Kombrinck KW, Talmage KE, Shaw MA, Witte DP, Ullman JM, et al. Genetic elimination of prothrombin in adult mice is not compatible with survival and results in spontaneous hemorrhagic events in both heart and brain. *Blood* 2009;**113**:696–704.
9. Nierodzik ML, Karpatkin S. Thrombin induces tumor growth, metastasis, and angiogenesis: evidence for a thrombin-regulated dormant tumor phenotype. *Cancer Cell* 2006;**10**:355–62.
10. Folkman J. Antiangiogenesis in cancer therapy–endostatin and its mechanisms of action. *Exp Cell Res* 2006;**312**:594–607.
11. Coughlin SR. Protease-activated receptors in hemostasis, thrombosis and vascular biology. *J Thromb Haemost* 2005;**3**:1800–14.
12. Qian Y, Pan J, Zhou X, Weiser P, Lu H, Shih FF, et al. Heparin and oversulfated heparin byproduct induce thrombin generation through contact system activation in plasma of patients with HIT. *Clin Appl Thromb Hemost* 2010;**16**:251–60.
13. Di Cera E. Thrombin. *Mol Aspects Med* 2008;**29**:203–54.
14. Huber-Lang M, Sarma JV, Zetoune FS, Rittirsch D, Neff TA, McGuire SR, et al. Generation of C5a in the absence of C3: a new complement activation pathway. *Nat Med* 2006;**12**:682–7.
15. Orfeo T, Brummel-Ziedins KE, Gissel M, Butenas S, Mann KG. The nature of the stable blood clot procoagulant activities. *J Biol Chem* 2008;**283**:9776–86.
16. Qian Y, Pan J, Zhou X, Weiser P, Hourcade D, Lizsewski KM, et al. Oversulfated heparin byproducts induce thrombin generation through contact system activation. *Clin Appl Thromb Hemost* 2010;**16**:244–50.
17. Stief TW. Kallikrein activates prothrombin. *Clin Appl Thromb Hemost* 2008;**14**:97–8.

18. Colman RW, Clowes AW, George JN, Goldhaber SZ, Marder VJ. Overview of hemostasis. In: Colman RW, editor. *Hemostasis and thrombosis*. Philadelphia: Lippincott Williams & Wilkins; 2006. p. 3–16.
19. Lee AY. Cancer and thromboembolic disease: pathogenic mechanisms. *Cancer Treat Rev* 2002;**28**:137–40.
20. Furie B, Furie BC. Mechanisms of thrombus formation. *N Engl J Med* 2008;**359**:938–49.
21. Schaffner F, Ruf W. Tissue factor and protease-activated receptor signaling in cancer. *Semin Thromb Hemost* 2008;**34**:147–53.
22. Colman RW. Contact activation (kallikrein-kinin) pathway: multiple physiologic and pathophysiologic activities. In: Colman RW, editor. *Hemostasis and thrombosis*. Philadelphia: Lippincott Williams & Wilkins; 2006. p. 107–30.
23. Waaler BA. Contact activation in the intrinsic blood clotting system; studies on a plasma product formed on contact with glass and similar surfaces. *Scand J Clin Lab Invest* 1959;**11**:1–133.
24. Kishimoto TK, Viswanathan K, Ganguly T, Elankumaran S, Smith S, Pelzer K, et al. Contaminated heparin associated with adverse clinical events and activation of the contact system. *N Engl J Med* 2008;**358**:2457–67.
25. Pan J, Qian Y, Zhou X, Pazandak A, Frazier SB, Weiser P, et al. Oversulfated chondroitin sulfate is not the sole contaminant in heparin. *Nat Biotechnol* 2010;**28**:203–7.
26. Pan J, Qian Y, Zhou X, Pazandak A, Frazier SB, Weiser P, et al. Identification of chemically sulfated/desulfated glycosaminoglycans in contaminated heparins and development of a simple assay for the detection of most contaminants in heparin. *Glycobiology Insights* 2010;**2**:1–12.
27. Pan J, Qi Y, Zhou X, Lu H, Ramacciotti E, Zhang L. Chemically oversulfated glycosaminoglycans are potent modulators of contact system activation and different cell signaling pathways. *J Biol Chem* 2010;doi:10.1074/jbc.M109.063735.
28. Studelska DR, Giljum K, McDowell LM, Zhang L. Quantification of glycosaminoglycans by reversed-phase HPLC separation of fluorescent isoindole derivatives. *Glycobiology* 2006;**16**:65–72.
29. Zuraw BL, Curd JG. Demonstration of modified inactive first component of complement (C1) inhibitor in the plasmas of C1 inhibitor-deficient patients. *J Clin Invest* 1986;**78**:567–75.
30. Seidel C, Sundan A, Hjorth M, Turesson I, Dahl IM, Abildgaard N, et al. Serum syndecan-1: a new independent prognostic marker in multiple myeloma. *Blood* 2000;**95**:388–92.
31. Filmus J, Capurro M. Glypican-3 and alphafetoprotein as diagnostic tests for hepatocellular carcinoma. *Mol Diagn* 2004;**8**:207–12.
32. Sarrazin S, Adam E, Lyon M, Depontieu F, Motte V, Landolfi C, et al. Endocan or endothelial cell specific molecule-1 (ESM-1): a potential novel endothelial cell marker and a new target for cancer therapy. *Biochim Biophys Acta* 2006;**1765**:25–37.
33. Osman I, Bajorin DF, Sun T-T, Zhong H, Douglas D, Scattergood J, et al. Novel blood biomarkers of human urinary bladder cancer. *Clin Cancer Res* 2006;**12**:3374–80.
34. Campoli MR, Chang CC, Kageshita T, Wang X, McCarthy JB, Ferrone S. Human high molecular weight-melanoma-associated antigen (HMW-MAA): a melanoma cell surface chondroitin sulfate proteoglycan (MSCP) with biological and clinical significance. *Crit Rev Immunol* 2004;**24**:267–96.
35. Vergilis IJ, Szarek M, Ferrone S, Reynolds SR. Presence and prognostic significance of melanoma-associated antigens CYT-MAA and HMW-MAA in serum of patients with melanoma. *J Invest Dermatol* 2005;**125**:526–31.
36. Skandalis SS, Kletsas D, Kyriakopoulou D, Stavropoulos M, Theocharis DA. The greatly increased amounts of accumulated versican and decorin with specific post-translational modifications may be closely associated with the malignant phenotype of pancreatic cancer. *Biochim Biophys Acta* 2006;**1760**:1217–25.

37. Horai T, Nakamura N, Tateishi R, Hattori S. Glycosaminoglycans in human lung cancer. *Cancer* 1981;**48**:2016–21.
38. Grigoriu BD, Depontieu F, Scherpereel A, Gourcerol D, Devos P, Ouatas T, et al. Endocan expression and relationship with survival in human non-small cell lung cancer. *Clin Cancer Res* 2006;**12**:4575–82.
39. Abid MR, Yi X, Yano K, Shih S-C, Aird WC. Vascular endocan is preferentially expressed in tumor endothelium. *Microvasc Res* 2006;**72**:136–45.
40. Duffy MJ. CA 19-9 as a marker for gastrointestinal cancers: a review. *Ann Clin Biochem* 1998;**35**(Pt 3):364–70.
41. Yin BW, Dnistrian A, Lloyd KO. Ovarian cancer antigen CA125 is encoded by the MUC16 mucin gene. *Int J Cancer* 2002;**98**:737–40.
42. Pineo GF, Brain MC, Gallus AS, Hirsh J, Hatton MW, Regoeczi E. Tumors, mucus production, and hypercoagulability. *Ann NY Acad Sci* 1974;**230**:262–70.
43. Biggee BA, Blinn CM, McAlindon TE, Nuite M, Silbert JE. Low levels of human serum glucosamine after ingestion of glucosamine sulphate relative to capability for peripheral effectiveness. *Ann Rheum Dis* 2006;**65**:222–6.
44. Wahrenbrock MG, Varki A. Multiple hepatic receptors cooperate to eliminate secretory mucins aberrantly entering the bloodstream: are circulating cancer mucins the "tip of the iceberg"? *Cancer Res* 2006;**66**:2433–41.
45. Harris EN, Weigel JA, Weigel PH. The human hyaluronan receptor for endocytosis (HARE/Stabilin-2) is a systemic clearance receptor for heparin. *J Biol Chem* 2008;**283**:17341–50.
46. Harris EN, Kyosseva SV, Weigel JA, Weigel PH. Expression, processing, and glycosaminoglycan binding activity of the recombinant human 315-kDa hyaluronic acid receptor for endocytosis (HARE). *J Biol Chem* 2007;**282**:2785–97.
47. Stanley FE, Warner AM, Gutierrez SM, Stalcup AM. Heparin-induced circular dichroism of chloroquine. *Biochem Biophys Res Commun* 2009;**388**:28–30.
48. Binstadt BA, Patel PR, Alencar H, Nigrovic PA, Lee DM, Mahmood U, et al. Particularities of the vasculature can promote the organ specificity of autoimmune attack. *Nat Immunol* 2006;**7**:284–92.
49. Duvall WL. Antithrombotic therapy. *Curr Mol Med* 2006;**6**:603–19.
50. Bartnik M, Norhammar A, Ryden L. Hyperglycaemia and cardiovascular disease. *J Int Med* 2007;**262**:145–56.
51. Grand'Maison A, Charest AF, Geerts WH. Anticoagulant use in patients with chronic renal impairment. *Am J Cardiovasc Drugs* 2005;**5**:291–305.
52. Swigris JJ, Fischer A, Gillis J, Meehan RT, Brown KK. Pulmonary and thrombotic manifestations of systemic lupus erythematosus. *Chest* 2008;**133**:271–80.
53. Khorana AA. The NCCN clinical practice guidelines on venous thromboembolic disease: strategies for improving VTE prophylaxis in hospitalized cancer patients. *Oncologist* 2007;**12**:1361–70.
54. Linkins L-A, Warkentin T. The approach to heparin-induced thrombocytopenia. *Semin Respir Crit Care Med* 2008;066–74.

Index

D

E

H

O

P

R

S

T

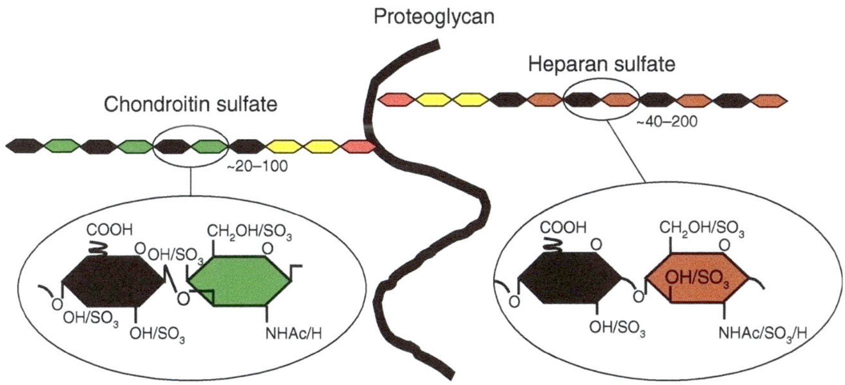

LIJUAN ZHANG, FIG. 1. Diagrammatic representation of GAG assembly on a proteoglycan core protein. Both heparan sulfate (HS) and chondroitin sulfate (CS) are attached to specific serine residues of proteoglycan core protein by the linkage tetrasaccharide GlcA (black)-Gal (yellow)-Gal (yellow)-Xyl (pink). Biosynthesis starts with the transfer of xylose from UDP-xylose to a serine residue of a core protein catalyzed by two xylosyltransferases. The linkage region is then synthesized by the sequential addition of two galactose residues (by galactosyltransferase I and II), and glucuronic acid (by glucuronosyltransferase I) from the corresponding UDP-sugars. After completion of the linkage tetrasaccharides, the addition of first HexNac residue occurs. Addition of GalNAc from UDP-GalNAc by *N*-acetylgalactosaminyl transferase I to the nonreducing terminal GlcA commits the intermediate to CS sysnthesis, which occurs subsequently through alternating addition of GlcA and GalNAc (green) by chondroitin synthase. If GlcNAc is added to the linkage tetrasaccharide instead by *N*-acetylglucosaminyltransferase I, HS synthesis occurs. Alternating GlcA and GlcNAc (red) residues are then added by HS copolymerases (EXT-1 and EXT-2) from their corresponding UDP-sugars.

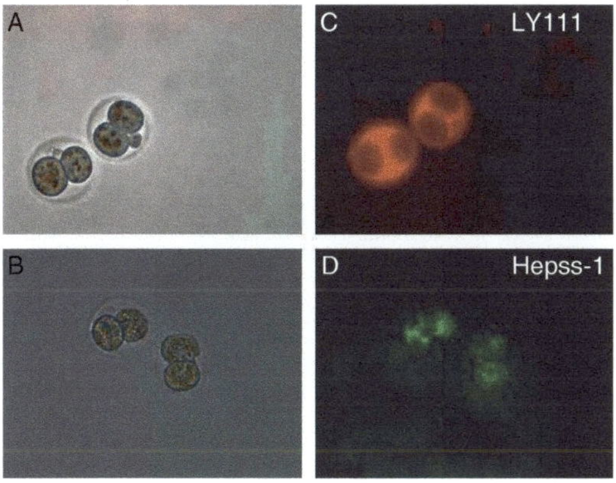

IZUMIKAWA AND KITAGAWA, FIG. 5. Immunocytohistochemistry of mouse 2-cell embryos using an anti-CS or anti-HS monoclonal antibody.[20] Wild-type embryos at E1.5 were collected and stained. Left panels (A, B) show embryos examined by phase-contrast microscopy. Immunofluorescent staining by anti-CS (LY111) or anti-HS (Hepss-1) antibody is shown in right panels (C, D).

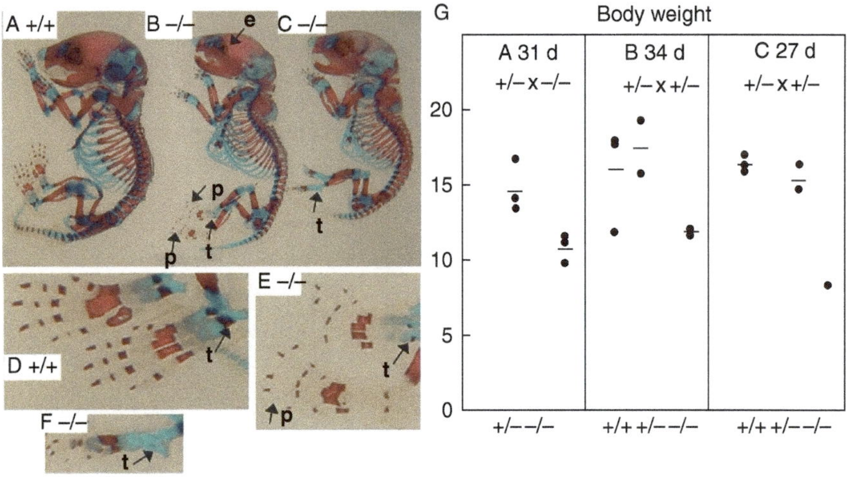

HABUCHI AND KIMATA, FIG. 7. Growth retardation of the HS6ST-1-deficient mice. The skeletal samples of newborn mice were stained with alcian blue and arizarin red. On the whole, the skeletons of the HS6ST-1-deficient mice (B, C) were smaller than those of wild-type mice (A). They often lacked a hind limb phalange, and either possessed a smaller tarsus in the hind limb compared to the tarsus in wild-type mice, or were completely lacking a tarsus. (D–F) Higher magnification of the respective hind limbs shown in (A)–(C). *Arrows* indicate the tarsus (t), phalanges (p), and eyes (e). Right panel (G) shows the body weight of about 4–5-week-old mice. (Redrawn, with permission, from Ref. 41.)

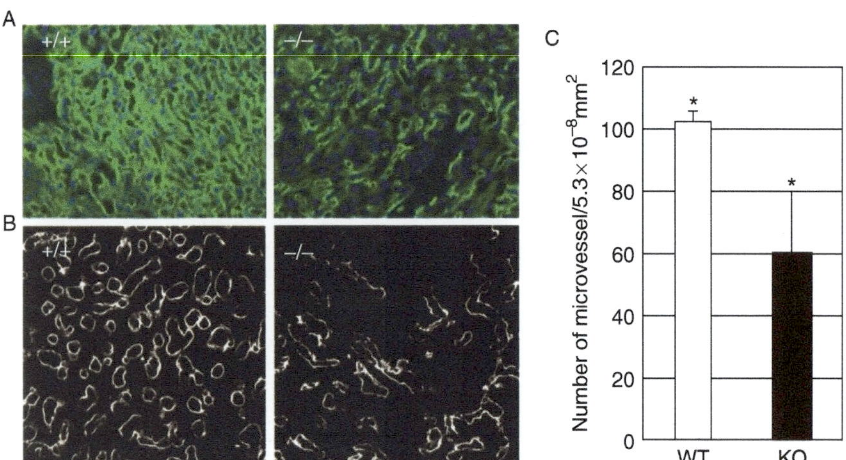

HABUCHI AND KIMATA, FIG. 9. Reduced microvessels in the placenta of a homozygous embryo from a heterozygous mother. (A) Frozen sections from the E15.5 placentas were stained with an anti-CD31 antibody (green). (B) The left panel shows the photographs of a wild-type embryo, and the right panel shows the photographs of a homozygote embryo analyzed by confocal microscopy. (C) Bars indicate the mean number of microvessels in the labyrinthine zone per 5.3×10^{-8} mm^2 counted from the photographs obtained from the confocal microscopic analysis. $^*p < 0.03$. (Redrawn, with permission, from Ref. 41.)

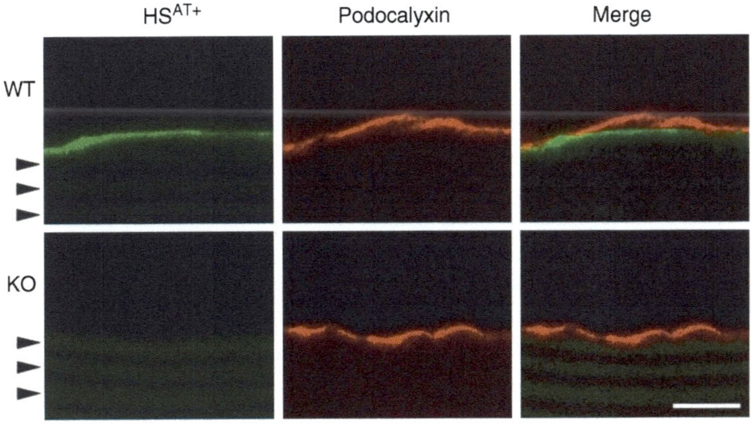

SHWORAK *ET AL.*, FIG. 6. Expression of HS^{AT+} on *Hs3st1*$^{-/-}$ aortic ECs is reduced to undetectable levels. Aorta cryosections (~5 μm) from *Hs3st1*$^{+/+}$ (*WT*) and *Hs3st1*$^{-/-}$ (*KO*) mice were costained with AT-647 to detect HS^{AT+} (*green*), and antipodocalyxin to detect ECs (*red*). The yellow color indicates coexpression of podocalyxin and HS^{AT+}. The *bar* indicates 10 μm. *Arrow heads*, elastic lamina; autofluorescence of elastin.

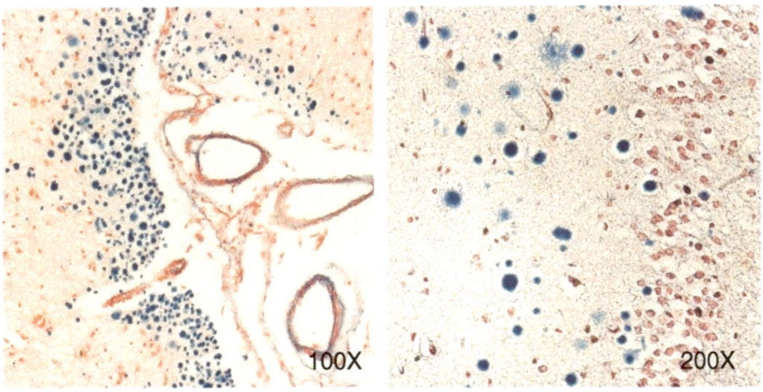

ZHANG AND LI, FIG. 1. Sulfated Alcian blue staining of cerebral corpora amylacea in human brain (courtesy of Paul O'Callaghan, Department of Public Health and Caring Sciences, Molecular Geriatrics, University of Uppsala, Uppsala Sweden). Corpora amylacea accumulate in the brain during normal aging, and can also be found in other organs, for example, heart, liver, prostate, lung, thyroid, and muscle.[167] In 1853, Rudolph Virchow demonstrated the reaction of cerebral corpora amylacea of the brain to the iodine test on the assumption that they may be cellulose derivatives. He also applied the test to what appeared to be similar to corpuscles in a "waxy" spleen. These experiments convinced him that the components of cerebral corpora amylacea and waxy spleen could be identical to starch; he preferred "amyloid" to commonly used terms "lardaceous" or "waxy" changes.[7] A few years later, the presence of protein was demonstrated in waxy spleen; this finding switched attention to the protein moiety in amyloid research thereafter. The term "amyloid" has become the common name for the protein components in amyloid deposits (For review see Puchtler et al.[168]).

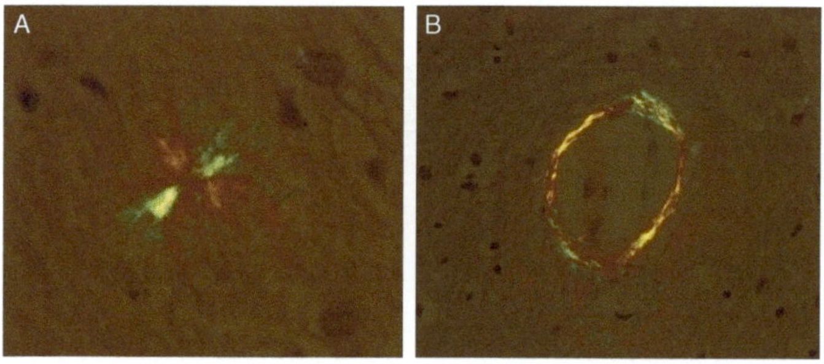

ZHANG AND LI, FIG. 2. Congo red staining of Aβ deposits in the brain of patient with Alzheimer's disease (courtesy of Paul O'Callaghan, Department of Public Health and Caring Sciences, Molecular Geriatrics, University of Uppsala, Uppsala Sweden). (A) A neuritic plaque stained with Congo red showing typical apple-green birefringence with a "Maltese cross" morphology as viewed under polarized light. (B) Apple-green birefringence of Congo red staining of a blood vessel with Aβ deposits.

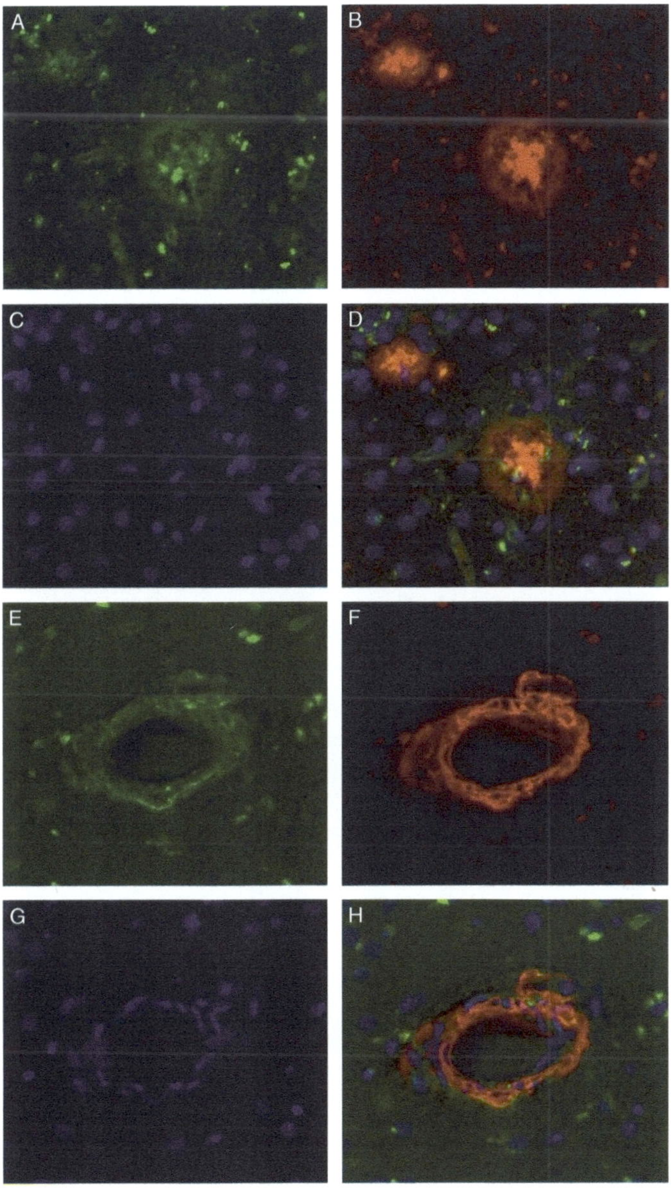

ZHANG AND LI, FIG. 3. Codeposition of amyloid-β (Aβ) and heparan sulfate (HS) in the brain of a patient with sporadic Alzheimer's disease. Double immnostaining was performed with an anti-HS antibody in combination with anti-Aβ42 antibody for plaques and Aβ40 antibody for blood vessels. Codeposition of HS with Aβ in neuritic plaques: (A) HS; (B) neuritic plaques; (C) DAPI counterstaining for nuclei; (D) merge of (A)–(C). Codeposition of HS with Aβ in blood vessel wall: (E) HS deposits; (F) Ab deposits; (G) DAPI counterstaining for nuclei; (H) merge of (E)–(G).

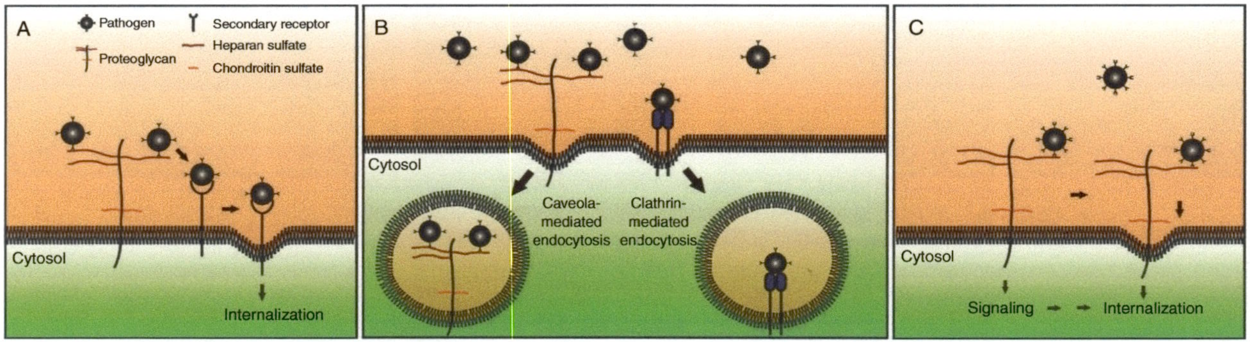

AQUINO *ET AL.*, FIG. 1. GAGs in pathogen attachment and invasion. (A) GAGs as cell surface coreceptors: pathogens bind to cell surface GAGs to increase their concentration on the cell surface and facilitate their interaction with secondary internalization receptors. (B) Viral internalization by endocytic mechanisms of host cells: the initial interaction of FMDV with GAGs or other receptors specifies whether FMDV virions will be internalized via a caveola-mediated or clathrin-mediated endocytic pathway. (C) Bacterial internalization by activation of host cell signaling: binding of *N. gonorrhoeae* to cell surface HS activates an intracellular mechanism that leads to its internalization.

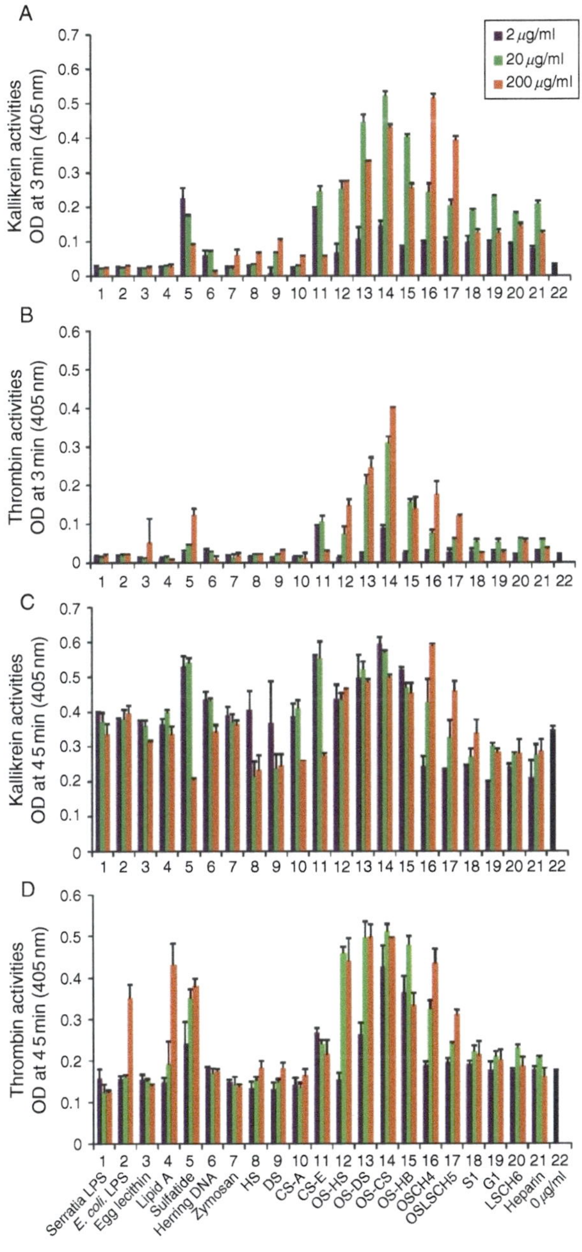

QIAN *ET AL.*, FIG. 1. Negatively charged molecule induced kallikrein- and thrombin-like activities. Normal pooled human plasma (30 μl) was incubated with 20 μl negatively charged molecules at a final concentration of 0, 2, 20, and 200 μg/ml at 37 °C for 5 min. Kallikrein-like activities were assessed by addition of 150 μl of the kallikrein chromogenic substrate, 0.5 mM *N*-benzoyl-Pro-Phe-Arg-*p*-nitroanilide hydrochloride dissolved in TBS. OD readings were taken every 90 s for 40 mins (Fig. S3). Thrombin-like activity was assessed by addition of the thrombin chromogenic substrate, 1 mM Sar-Pro-Arg-*p*-nitroanilide dihydrochloride dissolved in TBS. OD readings were taken every 90 s for 40 min. (A) The OD readings of kallikrein-like activities at 3 min from two independent experiments. (B) The OD readings of thrombin-like activities at 3 min from two independent experiments. (C) The OD readings of kallikrein-like activities at 40 min from two independent experiments. (D) The OD readings of thrombin-like activities at 40 min from two independent experiments. The OD readings of kallikrein activity induced by 20 μg/ml of HS, DS, CS, OSHS, OSDS, and OSCS at 3 and 40 min have been reported previously.[50]

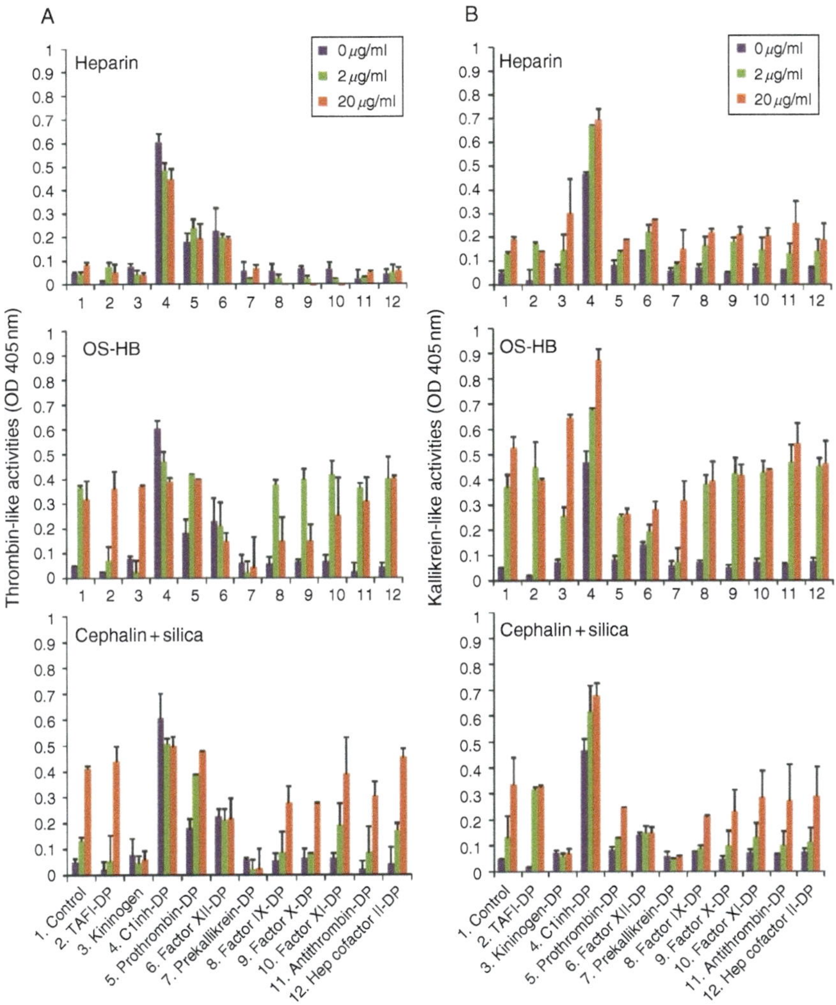

QIAN *ET AL.*, FIG. 3. Induction of thrombin and kallikrein activities. Human control plasma, and plasmas depleted in kallikrein, kininogen, antithrombin, heparin cofactor II, C1 inhibitor, FXII, FXI, FX, FIX, FII (prothrombin), and TAFI were incubated with 0, 2, and 20 μg/ml heparin, OS-HB, or cephalin/silica for 5 min at 37 °C. (A) Kallikrein activity was assessed by addition of 150 μl of the kallikrein chromogenic substrate, 0.5 mM *N*-benzoyl-Pro-Phe-Arg-*p*-nitroanilide hydrochloride dissolved in TBS. OD readings were taken every 90 s for 40 min. The average OD readings of kallikrein activities at 40 min from two independent experiments were presented. (B) Thrombin activity was assessed by addition of the thrombin chromogenic substrate, 1 mM Sar-Pro-Arg-*p*-nitroanilide dihydrochloride dissolved in TBS. OD readings were taken every 90 s for 40 min. The average OD readings of thrombin activities at 40 min from two independent experiments were presented.

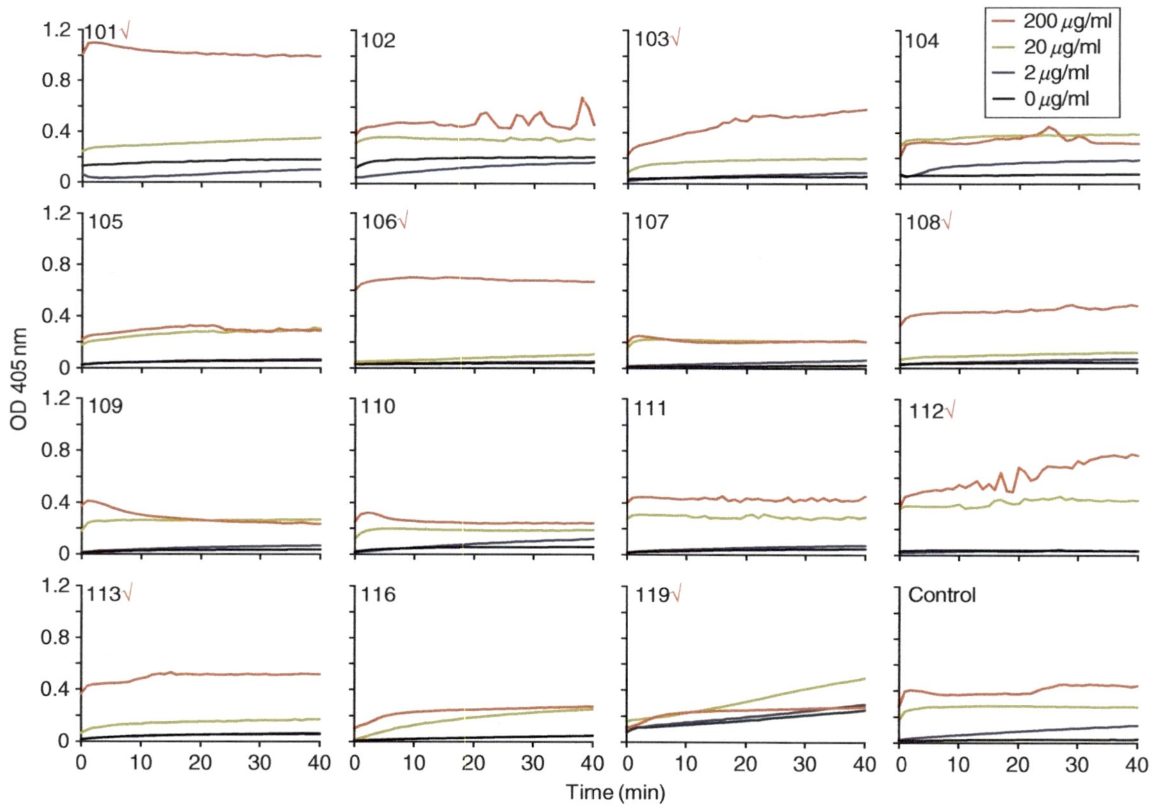

Qian *et al.*, Fig. 6. HIT diagnosis by measuring OS-HB-induced thrombin activities in HIT patient plasmas. Normal pooled human plasma or HIT patient plasma (30 μl was incubated with 20 μl OS-HB with a final concentration of 0, 2, 20, and 200 μg/ml at 37 °C for 5 min. Thrombin activity was assessed by addition of 150 μl of 100 mM Sar-Pro-Arg-*p*-nitroanilide dihydrochloride chromogenic substrate. OD readings were taken every 90 s for 40 min.

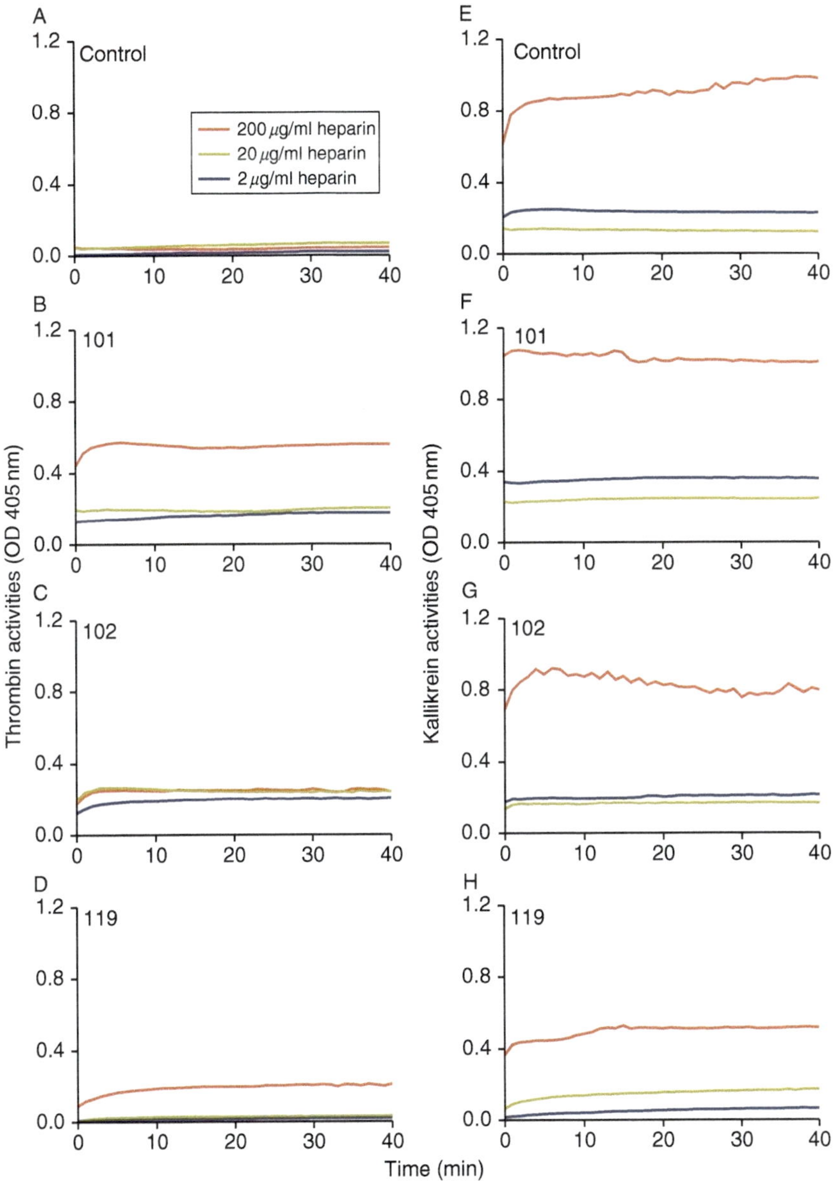

QIAN *ET AL.*, FIG. 8. Heparin-induced thrombin and kallikrein activities in HIT patient plasma. Patient or normal pooled human plasma (30 μl) was incubated with 20 μl heparin, with a final concentration of 0, 2, 20, and 200 μg/ml at 37 °C for 5 min. *Left panel*: Thrombin activity was assessed by addition of the thrombin chromogenic substrate, 1 mM Sar-Pro-Arg-*p*-nitroanilide dihydrochloride dissolved in TBS. OD readings were taken every 90 s for 40 min. The experiment was repeated twice with similar results. *Right panel*: Kallikrein activity was assessed by addition of 150 μl of the kallikrein chromogenic substrate, 0.5 mM *N*-benzoyl-Pro-Phe-Arg-*p*-nitroanilide hydrochloride dissolved in TBS. OD readings were taken every 90 s for 40 min.

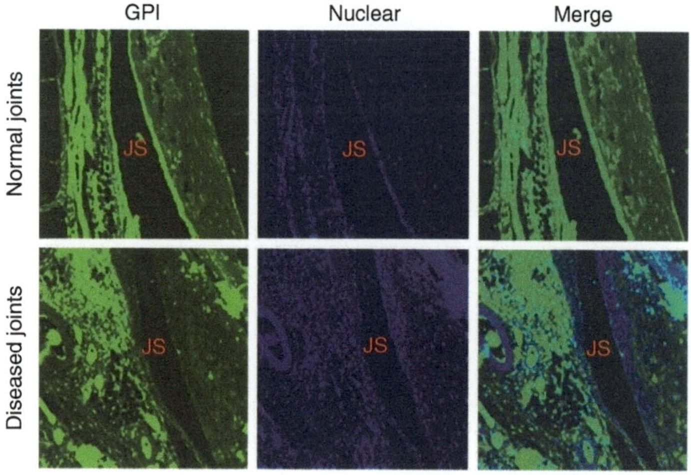

ZHOU *ET AL.*, FIG. 1. Diseased cartilage surface did not bind to GPI. Normal and diseased joint sections were deparaffinized and air-dried. Sections were fixed with 50% acetone in PBS, incubated with PBS containing 2 M NaCl to remove GAG binding proteins. Sections were incubated with 20 nM recombinant mouse GPI, followed by incubation with 1:50 diluted sera of arthritic K/BxN mice. After wash, sections were incubated with Alexa-fluor 488 conjugated goat anti-mouse secondary antibody. Nuclear was stained by To-Pro-3. Sections were examined by confocal fluorescence microscopy. The experiments were repeated twice with similar results.

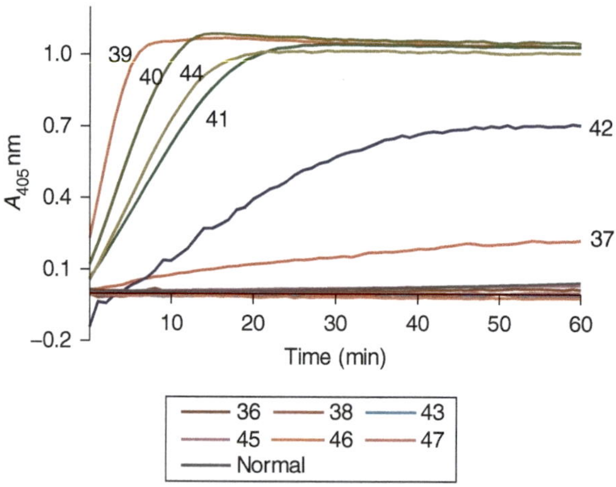

ZHOU *ET AL.*, FIG. 4. (A) Kinetics of thrombin activities of normal and RA patient plasmas. The thrombin amidolytic activities of pooled normal human plasma and 12 RA patient plasmas (30 ml) were assessed by addition of 150 μl of 100 mM Sar-Pro-Arg-*p*-nitroanilide dihydrochloride substrate. OD readings at 405 nm were taken every minute for 60 min. The experiments were repeated twice with similar results.

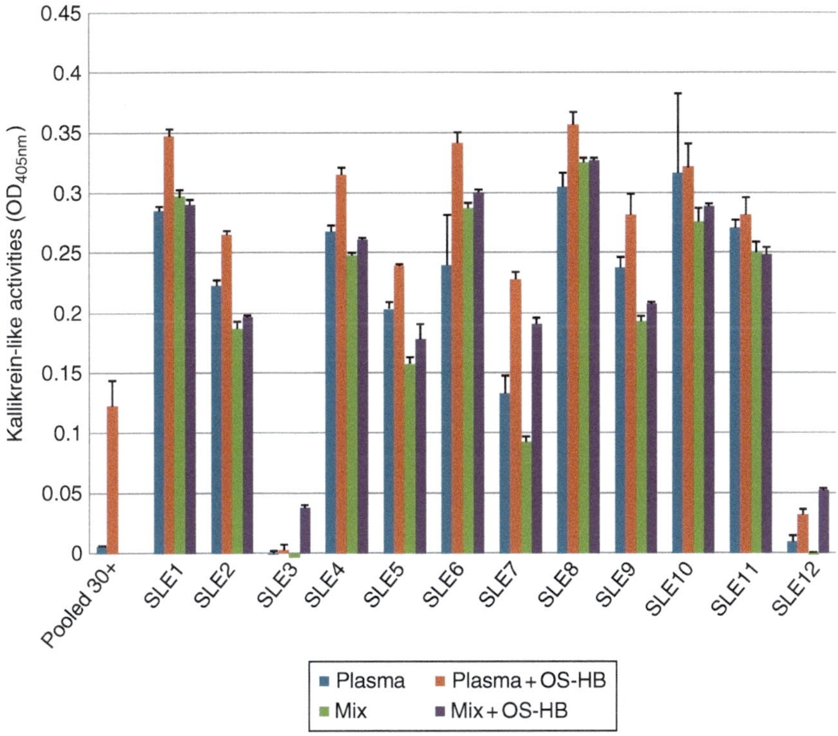

WEISER *ET AL.*, FIG. 1. Abnormal kallikrein activity in SLE patient plasmas. Plasmas were incubated with 20 μl TBS or 20 μl OS-HB at a final concentration of 200 μg/ml at 37 °C for 5 min. Kallikrein activity was assessed by addition of 150 μl of the kallikrein chromogenic substrate, 0.5 mM *N*-benzoyl-Pro-Phe-Arg-*p*-nitroanilide hydrochloride dissolved in TBS. Pooled 30+ indicates the normal plasma control obtained from 30+ healthy individuals. Detailed information about the SLE patients are shown in Table I. Bars represent arithmetic mean of the change of the $OD_{405\ nm}$ at 60 min from starting $OD_{405\ nm}$; each value was established from three independent measurements. The error indicated is the standard deviation. Blue bar, kallikrein activities in normal and SLE patient plasmas. Red bar, OS-HB-induced kallikrein activities in normal and SLE patient plasmas. Green bar, kallikrein activities in a 1:1 mixture of normal and SLE patient plasmas. Purple bar, kallikrein activities in an OS-HB-induced 1:1 mixture of normal and SLE patient plasmas.

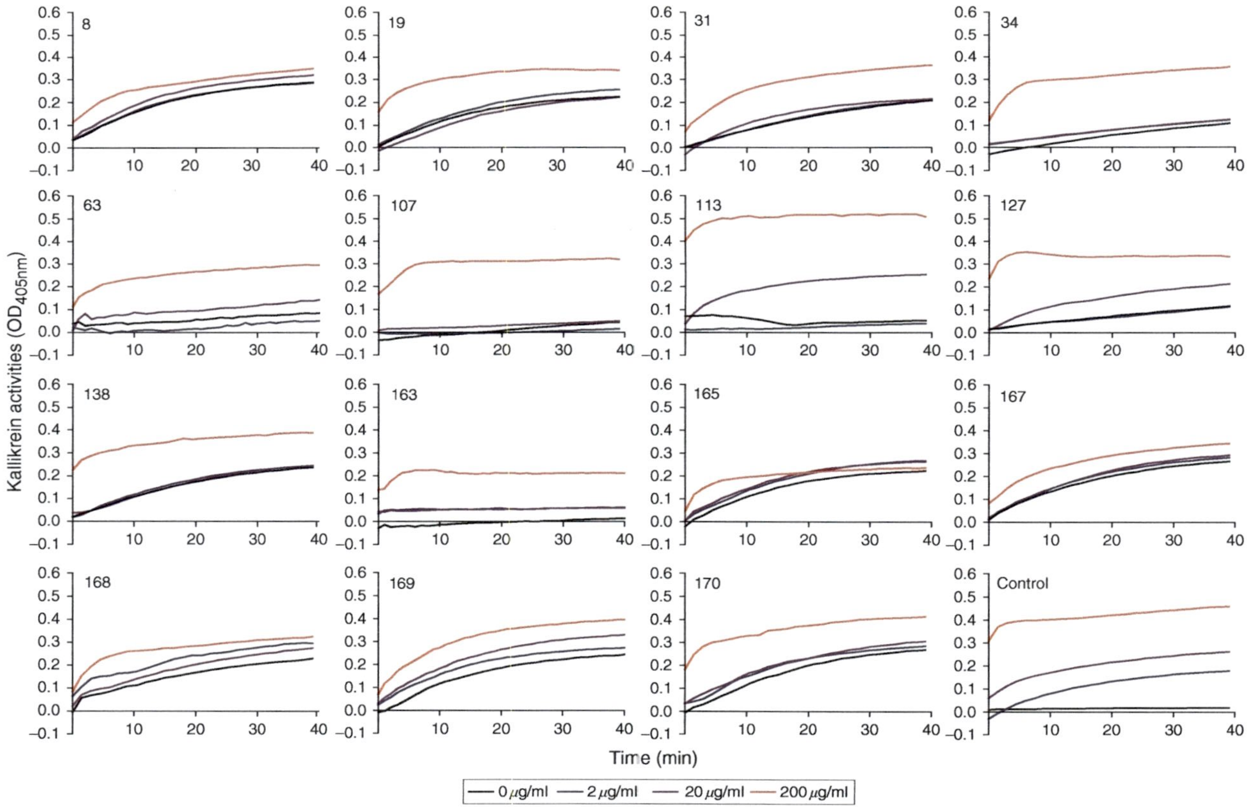

WEISER *ET AL.*, FIG. 3. High naturally occurring and low OS-HB-induced kallikrein activity in teenage SLE patient plasmas. Normal pooled human plasma or 15 SLE patient plasmas were incubated with 20 μl OS-HB with a final concentration of 0, 2, 20, or 200 μg/ml at 37 °C for 5 min. Kallikrein activity was assessed by addition of 150 μl of the kallikrein chromogenic substrate, 0.5 mM *N*-benzoyl-Pro-Phe-Arg-*p*-nitroanilide hydrochloride dissolved in TBS. OD readings were taken every 90 s for 40 min.

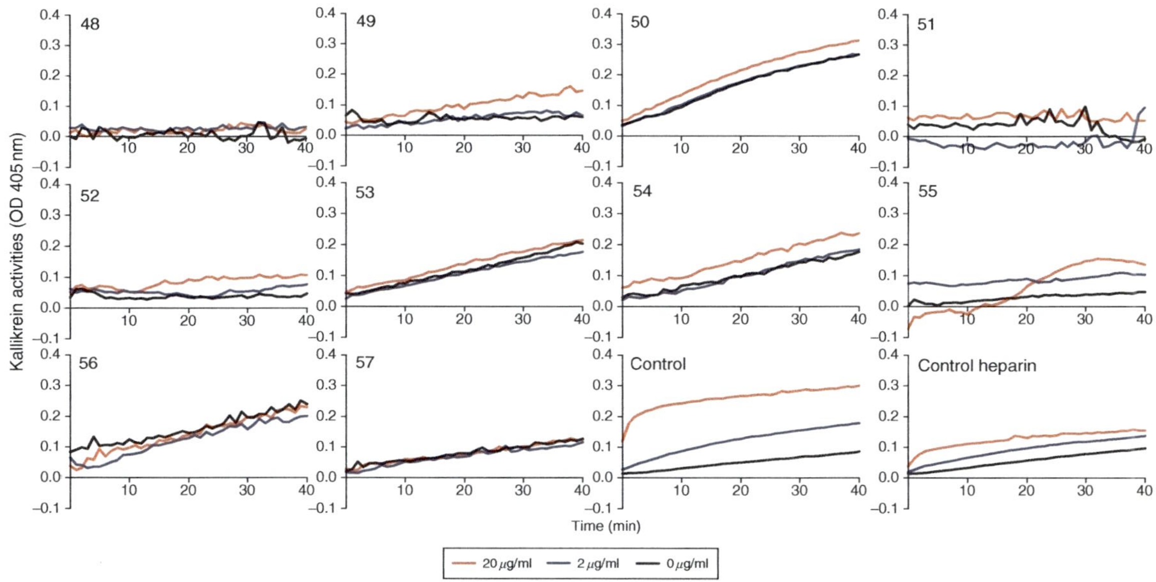
48
49
50
51
52
53
54
55
56
57
Control
Control heparin
Kallikrein activities (OD 405 nm)
Time (min)
20 μg/ml
2 μg/ml
0 μg/ml

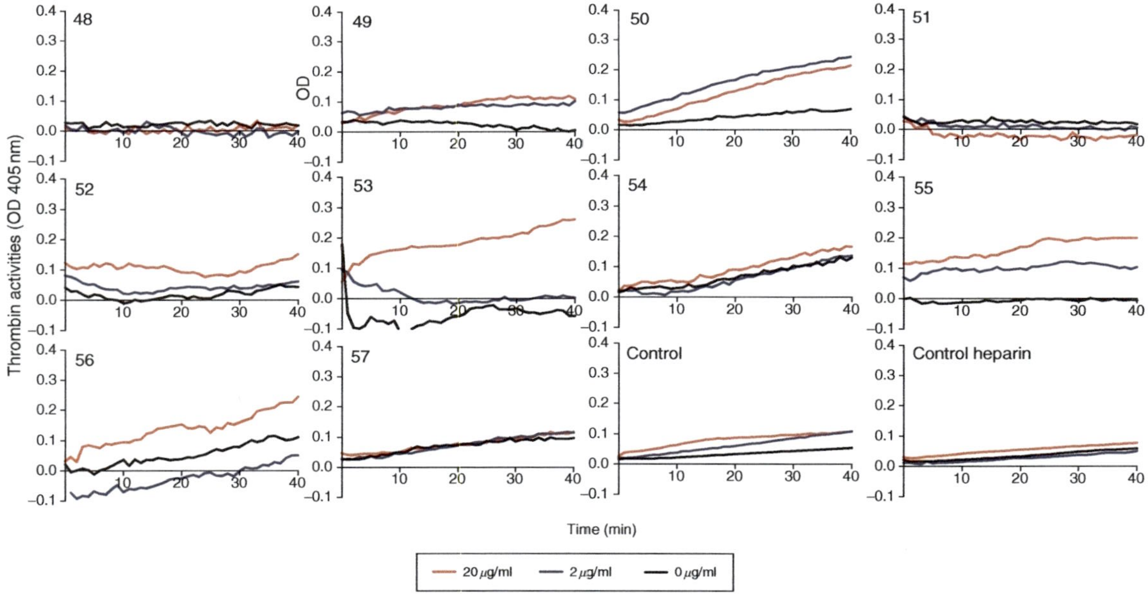

PAN *ET AL.*, FIG. 5. Abnormal kallikrein and thrombin activities in lung cancer patient plasmas. Normal pooled human plasma (30 μl) was incubated with 20 μl negatively charged molecules at a final concentration of 0, 2, and 20 μg/ml at 37 °C for 5 min. Kallikrein activity was assessed by addition of 150 μl of the kallikrein chromogenic substrate, and 0.5 mM *N*-benzoyl-Pro-Phe-Arg-*p*-nitroanilide hydrochloride dissolved in TBS. Absorbance readings were taken every 90 s for 40 min (Fig. 2A). Thrombin activity was assessed by addition of the thrombin chromogenic substrate, 1 mM Sar-Pro-Arg-*p*-nitroanilide dihydrochloride dissolved in TBS. Absorbance readings were taken every 90 s for 40 min (Fig. 2B).

FSC
www.fsc.org
MIX
Papier | Fördert gute Waldnutzung
FSC® C083411

Kolibri 360 GmbH
Ferdinand-Jühlke-Straße 7
99095 Erfurt, Deutschland
produktsicherheit@kolibri360.de